ECONOMIC GEOGRAPHY

ECONOMIC GEOGRAPHY

A Contemporary Introduction

Second Edition

Neil M. Coe, Philip F. Kelly and
Henry W.C. Yeung

WILEY

VP & EXECUTIVE PUBLISHER:	Jay O'Callaghan
EXECUTIVE EDITOR	Ryan Flahive
EDITORIAL ASSISTANT:	Julia Nollen
MARKETING MANAGER:	Margaret Barrett
PHOTO EDITOR:	Sheena Goldstein
DESIGNER:	Kenji Ngieng
SENIOR PRODUCTION MANAGER:	Janis Soo
ASSOCIATE PRODUCTION MANAGER:	Joyce Poh
COVER PHOTO:	Adnan/Age Fotostock America, Inc.

This book was set by Laserwords Private Limited. Cover and text printed and bound by Courier Kendallville.

This book is printed on acid free paper.

Founded in 1807, John Wiley & Sons, Inc. has been a valued source of knowledge and understanding for more than 200 years, helping people around the world meet their needs and fulfill their aspirations. Our company is built on a foundation of principles that include responsibility to the communities we serve and where we live and work. In 2008, we launched a Corporate Citizenship Initiative, a global effort to address the environmental, social, economic, and ethical challenges we face in our business. Among the issues we are addressing are carbon impact, paper specifications and procurement, ethical conduct within our business and among our vendors, and community and charitable support. For more information, please visit our website: www.wiley.com/go/citizenship.

Evaluation copies are provided to qualified academics and professionals for review purposes only, for use in their courses during the next academic year. These copies are licensed and may not be sold or transferred to a third party. Upon completion of the review period, please return the evaluation copy to Wiley. Return instructions and a free of charge return mailing label are available at www.wiley.com/go/returnlabel. If you have chosen to adopt this textbook for use in your course, please accept this book as your complimentary desk copy. Outside of the United States, please contact your local sales representative.

Library of Congress Cataloging-in-Publication Data

Coe, Neil M.
  Economic geography : a contemporary introduction / Neil M. Coe,
Philip F. Kelly, Henry W.C. Yeung. – 2nd ed.
    p. cm.
  Includes index.
  ISBN 978-0-470-94338-0 (pbk.)
  1. Economic geography.   2. Economic development.   I. Kelly, Philip F.,   1970-
II. Yeung, Henry Wai-Chung.   III. Title.
HF1025.C73 2012
330.9 – dc23

2012027173

Printed in the United States of America

10 9 8 7 6 5 4 3 2 1

CONTENTS

List of Figures   xi
List of Tables   xv
List of Boxes   xvii
Preface   xx
Acknowledgements   xxvi

Part I   Conceptual Foundations   1

1   Thinking Geographically   3
2   The Economy: What Does It Mean?   27
3   Capitalism In Motion: Why Is Economic Growth So Uneven?   55

Part II   Making the (Spatial) Economy   81

4   The State: Who Runs The Economy?   83
5   Environment/Economy: Can Nature Be A Commodity?   123
6   Labor Power: Can Workers Shape Economic Geographies?   154
7   Making Money: Why Has Finance Become So Powerful?   187

Part III   Organizing Economic Space   221

8   Commodity Chains: Where Does Your Breakfast Come From?   223
9   Technological Change: Is The World Getting Smaller?   261
10   The Transnational Corporation: How Does The Global Firm Keep It
     All Together?   294
11   Spaces of Sale: How And Where Do We Shop?   333

Part IV   People, Identities, And Economic Life   369

12   Clusters: Why Do Proximity And Place Matter?   371
13   Gendered Economies: Does Gender Shape Economic Lives?   402
14   Ethnic Economies: Do Cultures Have Economies?   432
15   Consumption: You Are What You Buy   466

Part V   Conclusion   497

16   Economic Geography: Intellectual Journeys And Future Horizons   499

Index   521

DETAILED CONTENTS

List of Figures		**xi**
List of Tables		**xv**
List of Boxes		**xvii**
Preface		**xx**
Acknowledgements		**xxvi**
Part I	Conceptual Foundations	1
1	Thinking Geographically	3
	1.1 Introduction	3
	1.2 Location and Distance	6
	1.3 Territory	12
	1.4 Place	14
	1.5 Scale	17
	1.6 Summary	23
2	The Economy: What Does It Mean?	27
	2.1 Introduction	27
	2.2 The Taken-for-Granted Economy	29
	2.3 A Brief History of "the Economy"	30
	2.4 Basic Economic Processes	38
	2.5 Beyond the Assumptions of Economics	43
	2.6 Summary	50
3	Capitalism In Motion: Why Is Economic Growth So Uneven?	55
	3.1 Introduction	55
	3.2 Uneven Development – Naturally!	57
	3.3 Fundamentals of the Capitalist System	58
	3.4 Inherent Uneven Geographies of Capitalism	64
	3.5 Placing and Scaling Capitalism	67

3.6   Going Beyond National Capitalism: A "Global California"?   72
3.7   Summary   76

Part II   Making the (Spatial) Economy   81

4   The State: Who Runs The Economy?   83
     4.1   Introduction   83
     4.2   The "Globalization Excuse" and the End of the State?   86
     4.3   The State as the Architect of the National Economy   88
     4.4   Varieties of States   100
     4.5   Rescaling the State   108
     4.6   Hollowing-Out the State?   117
     4.7   Summary   119

5   Environment/Economy: Can Nature Be A Commodity?   123
     5.1   Introduction   123
     5.2   How Is Nature Counted in Economic Thought?   126
     5.3   Incorporating Nature: Commodification and Ownership   128
     5.4   Valuing Nature: The Commodification of Environmental
            Protection   141
     5.5   Human Nature: The Body as Commodity   146
     5.6   Summary   150

6   Labor Power: Can Workers Shape Economic Geographies?   154
     6.1   Introduction   154
     6.2   Is Labor at the Mercy of Globally Mobile Capital?   157
     6.3   Geographies of Labor: Who Shapes Labor Markets?   159
     6.4   Labor Geographies: Workers as Agents of Change   170
     6.5   Migrant Labor   177
     6.6   Beyond Capital versus Labor: Toward Alternative Ways
            of Working?   180
     6.7   Summary   183

7   Making Money: Why Has Finance Become So Powerful?   187
     7.1   Introduction   187
     7.2   Is Global Finance Placeless?   191
     7.3   Financing Production: The Evolution of Banking   193
     7.4   The Rise of Global Finance   201
     7.5   Circulating Capital: Financialization   211
     7.6   Summary   216

Part III   Organizing Economic Space   221

8   Commodity Chains: Where Does Your Breakfast Come From?   223
     8.1   Introduction   223

8.2	Capitalism, Commodities, and Consumers	225
8.3	Linking Producers and Consumers: The Commodity Chain Approach	229
8.4	Re-regulating Commodity Chains: The World of Standards	244
8.5	Where Does a Commodity Chain End? From Waste to Commodities Again	255
8.6	Summary	256

9	Technological Change: Is The World Getting Smaller?	261
9.1	Introduction	261
9.2	The Universalization of Technology?	263
9.3	The Space-Shrinking Technologies	266
9.4	Product and Process Technologies	278
9.5	The Uneven Geography of Technology Creation	288
9.6	Summary	290

10	The Transnational Corporation: How Does The Global Firm Keep It All Together?	294
10.1	Introduction	294
10.2	The Myth of Being Everywhere, Effortlessly	296
10.3	Value Activity and Production Networks: The Basic Building Blocks of TNCs	298
10.4	Organizing Transnational Economic Activities 1: Intra-firm Relationships	302
10.5	Organizing Transnational Economic Activities 2: Inter-firm Relationships	312
10.6	Are There Cultural Limits to Global Reach?	324
10.7	Summary	329

11	Spaces of Sale: How And Where Do We Shop?	333
11.1	Introduction	333
11.2	Explaining Retail Geographies: Central Place Theory and Beyond	335
11.3	The Shifting Geographies of Retailing	338
11.4	The Configuration of Retail Spaces	353
11.5	Constructing Needs and Desires: The Advertising Industry	361
11.6	Summary	365

Part IV	People, Identities, And Economic Life	369

12	Clusters: Why Do Proximity And Place Matter?	371
12.1	Introduction	371
12.2	Industrial Location Theory	373
12.3	Binding Clusters Together: Agglomeration Economies	376

12.4   Untraded Interdependencies and Regional
       Cultures of Production   380
12.5   Toward a Typology of Clusters?   389
12.6   Rethinking Proximity   391
12.7   Summary   398

13 Gendered Economies: Does Gender Shape Economic Lives?   402
    13.1   Introduction   402
    13.2   Seeing Gender in the Economy   404
    13.3   Gendered Patterns of Unpaid Work   406
    13.4   Gendering Jobs and Workplaces   410
    13.5   Home, Work, and Space in the Labor Market   422
    13.6   Entrepreneurship and Livelihood Strategies   423
    13.7   Toward a Feminist Economic Geography?   426
    13.8   Summary   428

14 Ethnic Economies: Do Cultures Have Economies?   432
    14.1   Introduction   432
    14.2   "Color Blind" Economics   434
    14.3   Ethnic Sorting in the Workforce   436
    14.4   Ethnic Businesses and Clusters   445
    14.5   The Economic Geographies of Transnationalism   453
    14.6   The Limits to Ethnicity   460
    14.7   Summary   462

15 Consumption: You Are What You Buy   466
    15.1   Introduction   466
    15.2   Interpreting the Consumption Process   468
    15.3   The Changing Global Consumption Landscape   471
    15.4   Cultures of Consumption, Place, and Identity   476
    15.5   Toward an Ethical Consumption Politics?   484
    15.6   Consuming Places: Travel and Tourism   487
    15.7   Summary   493

Part V   Conclusion   497

16 Economic Geography: Intellectual Journeys And Future Horizons   499
    16.1   Introduction   499
    16.2   A Changing Field   501
    16.3   A Changing World   513
    16.4   Summary   517

Index   521

LIST OF FIGURES

1.1	Facebook – the transcendence of geographical distance?	4
1.2	Von Thünen's idealized model of land use	9
1.3	The Facebook friend map – global intensities of Facebook friend relationships, 2010	13
1.4	Spatial scales	19
1.5	Facebook's data center in Prineville, and an extension facility under construction	20
1.6	The location of Prineville, Oregon	21
1.7	Space, territory, place, and scale	24
2.1	The economy as an organic entity	29
2.2	The components of Gross Domestic Product	31
2.3	Irving Fisher's lecture hall apparatus, simulating the economy, ca 1925	35
2.4	The supply and demand curves	39
2.5	Many consumers, many sellers in the Tsukiji Fish Market, Tokyo – but a perfectly competitive market is hard to find	40
2.6	The circuit of global financial centers in the Islamic banking system	45
2.7	The economic iceberg and the submerged non-economy	47
2.8	Air pollution in China	48
3.1	A landscape of contemporary capitalism: an industrial park in Suzhou, China	64
3.2	Spatial divisions of labor	66
3.3	Waves of industrialization in East, Southeast, and South Asia, 1950–present	69
3.4	Industrial restructuring during the 1970s in the United States	70
3.5	Redundant industrial landscape in urban Philadelphia, U.S.	71

3.6 Galleries and apartments now occupy 19th-century industrial
      infrastructure in Liverpool, England   72
4.1 The border crossing between China's Shenzhen and Hong Kong's
      Lo Wu   91
4.2 The global landscape of sovereign wealth funds, January 2011   97
4.3 Institutions of global governance?
      (a) The IMF, 700 19th Street, Washington, D.C.
      (b) The World Bank, 1818 H Street, Washington, D.C.   109
4.4 Map of the G20 countries   112
4.5 The second G20 leaders' summit, London, April 2009   113
4.6 The expansion of the European Union since 1957   115
5.1 The Deepwater Horizon oil rig, April 2010   124
5.2 The economy as a system of material flows   130
5.3 The Rosia Poieni Copper Mine in Romania   131
5.4 The oil sands of Alberta, Canada   133
5.5 Grain wagons operated by the federal and provincial
      governments in Canada   134
5.6 Shifting cultivation of glutinous rice in Khammouane
      Province, Laos   136
5.7 Modes of urban water supply provision   140
6.1 Labor control regimes in Southeast Asia   169
6.2 Worker dormitories in Batam, Indonesia   169
6.3 The Toronto–Windsor corridor of auto assemblers, 2012   172
6.4 The Ford plant in Oakville, Ontario, flying the flags of Canada,
      Ford, and the CAW   173
7.1 Global finance gone mad in 2007–2008, from Wisconsin
      and Toronto to Dublin, the Cayman Islands, and Germany   188
7.2 Concentration and consolidation in the U.S. banking
      sector, 1998–2009   199
7.3 Relative importance of different types of global capital
      inflows since the 1980s (percent of total)   202
7.4 The global network of financial centers and global cities   208
7.5 The Occupy Wall Street movement in New York City   210
7.6 Global finance and the shifting relationship with local
      mortgage lending   213
8.1 The jewelry shop window – the start or the end
      of a complex commodity chain?   228
8.2 Geography is a flavor …   229
8.3 The basic commodity chain of our breakfast   229
8.4 The coffee commodity chain: who gains most?   231
8.5 The catfish commodity chain   232
8.6 Producer-driven and buyer-driven commodity chains   235

8.7 The top twelve coffee-producing (exports) and consuming (imports) countries, 2009 242

8.8 The coffee commodity chain: the changing institutional framework 243

8.9 The regulation of the dolphin-safe tuna packaging industry by nongovernmental organizations 247

9.1 Kondratiev long waves and their characteristics 266

9.2 The top 30 cargo airports in 2009 267

9.3 The top 30 container ports in 2009 268

9.4 Interregional internet bandwidth, 2005 and 2010 272

9.5 Dell's global operations, mid-2009 284

9.6 Industrial districts in Italy 286

9.7 Location of Toyota Motors, large-parts suppliers and third-party logistics providers around Toyota City in 2006 287

10.1 Different forms of organizing transnational operations 305

10.2 The BMW headquarter office in Munich, Germany 305

10.3 BMW's global production networks 306

10.4 Geographies of transnational production units 307

10.5 Making a BMW: Munich and beyond 311

10.6 Apple iPhone 3G: its components and key suppliers 313

10.7 The automotive cluster, Rayong Province, southern Thailand 319

10.8 Fast-food chains in the Caribbean 323

10.9 "Chamber of Fear" and the cultural limits to global reach: China's ban on a Nike television commercial in December 2004 328

11.1 Christaller's hexagonal central place theory pattern 336

11.2 The global distribution of Tesco and Wal-Mart stores in 2011 340

11.3 Tesco Lotus in Thailand 343

11.4 The development of Chicago's suburban shopping centers, 1949–1974 346

11.5 Britain's largest shopping centers, 2011 348

11.6 Cheshire Oaks outlet mall 349

11.7 Akihabara – Tokyo's electronics shopping district 354

11.8 Different retail streetscapes in Manchester.
(a) The Avenue, Spinningfields
(b) Market Street
(c) The Northern Quarter 355

11.9 Vegetable market in L'Isle sur la Sorgue, Provence, France 359

11.10 WPP group's organizational structure 364

12.1 Venture capitalists on Silicon Valley's Sand Hill Road 372

12.2 Weber's industrial location theory 375

12.3 The Hollywood film production cluster: location of production companies and major studios 378

12.4   Schematic representation of the Hollywood film production
       agglomeration in Los Angeles   379
12.5   Motorsport Valley in the United Kingdom, late 1990s   383
12.6   Silicon Valley and Route 128 corridor, United States   386
12.7   Fusionopolis and one-north, Singapore   387
12.8   A multifaceted cluster? High-tech business in Silicon Valley,
       California   391
12.9   Local "buzz" and global pipelines   396
12.10 Temporary clusters?   396
13.1   Main categories of unpaid work in various countries, ranked
       according to time spent on unpaid work (in minutes per day)   407
13.2   Minutes spent on unpaid work per day in various countries,
       ranked according to male–female disparity   408
13.3   Female labor force participation in selected countries,
       1980 and 2009   411
13.4   Women wait for their shift to start at an Indonesian
       electronics factory   412
13.5   Flight attendants: feminine work that becomes part
       of the product   418
13.6   Women and children gathering resources in mangrove forest
       areas of Aklan Province, the Philippines   425
14.1   Population of Urumqi in China's Xinjiang autonomous
       region by ethnicity and district, 2009   440
14.2   Geography of Chinese immigrants in the San Francisco
       Consolidated Metropolitan Statistical Area, by type
       of concentration pattern   443
14.3   Singapore's Little India   445
14.4   Ethnic businesses on San Francisco Street, Bilbao, Spain   446
14.5   How does ethnicity matter in the formation of business
       clusters and markets?   450
14.6   Chicago's Andersonville   451
14.7   Remittances into the Philippines by top 10 source countries,
       January–November 2011   456
15.1   A consumption landscape – The Strip, Las Vegas, Nevada   470
15.2   Tata's Nano car   475
15.3   Real (on the left) and fake (on the right) iPhones available
       in the Chinese market   476
15.4   Hybrid cultures? Bollywood dancing outside the National
       Gallery, London   482
15.5   Poster for "Buy Nothing Day"   485
15.6   World tourist flows, 2010   489
15.7   Culture-led redevelopment – the waterfront in Bilbao, Spain   491
15.8   Reclaimed industrial heritage – Legoland in Duisburg, Germany   491
15.9   Magical Kenya   493

LIST OF TABLES

4.1 World's twelve largest sovereign wealth funds in 2010 (US$ billion)   96

4.2 Varieties of states in the global economy   101

4.3 Major regional economic blocs in the global economy   114

6.1 The world's biggest employers, 2010   161

6.2 Different national labor conditions: two ideal types   165

7.1 The changing regimes of financial regulation in the global economy   204

7.2 Shanghai's position in global financial markets, 2008–2009 (in US$ billion)   210

8.1 Characteristics of producer-driven and buyer-driven chains   237

8.2 The world of standards   245

8.3 Certification schemes for sustainable coffee   252

8.4 Regional share of ISO9001:2000/2008 certificates (December 2009)   253

9.1 Expansion of submarine cable capacity, 1979–2005   270

9.2 Leading logistical providers – key facts and figures in 2008   277

9.3 Comparing contemporary production systems   282

9.4 The characteristics of "just-in-case" and "just-in-time" systems   283

10.1 Non-Equity Modes (NEMs) of cross-border activity by TNCs in selected industries, 2010 (in billions of dollars and millions of employees)   314

10.2 Subcontracting of the world's top notebook brand-name companies to Taiwanese firms, 2010   315

10.3 Different forms of organizing transnational operations – costs and benefits   325

11.1 Leading transnational retailers, ranked by sales outside home market, 2010   339

11.2 Top grocery retailers in Poland, 2010   341
11.3 Top grocery retailers in Thailand, 2010   342
11.4 The leading global communications groups and their
      advertising agencies   363
13.1 Contrasting views on the emancipatory potential of industrial
      employment for women   413
13.2 Top 10 occupations of women employed in the United States,
      ages 16 and over, 2010   416
14.1 Distribution of ethno-racial groups in various occupations,
      Los Angeles, 2009   438
14.2 The two sides of ethnic enterprise   452
14.3 Top twenty remittance-receiving countries in 2010   455
15.1 Asia's burgeoning middle class?   472
15.2 Mass consumption and post-Fordist consumption compared   477
15.3 International tourism receipts and expenditure – top 10
      countries in 2010   489

LIST OF BOXES

1.1   J. von Thünen, land use, and bid-rent curves   8
1.2   Paul Krugman and geographical economics   10
1.3   Why Prineville?   20
2.1   Metaphors of economy   34
2.2   Divisions of labor   42
2.3   The place of markets   51
3.1   Regulation theory   63
3.2   Sustaining the global capitalist system: geopolitics and domination   74
4.1   Unpacking the state   85
4.2   BHP Billiton's failed bid to buy PotashCorp   93
4.3   National business systems   99
4.4   Neoliberalism   102
4.5   Graduated sovereignty?   106
4.6   Shock therapy   110
5.1   Evaluating nature through the ecosystem services approach   129
5.2   Primitive accumulation and accumulation by dispossession   138
5.3   A global land grab?   139
5.4   Actor-Network Theory (ANT)   147
5.5   Selling cells   149
6.1   Chicago's prisons as labor market institutions   162
6.2   The temporary staffing industry   163
6.3   Lean production   166
6.4   Local labor control regimes   167
6.5   The Mondragon Cooperative Corporation   182
7.1   A glossary of common financial terms   189
7.2   Why did Citi expand beyond the United States?   196

7.3	Global cities	207
7.4	The Cayman Islands as an Offshore Financial Center (OFC)	212
7.5	Subprime and the crisis of global finance in 2008/2009	215
8.1	Coffee, Cafés, and Connections	226
8.2	Upgrading strategies in global commodity chains	233
8.3	Trading giants – the Japanese Sogo Shosha	238
8.4	Environmental certification of the dolphin-safe tuna commodity chain	246
8.5	The limits to standards?	254
9.1	The Box	269
9.2	Offshore Services in India	274
9.3	Fordism – More than a Production System?	281
9.4	Production Process Innovation: The Case of Dell Computer	284
10.1	From commodity chains to Global Production Networks (GPNs)?	298
10.2	Corporate cultures	302
10.3	Transnational production in the *maquiladoras* of northern Mexico	309
10.4	Transnational corporations and the new international division of labor	317
10.5	Global production networks and macro-regional integration in Southeast Asia	319
10.6	Transnational corporations and anti-globalization movements	326
11.1	Success and failure in global retailing	344
11.2	Asos – online clothing retailer	352
11.3	Geographies of branding	362
12.1	The creative class and spatial clustering	381
12.2	High-tech clusters in Asia	387
12.3	Clusters – the mesmerizing mantra?	392
12.4	Project working	395
13.1	Patriarchy	409
13.2	Devaluing the third world woman	414
13.3	Theories of labor market segmentation	417
13.4	Redundant masculinities	421
13.5	A queer Economic Geography?	427
14.1	What is Ethnicity?	436
14.2	Ontario's South Korean convenience stores	447
14.3	Transnationalism	454
14.4	Ethnic Turks in Germany	458
15.1	Viva Las Vegas!	469
15.2	Bottom of Pyramid markets	474

15.3 Consumption work 478
15.4 Bourdieu's cultural capital 480
15.5 McDonald's in Asia 483
15.6 The world's biggest industry? 488
16.1 Ontology, Epistemology, and Methodology 502
16.2 The "Cultural Turn" in Economic Geography 507
16.3 What is discourse? 509
16.4 Economic Geography beyond the Anglosphere 511

PREFACE

When the first edition of this book was published in 2007, there was a dearth of material available for students that reflected the contemporary state of economic geography. Since then, some excellent volumes have appeared, but we believe that the model developed for this textbook remains distinctive in several ways and we have retained these features in this second edition:

- First, the book is structured on the basis of topical issues that are tackled using a geographical perspective, rather than on the basis of intellectual history or academic debates. We believe this is still the best way to engage students, many of whom come to our courses with a curiosity about the world around them, but not necessarily a commitment to Geography as a discipline, or any prior knowledge of the field.
- Second, the book is written in what we hope is a clear and engaging style. For many students Economic Geography is not a field with a great deal of immediate appeal, but we have tried to present it in an accessible and interesting way. To avoid overcrowding the text and reading, we use only sufficient data and facts to support our arguments and explanations.
- Third, although this is not a book about the global economy *per se*, we have made a deliberate effort to ensure that it addresses the major issues confronting the global economy today and it draws examples from around the world, reflecting the varied contexts in which the book is used. As such, we make no distinction between Economic Geography and the often-separate subfield of 'Development Geography'.
- Fourth, the book reflects the range of topical and theoretical approaches that exist in contemporary Economic Geography. Instructors will recognize that political–economic and institutional approaches underpin much of the book, but at the same time post-structural thinking and efforts to explore the economic implications of culture and identity are also taken seriously.

In short, this book aims to present a conceptually rich and yet readable introduction to the field of Economic Geography that showcases the different ways through which economic geographers understand economic processes. It is designed to appeal to students who are coming to Economic Geography for the first time, while also offering depth to those more familiar with the field.

Changes in the second edition

In writing the first edition, we were conscious of the need to avoid publishing a book that was 'of the moment' – in other words, a compendium of contemporary economic patterns that would be outdated as soon as it was published. We opted instead for a pedagogical model that was based on developing geographical arguments about the empirical issues at hand. These were substantiated with data, but the arguments, we believed, would outlast the data. Eventually, however, the ongoing dynamic of change in economic processes, and the shifting directions of scholarship in Economic Geography, mean that both topics and arguments need to be revisited. In this second edition we have made a number of changes to reflect our changing times.

- An entirely new chapter on finance has been added. Although we have consciously avoided structuring the book around a list of economic sectors, we believe that the global financial system today represents not just a sector in its own right but also an essential input to all other economic activities. As the financial crisis of 2007–09 demonstrated for North American and European economies (and the earlier crisis of 1997–98 demonstrated for East Asia), a clearer understanding of how the financial system works and how it is integrated with the rest of the economy must now be central to the study of Economic Geography.
- Four new chapters replace parts of chapters in the first edition. The new chapters relate to Technology, Retail, Clusters, and Consumption. This allows us: (a) to tackle influential ideas concerning innovation and economic clusters in far more detail; (b) to examine closely the changing geographies of retail activity that now account for such a large proportion of employment and value-added in many economies; and (c) to explore issues of consumption in more detail, including the consumption of places through tourism – a sector that was largely missing in the first edition.
- Chapter 1 introduces geographical concepts of location, territory, place, and scale in an entirely new way, and using an example that is literally at the fingertips of every student: Facebook.
- The first edition of the book was vigorous in its critique of approaches in Economics, which were sometimes presented in rather caricatured form. In this edition, we have avoided adversarial confrontation of this kind, preferring instead to focus on the positive appeal of a geographical approach. At the same

time, we have sought in Chapter 2 to provide more coverage of fundamental economic concepts so that students will be equipped with the vocabulary necessary to understand the different approaches offered by Geography and Economics.

- Although, as noted above, we generally avoid mapping out disciplinary issues, this edition features a new concluding chapter that will help students understand how the material they have read fits into the context of Economic Geography as an academic field. This is targeted especially at those intending to pursue higher-level studies in the field and who want to know how approaches in Economic Geography correspond with wider intellectual trends in Geography as a whole.

- Finally, data, examples, and references have been extensively updated throughout. In both developed and developing countries the last five years have seen turbulent economic upheavals and structural shifts of various kinds. It is, in some respects, a different economic world. Where relevant, we have reflected this in the data and examples presented. Economic Geography has also moved ahead intellectually, and we have tried to incorporate this as well.

Audience

The book is designed for introductory courses on Economic Geography in an undergraduate degree programme. The text is written in an accessible way, but some of the processes and ideas that it discusses are inevitably complicated. The ways in which the text is used will therefore depend on instructors' assessments of their students' background and preparation. Students who already have some familiarity with the concepts and arguments presented in this book could likely use the chapters as their starting point for further exploration of a given topic through articles from the research literature in Economic Geography, including those suggested in the reading notes for each chapter. But for those with little background in Geography, or even in the social sciences in general, the chapters in this book might be better approached through an initial reading pitched at a popular audience, from a news magazine or website. In other words, the chapters in this book may be the starting point or the end point, depending on the students involved. The book is designed so that it could serve either purpose.

While the pitch of this book is intended for a particular audience, it is also worth noting that a specific conception of what constitutes Economic Geography is implicit in our selection and treatment of topics. The text is therefore targeted to those instructors who share, or wish to adopt, this approach. A few points are worth making in this regard:

- First, this is a book that explores the multiple scales of economic processes and is not, therefore, focused exclusively on larger processes at global or national scales. For example, we believe that Economic Geography has as much to contribute in thinking about how gender roles in the household play out within the spaces of the urban labour market, as it does in understanding the globalizing organizational forms of transnational corporations (and so we cover both).

- Second, ours is a largely qualitative vision of the field, in the sense that we do not emphasize formal analytical techniques in the book. Rather than providing exercises in quantitative analysis, we focus instead on stimulating students with critical perspectives and arguments. For example, in thinking about ethnically-structured labour markets, we are more interested in inviting students to think about the processes that lie behind such phenomena than in explaining how to demonstrate statistically that such patterns exist. That said, statistical exercises can, of course, be used as supplementary assignments alongside this text.

- Third, we focus on what we see as some of the best of recent scholarship in Economic Geography. Although classic models and approaches are covered, our goal is to expose students to the insights that *contemporary* Economic Geography can provide in making sense of the world around them. We tend, therefore, to cover locational analysis and other classical approaches in detailed boxes rather than the main text.

- Fourth, we do not seek to establish impervious boundaries between Economic Geography and other cognate fields concerned with social, cultural and political processes. Our vision of the discipline is a porous one and we take seriously the need to view the economy as embedded in other spheres of life. For example, we see consumption not 'just' as an economic act, but also a political engagement through fair trade and other certified products, and as a component of identity formation. In this sense, the book is very much in tune with what geographers have called the 'new economic geography' (not to be confused with the approaches in Economics that are often given the same label). The audience for this book is, then, among those who share this ecumenical vision of Economic Geography.

Organization of the book

This book takes the form of a series of linked chapters on topical issues and contemporary debates that draw upon, and showcase, some of the best research in Economic Geography. These issues are drawn from contemporary economic life, which is increasingly constituted at a global scale – from uneven development, space-shrinking technologies, and environmental degradation, to powerful global corporations, organized labour, and ethnic economies. We see each of these as issues rather than just phenomena, i.e. they are processes to be debated rather than factual realities to be described. Each chapter thus seeks to answer a significant contemporary question that a curious and well-informed reader might reasonably be expected to ask about the world around them.

This, then, is not a conventional text: our aim is to develop well-grounded *arguments* from an economic geography perspective, not necessarily to present simplifications of multiple viewpoints or collections of facts and data. We are, however, trying to develop these arguments in straightforward and accessible ways.

The book is organized into four parts:

Part 1: Conceptual Foundations – This section introduces the basic building blocks of geographical analysis and our understanding of the economy. These are brought together in a geographical understanding of the dynamics of the capitalist system. Chapter 1 examines location, territory, place, and scale as core geographical concepts, while Chapter 2 explores where the idea of 'the economy' comes from historically and some of the common concepts used in economic analysis such as demand, supply, production, markets, and firms. Chapter 3 then mobilizes these geographical and economic concepts into a dynamic and structural account of uneven development in a capitalist economy.

Part 2: Making the (Spatial) Economy – having explained the dynamics of capitalism, here we introduce the inputs and actors, besides capitalists, that make the system work: the state, nature, labour, and finance capital. Chapter 4 discusses the state in its varied manifestations and the ways in which it shapes economic activity. Nature is examined, in Chapter 5, as an input to the economy through processes of commodification. Chapter 6 considers labour as a factor of production, but more particularly as an active agent in shaping the economic landscape. Chapter 7 looks at the financial system both as a source of productive capital and an economic sector in its own right.

Part 3: Organizing Economic Space – here we explore the ways in which economic relationships across space are established, organized and maintained. Chapter 8 introduces the commodity chain concept, which provides the overall framework for this part of the book, and we will ask how we might seek to regulate and control the production processes behind the commodities we consume. The commodity chain is fundamentally shaped by technologies of movement, communication and product innovation that are examined in Chapter 9. The role of transnational corporations in organizing global production networks is explored in Chapter 10, while Chapter 11 focuses on restructuring in the retail business as the end point of the commodity chain.

Part 4: People, Identities, and Economic Life – The final part of the book explores the blurred line between economic processes and the social and cultural contexts in which they are embedded. Chapter 12 highlights the very social process of economic cluster formation, and the benefits of learning and innovation that result. Chapter 13 examines the role that gender plays in economic life, in reproductive work, in waged workplaces, and in entrepreneurship. Ethnicity is the focus of Chapter 14, which explores how labour markets are 'colour coded', how ethnic entrepreneurship works, and how transnational economies result from migration. Chapter 15 then examines the intersections between who we are and what we consume, i.e. how identity is tied up with consumption and how places themselves become consumable through tourism. In concluding the book, Chapter 16 opens up two broader issues for students interested in pursuing the subject further: understanding the intellectual development of Economic Geography as an academic discipline, and thinking about how future economic trends may present new challenges to the field.

Pedagogical Strategies

Each chapter in this book follows a similar structure. In most cases the chapter title is worded as a fairly intuitive question, reflecting our attempt to engage with questions that students might have of their own economic worlds. Although the topic for each chapter also lends itself to coverage of a defined field within Economic Geography, we have deliberately avoided framing chapters in disciplinary terms in this way.

The chapters open with what we call the 'hook', i.e. a (hopefully engaging) contemporary example or issue used to introduce the key theme of the chapter. In the second section we tackle a commonly held myth or misapprehension about the topic at hand (e.g. the nation state is dead, or transnational corporations are all-powerful) and illustrate how these myths often rest, in large part, on a non-geographical understanding of the world around us. The main body of each chapter then serves to illustrate the necessity and effectiveness of taking an explicitly geographical approach for understanding different aspects of the economy.

Our aim is to make these arguments in a clearly understandable, lightly referenced, jargon-free manner, drawing on a wide range of examples from across different sectors of the economy, and from around the world. Boxes within the text are labelled as 'key concepts', 'case studies', and 'further thinking', and they offer more detailed elaborations on specific ideas or examples.

The penultimate section of the chapter is designed to add a 'twist' to the arguments that have preceded it; or, in other words, to probe somewhat more deeply into the complexity of contemporary economic geographies. Additional nuances and insights are offered in these twists to encourage students to avoid simplistic views of economic activity. Each chapter then concludes with a short summary of the main themes covered.

What lies after the summary is also important. First, for ease of use, the reference list is included on a chapter by chapter basis. Second, the reading notes in each chapter guide the student towards what we identify as the most engaging and accessible literature on the topic. Some of these readings identify the sources of well-known case studies we have drawn from the geographical literature, enabling students to 'flesh out' the brief summaries offered in the chapter. We also identify up to five online resources per chapter that can also be used to supplement the chapters. In most cases, these include web resources that can be consulted for updated examples and data.

Overall, our intention is to offer an exploration of Economic Geography rich in examples and case studies that can, on the one hand, expose students to economic life and practices in various parts of the world, and, at the same time, introduce concepts that can be 'put to work' in their own local contexts. Hence the text can be used alongside local literature and case studies wherever the book is adapted.

ACKNOWLEDGEMENTS

This second edition has been a long time in the making! Throughout the writing process, however, we have been sustained by the positive feedback and encouragement of users and readers of the first edition. We hope that this new edition meets their expectations. At the start of the process, we benefited tremendously from the input of Yuko Aoyama, Geoff Mann, Sam Scheuth, Eric Sheppard and Dick Walker, who provided close and critical readings of the first edition. Their detailed and thoughtful comments were crucial in shaping how we approached this book. Our plans for the second edition were in turn sharpened by the insights of eight proposal reviewers, while the final version itself has benefitted greatly from the feedback of the ten reviewers who looked at various draft chapters. We thank all these commentators for the time and care taken in helping us to make this a better book. At the same time, this edition still benefits from the wise advice that shaped the first edition, and for that we are grateful to Trevor Barnes, Gavin Bridge and Tim Bunnell. Above all, Peter Dicken remains our collective primary inspiration in striving to improve the accessibility and visibility of Economic Geography.

At John Wiley, our editor Ryan Flahive has been patient and supportive as this project has slowly and steadily moved towards completion. We are very grateful for his ongoing encouragement and confidence in our project, and for the excellent editorial assistance of Julia Nollen and Joyce Poh. Clive Agnew, Keith Barney, Martin Hess, George Lin and Claire Major gave us permission to reproduce their photos in the book. Adam Lukasiewicz did an excellent job of preparing the web-based materials for this edition. Finally, we offer a huge thanks to Graham Bowden at the University of Manchester, who once again has done a superb job in designing and producing all the graphics for the book, while remaining calm and assured despite the erratic workflow we have thrown at him! His contribution is absolutely central to the look and feel of this volume.

On an individual level, we all would like to thank our respective colleagues, friends and families for their support and understanding. Neil's contributions were completed before his move to the National University of Singapore in August 2012. As such he would like to thank his University of Manchester Economic Geography colleagues – Gavin Bridge, Noel Castree, Martin Hess, Erik Swyngedouw and Kevin Ward – for their collegiality and for making Manchester such a great environment in which to 'do' Economic Geography. The golf gang – Tim Allott, Martin Evans and Chris Perkins – again deserve thanks for providing increasingly sporadic but much needed distraction. More broadly, he is extremely grateful for the ongoing support offered by Trevor Barnes, Meric Gertler, Jamie Peck, Jessie Poon, Roger Lee, Anders Malmberg, John Pickles, Adam Tickell and Neil Wrigley. Thanks are also due to the Yeung family for their fantastic hospitality during book meetings in Singapore. Closer to home, Laura and Adam were very excited to see their names in the first edition and they once again deserve recognition in print! They are a constant source of joy and inspiration, as is Emma, whose calm and unflagging support is invaluable and much appreciated.

Philip is grateful to colleagues in Toronto and elsewhere who have supported this endeavour, knowingly or unknowingly, along the way. Graduate students have played a key role as colleagues, collaborators, and teaching assistants in Economic Geography: Jean-Paul Addie, Keith Barney, Simon Chilvers, Nel Coloma-Moya, Anne-Marie Debbané, Conely de Leon, Veronica Javier, Adam Lukasiewicz, Julia Mais, Claire Major, Elliot Siemiatycki, Ritika Shrimali, and Junjia Ye. Departmental colleagues in Economic Geography at York have provided a wonderfully collegial environment that has influenced this book in a variety of ways: Ranu Basu, Raju Das, Steve Tufts, Jennifer Hyndman, Lucia Lo, Glen Norcliffe, Valerie Preston and Peter Vandergeest. I also appreciate the help of Alicia Filipowich and Yvonne Yim, who have always been happy to show a brazen disregard for their job descriptions. I am grateful for grant funding provided by Wiley-Blackwell, which allowed me to travel to work with Neil and Henry on this project; and thanks to the Coe and Yeung families who provided hospitality that funding just can't buy! Finally, and for everything else: Hayley, Alexander, Jack and Theo.

Henry has used the first edition as the primary text in his GE2202 Economy and Space class for over four years. Over this period, several hundred students on this course have delved deep into the text for inspiration and, just as often in the hot tropical climate, perspiration! He is most grateful for their candid and positive feedback on what the text can offer. This tangible impact on the "final consumer" makes the writing immensely satisfying. Meanwhile, his colleagues in the Department of Geography, National University of Singapore, have been most engaging and encouraging, particularly those in the Politics, Economies, And Space (PEAS) group. Special thanks go to Godfrey Yeung, Zhang Jun, Karen Lai, James Sidaway, Liu Yi, and Chen Rui for their ongoing contributions to this book's preparation. During the past four years, the three of us gathered in Glossop and Singapore several times to work on the proposal and book manuscript. Henry would like to thank Neil's family, Emma, Laura, and Adam, for their forbearance of the noisy and demanding visitor from Singapore! Valerie and Peter Dicken always make him very at ease in his Manchester "second home" and their English breakfast is simply THE best. Back in Singapore, his family has come to terms with the sustainability of this text and thus seeing another volume (with the same title!) added to the shelf area in the living room. Weiyu has been a great spouse who supports every aspect of this project. Kay and Lucas have grown up with the first edition and, hopefully, learnt a thing or two about/from Uncle Neil and Uncle Phil! Henry would like to dedicate this edition to their continual learning journey to become better students (of Economic Geography??).

NC, PK and HY
Glossop, Toronto and Singapore
June 2012

PART I

CONCEPTUAL FOUNDATIONS

CHAPTER 1

THINKING GEOGRAPHICALLY

Goals of this chapter

- To understand the distinctive elements of a geographical mode of thinking
- To elaborate on key concepts such as location, distance, territory, place, and scale
- To apply these concepts to economic phenomena

1.1   Introduction

In February 2004, while still a 19-year-old undergraduate student at Harvard, Mark Zuckerberg created a social networking website called Facebook. Six months later, his company had established itself in California's Silicon Valley, just outside the campus of Stanford University. By the end of 2010, *Time* magazine had lauded Zuckerberg as its Person of the Year, a critically-acclaimed movie had told the story of the company's creation, and by 2012 around 850 million people worldwide were Facebook subscribers. In other words, more than one-tenth of the world's population had an account. When Facebook shares were floated in 2012, the company was valued at US$80–100 billion – more than many long-established media corporations, like News Corporation or Time Warner, and more than many of the iconic corporate entities of a previous generation, such as Sony or General Motors.

Facebook's value was not, however, based on its income, assets, or profits. In 2011, its profits amounted to about $1 billion, largely generated through advertising revenue. Rather, the price tag reflected recognition of the potential value attached to its 850 million subscribers and its dominance in terms of web traffic (*The Economist*, 2012). In 2010, Facebook accounted for 8.9% of all web visits in the United States (even exceeding Google's 7.2%). Subscribers to Facebook spent about 700 billion minutes interacting with its site every month (an average of almost 17 hours per month for each user) (Facebook, 2011). This amounts to a huge potential for earnings, through advertising, linked applications, and online games.

Facebook might seem an odd example to use in introducing the field of Economic Geography. To most of its users, after all, Facebook is primarily a venue for social networking, for staying in touch with friends and sharing photographs, thoughts, invitations, and so on, rather than engaging in economic activity. Its product is free to use and does not require any economic transactions on the part of users, beyond their prior purchase of a computer or mobile phone and a contract with an internet service provider. Furthermore, communications on Facebook seldom concern "work-related" activities, as many employers are acutely aware!

In addition to being apparently noneconomic, Facebook might also seem to make geography itself irrelevant. Time spent on Facebook appears to involve no interaction with the earth's physical systems, and the company's product is, to its users, weightless and virtual. Facebook would also seem to eradicate the impediments created by geographical space. Its own home page depicts a graphic of social linkages transcending a map of the globe (see Figure 1.1). Distances, places, borders, and other markers of geographical space appear to offer little hindrance to the spread of an individual's network of friends through the site, and the possibilities for communication. Users can stay in touch with a network of friends almost regardless of where on earth they are located, and the difference in connection times between various locations is almost imperceptible. Perhaps more than any other business, then, Facebook would seem to exemplify the end of geographical space.

Figure 1.1   Facebook – the transcendence of geographical distance?

If we look a little closer, though, Facebook and its business operations provide both an economic and a geographical story. Clearly Facebook is an economic phenomenon. As a business itself, Facebook employed around 3,000 people in 2011 and generated sales of nearly US$4 billion. But its product also represents an economic tool used by many other businesses to reach and connect with clients and customers, both as a networking device in its own right and as a venue for advertising that might once have appeared in newspapers or magazines. Facebook also represents a platform on which developers are creating software applications, thereby generating economic activity well beyond the company itself. Facebook estimates that entrepreneurs and software developers in 190 countries work with its platform. Looked at from a different angle, Facebook and the internet more generally provide channels through which networks of solidarity and struggle might be created between workers around the world (a theme we will return to in Chapter 6).

Facebook is also a profoundly geographical story. Here we take "geographical" to mean the patterning of activities (in this case, economic activities) on the earth's surface. The central questions asked by economic geographers are how economic patterns across space are configured, and why things happen where they do. But space itself is not a straightforward idea, as it is more than just the canvas on which such patterns are imprinted. If that were the limit of our interest in space, then we would simply be describing patterns rather than trying to explain them. Rather, space is an active part of explaining geographical patterns, meaning that economic activities are shaped by spatial relations – space is not just *where* things happen, it is also *why* things happen where they do. Seeing space as an explanatory factor in this way involves thinking carefully about what we mean by space. In this chapter we develop four conceptions of space and illustrate their meaning with reference to the Facebook example.

The first concept is location (Section 1.2), which involves the positioning of people and objects relative to each other. A key variable here is distance. Overcoming distance requires time and money and so it determines a great deal about how the economic landscape is configured. This is true whether we are thinking about traded goods being moved, people commuting to work, or shoppers traveling to retail outlets. We will examine, in particular, some of the classic models that economic geographers have used to demonstrate how the cost of distance (sometimes called the "friction of distance") affects location in space.

A second geographical conception of space is territory (Section 1.3). If location and distance are about coordinates on a map and the physical space that separates them, then territory is about carving out defined portions of space and exercising power over them. The primary form of territorial power is exercised by governments, who can affect both economic activities within their territories and economic flows across their borders.

Third, we consider the concept of place (Section 1.4). Places are formed when space takes on certain unique characteristics that are meaningful to the people who interact with them. Places may have cultural or political significance, but they also shape economic patterns in important ways. Economic places do not, however, just create themselves internally – rather they are the product of various flows across space that intersect differently in different places to generate one-of-a-kind outcomes. It is the uniqueness of those outcomes that plays a part in determining where economic activities will "take place."

Finally, we will think about scale (Section 1.5). Both territories and places are defined areas of space, but they might represent a range of different scales – state territories vary widely in size, and while a house is a place, so too is Beijing or South Wales. This might seem a rather straightforward idea, but scale becomes complicated when we are thinking about space in economic life. Which scale should we focus our attention upon? How do different scales relate to each other?

1.2   Location and Distance

Perhaps the most basic way of thinking about space is as a grid of points that we can describe using a system of coordinates. Detailed maps do this quite effectively using longitude and latitude. The lecture hall where your class takes place could easily be described in these terms. It could be given precise global coordinates and you could then also describe your own location in that classroom through a regular pattern of rows and seats (3rd row, 4th seat from the end, etc.). This conception of space, based on some kind of definable measure of position, is often referred to as absolute space. It is the space of geometry and mathematics, and allows the specification of point locations, lines, and areas. Space in this instance is simply a grid for defining an absolute position.

Taking space as a grid of coordinates also allows us to go one stage further. We can examine the location of people and things in space in terms of their position in relation to each other. This is, in fact, far more important than simply knowing the coordinates of something on a map, as it allows us to start thinking about patterns in economic space – what kinds of things are happening where, and why this might be so. The absolute space of distance helps us with this explanation because overcoming distance requires time and money and so is nearly always a factor in determining the location of economic activities. As raw materials and finished goods are transported for longer distances, for example, the more expensive they become or the more they deteriorate in quality. As people decide where to shop, they must factor in the costs in time and money of traveling to retail outlets at varying distances away. In an urban center, a labor market can usually only be created out of those people who are close enough to commute to work at the beginning and end of the work day. This means that most cities have a labor market represented by areas within 1–2 hours of commuting time.

It is also important to note that the distance between points on a map is not necessarily the most significant factor in determining the effect of location on economic activities. It may, for example, be quicker to get from New York to Amsterdam than to get from Green Bay, Wisconsin, to Austin, Texas, even though the trans-Atlantic journey is much longer in terms of absolute distance. Likewise, it may be less expensive to ship large quantities of manufactured goods from China to California than it is to move them from California to urban centers across the United States. What matters, then, is not absolute distance measured on a map, but rather *relative* distance, measured in freight/transport costs or travel time. This is, in turn, dependent on the configuration of transportation modes and intersections. Relative space is, then, frequently more important in economic terms than absolute space.

For many years, absolute and relative locations were the key conceptions of space used by economic geographers as they sought to understand how the "friction of distance" affected the spatial pattern of economic activities. A classic early version of this idea was provided in the early 19th century by Johann von Thünen, who examined the way distance affected agricultural land use patterns (see Box 1.1). Another German theorist, Walter Christaller, applied a similar style of thinking to the patterns of cities, towns, and villages that develop across economic space (to be discussed further in Chapter 11).

The key point to note in both von Thünen's and Christaller's ideas was that they saw space in terms of relative location – between producers and marketplaces, and between marketplaces and consumers. By establishing certain assumptions and by analyzing how distance affected relationships, they were able to develop models that could predict how economic activities might be arranged in space. This was a powerful way of thinking and inspired a great deal of subsequent research in Economic Geography, much of it developing a more sophisticated mathematical version of the analyses produced in these early models. Such approaches are often labeled "locational analysis" or "spatial science," indicating their ambition to find principles that underpin the arrangement of economic activities in space. The effects of distance and transport costs on patterns in economic space also formed the basis for studies by economists developing what they term "the new economic geography" (see Box 1.2).

If we return to the example of Facebook, we might at first imagine that the study of location and distance has little to offer. It is certainly true that relative space has been greatly disrupted by the networks created through internet connectivity. It is, for example, just as easy to connect on Facebook whether a person is in New York, New Delhi, or New Zealand. And to a certain extent, the internet allows real economic transactions to take place without any "friction of distance." Buying applications, music/video downloads, or software are all possible regardless of where the consumer, or the business selling those products, is located. Other kinds of production may also be facilitated by the internet. Even physical products, like this book, can be created using the internet to overcome distance (for example, as draft chapters are circulated by email between authors located in the U.K., Canada, and Singapore, and publishers in the U.S.).

KEY CONCEPT

Box 1.1   *J. von Thünen, land use, and bid-rent curves*

One of the earliest examples of an attempt to analyze the effect of distance on patterns of land use was developed by a German landowner almost 200 years ago. Johann-Heinrich von Thünen (1783–1850) was trying to find a logical way of understanding how to use land on his estate in order to maximize economic benefits. Using a very simple model he showed that the different value and transportation costs of various crops meant that they would "bid" for the use of a particular parcel of land at different rates as distance from a market town increased. The result was a pattern of rings around a town in which distance dictated agricultural land use. Some land uses, such as dairy cows, produced a commodity that was heavy to transport, easily perished, and needed on a daily basis.

Under these circumstances this land use would "bid" a high price for land close to the market center, but this would rapidly fall away since beyond a certain distance it would not be possible to transport the milk to the market without it spoiling.

Agricultural crops, on the other hand, yielded less daily income from a given area of land, and would only require seasonal transportation. They would thus bid less for land near the center, but could continue at much greater distances (described by a "bid-rent curve," Figure 1.2a). Once a variety of products were assessed in this way, von Thünen showed that a pattern of concentric land use rings would result around the market center, all calculated based on the transport cost of the products (Figure 1.2a). This model made a variety of assumptions – for example, that transport costs were directly proportional to distance, that the landscape, soil fertility, etc. were uniform in all directions, and that there was only one market center for agricultural products.

Obviously, the pattern would begin to get more complicated once these assumptions were relaxed (Figure 1.2b). If we think about contemporary food systems, in which for example Eastern Canada is supplied with fresh salad from California, Mexico, and other places, von Thünen's scheme begins to look very dated. Nevertheless, the model does capture the essence of how the costs associated with distance can affect the spatial structure of economic activity.

Although our technological ability to overcome some of these distances greatly expands the size of the "circles," a similar calculation is still being made. Another context in which bid-rent curves have been applied is in the urban setting, where prices and land use patterns (e.g., of office, commercial, residential, or industrial activities) are often directly related to their distance from a central business district.

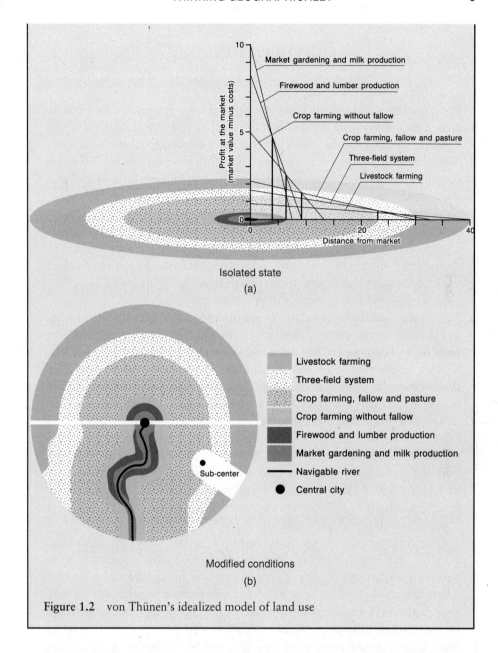

Figure 1.2   von Thünen's idealized model of land use

Nevertheless, location is still surprisingly important even for internet applications, and several examples illustrate this point. First, Facebook's business is based on selling advertising space. This advertising is targeted, including by geographical area, so that the ads that users see on their Facebook pages will generally be those that relate to services or products in their immediate geographical area.

FURTHER THINKING

Box 1.2   Paul Krugman and geographical economics

Paul Krugman was awarded the Nobel Prize for Economics in 2008 in recognition of his work on economic geography and international trade.

Although Krugman has achieved fame as a columnist for the *New York Times*, writing for a popular audience, the work for which he was awarded the prize was quite abstract. It involved the development of a model that showed why an economy will, over time, tend to develop core regions of manufacturing and peripheral regions based an agricultural production. Of particular interest here are the ways in which Krugman (1991a and b) treats transportation costs. He assumes that transport costs are proportional to distance and operate like an "iceberg," meaning that the greater the distance traveled the less value a product retains (because it "melts away").

He then shows that a number of processes take place.

First, if a manufacturer locates in proximity to other manufacturers (e.g., in the same urban center), then it will benefit from the lower transport costs for its components that are being supplied from other local factories. Second, he shows that because manufactured goods are less expensive in that location (due to lower transport costs), the wages of workers who are buying them will be worth relatively more, and thus it will be more attractive as a place to work. Third, transport costs to reach a large market will also be reduced by locating in proximity to that market – and as the attractiveness of that market draws more manufacturers, their workers will also become part of the market. Finally, he shows that if transport costs across space are low enough, then there will be a tendency for manufacturing producers (and population) to concentrate in a core area and from there, they will supply a population dispersed across space.

For all of these reasons, a process can be triggered in which manufacturing activities and population will concentrate in one area, creating a core and periphery across economic space. This will not happen in a context where transport costs are very high, but will become feasible as they fall. Which area will be the core, and which the periphery, can be affected by quite small differences at key points in time, but will then lead to large differences as these processes play out – a widening gap between core and periphery due to what can be termed "cumulative causation." To derive this result, Krugman used what he later termed "aggressively unrealistic" assumptions. His purpose was not to describe the real world, but to characterize mathematically the operation of an idealized model. He does point out, however, that the model makes intuitive sense and that real-world cases provide some support for his ideas. For example, manufactured goods were produced in small towns in the United States as farming settlement spread westward in the 18th and 19th centuries. But as the railroad reduced transport costs it became possible for manufacturing to concentrate in a few urban centers, which grew rapidly and whose predominance has been largely maintained. Krugman's model initiated a generation of research in Economics, known as Geographical Economics, that has sought to incorporate the role of space into the modeling of economic development and trade flows.

The process for placing advertisements does not involve paying a fixed price, but instead bids are accepted by Facebook based on either each "view" of an advertisement or each "click" on the hyperlink embedded in the advertisement. Bidders decide what they can afford to pay to reach their desired population and market area. Thus, proximity still matters, because many advertisers are ultimately more interested in reaching people who are nearby.

Distance matters in another sense too. To get almost instantaneous connections, the location of computer servers (where data are stored and processed) is important. Although Facebook still has all of its servers in the United States, this is likely to change in the coming years, and Google has already recognized the need to locate servers around the world in order to deliver the response times that users have come to expect. But computer servers must also be placed with certain locational requirements in mind. For example, server farms or data centers need huge amounts of electrical power and so it pays to be near a cheap supply of electricity, such as a hydroelectric power plant. The heat generated by this kind of power consumption also means that large amounts of cooling water are needed, and so this too shapes locational decisions. Finally, and perhaps most obviously, the locations of such data centers need to be in areas that have very good connectivity to the internet. In each of these ways, then, essential elements of internet infrastructure do in fact have locational requirements to be near certain key inputs for their "production" process.

Surprisingly perhaps, these locational requirements are similar to some of the factors that would have shaped the very earliest geographical patterns of manufacturing industries. In 18th century Europe, a textile or flour mill would have located next to a river in order to generate power (using a water wheel) and to serve as a transportation corridor for raw materials and finished products. In principle, these requirements are not so different from a data center's need for electrical power and a good location on a network for moving data. In both cases, the spatial pattern of economic activity is shaped by distance and proximity to certain production requirements. Space as distance, then, can be as important now for Facebook as it was 300 years ago for a flour mill in determining where in space activities will happen.

1.3   Territory

So far, the way in which we have thought about space has largely reduced it to the role of physical distances in explaining the pattern of economic activities across the landscape. As we have seen, there are certainly merits to this approach – as the time and cost associated with overcoming distance are undoubtedly important factors in determining where things are located.

Space is, however, more complicated than just the existence of distance. It can also be viewed as having qualities that distinguish one area from another. An especially important attribute of spaces is the power to control them in some way. This brings us to the concept of territory, which can be defined as "a bounded space under the control of a group of people, usually a state" (Elden, 2009, xxv). Thus, while locations and distances are indicated on a map of the world by gridlines and a scale, territories are defined by the lines and shading that indicate the entities that control various areas.

Control over territory can have important implications. National governments have the power to police, guard, and enforce the boundaries of their territory. This means that they can control what flows across those boundaries, including workers, products, and money. Even in a globalizing world there is little doubt that national governments still exercise many powers in this respect (as we will discuss in greater detail in Chapter 4). Where states surrender such powers, it is often to other territorial entities that are themselves state-like, such as the European Union.

Territorial space also implies the power to exert some degree of control over certain processes within those boundaries. In the case of national states, this is clearly manifested in a range of ways – from education and training programs that are run by governments, to laws concerning property, pollution, contracts, and employment relations, to welfare, unemployment benefits and tax policy. In these and many other ways, governments shape the nature of economic activity within their borders quite fundamentally. Governments are also usually the single largest employer and purchaser of goods and services. It is not hard to see, then, why control over borders, and over economic activities within them, would make national territories a key dimension that shapes where economic activities happen.

National states, may, however surrender part of their absolute control over territory to others. City governments, for example, often take on the role of land use planners and thus decide what kinds of activities will be allowed in which locations. Through inspectors, enforcement officers, and of course police departments, the location and conduct of economic activity is also closely regulated.

Ownership of land may also allow private entities to exercise control in economically important ways. Consider, for example, a shopping mall, which in many respects is a territorial entity carved out in space. The managers of the mall are permitted, by virtue of their ownership of the facility, to exercise some degree of territorial control over it. They can, for example, shape what

Karen Bleier/AFP/Getty Images, Inc.

Figure 1.3 The Facebook friend map – global intensities of Facebook friend relationships, 2010

kinds of activities occur within the mall, including decisions about the type and mix of retail tenants (discussed further in Chapter 11). They can also "police" the behavior of people within the mall, by denying access to those they see as undesirable and by disallowing certain activities – try lying down for a nap or selling some home-baked cookies in a shopping mall, and you will likely find out fairly quickly that it is a closely controlled territory!

Returning to the example of Facebook, we can see several ways in which territorial control is influential. Figure 1.3 shows a map of Facebook linkages around the world. Created in 2010, it plots the connections between Facebook "friends" across global space. The map became an object of some fascination in China when it was released as it clearly shows the country as a "dark" area without Facebook friend connections. In reporting on the map, the *Wall Street Journal* quoted one Chinese blogger as asking, "Here is the Facebook map. This octopus-like creature hasn't been able to invade our territory, should we be happy and content with this?" The reason that Facebook is absent from China is because the national government has exercised its territorial power to block flows across its national borders – in this case blocking the Facebook site within China. This occurred in July 2009 after the national government in China became fearful that political dissidents in some peripheral regions were using social networking sites to organize their resistance.

Whereas China blocked Facebook from engaging in business within its territory, national governments may also actively seek to attract certain economic activities. We noted earlier that internet data centers have certain locational requirements, but they may also be lured by the promotional activities and incentives that governments offer. Since the collapse of its financial system in 2008, Iceland has attempted to promote itself as a prime location for data centers. With a surplus of very low cost electrical power, much of it generated geothermally, and a cool climate, the country has certain natural advantages for the location of data centers. But the government has also moved to ensure that its taxation regime is attractive to locators of data centers, has invested in high-capacity underwater cables to ensure fast connections to the global internet, and has mounted a campaign to "sell" the various benefits of locating within its territory: its promotional campaign sells Iceland as "the coolest location for data centres." The Icelandic state, as a territorial entity, has thus sought to render itself attractive to this growing sector in a variety of ways through its regulatory power over territory.

1.4   Place

Place represents a third conception of space and is distinct from both location and territory in important ways. In one sense a place is indeed a location, as it occupies specific coordinates on the earth's surface. But it is not the occupation of a singular location that defines a place. Rather, a place is a unique ensemble of human and physical features on the earth's surface, including environmental conditions, physical and human landscapes, cultural practices, social life, and economic activities.

Where do these distinctive features of a place come from? Some are part of the physical environment and owe little to the impacts of human activity. These should not be forgotten as they matter a great deal in determining the resources that form the basis for a place's economic development (as we will discuss in Chapter 5). They also matter in terms of the attractiveness of places for human settlement. But the characteristics of a place are also fundamentally shaped by human activity. Forms of government, religious traditions, linguistic groups, norms relating to gender roles, architecture, artistic expression, ways of interacting with other people, levels of wealth and inequalities of wealth, the types of work that people do, the shops that exist and the things that they sell – these are all human activities that give rise to unique characteristics of particular places, and they may vary greatly even within the same territorial spaces. New York City, for example, is a quite different place from the capital of New York state in Albany. Vancouver is a distinctive place in itself, but the Kitsilano neighborhood is a quite different place from New Westminster.

These differences in human society between places are to some extent created in situ – that is, they are created internally within such places. Traditions of art, different styles of architecture, and styles of social interaction might emerge as distinctive and quirky features of a particular place. They are also reproduced through the everyday behavior of people in that place. There would, after all, be no such thing as a distinctive New York attitude and way of behaving unless New Yorkers themselves performed that style every day! The characteristics of a place may also emerge because of chance happenings. Detroit, for example, may not have become North America's 20th-century center of auto manufacturing if Henry Ford had not been born nearby, nor would Rochester, New York, have been the headquarters of the Eastman Kodak company if George Eastman had not been raised in upstate New York. On the other hand, we could equally argue that the conditions for these innovators to establish new industries were created by precisely those places in which they were raised.

For the most part, though, differences between places are created as much by the connections between places as they are by processes that are internal to them. We can therefore think of any given place as a unique knot of linkages tying it to many other places. The role that a place plays in larger structures explains a lot about its characteristics. New York City and Hong Kong are wealthy cities, but they remain so in large part because of the paramount role that they play in a globalized financial system. Each has concentrations of major banks and financial institutions with global reach, and billions of dollars worth of transactions are completed daily among these global financial centers (described in greater detail in Chapter 7). Sydney (Australia) and Vancouver (Canada) have economies that are shaped in part by their centrality in flows of migration to their respective countries. Oxford, England, is a city defined by its role as a major center of higher education and research, but this role exists because the university represents a node in various networks of research funds, students, faculty, and knowledge exchange. Places, then, can be seen as the "coming together" of flows across space to create unique intersections in particular locations. This conception of place is what Doreen Massey (1994) calls a "global sense of place."

It is, however, important to remember that places are not just the outcome of contemporary connections. They are also the result of historical place-making from different periods of time, each layered on top of the period before. This historical layering process is another reason why no two places are alike. We cannot understand the grandeur and wealth of London without acknowledging its role at the heart of a global empire over the last few centuries. Nor can we fully understand Manila or Mumbai without thinking about the ways in which centuries of colonialism shaped their societies, cultures, and economies. Even the most remote village is shaped by these historical connections, and their contemporary consequences, in a most profound way.

Understanding the uniqueness of places is, therefore, complicated. It requires us to think about both historical and contemporary processes and how they have shaped a place, and how the characteristics of a place in the past shaped (without actually *determining*) what it could become in the future. It also requires us to examine not just what the place itself is like, but also the part it plays in larger structures and processes – studying a place is not therefore about studying *just* that place. Furthermore, studying places is about recognizing how diverse factors, from the natural environment, to cultural practices, to economic activities, are all interconnected. This encounter with the complexities of a specific place is, in many ways, a quintessentially geographical undertaking.

It is also worth pointing out that this geographical approach to the world requires a somewhat different mindset than we might find in other disciplines, and especially Economics. Taking a geographical approach to patterns of economic activity requires us to see them in all the complexity and messiness of the real-world places in which they are situated. This does not mean that there are no bigger forces operating that shape these activities, but it does mean that we are seeking to understand them as unique instances in real and lived places, rather than as ways of deriving model "laws" or principles. While economists are often seeking universally applicable generalizations (a "science" of economic processes), economic geographers are usually going in the opposite direction – trying to understand why certain things happen in specific places in the context of all the richness and complexity of that place.

We can illustrate this approach by returning to the example of Facebook. We noted at the beginning of this chapter that Mark Zuckerberg started the website while still a student at Harvard University (near Boston), but within months he had moved to California's Silicon Valley. Palo Alto, just south of San Francisco, became the company's new home, situating it at the heart of a major global center for software development. The qualities of that place are an important part of the Facebook story. Facebook's first home was in the Stanford University Business Park – an environment established precisely to assist new start-up companies needing office space, and particularly those established by the university's faculty and graduates. The company's first significant investment, in June 2004, came from Peter Thiel, who several years earlier had cofounded the online payment system Paypal with a group of fellow Stanford graduates. An important part of Palo Alto's quality as a place, then, was the presence of both an entrepreneurial atmosphere and a willingness to bet venture capital on new companies. Silicon Valley also represented a rich pool of workers for a new high-tech company. Facebook's employees were drawn from various sources, but many had previous experience with new start-up ventures in Silicon Valley.

The important point to note here is that Silicon Valley represents a place where information, ideas, capital, and highly skilled workers all come together in a way that fosters the development of firms like Facebook. To understand truly the emergence of Silicon Valley as the place it is today, we would also have to examine the role of American government defence expenditure in the area in the second half of the 20th century, and various other features of California's historical development (for more on California, see Chapter 3). The development of Facebook needed these unique features of Silicon Valley, and, as Facebook has developed, it further enhances these qualities of Silicon Valley as a place (for more on high-tech clusters like Silicon Valley, see Chapter 12).

A second, but quite different, example from the Facebook story also illustrates the distinctiveness of places and how they might shape economic opportunities. We noted earlier that the global map of Facebook "friend" connections (Figure 1.3) shows a large dark area over China, due to the government's application of its territorial power to exclude. It is also worth noting, however, that Japan is also rather underrepresented. While 60% of internet users in the United States are now Facebook subscribers, the proportion in Japan in less than 2%

(Tabuchi, 2011). Instead, Japanese users have been drawn to other social network sites, especially Mixi. The near absence of Facebook in Japan is not because of language barriers (it is available in Japanese) or because of the actions of a territorial government. Rather, we see a distinctively place-based explanation for why Facebook has not captivated Japanese users. This revolves around the culture of internet usage in Japan that rejects the openness of Facebook, where real names and pictures are used, and favors instead the use of pseudonyms or false identities when connecting to others online. The idea that strangers might see personal information or even photographs is quite alien to Japanese internet culture. The result is that although there was an US$8.5 billion online advertising industry in Japan in 2009, Facebook captured relatively little of it. Here we see place operating in a quite different way. A company's product that has been so successful in so many places was found to be "out of place" in a different cultural context. This form of place-based cultural limit to the global spread of transnational corporations is discussed further in Chapter 10.

1.5   Scale

In the discussions above we have discussed territories that range from national states to shopping malls. We have also discussed places from the suburb of Palo Alto in California, to New York City, to Japan as a whole. This implies that we need to give some explicit consideration to the issue of scale.

When used in relation to a map, scale refers to the way in which distances on the ground are represented. A map scale of 1:50,000, for example, would represent 500 meters of real-world distance with one centimeter on the map. But scale is more commonly discussed in relation to different levels or sizes, and when these are applied to space we are usually talking about variously sized areas. We can identify a long list of scales that have already been used in our discussion of Facebook (see Figure 1.4):

- A *global* scale allows us to understand the full scope and reach of Facebook's client base and business. Around 80% of the company's subscribers are outside the United States, and it is the dominant social network site in many countries in the Americas, Europe, Africa, and Asia.
- The *macro-regional* scale is larger than the national scale, but more defined than the global. It usually refers to a group of countries that should be considered together – for example, because of some kind of economic commonality. The European Union would be one example. Thus, one part of Facebook's global corporate structure is a European headquarters located in Dublin, Ireland.
- The *national* scale is the most important scale at which Facebook must comply with the regulations imposed by territorial states. It is also often the

scale at which internet service providers (crucial intermediaries in Facebook's business) usually operate. And, as we have seen in Japan, place-based cultures that demand certain types of internet applications may also be nationally defined.

- The *regional* scale may represent a territorial unit of control (for example a subnational state or province with jurisdiction over certain issues); but a region may also represent the scale at which advertisers on Facebook are targeting their potential customers; or we might think about Facebook's headquarters in the regional cluster of high technology development stretching along the Santa Clara Valley south of San Francisco.

- The *urban* scale is where most of our everyday economic activities take place, including the daily journey to work of around 1,000 Facebook employees at their Palo Alto headquarters. A city may be significant in other ways too – when Facebook sought to raise very large amounts of capital in early 2011, it turned to the New York investment bank Goldman Sachs. The bank's location in downtown Manhattan among a major cluster of financial firms is just as important as Facebook's location in the Silicon Valley high-tech cluster.

- The *local* scale might reflect the immediate neighbors located close to Facebook's headquarters, such as Google and Stanford University, creating flows of people, ideas, and competitive pressures that drive new innovation.

- The *workplace*, or the *home*, are scales where many of the micro-processes of everyday life are played out. Facebook's headquarters features open-plan desk space designed for frequent interaction and innovation. The home lives of its employees will shape why they work there in many complex ways (while the home lives of other people may dictate that employment in a high-pressure environment with long hours of work is not possible).

- Finally, the *body* is the scale at which we all occupy, navigate, and experience space. And the "coding" that we carry around through our bodies (our gender, ethnicity, sexuality, citizenship, linguistic skills, education, cultural style, etc.) often shapes the ways in which we are treated in our economic transactions. The body may be important in another sense too – an individual or small group of individuals (Mark Zuckerberg, Henry Ford, and George Eastman have all been mentioned in this chapter) may be a critical player whose personality and background shapes an entire economic sector.

These scales are useful frameworks for thinking about economic processes, but there are three important points to remember that refine our understanding of scale.

The first point is that scales are not hierarchical. It is tempting to assume that larger scales determine what goes on at smaller scales. The economic world is, in fact, more complicated than this. Clearly even a global firm like Facebook is unable, in this instance, to override the national territorial power of a state,

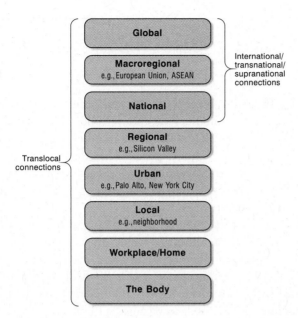

Figure 1.4   Spatial scales
Source: Adaptation from Castree, N., Coe, N. M., Ward, K. and Samers, M. (2004) Spaces of Work: Global Capitalism and, Geographies of Labour, figure 0.1.

such as China. Likewise, the wealth and power of Wall Street financiers may play a significant role in shaping the national economic policy of the United States. Finally, we can note that while the high-technology sector (Facebook, Apple, Google, etc.) may be truly global in scale, its important functions are played out in specific regions such as Silicon Valley in California. And, of course, innovative new ideas that will change the world occur in the workplaces, kitchens, bedrooms, and coffee shops of the world. Thus, scales may vary in size, but this does not necessarily mean that greater size implies greater power.

A second point to make is that economic processes work at multiple scales simultaneously. To try to understand a set of processes at one scale alone will inevitably produce a very incomplete picture of what is happening. In 2011, when Facebook opened its first self-owned data center in Prineville, Oregon (see Box 1.3), decisions would have been made at many different scales. On a global and continental scale, it was clearly decided that the servers should be in the United States, while on a national scale the Northwest of the country was chosen as somewhere with a climate that would not exacerbate cooling costs. Incentives from local governments and the availability of a suitable site also played a part at a yet smaller scale. In other words, the dynamic behind the geography of this particular economic activity is working at several scales simultaneously. The purpose of geographical analysis is not, therefore, to pick the "correct" scale to focus upon, but rather to the keep multiple scales in mind at the same time. To think at just one scale usually provides a very partial and misleading understanding.

CASE STUDY

Box 1.3   Why Prineville?

In April 2011, Facebook opened its own data center (see Figure 1.5) in Prineville, Oregon, a community of about 9,000 (see Figure 1.6). Until that point, the company had leased data center capacity from other providers. The facility in Oregon reportedly has the capacity to handle all of Facebook's current data traffic and storage needs, but the company has plans for further expansions. Why would Prineville represent the chosen location for such a facility?

Data centers are vast warehouses of computing power, and the conventional factors explaining their location include the need for a large electrical power supply, a copious supply of cooling water, and good internet connectivity. As noted earlier in this chapter, these kinds of requirements can be accommodated within a locational analysis – one that seeks to understand what an economic activity needs to be near.

A locational analysis of Facebook's Prineville site provides some indications of why the facility was located there, although interestingly the explanations are rather different from the ones we might expect. Rather than be near plentiful supplies of cheap electricity – for example, from hydro generation – the facility will draw power from a local utility company that uses mostly coal. However, the utility was able to meet Facebook's

Faith Cathcart/The Oregonian/Landov LLC

Figure 1.5   Facebook's data center in Prineville, and an extension facility under construction

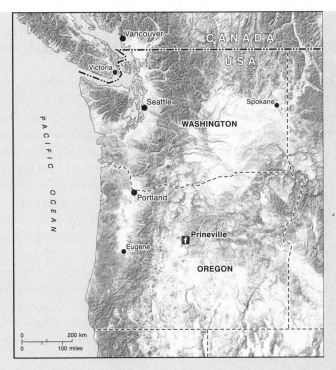

Figure 1.6   The location of Prineville, Oregon
Source: Original drawing, authors

needs, perhaps due to excess capacity created by local deindustrialization. Formerly home to a thriving wood products industry, Prineville had seen a long-term decline, with the last mill and another major employer (a tire manufacturer) having left by 2008. By late 2010, unemployment rates were close to 20%. Nor did Prineville provide plentiful quantities of cooling water. In fact, central Oregon is known as the "high desert" as it sits in a rain shadow zone, and local water supply is far from plentiful. Prineville does, however, offer other physical locational advantages. A relatively dry climate helps with the cooling mechanism used in the data center, which involves evaporative cooling rather than air conditioning.

But such locational considerations of proximity and distance only take the explanation so far. Another dimension of Facebook's locational decision concerned the incentives on offer from local governments. Prineville is in one of the state of Oregon's "long-term rural enterprise zones," which provide for up to 15 years of exemption from property and corporate income tax. The ability to provide security for the facility was also a major consideration. Here, the location of the site in a very remote location is an advantage – on a bluff above a valley surrounded by forested hills in the interior of Oregon. Thus, the territorial space of government economic jurisdiction was important, and so was the ability to secure territorial control over the site itself.

Had Facebook been seeking to lure highly mobile software engineers to work at their data center, they would likely have looked elsewhere, to places like Silicon Valley. But attracting large numbers of highly technical employees was not an issue. When fully operational, the data center requires only around 35 full-time workers, from security guards, to maintenance workers, to computer technicians. The local site manager for Facebook was quoted as saying that "We're the blue collar guys of the tech industry, and we're really proud of that... This is a factory. It's just a different kind of factory than you might be used to. It's not a sawmill or a plywood mill, but it's a factory nonetheless." In that sense, Facebook identified a place with the right kind of local work culture.

As this brief example suggests, understanding the location of an economic activity requires several conceptions of space. Location and distance are important in understanding why proximity to certain requirements will always matter. But territorial control was also clearly a factor. Finally, place matters, even where it apparently does not. In this case there was no imperative to locate somewhere to which people would be attracted, as employment requirements were small, and there would be advantages to creating a stable workforce in a small town – quite different from the frenetic mobility characterizing Facebook headquarters.

Finally, we need to avoid the temptation to see these scales as somehow naturally occurring. Each of the scales listed above is humanly created in two senses. First, and most obviously, a scale such as "the national" or "the urban" refers to entities that we have collectively created. Nations and cities are not naturally occurring phenomena! Second, and less obviously, each of these scales is being actively constructed and reconstructed with ongoing changes in our economies and societies – a process called the "production of scale." There was a time, for example, when the national economies of the industrialized world were reasonably self-contained and governments could think about economic policies at the national scale. Many would argue that Western European and North American economies could be thought of in these terms from the 1940s to the 1970s. It is important to point out that this was not the experience of many former European colonies for whom national autonomy was much more constrained – here we might have to think about "empire" as a scale of analysis. But the national scale also proved unstable in Europe and North America. The expansion of trade, flows of investment capital, and the reach of transnational corporations since the 1970s have facilitated a globalizing era in which it is very difficult to think about economic management operating solely at the national scale. The scale of our capitalist economy has thus been reworked, and in this sense, scale is constantly being produced.

Alongside the rise of global processes, many would also argue that we now live in an economic world in which it is urban centers and regions that are the nodes of growth and innovation (Scott, 1998). It is networks of world cities that drive the global financial system (New York, London, Tokyo, Frankfurt, Hong Kong, Singapore, etc.) and major manufacturing centers are located in regions such as Guangdong province in China, and export processing zones across many developing countries. It is, therefore, increasingly the characteristics and connections of these urban regions that shape the geography of global economic activity. Again, then, we see scale being restlessly restructured in the context of a dynamic capitalist economy (an issue we return to in Chapter 3).

1.6   Summary

There are many popular conceptions about what geographers study. If you are a student of geography, you will have heard most of them. Geography, it is often assumed, is about maps (knowing where things are located, and how borders are configured), the natural environment (rivers, mountains, volcanoes, glaciers, etc.), and places (going there for field trips, knowing capital cities, etc.).

This chapter has shown that these preconceptions of Geography hold elements of truth, but in fact geographical analysis is about much more. The old stereotype about geographers and their maps is accurate in so far as we are certainly concerned with questions about where things are located and how physical distance affects patterns of economic activity. The notions of space that arise from thinking geographically in this way are also central to the geographical analyses that have become popular in the discipline of Economics. In Figure 1.7, this kind of space is represented by grid lines across which locations and distances can be measured.

The lines on a map that define the territories controlled by states or other entities are also key elements of geographical analysis. But we are less interested in where exactly a border is located than what its effects might be on economic activities within it, or economic flows across it. It is therefore the carving up of space into controlled or managed units that makes territory an important concept. The zones shaded on Figure 1.7 indicate how territory may "color in" space in this way.

If location gives us lines and dots on a map, and territory shades them in and indicates who controls them, then places – in all of their richness, uniqueness, and complexity – are utterly impossible to convey using a map. They are fashioned out of every possible dimension that differentiates space: the natural environment, landscape, cultural life, political processes, types of work, the things that get produced, consumption patterns, and so on. This does not, however, imply that

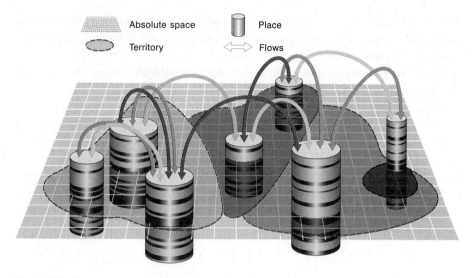

Figure 1.7 Space, territory, place, and scale

places cannot be understood in relation to larger processes or ideas. When we study a place we certainly need to think about big ideas such as capitalism, gender ideology, and identity. But we do so in order to understand the place, not in order to find ways of making it fit with certain universal principles. Nor does the study of place imply that the place alone is studied – places are created out of historical and contemporary connections with other places across space. In Figure 1.7, the complexity of place is depicted using a layered pattern – one that suggests the unique sedimentation of characteristics over time, shaped both internally and by relationships or flows across space.

Finally, all of these spatial concepts are discussed with an implicit framework of scale that structures them. The global, the national, the regional, and the urban are all scales that can be used to "frame" our understanding of the economic. But we have to remember to pay attention to multiple scales at the same time – to take in the whole canvas as well as the detailed brushstrokes. We also have to remember that scales are humanly produced and much of the restructuring of our economic lives is a reworking of the scales at which they are constituted.

By now it should be clear that space is far more than just the canvas on which geographical patterns are mapped out. It is space that creates geographical patterns through the influence of distance, territory, and place upon economic activities. In other words, space is an active agent in shaping economic geography. At the same time, however, every geographical pattern that emerges is itself a reshaping of space; when a place like Silicon Valley is created, it is more than just a pattern of existing economic activity – it is also a magnet for future activities as well. Space has been altered for the future. When a newly scaled territory is created, like the European Union, the opportunities for business and for labor are reworked. And when a new subway or mass transit line reaches an outer suburb, the relative distance to that location has been permanently changed. We are, then, always producing space, and it is the role of geographers to understand not just how patterns in space are changing, but also how space itself is changing.

Notes on the references

- Alistair Bonnet's (2008) *What Is Geography?* provides a short and accessible introduction to the question that it poses.
- Doreen Massey's (1994) collection of essays remains some of the best and clearest thinking on the subject of place, but Cresswell (2004) also provides an excellent overview.
- Although not focused on economic issues, Stuart Elden's (2009) *Terror and Territory* provides a clear and sophisticated meditation on the implications of territory.
- Keil and Mahon (2009) provide an excellent review of the geographical literature on scale.
- Barnes et al. (2012) and Leyshon et al. (2011) bring together excellent collections of statements by economic geographers about the state of the field.
- Hudson (2004) provides a challenging geographical conceptualization of the spaces, flows, and circuits of economic activities.

Sample essay questions

- Is space the canvas on which economic geographical patterns are mapped out, or does it play a role in shaping those patterns?
- What is territory, and how does it affect the conduct of economic activity?
- How is "place" different from "location"?
- Why is scale more than simply the physical space in which social processes happen?

Resources for further learning

- http://people.hofstra.edu/geotrans/index.html. The Geography of Transport Systems website at Hofstra University has extensive materials that explain the role of distance in the formation of geographical patterns.
- http://faculty.washington.edu/krumme/ebg/contents.html#location. Gunter Krumme's website on "Economic and Business Geography" at the University of Washington has a wealth of information about key economic geographical ideas.

- A complete lecture series on the Economic Geography of the Industrial World, by Richard Walker University of California, Berkeley, is available as a free download on iTunes.

References

Barnes, T., Peck, J., and Sheppard, E., eds. (2012). *The Wiley-Blackwell Companion to Economic Geography*. New York: Wiley-Blackwell.

Bonnett, A. (2008). *What Is Geography?* London: Sage.

Castree, N., Coe, N. M., Ward, K., and Samers, M. (2004). *Spaces of Work: Global Capitalism and Geographies of Labour*. London: Sage.

Cresswell, T. (2004). *Place: A Short Introduction*. Oxford: Wiley-Blackwell.

Elden, S. (2009). *Terror and Territory: The Spatial Extent of Sovereignty*. Minneapolis: University of Minnesota Press.

Economist, The. (2012, February 4). Floating Facebook: the value of friendship.

Facebook. (2011). Facebook statistics. www.facebook.com/press/info.php?statistics (accessed on January 15, 2011).

Hudson, R. (2004). *Economic Geographies*. London: Sage.

Keil, R., and Mahon, R., eds. (2009). *Leviathan Undone? Towards a Political Economy of Scale*. Vancouver: UBC Press.

Krugman, P. R. (1991a). *Geography and Trade*. Cambridge: MIT Press.

Krugman, P. R. (1991b). Increasing returns and economic geography. *Journal of Political Economy*, **99**, 483–499.

Leyshon, A., Lee, R., McDowell, L., and Sunley, P., eds. (2011). *The Sage Handbook of Economic Geography*. London: Sage.

Massey, D. (1994). *Space, Place and Gender*. Minneapolis: University of Minnesota Press.

Scott, A. J. (1998). *Regions and the World Economy: The Coming Shape of Global Production, Competition and Political Order*. Oxford: Oxford University Press.

Sheppard, E., and Barnes, T., eds. (2000). The Blackwell Companion to Economic Geography. Oxford: Wiley-Blackwell.

Tabuchi, H. (2011). Facebook wins relatively few friends in Japan. *New York Times*, January 10, B1.

CHAPTER 2

THE ECONOMY
What does it mean?

Goals of this chapter

- To show that "the economy" is an idea that arose in a particular time and place
- To introduce some concepts that are used in understanding economic processes
- To explore how conventional notions of "the economy" might be expanded

2.1   Introduction

I come from down in the valley where mister when you're young
They bring you up to do, just like your daddy done [...]
Then I got Mary pregnant, and man that was all she wrote
And for my nineteenth birthday I got a union card and a wedding coat [...]
I got a job working construction, for the Johnstown company
But lately there ain't been much work, on account of the economy

I've seen a lot of bailouts in my life.
But why is it I never see a bailout for the homeless and the poor?[...]
Do you want to live your life a slave? In chains from the cradle to the grave.
The economy is suffering, let it die. Yeah!

"The Economy is Suffering, Let it Die", Anti-Flag, The People
or the Gun, 2009 (Sideonedummy Records)

The economy is perhaps an unlikely theme to inspire the lyrics of a song. But when economic themes do enter popular musical culture, the way in which they are understood can be interesting. Bruce Springsteen's lyrics from his 1980 album, *The River*, mourn the effects of economic decline on an industrial town in Pennsylvania. The song describes a social context in which young men are raised to work in blue-collar jobs, just like their fathers, but the unemployment that characterized many such areas in the 1970s means that "lately there ain't been much work." The lyrics and music lament the effects of economic change on everyday lives, but there is little indication that there is much that can be done about it. The contemporary punk band Anti-Flag, on the other hand, presents an altogether angrier and more resistant response to current economic problems. They rail against the way in which governments in the United States, the United Kingdom, and elsewhere rescued major banks and financial institutions during the credit crisis of 2008–09, while neglecting the poor and the homeless. Their response to the suffering of the economy is to "let it die."

While the two songs offer quite different reactions to economic processes, they share an important and widespread understanding of something called "the economy." In both cases, the economy is seen as external to everyday life – in Springsteen's case it is a force that has brought recession and despair to a locality and its people, while for Anti-Flag it is so external to everyday concerns that letting it "die" is presented as an option. This strategy is not something mainstream commentators would advocate, but it is certainly very common to hear that the economy is "healthy," "growing," "sick," or "weak." These terms imply that the economy is a body that can become stronger or weaker over time, can grow or shrink, and can be healthy or sick. In that sense, "let it die" is simply taking the metaphor one stage further. The economy is thus being treated as an entity – perhaps organic, perhaps mechanical – but clearly something separate, distinct, and identifiable. The sense that the economy is "out there" is common to both of these songs and to almost all other academic, political, or popular discussion that we often hear. What does it mean to understand the economy in this way? Why should we separate out particular aspects of our lives and label them economic? And in doing so, what are we excluding? These are some of the key questions that we will examine in this chapter. Just as Chapter 1 sought to unpack and explain what we mean by geography, this chapter examines the implications of adopting a subject of analysis called the economy.

The chapter starts in Section 2.2 by considering the conventional ways in which we understand the economy. In particular, we examine how indicators such as gross domestic product are used to define and measure all things economic. While Gross Domestic Product treats the economy as a measurable entity, we show in Section 2.3 that this understanding of the economy is a surprisingly recent development. Even in the early 20th century, just a few generations ago, the notion of an economic sphere implied something quite different. It was during a few crucial years in the mid-20th century that the economy, as we know it today, became embedded in the academic and popular imagination. In Section 2.4 we then lay out some of the basic concepts of supply, demand, prices, and markets that are usually assumed to be the fundamental processes of economic life. Having examined how the economy came about historically and how it is analyzed, we then in Section 2.5 show how much is excluded in our conventional view of the economy. The implication, in short, is that we need to have a less "economic" understanding of the economy. This idea will underpin much of this book, as we explore the geographies of economic life as it is *actually* lived, rather than how it *should* be understood in theory.

2.2   The Taken-for-Granted Economy

It is a staple of our daily media diet to be informed about the state of the economy at various geographical scales: the local economy, the national economy, or the global economy. Political leaders describe their plans for the economy and take credit for making it grow. With increasing frequency, gatherings of global leaders take place at which rescuing, saving, or resuscitating the economy is the prime goal. But even beyond the rhetoric of electoral politics, "the economy" is a notion that we constantly face as we digest the daily news. Major events such as terrorist attacks, natural disasters, political tensions, and so on are assessed in relation to their impact upon the economy. A country's economy is examined for signs of health or weakness, almost like a patient being diagnosed. Figure 2.1 illustrates this idea: an obese "Mr. Economy" is on the weighing scales and looks likely to push up interest rates due to his increasing size.

Figure 2.1   The economy as an organic entity

The collective lack of questioning that exists concerning the economy perhaps reflects an assumption that it is such a solid and tangible "thing" that its existence could hardly be in doubt. This sense of a tangible thing out there is reinforced by the various ways in which the economy is defined and measured. Most commonly, this measurement takes the form of Gross Domestic Product (GDP), which adds together the total market value of the production of goods and services – that is, the total output – in a particular economy. There are several ways of calculating Gross Domestic Product (see World Bank, 2011). The most straightforward is to add together all recorded final expenditures on goods and services in a particular territory. This expenditure would include the private consumption expenditure of individuals (all the goods and services bought on a daily basis), the consumption expenditure of governments (supplying hospitals, schools, armed forces, etc.), and the expenditure directed toward investment. In addition, the calculation would also include money entering the economy from abroad. This can be done by taking the value of output that is exported, minus the value of imports. The measure deliberately excludes money that changes hands for intermediate products and services that are used to create the end product, as the value of this output is ultimately factored into the value of the end product that is bought. Figure 2.2 summarizes the components of GDP. The final figure for GDP gives a total amount of wealth that is being generated in an economy and gives a sense of how different economies compare, especially when the total wealth is given on a per capita basis.

GDP is most often calculated for national economies, but figures are also sometimes quoted for subnational units, such as states, provinces, or regions, and also for supranational entities like the European Union. Although there are many problems with GDP as a measure of economic activity and well-being, as we will discuss later, it does give an immediate sense of the geographical unevenness of global wealth and power.

GDP gives us a way of describing the economy as a whole, and in many ways it reflects a popular conception of what constitutes the economy – a territorially bounded entity that consists of a complex set of wealth-generating processes. In the next two sections, we examine more closely where this idea of the economy as an entity comes from and how we go about understanding its internal processes. Only then can we start to think carefully about what the concept's limitations might be.

2.3   A Brief History of "the Economy"

In the English-speaking world of the early 18th century, the word "economy" would have referred to the management of a household or a family budget. In some contexts we still use the word in this sense today. When we talk about economizing we are usually referring to personal financial budgeting. If we drive

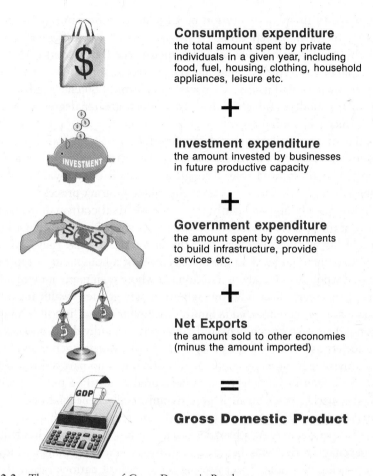

Consumption expenditure
the total amount spent by private
individuals in a given year, including
food, fuel, housing, clothing, household
appliances, leisure etc.

+

Investment expenditure
the amount invested by businesses
in future productive capacity

+

Government expenditure
the amount spent by governments
to build infrastructure, provide
services etc.

+

Net Exports
the amount sold to other economies
(minus the amount imported)

=

Gross Domestic Product

Figure 2.2   The components of Gross Domestic Product

an economy-size car, or buy an economy-class air ticket, the implication in each case is that they are smaller and more frugal options.

The 18th century, however, was a period of rapid change, and its latter decades in particular marked the beginning of the Industrial Revolution in England. Before then, when the production and use of goods and services was on a very small scale, there was little need to think about a set of economic processes that exceeded the scale of the family or household. In pre-industrial societies, most agricultural or craft production was for subsistence purposes, perhaps with some surplus paid to a feudal aristocrat or exchanged in local markets. The Industrial Revolution, however, saw Western Europe undergoing significant changes in the nature of production and the social relations that surrounded it. In particular, larger scale agricultural production emerged along with factory-based manufacturing. With production on a much grander scale and increasing levels of work specialization, it was possible to think of a *division of labor* for the first time. This division refers to the ways in which different people carried out different tasks and were thereby dependent on each other for their needs (Buck-Morss, 1995). The division of labor, nevertheless, had always existed in some sense. Each village would probably have had its own carpenter, blacksmith, butcher, and so on. But the birth of the modern industrial era saw a much greater degree and scale of interdependence and specialization.

It was this division of labor that Adam Smith famously identified in his book *The Wealth of Nations*, published in 1776. Using the example of a factory making pins, Smith showed that through a "division of labour," a group of people specializing in different stages of the manufacturing process could make pins far more quickly and efficiently than an individual craftsman carrying out every stage of the process himself. Smith's insights were, however, much broader than this. In his writing there was, for the first time, a sense that "economy" concerned something larger than the management of a household. It represented an integrated whole at the scale of a nation – a whole with many individual parts that worked together, albeit unknowingly, to create greater wealth for all. This interdependence was summarized in Smith's metaphor of the "invisible hand." By seeking only their own enrichment and advancement, individuals unconsciously benefited society as a whole in the process. In Smith's words, each individual was "led by an invisible hand to promote an end which was no part of his intention" (1976, p. 477). Smith's ideas have since been used to justify a political ideology in which the market mechanism is seen as universally beneficial and the most appropriate way of organizing society in all contexts. This is a rather partial reading of his perspective. At a broader level, Smith was perceptively identifying the interdependence that was being created by an emerging modern industrial society. By seeing "economy" as fostering the wealth of nations rather than the management of households, he was pointing to the development of an integrated whole in which people were participating as economic actors.

The analysis of national production and consumption, which Adam Smith developed, was then known as *political economy*. It was essentially concerned with the management of resources at a national scale. In Britain, institutions such as the Bank of England and the London Stock Exchange were emerging at about the same time. They provided an organizational basis for understanding economic processes on a larger scale. By the early 1800s, economic processes were understood as matters of national significance and not simply practices of household management. This was a significant step, but it was still some way from the notion of the economy as we use the concept today, or as we saw it used in the opening quotes from Bruce Springsteen and Anti-Flag.

To find the beginnings of the economy being understood as a separate and distinct entity, we have to move forward to the end of the 19th century. Until that time, it was political economy in the tradition of Smith, and later in the more radical version developed by Karl Marx writing in the mid-19th century, that generated understandings of collective production, consumption, trade, and wealth in society. The late 19th century was, however, a period of profound scientific, technological, and intellectual change. European societies were expanding into colonial territories, large industrial cities were growing rapidly, and great strides were being made in fields such as electrical engineering, medicine, and the natural sciences. It was also during this period that modern Economics, as a profession and an academic discipline, was being established (Mitchell, 1998).

From about the 1870s onwards, economists started to think less in terms of *national* political economy – that is, the marshalling and management of a country's wealth – and more in terms of *individual* economic decisions and outcomes. These individual economic actions could, in turn, be aggregated together in order to develop an analytical model of how the national economy as a whole would work as a system. It was no coincidence that such thinking emerged at the same time as physics and chemistry were developing scientific models of the natural world. Economists adopted very similar approaches and even used terminology that was borrowed from the physical sciences. Underlying the new field of physics was an understanding of energy as the unifying force that connected all matter and processes in the universe together. Economists found their equivalent in the concept of "utility" through which anything "economic" could be measured in terms of its value or its usefulness to individuals. Like energy in physics, utility was assumed to represent a common element in all economic transactions and rendered them comparable and quantifiable. With this assumed basis for a unified understanding of economic processes, individual producers and consumers could be viewed as analogous to the predictable behavior of atoms and molecules, and the economic processes they created could be likened to the forces and dynamics studied by physicists. Indeed, the concepts that economists used to understand economic behavior drew directly on physical processes such as equilibrium, stability, elasticity, inflation, and friction (Mitchell, 1998). These kinds of metaphors are still very much in use today by economists and just about everyone else, as noted in Box 2.1.

For these natural science metaphors to be meaningful, it was necessary to make some quite significant assumptions about the behavior of human economic actors. These assumptions are still largely applied in the field of Economics and are described in more detail in Section 2.5. Essentially, people had to be assumed to behave in a similarly rational and consistent manner in *all* circumstances. This rationality was based on maximizing their own economic rewards (or utility) that could be quantified and counted. Equipped with such assumptions, the economists of the later 19th century could start to predict how different stimuli would affect economic behavior of individuals that could be aggregated in statistical terms. They began to develop an understanding of the economy as not simply the

FURTHER THINKING

Box 2.1   *Metaphors of economy*

Metaphors are required any time we need to *reduce* something that is complicated and difficult to *grasp* into a more conceptually manageable *picture*. The fact that the previous sentence contains three (*italicized*) metaphors to make its point illustrates how often we have to resort to them. But these are relatively minor metaphors. Economics employs much bigger conceptual metaphors in trying to make the economy comprehensible. Conventionally, these have involved drawing upon the scientific language and models of physics and biology to represent economic processes.

Sometimes the language of Newtonian physics is used to conceptualize economic processes as interactions of objects, flows, forces, and waves. Some everyday examples illustrate this point: the market provides a *mechanism* for bringing together buyers and sellers and for allocating resources; the location of raw materials exerts a *gravitational* attraction upon producers; and distance provides a *frictional* force preventing consumers from shopping in far-off places. Participants in the economy are therefore reduced to rational economic beings, obeying the "laws" of economics, just as atoms obey the laws of physics. More complex understandings of the economy may also resort to physical metaphors – for example, when we talk about economic *cycles* and *waves* of investment.

An even more *fertile* source of metaphors is found in biological processes. To talk of the *health* and *growth* of an economy is to represent it in terms of an organism or body. This metaphor is also found in ideas such as the *circulation*, *reproduction*, *herd instincts*, and the *contagion* of economic crises. Larger contexts of logic are also represented through biological metaphors – for example, when the market is seen as providing a process of *natural selection* in which only the *fittest* and *strongest* (by implication, the most efficient) will survive, and when crises are understood to *weed out* the least efficient producers. *Development* too is essentially a biological metaphor for understanding growth and change as a natural process, as in the maturing of a human body.

Clearly, we cannot do away with metaphors. We will always need them to facilitate our thinking. But we do need to be aware of the implications of understanding economic processes by resorting to these images, as they will inevitably only partially fit the context in which they are being applied. Thus, while in some circumstances it will be useful to think about economic actors as law-abiding "atoms," it is easy to see that this will exclude much that is interesting in the economic world. More broadly, by transposing the laws of nature onto the social world, we are closing off many possibilities of alternative thinking and alternative economic arrangements.

management of national wealth, but as a system of inputs, outputs, and decisions that occupied the realm of rationality and predictability entirely separate from issues of government, culture and society.

Perhaps the most graphic illustration of this new perspective on the economy was provided by the celebrated American economist Irving Fisher. In his doctoral dissertation, completed in 1892, Fisher designed a mechanical model of an economy with an intricate system of water tanks, levers, pipes, pivots, and stoppers. The following year, Fisher actually built the model and used it during his lectures at Yale University for several decades. By adjusting various flows and water levels, Fisher claimed that he had developed a predictive model of an economy that could be used to experiment with, and predict, the effects of various changes in market circumstances (Brainard and Scarf, 2005). Figure 2.3 shows the laboratory apparatus that Fisher built in 1925 to replace the first model.

While Fisher continued to experiment with mechanical analogies of the economy, others were developing mathematical models to study the economy as a coherent and logical entity. The 1930s, in particular, saw the emergence of *econometrics* as a field of study in which complex mathematical techniques were used to capture economic processes. These models were important not just because

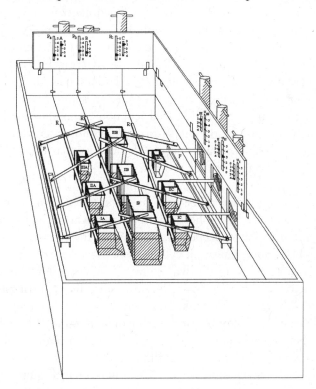

Figure 2.3   Irving Fisher's lecture hall apparatus, simulating the economy, ca 1925
Source: Brainard, W. and Scarf, H. (2005) How to compute equilibrium prices in 1891. *The American Journal of Economics and Sociology*, 64(1), 57–83. Image: Irving Fisher's lecture hall apparatus.

of their sophistication and the appearance of mathematical scientific rigor that they gave to the field of Economics, but also because they enabled economists to go beyond the notion of the economy as a set of mechanical levers, tanks, and pulleys. Models such as those constructed by Fisher could generate ever more complicated depictions of economic processes in the economy, but they were essentially static. They could not handle any kind of expansion in the economy as a whole, or any kind of external change. What would happen if the water in Fisher's apparatus started to evaporate, or started to leak? Or if the tanks changed in size and number over time? Or if some kind of external shock disrupted the apparatus? Such a model could not accommodate these kinds of dynamic changes, but the mathematical versions being developed by the 1930s could analyze such dynamism. Growth and change in the economy *as a whole*, driven by both internal and external forces, could be the subject of analysis for the first time.

While econometrics delivered the analytical tools to study the economy as a whole, the notion of national economic management was emerging at the same time. In his *General Theory of Employment, Interest and Money*, published in 1936, the British economist John Maynard Keynes established the idea that a national economy was a bounded and self-contained entity that could be managed using particular policy tools. These tools included controls over interest rates, price levels, and consumer demand. While Keynes developed his ideas before the Second World War, in a context of depression and unemployment, it was after the war that his ideas really took hold and governments started to engage in the kinds of economic management that he had advocated. For almost three decades from the 1940s until the early 1970s, Keynesianism was the orthodoxy of economic management in the industrialized, non-communist world.

It was, therefore, only by the 1940s that the economy as we know it today had really emerged as a popular concept. The idea by that time implied several important features about economic processes:

- First, the economy came to be seen as an *external sphere*, separate from the rest of our lives. Metaphors based on physical or biological processes greatly assisted in this conceptualization. When the economy is imagined as a machine or as an organism, it can be more easily seen as an external force bearing down upon us. In this way, it was possible to think about someone, or some place, being affected by the economy. Bruce Springsteen's lyrics, which would have been unthinkable when the word "economy" referred to household management or the stewardship of national wealth, could therefore only have been written in the second half of the 20th century. Understanding the economy as a machine or organism also enables us to think of its health and robustness – again, an idea that would have seemed quite odd a century ago. The addition of the word "the" to the word "economy" is an important change that indicates this new meaning. Whereas before economy was an activity or an attitude, it is now a thing.

- Second, because of the notion that the economic sphere operates according to its own internal logic, it has become seen as *independent* of social, political, and cultural processes. While political systems, ideologies, and parties might vary across time and space, and while cultures and social practices might also display rich geographical variation, our understanding of the economic realm often presents it as operating according to a separate and universal logic of its own, such as profit or utility maximization.

- Third, the economy, as it emerged in the 1940s, was primarily a concept that focused on the *national* scale. It was the national economy that was analyzed and modeled using the new techniques and approaches developed in the 1930s. It was also in the 1940s that measurements of national economic well-being, such as Gross Domestic Product, were first developed. Imagining the economy as constituted at a national scale allowed the global economy to be organized as an *inter*-national order. The institutions established in the 1940s, including the International Monetary Fund and the World Bank, were explicitly coalitions of national governments coming together in order to coordinate their economies.

- Finally, the idea of a national economy as a machine that could be maintained and *managed* by government intervention introduced the notion that continual economic growth was not only possible, it was the responsibility of national governments to deliver it. Politicians around the world recognize this and usually see economic growth as their first priority. And when they fail to deliver, their critics are often quick to say how they could do it better. This is, however, a relatively novel idea that is a product of the second half of the 20th century.

We have seen, then, that much of what we take for granted in our notion of the economy is actually a set of ideas that are relatively recent. Out of the Industrial Revolution in Europe there emerged a sense of a national division of labor as a set of economic relations that constituted a sum of many parts. But it was not until the late 19th century that the idea of an economy as a regulated organism or mechanism started to emerge. It was as recently as the 1940s that this idea was widely adopted and started to become part of political rhetoric and popular understanding in the public domain.

By examining the historical emergence of the idea of the economy in this way, we have also implicitly made an important point: that there is nothing natural or fundamental about the way in which we understand, measure, and manage the economy. Instead, current ways of thinking are a product of particular historical circumstances in particular places.

2.4 Basic Economic Processes

We have seen that the economy tends to be treated as an entity that is "out there," and where this perspective comes from historically. In this section, we will examine more closely the mechanisms that are usually assumed to drive economic processes, including demand and supply, firms, production, markets, and prices.

Economics is normally defined as the study of the allocation of scarce resources. The starting point for economic analysis is usually, then, the existence of **demand** for a product or a service and the process of matching that demand to a supply. Demand could range from fundamental human needs for food, shelter, and clothing, to the desire for luxuries such as expensive holidays, fine dining, and art collections. We could, of course, all have the desire for these items. But to have economic relevance these desires would have to become *effective* demand – that is, desires by individuals backed up by their ability to pay for them. Demand is not, however, fixed. It will vary depending on the cost of purchasing something and depending on the other factors such as fashion, taste, age, gender, and location. For example, demand for a new electronic device can be created using advertising. It may be coming largely from those in a particular demographic cohort, and it may be concentrated in certain places where incomes and interest are sufficient. We should also remember that much of the demand for products and services does not come from individual end-users, but comes instead from various points within the production system. A piece of heavy machinery might be purchased by a copper mining company, whose copper might be demanded by a maker of cash registers, whose product might be bought by a chain of grocery stores. We may, as the final consumers, shop for groceries, but we would never directly demand the mining machinery, the copper, or the cash register. It is our purchases at the grocery store that create the demand for those products.

On the other side of the equation, a process of **supply** must be in place to meet demand – that is, someone must be willing to provide the product or service. This might be a relatively simple process in which an individual or group of people cook a meal, provide a haircut, or fix a car. Or it might be a highly complex operation in which thousands of people are involved in designing and manufacturing a sophisticated piece of machinery such as a computer, or providing a complex service such as a credit card payment system. This is where **firms** enter the picture, as organizing units that coordinate the production of almost every conceivable product or service. While firms are often assumed to be motivated purely by an imperative to make as much profit as possible, this may be manifested in varied ways: some may wish to expand market share rather than their immediate profit on each unit produced; others may be seeking to establish or maintain a reputation, even at the expense of lower profitability; while others may forego short-term profitability in order to meet goals on a longer time horizon (e.g., family business). The goal is always ultimately profitability, but this may not be the immediate imperative upon which a company is acting.

It is also important to note that firms come in many ownership or organizational forms, from single proprietor companies (e.g., a lone web-page designer), to partnerships (e.g., a law firm), to privately-held companies (in which ownership rests with an individual or group), to public companies whose shares are bought and sold on stock markets (and whose ownership is often mostly separated from its management). We must also recall that a great deal of consumption and production ranging from healthcare, to education, to policing is conducted by public entities, usually governments. Not-for-profit and charitable organizations also play a significant role, for example in the provision of private education from kindergarten up to some of the world's most prestigious research universities. The organizers of economic activity are therefore diverse.

Whatever the source of demand and whoever the supplier of a product or service, conventional economic analysis assumes that they will come together in a **market**. In its purest form a market operates to unite buyers and sellers and to fix a **price** for the product or service. Both demand and supply are sensitive to prices. The higher the price for something, the less we will demand, but the more someone somewhere will be willing to supply. As prices decrease, demand will go up, but firms will have less incentive to sell a product or service. These tendencies of buyers and sellers can be represented in what is perhaps the most iconic diagram for students of Economics – the meeting of demand and supply curves (see Figure 2.4). Where the two lines intersect, demand and supply have reached an **equilibrium** and this is the price that will be set.

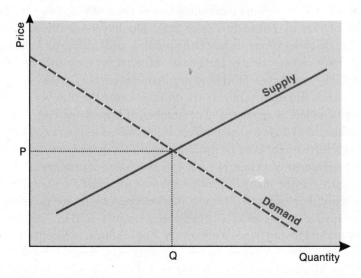

Figure 2.4   The supply and demand curves

The authors

Figure 2.5 Many consumers, many sellers in the Tsukiji Fish Market, Tokyo – but a perfectly competitive market is hard to find

The most basic and idealized model of the economic process involves many consumers and many producers coming together in a market. All have complete knowledge of their options in terms of who is available as a buyer or seller and so all are making fully informed decisions. This is a situation known as perfect competition. Perhaps the closest approximation to this scenario would be a fresh produce market in which many vendors are selling fruits and vegetables, fish, and so on, and many shoppers are examining the wares available (see, for example, Tsukiji fish market in Tokyo in Figure 2.5). The buyers in this scenario have many choices, such as where to buy their prawns, or whether to substitute tuna for tilapia. They can see clearly the quality of what is being sold and the price being asked. They may even be able to negotiate for a lower price.

In fact, such a situation rarely exists. In most cases the number of buyers or the number of sellers is restricted. The number of buyers for large aircraft, for example, is limited to a very small number of purchasers known as an oligopsony. In some cases there might even be just one buyer or a monopsony – for example, where the government of a country is the sole purchaser of military aircraft. More commonly there may be many buyers for a product but just a few suppliers – that is, an oligopoly. For example, in most market areas there are no more than a handful of suppliers of gas for cars or cell phone services. And in a few cases, there might be only a single supplier or a monopoly. For example, in Sweden wine is sold exclusively in government liquor stores. Markets may, therefore, take on many forms, and the idealized case of many buyers and many sellers is seldom seen.

It is also important to remember that even if there are many suppliers in a marketplace, those buying a product or service may not have complete information about what is out there. The explosion of online information in the last ten years has perhaps brought us closer to full awareness of the marketplace, but many consumers are still limited in their ability to make choices. Here, geography becomes an important factor. For example, a small town with a single grocery store is essentially a monopoly market, even if the internet allows consumers to see the price and selection of goods in a city far away. In short, the model of "perfect competition," which is so often assumed to determine a "fair price" in the marketplace for a product or service, is actually very difficult to find.

One market in particular deserves special mention, as it is perhaps the most important market of all: the demand for, and supply of, human labor. Ultimately all products and services are the result of collective human labor, and the prices they fetch in the marketplace are to a large extent determined by how the labor that created them is valued. The usual assumption in economic analysis is that the market for human labor is a competitive process just like any other market. It is also assumed that people are evaluated in this marketplace based on the education, training, and skills that they bring to a production process. In this way, it is their contribution to the efficient delivery of a product or service that determines their value in the form of a wage. The labor market, as we call it, is very complex because it comprises many connected but distinct specializations that come together in a division of labor. We have already noted a technical division of labor identified by Adam Smith earlier in Section 2.3. Box 2.2 examines the various kinds of social and spatial division of labor in more detail.

In markets of all kinds, prices are determined through the use of **money** as a universal measure of value. Although we tend to take for granted the use of money in a system of market exchange, it is in fact a remarkable invention. Currency allows products, services, and people of all kinds and in all places to be held up against a common yardstick. It also allows value to be stored and saved in forms such as coins, notes, and bank balances that otherwise have very little intrinsic value. Not only is money essential to the operation of a market, but there is also a market in money itself. Foreign exchange markets determine the price of one currency against another, while credit markets determine the cost of borrowing money, with its price being reflected in rates of interest.

While it is easy to lay out the basic parameters of a market-based economy, the actual operation of the system is, of course, very complicated. It is this complexity that is studied by the field of Economics. Complexity may arise in many ways:

- There may be difficulties in matching, in the same place and time, the skills and workforce needed, with the producers who are employing them.
- Information about prices, products, and services may be imperfect, thereby creating anomalies.
- The market for money itself may have effects upon the markets in which money is used as a unit of exchange. The cost and supply of money may be affected by external factors such as foreign traders or government policies.

KEY CONCEPT

Box 2.2   *Divisions of labor*

Divisions of labor can take many forms, but they all essentially refer to the separation of roles in the process of production and in society at large. The **technical** division of labor is the differentiation of specific tasks in a production process, so that individual people contribute just one small part to the larger goal. A classic example is the production line of a manufacturing plant in which individual workers repetitively add just one component to a product as it passes along an assembly line. Beyond the physical act of manufacturing, any large organization will have a division of labor in which tasks are specialized into finance, accounting, public relations, senior management, human resource management, procurement, and so on.

The wider economy is also characterized by an interdependent **social** division of labor. This might refer to the different industrial sectors of an economy such as retail, manufacturing, banking, shipping, and so on, but it can also imply the division of society into different classes of people. In particular, there are classes that own firms, or occupy highly privileged positions within them, and there are those who are paid to contribute their labor to such firms. Both gendered and ethnic divisions of labor may also exist in an economy. **Gender divisions of labor** occur within households as certain tasks tend to fall to men or to women, but it is also evident in society at large, as certain forms of waged employment become associated with masculinity or femininity. **Ethnic divisions of labor** are also apparent where ethnic or racial groups are concentrated in particular lines of work. They may arise out of choice, to some extent, but are more likely the product of current or historical forces of exclusion by the state or by a majority population.

Spatial divisions of labor arise where particular types of economic activity or production tasks are concentrated in specific geographical areas. For example, head offices might be concentrated in capital cities, manufacturing in other urban centres, and mining in the resource-rich periphery of a country. The **international division of labor** is an example of a spatial division of labor at a larger spatial scale, whereby countries specialize in particular kinds of activity. In the colonial era this typically involved extracting resources from colonized foreign territories for manufacturing sectors in the colonial metropolitan core. The **New International Division of Labor** refers to the restructuring of this arrangement, so that manufacturing is now increasingly carried out in places that were once colonial peripheries.

- More broadly, governments and other entities may not demand or supply things based on a self-interested rationality and may therefore distort the "perfect" market. For example, most governments limit the number of cell phone companies that can operate in a given market.
- The market in one set of products may affect the prices of others. A rise in the price of oranges, for example, might cause an increase in the demand for apples, while an increase in the price of oil has an effect on many other products and services.
- Uncertainties about the future might cause demand and supply to move in ways that do not reflect current needs and resources.
- Distortions in markets may also occur when information about a product is misleading. As we will see in Chapter 7, misinformation in the market for financial investment products was a major cause of the 2008–09 credit crisis.
- New technologies of production, new demand, or new regulations may bring about unexpected shifts in the process through which a price is determined, and may even lead to entirely new sectors. The auto and computer sectors are classic examples of unanticipated new economic sectors of historic importance.

All of these complications are often assumed to be the analytical domain of economists, and they have developed highly sophisticated statistical tools of analysis to examine the relationships between movements in prices, wages, interest rates, exchange rates, and other indicators. But they are ultimately complications that can be accommodated within the economists' peculiar worldview. In the next section, we will explore the ways in which the real world often confounds the foundations of the economists' model.

2.5   Beyond the Assumptions of Economics

We have seen that the economy was being re-imagined in the mid-20th century as a separate and identifiable entity. We have also seen that the key process understood to underpin the functioning of the economy was the market mechanism establishing a monetary value for all economic transactions. Economic analysis is now a highly sophisticated field and proclaims to accommodate all sorts of complications that arise in market operations, but it still has its limitations when we start to question its boundaries and assumptions. In this section we address three areas in which the assumptions of economic analysis are often challenged and where a geographical approach that is sensitive to real people and places might have important things to say.

Beyond Rationality

Economic analysis usually assumes two things about how people behave individually. The first is that people act through self-interest and therefore seek to maximize their personal benefit, or "utility," in any economic transaction. In other words, they are acting on their own behalf and not out of altruism and sympathy for someone else. The second assumption is that people enter into economic transactions with full knowledge and information about both their own preferences and about the ways in which those preferences can be met. In other words, they can fully assess their own needs and desires in terms of what they are willing to pay or charge for a product or service, *and* they have full access to information about alternative possibilities.

In reality, the decisions people make are far more complicated and irrational. Self-interest may be mixed with altruistic concerns – for example, by choosing to reduce personal consumption out of consideration for the effects on the natural environment. The maximization of personal gains may also be called into question where we pay a premium for "fair trade" products, for example. Even self-knowledge cannot be assumed. While the market mechanism assumes people have a threshold price that they are willing to pay for a product or service, in practice we seldom make such a careful calculation. Knowledge of the full set of options offered by the market is often unrealistic in many contexts.

All of these exceptions could in some ways be incorporated into conventional economic analysis, for example, by inferring an amount for the utility that a person derives from being altruistic or ethical, or by looking at collectively-determined prices rather than individual willingness to pay. But people also act in ways that cannot be reduced to a calculation of how their personal utility will be maximized. Many consumption and production decisions are based on the symbolic or cultural value that is embodied in a product or based on the norms, traditions, and expectations of their cultural group. Different ethnic groups, for example, might have quite different profiles of demand depending on their cultural background. On the supply side, companies too might make production decisions that are driven by their traditions or culture rather than purely rational economic calculation.

The wider point here is that the assumption of economic rationality may be less valid than is often believed. Much of the edifice of economic analysis is based on the idea that people are rational economic persons, the so-called *homo economicus*. In other words, they are always trying to behave in ways that maximize their satisfaction, wealth, or enjoyment. In reality, people nearly always fall short of this ideal. A contemporary example is found in the practices of Islamic banking and finances. A core tenet of Islamic belief is that money cannot be used to make further wealth outside of a productive process. While it is permissible to invest money in manufacturing, services, or trade, the charging or paying of interest (*riba*) on a loan is not allowed under Sharia law (Pollard and Samers, 2007). In addition, investment in activities that are considered undesirable, such as gambling, alcohol, and pork-related products, are also shunned. This religious/cultural belief system, which requires behavior that lies outside of the profit-maximizing imperative of most conventional banks, has fostered the development of an alternative banking system. Given the wealth of several Islamic countries in the Middle East (based largely on oil) the result has been the creation of an alternative circuit of significant financial centers with a different geographical structure than the mainstream network of global cities. While mainstream global finance moves between cities such as New York, London, Frankfurt, Tokyo, and Hong Kong (see Chapter 7), Islamic banking has established a circuit centered on Manama (Bahrain), and includes Dubai, Tehran (Iran), Karachi (Pakistan), Kuala Lumpur (Malaysia), and London (U.K.) (Bassens et al., 2010). Figure 2.6 illustrates this map of Islamic financial centers.

While "irrational" behavior by individuals represents one problem for economic analysis, a related problem on a much larger scale is found in the collective irrationality of markets as a whole. Sometimes financial markets, in particular, display patterns or trends that are clearly irrational but reflect the mechanisms through which markets work. For example, an ever deepening risk associated with U.S. mortgage loans and investment products associated with them became apparent in the wake of the financial crisis of 2008–09. Yet banks continued to lend mortgages to risky borrowers because competition in the marketplace dictated that they must follow the "herd." A similar phenomenon occurred during the Asian financial crisis in 1997–98 when the bubble of the massively overinflated property market finally burst. One could argue that such crises (which we will examine in Chapter 7) represent an effective correction in an open marketplace, but clearly markets had been moving with "irrational exuberance" up until that point.

Another instance of ongoing market irrationality is found in the labor market. When markets decide on the economic value of something, it is generally assumed that they are working without cultural prejudices. This is particularly problematic when we look at how human labor is bought, sold, and valued. The assumption

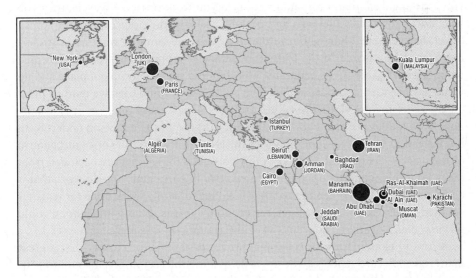

Figure 2.6 The circuit of global financial centers in the Islamic banking system
Source: Adapted from Bassens, D., Derudder, B. and Witlox, F., 2010. "Searching for the Mecca of finance: Islamic financial services and the world city network", *Area*, 42(1). pp. 35–46.

leads us to believe that the market simply and impartially assesses the quantitative value of a person's labor time in terms of the economic value that it generates. Appropriate skills are thus matched with appropriate jobs in a dispassionate and objective manner. As we will show in more detail in Chapters 13 and 14, however, gender and ethnicity often have profound effects on how individuals experience waged employment. Some jobs become associated with femininity and others with masculinity. In addition, some workplaces become highly masculine or feminine environments. Similarly, many urban labor markets show very clearly that processes of ethnic "sorting" are happening, with various jobs being disproportionately held by certain groups of people. Gender and ethnicity are not, then, non-economic factors. Instead, they are forms of identity through which economic resources are allocated. Furthermore, as culturally constructed ideas, gender and ethnicity have a very varied geography across global space, and so the economic processes that they influence are equally varied.

Beyond Calculability

A key feature of economic analysis is that, in general, something only counts if it can be counted. The rise of econometrics as the "science" of economic processes, and the role of a market based on money as the medium of exchange, have meant that a phenomenon has to be quantifiable in order to be included in models of the economy. If an activity takes place outside of the monetarized economy, or if a monetary value cannot readily be assigned to it, then it is not conventionally defined as an economic activity. This can lead to some odd conclusions. For example, if you drive or take a bus to your campus or to a place of work, then you have engaged in an economic act; but if you walk or ride a bicycle, then you have not. In one case, fuel bills and parking costs, or the price of a public transit ticket, have changed hands, but in the latter case the work of walking or cycling has been done outside of the formal monetary economy. Thus, in our accepted ways of understanding the economy, an economic act has not taken place and no value has been produced or added. The conceptual boundaries we tend to place on the economy are therefore rather arbitrary. The "economic," as it is usually understood, does not relate to the nature or purpose of the act being undertaken, but instead to whether or not money changes hands through the price mechanism or the market. As we will see later in this book, the things that are excluded from the measurable economy often follow a pattern. In particular, unpaid work that is commonly done by women in the home is excluded, while work done for wages is counted in full. At the outset then, the economy as it is usually understood could be viewed as a gendered concept (we will elaborate on this idea in Chapter 13).

A careful examination of economic processes reveals, of course, that a great many take place outside of the formal, quantifiable economy. In addition to household work, these transactions might be based on voluntary work, charitable donations of goods and services, exchange and barter arrangements, and even the underground economy where cash may change hands but not within the measurable economy. These activities have been likened to the submerged portion of an economic iceberg (see Figure 2.7). Production for a market, using cash transactions, and waged labor in a private company capture only a part of how people actually produce and exchange resources of various kinds in their daily lives. In families, in cooperatives, in neighborhoods, and in many other settings, people work, share, and exchange in ways that are seldom recognized as formal economic processes. The effect of this neglect is not just to *mis*-count the economy, but also to *dis*-count the work of certain people.

The miscounting of economic processes is especially important when it comes to our relationship with the natural world. As we will see in Chapter 5, it is often difficult to assign a quantitative monetary value for the "services" provided by the natural environment. Where this happens, nature is usually assumed to be outside of economic processes. And yet nature is absolutely fundamental to economic life. The tourism industry often relies on the benefits of unspoiled natural beauty, but these remain unquantified and outside of "the economic" (see

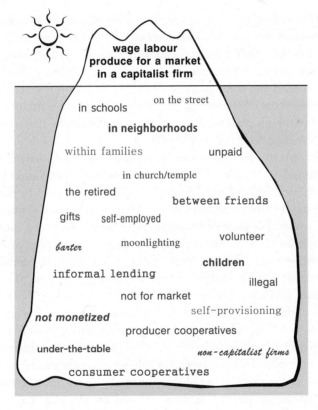

Figure 2.7   The economic iceberg and the submerged non-economy
Source: Gibson-Graham, J. K. (2006a) The End of Capitalism (as we knew it): A Feminist Critique of Political Economy, Second Edition. University of Minnesota Press. Drawing by Ken Byrne.

Yan Yan/Xinhua/Landov

Figure 2.8   Air pollution in China

Chapter 15). Other kinds of economic activities are also directly dependent on the natural environment for the disposal of waste products, but again the difficulties in quantifying the benefits of this service for individual users have meant that it is seldom factored into economic analyses. Figure 2.8 shows air pollution in China and gives a sense of the consequences when clean air is not given an adequate market value. In most of the world, it is only recently that attempts have been made to start putting a price on air pollution – for example, through the carbon credits and offsetting described in Chapter 5.

Even more directly, natural resources such as oil, gold, water, rubber, and so on are themselves bought and sold as commodities, and used as inputs in economic processes. Indeed, the economy is itself, in some ways, the process of transforming nature from one form to another, from rubber plantations to car tires, or from buried diamonds to engagement rings. At every turn, then, the economy is also an environmental process: as input, as process, and as impact. While this process does give a monetary value to nature, it is generally based on the present market value that does not reflect the long-run cost of depleting the earth's resources and life-support systems.

As a result of these silences in economic analysis, we often find an assumption that growth is unquestionably a good thing. A growing economy is considered healthy, robust, and thriving. Indeed a growing economy is considered a necessity such that new people, new resources, new efficiencies, new products, new wealth are all constantly needed. It would be unthinkable in most countries for a politician to mount an election campaign based on the idea that the economy has grown enough or is efficient enough already. Biological metaphors, of course, conveniently allow us to understand this kind of growth as an organic and natural process. But the separation of a "healthy" economy from its environmental consequences means that we often fail to take into account the full implications of a growing economy because not all of the consequences are unambiguously healthy. If more people use their cars rather than walking, then the economy has grown, but the environmental, health, and social consequences are far from positive. Even more bizarrely, if more people get involved in car accidents, resulting in repair bills, medical expenses, and so on, then again the economy registers growth, even if the social consequences are entirely negative.

The dependence of economic analysis on the quantifiability of economic transactions is therefore a major limitation. It leads almost inevitably to a neglect of work and transactions that do not occur through the monetarized market system. The silences that result often have a pattern to them manifested in the failure to account for the work done by women in particular, and to account for the true value of nature.

Beyond Private Property

A fundamental feature of a market economy is ownership, of products, of one's own labor, or of money to buy these things. In most societies private ownership of resources is indeed the norm, although there are exceptions. As we will see in Chapter 5, establishing private ownership of water, for example, has been attempted but it is difficult. A substance like air provides even more of a challenge to the system of private ownership. Private ownership of one's own labor power is also assumed, but not always practiced in reality. For example, where workers have been trafficked across international borders, where they are working in the homes of their employers, or where child labor is used, then the usual assumptions about employees selling their labor in a free market are not applicable.

Outside of mainstream capitalist societies, however, there exist groups where the assumption of private property is far from the norm within their economic culture. For example, indigenous people in Canada, the United States, Australia, Southeast Asia, and elsewhere often hold quite different views of property from the mainstream societies in which they are situated (Castree, 2004). While the modern Western view is that property can be owned, and bought and sold by individuals, in many indigenous cultures, individual ownership of this sort is an alien concept. Instead, the right to use the resources of the land or the sea is often collective and no individual has the right to sell them – and yet indigenous groups have historically been pressured to do just that. Countless disputes over drilling rights, fishing licenses, and logging concessions around the world hinge, in part, on this fundamental difference between market and cultural values, which Western companies have often done far too little to recognize.

Even in modern societies there are extensive forms of collective ownership that challenge the idea that private ownership is universal. Collective ownership by citizens happens especially, although not exclusively, through the agencies of the state. As we will show in Chapter 4, the state is a powerful economic entity in its own right, as both the supplier and consumer in many economic relationships. Much of a country's resources also lie within the control of the state, which may license them out to private companies. In Canada, for example, only about 10% of the country's land area is privately owned under freehold rights; the rest is all considered "Crown" land and is managed by federal or provincial governments. Indeed, in many contexts the state may be the *most* powerful force in economic life. And yet, decisions by the state may not be based upon any particular market logic, but rather upon a set of political calculations and national interests. To focus purely on market mechanisms is therefore to miss a great deal of what drives our economic lives and its variability across space.

In a variety of ways, then, the distinctions between the economic, the cultural, the political, and so on are quite arbitrary. Indeed, processes or objects that are conventionally assigned to one sphere have significant implications in another. Indeed it is the separation of these spheres in the social sciences today that is problematic. To understand economic processes it is necessary to place to them in the broader context of which they are a part, and once that context is added, then a world of geographical variability becomes apparent. We have come a long way from the treatment of markets as universalized mechanisms of rational behavior, to see them instead as geographically embedded. As Box 2.3 explains, this is an intellectual project that has a long heritage, as scholars have tried to grapple with the rightful place, both morally and analytically, of markets.

2.6   Summary

In this chapter we have critically examined some of the most cherished assumptions that are employed when economic issues are discussed either in the popular media or in academic analysis. We have seen that "the economy" is not such a natural and unproblematic concept as we might at first assume. It is, first of all, a notion that is of relatively recent origin and one that emerged in a particular historical and geographical context. This illustrates a wider point noted in the chapter. Rather than tending toward an ever more complete understanding of how the world works, knowledge in the social sciences is a product of its time and place. What makes sense when we discuss the economy now is a reflection of the circumstances in which we find ourselves. A few generations ago, the concept had different connotations, and perhaps a few generations from now its meaning may have shifted again.

We have also examined some of the fundamental ideas that are used to analyze and understand the economy: markets, prices, demand, supply, and so on. With increasing sophistication, economists claim to be able to model and predict these processes. And yet, if we limit ourselves to formal economic processes and to the dimensions of life that are quantifiable, then we arrive at a very partial

FURTHER THINKING

Box 2.3   The place of markets

The idea that markets should be the fundamental organizing principle in society is both widespread and deeply rooted. Starting with the political philosopher Friedrich von Hayek, there has been a body of economic thought that sees markets as both efficient and just. They are efficient because, in theory, they allocate resources to the most productive users; and fair because all relationships are reduced to their monetary fundamentals without attention to tradition, class, caste, gender, ethnicity, and other "nonmarket" forms of differentiation. These tenets, of efficiency and fairness, form the basis for a set of political beliefs often called neoliberalism, which argues for the expansion of market relations into all aspects of life such as healthcare, welfare, education, social services, and so on. Several geographers have traced the rise of neoliberalism into the mainstream of political life around the world (Harvey, 2007; Peck, 2011: see Box 4.4).

Another line of thinking takes a different view of markets and is often traced back to the political economist Karl Polanyi. Polanyi argued that markets do not provide a "pure" mechanism that ensures efficiency and fairness, but are instead *embedded* in society. Even those markets that seem to be freely operating must be created by social institutions that are not market-driven. It would, for example, be impossible to imagine the creation of transport and educational systems (upon which markets undoubtedly depend) without some forms of collective coordination and provision by state and nonstate actors. However, when "the market" is allowed to subordinate society, then it begins to undermine the very institutions that allowed it to exist in the first place. Thus, many would argue that the increasing introduction of a market mechanism into the provision of education, for example, is undermining the need of an advanced market economy for an educated workforce. This leads to what Polanyi calls counter-movements, where individuals and groups in society seek "social protection" from the forces of the market. The idea that markets are socially embedded makes a geographical approach to understanding how they might be rooted in specific places especially important. For a further discussion on the conceptualization of markets, and their embeddedness in place, see Rankin (2004).

understanding of how the economic world really works. The final section of this chapter thus explored some of the ways in which we need to broaden our understanding of economic processes. Our economic lives go beyond those processes that are quantifiable in monetary terms. They also spill over into all other forms of human activity such as social, cultural, and political practices. These other spheres might be considered as influences on the economy, or they might be seen as affected by the economy, but too often the idea that they are entirely separate still remains. As many of the chapters in this book will show, economic processes are very much embedded in other forms of human and/or environmental interaction, and it is this that gives them a rich geographical dimension that needs to be understood.

Notes on the references

- In 2008 the French government commissioned a major report on the measurement of economic performance and social progress amid dissatisfaction with GDP as the most commonly used indicator of development. The commission was led by Nobel Laureates Joseph Stiglitz and Amartya Sen, and the report provides an excellent discussion of the issues involved in measuring human progress. Available at: http://www.stiglitz-sen-fitoussi.fr/en/index.htm.
- Timothy Mitchell (1998, 2002) has written some excellent studies tracing the emergence of the economy as a concept, and examining the consequences of development discourse in Egypt.
- For early and perceptive discussions of the interface between economy and culture, see Thrift and Olds (1996) and Crang (1997). Castree (2004) provides an updated critique of this debate.
- Barnes (1996) and McCloskey (1998) examine the role of scientific metaphors in the concepts of economics. Kelly (2001) examines metaphors of economic crisis in Southeast Asia.
- For an innovative analysis of the political implications of seeing the economy as a discursive construct, see Gibson-Graham (2006b).
- Partha Dasgupta (2007) provides a very readable introduction to the thinking of Economics, and Chapter 4 explains the working of the market mechanism in particular. Silvia Nasar (2011) provides a readable and engaging history of economic thinkers, including Fisher and Keynes.

Sample essay questions

- Explain how, and why, the meaning of "the economy" has changed over time.
- When we discuss the economy, what activities of production, exchange, or consumption are usually included and excluded?
- Is it possible to understand economic processes without also understanding other dimensions of human society and the natural environment?

- Outline the assumptions of a conventional Economics approach to economic processes, and show how a geographical approach might challenge these assumptions.

Resources for further learning

- The History of Economic Thought (HET) website at the New School for Social Research university in New York provides some excellent materials on key economic thinkers and concepts: http://homepage.newschool.edu/~het.
- The United Nations Development Programme calculates a Human Development Index that provides an alternative measure of progress beyond the narrow confines of GDP: http://hdr.undp.org/hd.
- The Community Economies project, which includes the geographers Julie Graham and Kathy Gibson, is involved in thinking about and enacting alternative models of "the economy": http://www.communityeconomies.org/index.php.
- The Division of Labour website provides entertaining ways to learn and teach economics principles through song lyrics: http://divisionoflabour.com/music.

References

Barnes, T. (1996). *Logics of Dislocation: Models, Metaphors and Meanings of Economic Space*. New York: Guilford.

Bassens, D., Derudder, B., and Witlox, F. (2010). Searching for the Mecca of finance: Islamic financial services and the world city network. *Area*, **42**, 35–46.

Brainard, W., and Scarf, H. (2005). How to compute equilibrium prices in 1891. *The American Journal of Economics and Sociology*, **64**, 57–83.

Buck-Morss, S. (1995). Envisioning capital: political economy on display. *Critical Inquiry*, **21**, 434–467.

Castree, N. (2004). Economy and culture are dead! Long live economy and culture! *Progress in Human Geography*, **28**, 204–226.

Crang, P. (1997). Introduction: cultural turns and the re(constitution) of economic geography. In R. Lee and J. Wills, eds. *Geographies of Economies*. London: Arnold, pp. 3–15.

Dasgupta, P. (2007). *Economics: A Very Short Introduction*. Oxford: Oxford University Press.

Fischer, I. (1925). Mathematical Investigations in the Theory of Value and Price. New Haven: Yale University Press.

Gibson-Graham, J. K. (2006a). *The End of Capitalism (as We Knew It): A Feminist Critique of Political Economy*, Second Edition. Minneapolis: University of Minnesota Press.

Gibson-Graham, J. K. (2006b). *A Postcapitalist Politics*. Minneapolis: University of Minnesota Press.

Harvey, D. (2007). *A Brief History of Neoliberalism*. Oxford: Oxford University Press.

Kelly, P. F. (2001). Metaphors of meltdown: political representations of economic space in the Asian financial crisis. *Environment and Planning D: Society and Space*, **19**, 719–742.

McCloskey, D. (1998). *The Rhetoric of Economics*, Second Edition. Madison: University of Wisconsin Press.

Mitchell, T. (1998). Fixing the economy. *Cultural Studies*, **12**, 82–101.

Mitchell, T. (2002). *Rule of Experts: Egypt, Techno-politics, Modernity*. Berkeley: University of California Press.

Nasar, S. (2011). *Grand Pursuit: The Story of Economic Genius*. New York: Simon and Schuster.

Peck, J. (2011). *Constructions of Neoliberalism Reason*. Oxford: Oxford University Press.

Pollard, J., and Samers, M. (2007). Islamic banking and finance: postcolonial political economy and the decentring of economic geographies. *Transactions of the Institute of British Geographers* **32**, 313–330.

Rankin, K. (2004). *The Cultural Politics of Markets: Economic Liberalization and Social Change in Nepal*. Toronto: University of Toronto Press.

Smith, A. (1976, originally published 1776). *An Inquiry into the Nature and Causes of the Wealth of Nations*. Chicago: University of Chicago Press.

Thrift, N., and Olds, K. (1996). Refiguring the economic in economic geography. *Progress in Human Geography*, **20**, 311–337.

World Bank. (2011). World Bank Statistical Manual. http://web.worldbank.org/.

CHAPTER 3

CAPITALISM IN MOTION
Why is economic growth so uneven?

Goals of this chapter

- To understand the fundamentals of capitalism as an economic system
- To think structurally and systematically about capitalism and its uneven development processes
- To explore the integral role of space in capitalism
- To analyze different scales of capitalist geographies

3.1 Introduction

Imagine that we could go out into space and look back on the earth with a special kind of camera. The image produced by the camera would look like a thermal image in which hot spots show up bright, while cooler areas are darker. But now imagine that the camera is measuring economic activity instead of temperature. The bright areas on the image are those with rapid economic growth, rising incomes, and new enterprises being formed. The darker areas are those with declining economic activities, increasing unemployment and poverty, and businesses closing. The first thing we would notice about such an image is that it creates an uneven patchwork across global space – some areas are very bright, while others glow only dimly. The southern coastal provinces of China and the southern state of Karnataka in India would show up brightly, while crisis-hit parts of Europe such as Ireland, Spain, and Greece would be dimly lit.

We could then zoom in on parts of this economic image and notice that the geographical unevenness has further patterns at smaller scales. In the United Kingdom, the southeast of England around London shows up brightly, while the northern parts of the country much less so. Within Italy, the pattern is complicated because there are brighter patches in the north of the country, around Milan, but the south is relatively dark. In the United States, we see bright spots in places like Austin, Texas, but Detroit, Michigan, is noticeably struggling. Zooming in still further, we see cities that are patchworks of light and dark. Manila in the Philippines, for example, has residential neighborhoods and new shopping malls that are islands of brightly lit wealth and economic activity, but surrounding them are darker areas of poverty.

Next, imagine that these images from the "economic imaging camera" are not still pictures, but are instead different frames in a movie that has been filmed continuously for the past 100 years and is being played back at high speed. Now we would see the pattern of light and darkness shifting over time. Detroit is in darkness now, but it was burning bright in the 1950s (when the American automobile industry grew and brought great wealth to the city and the country). China was in darkness until the 1980s when the picture started to change rapidly. This imaginary example tells us that uneven development can occur in different places and at different geographical scales, from the global to the national, the regional, the urban, and the local. This geographical pattern can also change over time with the ebbs and flows of economic activity that constitute the global economic system. Overall, it is a structured pattern of wealth concentration, but one that is subject to a volatile process of boom and bust and boom again.

This chapter provides an important conceptual foundation for thinking about why the economic landscape at different geographical scales seems to be continuously changing. Our common sense tells us that the economic system in which we live, work, and play does not seem to be static and permanent; in fact, it is characterized by ups and downs in different times and places. This chapter will suggest that such dynamics of change are closely related to the way in which our economic system is organized. It is not the outcome of natural forces (e.g., the environment) or invisible hands (e.g., the market). The previous chapter has alluded to some common features of an economy as conceived in conventional Economics. In this chapter, we further examine the economic system often known as capitalism, and show how in its attempt to accumulate wealth it creates uneven development across space and through time. We will deal with more specific actors in capitalism such as the state, nature/resources, labor, and finance in the next four chapters.

The chapter starts with a view of uneven development that sees it as either the result of natural (i.e., environmental) conditions, or as a starting condition that market forces will even out over time (Section 3.2). In many ways, this corresponds to the popular perception of why some places are richer and more developed than others. In Section 3.3, we construct an alternative view using ideas drawn from geographical political economy. We explore the dynamics and mechanisms that produce uneven development in the capitalist system in particular – the system of wealth production and distribution that now prevails in all but a few communist states such as Cuba and North Korea. Having established these fundamental mechanisms, we then look at how the capitalist imperative has produced peculiar geographical outcomes (Section 3.4) and how particular places and scales are incorporated into the shifting landscape of capitalism (Section 3.5). Finally, we note in Section 3.6 that not all processes of accumulation are located within national capitalist systems, but are interconnected on a global scale.

3.2   Uneven Development – Naturally!

A common approach to the unevenness of economic development is to see it as a natural state of affairs – natural either because of the uneven geographical distribution of the bounty of nature (for example, oil fields can only be where oil is located), or because growth has to start somewhere and, under the right conditions, it will naturally spread. Both of these views are very much evident in contemporary debates about why economic development is spatially and socially uneven.

At first glance, the notion that human societies are uneven because nature is uneven is a rather tempting explanation. After all, the Industrial Revolution was initiated in the towns of northern England that had the appropriate endowments of water, coal, iron ore, and other raw materials in the second half of the 18th century. And today, nature has bestowed upon countries such as Australia, Canada, Russia, Sweden, and the United States an immense bounty of mineral, agricultural, and forest resources. There seems to be a case for using environmental factors to explain economic development outcomes – a line of thinking that can be termed "environmental determinism."

But if the list of examples is expanded, this argument for environmental determinism begins to unravel. Indonesia has a remarkable array of mining, agricultural, and forest resources, and yet its 237 million people enjoyed a per capita income of just over US$4,300 in 2010 (after adjustments are made for purchasing-power differences). In neighboring Singapore, a tiny island city-state with no natural resources at all, the per capita income was around US$54,700. Japan also has negligible oil, gas, and mineral resources, and relatively little arable land, and yet it too manages to exceed Indonesia's income almost nine-fold. Conversely, in 2010 Nigeria sat atop proven crude oil reserves estimated at 37.2 billion barrels (the eighth largest reserve in the world), and yet its 150 million people had only half of Indonesia's per capita income, and about 70% of them lived in poverty (hdr.undp.org, accessed on March 8, 2012). A territory replete with industrial raw materials is not, then, enough to guarantee the economic well-being of a nation's population – indeed, there is little correlation between a list of the world's wealthiest nations and a list of those with the greatest resource endowments. Clearly we have to look for other explanations of uneven development.

A second set of arguments starts not with the causes of unevenness, but with its presumed leveling out over time through processes of development that spread wealth across societies and spaces. This has been a key argument of mainstream approaches to economic development over the last 50 years. While intellectual fashions have changed, the argument has remained essentially the same: that all economies can develop if they adopt appropriate policies and strategies, and that uneven development is merely a temporary condition that will, naturally, be overcome. One of the earliest manifestations of this perspective was in modernization theory – a school of economic thinking prevalent in the 1950s and the 1960s that saw largely cultural and institutional barriers to development in the Third World (i.e., developing countries). Impoverished economies would develop toward the Western model of industrial production, a modern democratic society, and high mass consumption, if they first established certain preconditions. In early versions of the theory, developed by American economic historian Walt Rostow in 1960, these conditions included high savings and investment rates, and the removal of cultural resistance to modern science and industrial production. Such models implied that economic development will tend toward an equilibrium pattern in which differences are smoothed out over time.

More recent incarnations of this kind of economic thinking about development are found in the strong emphasis by powerful global institutions such as the International Monetary Fund, the World Bank, and the World Trade Organization on the role of free markets in generating wealth and development (see Chapter 4 for more discussion on these institutions). What remains constant, however, is the notion that underdevelopment is an aberration – a problem that will be naturally fixed if the capitalist system is allowed to operate fully. We will now turn our attention to the fundamentals of such a capitalist system in order to explore an alternative and more critical view – one that sees unevenness rather than market equilibrium as the normal state of affairs under capitalism, and suggests that any attempt to eradicate unevenness requires changes to the capitalist system itself.

· 3.3   Fundamentals of the Capitalist System

The conventional explanations of economic development in the previous section are deficient in one important respect. None of them actually seeks to explain *how* wealth is created, focusing instead on explanations for either relative differences in wealth between regions and countries (i.e., the resource endowment approach) or how these differences might be evened out over time (i.e., modernization theory). Taking one step backwards, we can instead think about the fundamental process through which wealth is generated. To do this, we need to consider the concept of value and the ways in which it is created and distributed in a system, or structure, of economic relationships.

Value Creation and Structures of Economic Life

Economic relationships involve the creation and distribution of value. Value is defined as either (i) the benefit we get from having or consuming something (use value) or (ii) the monetary worth of a good or a service traded in the market economy (exchange value). The creation of value is central to the question of economic development, and uneven development is a reflection of either a relative lack of the physical or organizational resources needed to create value (as in the explanations in Section 3.2), or a failure to capture and/or retain that value in the hands of the person, the household, or the community that created it. Ultimately, though, value is always created by people – that is, living human beings who are engaged in labor processes that make or transform a product (which may be a physical good or an intangible service).

There are, however, many ways of structuring the economic relationships of value creation. Peasant economies involve subsistence production so that value is almost entirely retained within the household; feudal economies involve the payment of tribute to a landlord; cooperative or collectivized economies involve sharing the value created in productive activities among the group; and capitalism, the system of economic organization that dominates our world today, involves the creation of value in waged labor processes, and the private ownership of property and assets. In identifying these different systems, we are thinking about the *structures* that shape economic processes.

Thinking structurally is a rather difficult thing to do, because it involves going behind the everyday processes that we participate in, and asking instead about the underlying logic of how our economic relationships are organized. But by considering the deeper ways in which our economic system is structured, we can start to discover its logics and fundamental features. This does not mean that we have some kind of grand theory that will explain every economic phenomenon around us, or predict future events, but it does mean that we can understand the imperatives that generally drive the system in which we live. Perhaps most importantly, thinking structurally implies going beyond the motivations or experiences of an individual person or firm participating in a system, and thinking about what motivates the system as a whole.

Examining the structural characteristics of a capitalist economy, we can see that it is a system in which a relatively small group of people own the tangible or intangible assets that are necessary for production and value creation. These owners of the means of production are termed capitalists, and they buy labor power from workers in order to execute the production process. Both the inputs and outputs of the production process, and the labor power that participates in it, are bought and sold through a market mechanism. Having established these structural foundations of a capitalist system, we can now examine its workings in more detail.

Driving Capitalism: Profit, Exploitation, and Creative Destruction

There are three fundamental logics that drive contemporary capitalism:

- Capitalism is profit-oriented;
- Growth in value rests on the exploitation of labor in the production process; and
- Capitalism is necessarily dynamic in technological and organizational terms.

First, the entire capitalist system is based on the incentive to profit from economic transactions. To maintain profit, it is necessary for capitalism as a system to grow continually. Without growth, profits decline. If new opportunities for profit are not constantly created, then existing profits will be whittled away through competition. Let us look, then, at exactly how profit gets created and distributed in a capitalist system. A capitalist can create profit by extracting surplus value from an employee, which is the amount of value that the person produces in excess of what he or she is paid. The price of labor will tend toward the amount needed to keep a worker clothed, fed, housed, and raising children (future workers!) – an amount that will vary across different places and historical eras. There will, of course, also be variation across different types and sectors of work, but for the system as a whole, it will be the income necessary to sustain and reproduce the workforce that defines the average wage.

The capitalist (employer) makes a profit by selling the commodity that is produced. He or she is able to keep the difference between the market value of whatever was produced by labor on the one hand and the wages of labor on the other. In everyday terms, the former represents revenue and the latter the total cost (bearing in mind all costs – materials and non-materials – are subsumed in this approach under the value of labor). This phenomenon of surplus value extraction is the second fundamental logic of capitalism, known as exploitation. This exploitation is only possible because the capitalist owns the means of production: machinery, land, raw materials, intellectual property, and so on. Private ownership rights, therefore, are a critical precondition for modern capitalism. At its core, capitalism is thus about a structural relationship between different social classes: a capitalist class that owns the means of production and a working class that owns the labor it "sells" to the capitalist. Wealth is generated through the extraction of surplus value, and this wealth accumulates with those who employ workers.

This is, of course, a very simplified picture of the reality of everyday life in contemporary capitalism. Today some employees (such as bankers, managers, accountants, lawyers, engineers, doctors, and software specialists) are paid exceptionally high salaries, and we would hesitate to label them as exploited. And the ownership of the firms (or the means of production) is often far more complicated today than it has ever been in the past. Most large enterprises are now owned by diverse shareholders, partnership arrangements are used in many professions, and every employee with a pension plan is effectively benefiting from shares that their pension fund owns. The distinction between employers and employees, or between classes, is thus often difficult to establish. Nevertheless, if our purpose is to identify the broad structural processes of a capitalist system (rather than categorizing every person into a specific class), the class relationship between capitalists and workers is an important and useful starting point.

The third fundamental logic of capitalism is its immense dynamism and creativity. The possibilities for accumulating more and more profit mean that system-wide incentives exist to create new products, new markets, new raw materials, new ways of organizing the production process, and new ways of saving on the cost of everything. This does not necessarily mean that every capitalist enterprise is buzzing with creativity and fresh thinking. But in a competitive environment, it is the entrepreneurs and innovators that the system will reward, while those who fail to innovate will be outcompeted. Only profitable firms will survive and provide the sources of growth in a capitalist system. There is therefore a fundamental logic or urge in capitalism toward "creative destruction," a term famously coined by the Austrian economist Joseph Schumpeter to describe the capitalist process of generating new growth through the destruction of old products, processes, and markets and the creation of new ones. For example, the capitalist system as a whole has continued to design and build better and better television sets over the last 50 years. But there are no longer any major American-owned TV manufacturers, despite the fact that Americans have one of the highest levels of TV ownership in the world. Instead, East Asian electronics firms (particularly those from Japan and South Korea) have dominated the market. American TV makers gradually went out of business as their Asian competitors surpassed them with product innovations, new manufacturing technologies, and lower costs. The same story can be told in many other industries, such as automobiles and clothing manufacturing. But at the same time the rise of new industries, for example in IT and finance, have created fresh frontiers of growth for American capitalism. This last point indicates that a national capitalist system can go through different rounds of crisis, restructuring, and recovery.

Contradictions, Crises, and Recovery in the Capitalist System

The approaches to economic relations outlined in Section 3.2 saw growth and development as spreading outward and ultimately tending toward equilibrium and evenness. However, a structural assessment of the capitalist system could lead to quite different conclusions. By looking further into the fundamental characteristics of the system that were just identified, we can see contradictions. These contradictions imply that crises, disequilibrium, and unevenness in the system may in fact be the norm, not the exception.

The first fundamental contradiction of capitalism is in its internal imperative for growth and profit. While growth increases the price of labor, a drive for profit requires that labor costs be minimized. This is a fundamental tension that is inherent in capitalism. One way of dealing with this problem is a technological fix whereby competing capitalist firms try to find technological ways of making their production more cost-efficient than their competitors. For these reasons, capitalists will try to push wages down or, where applicable and cost-effective, to invest in machinery and other technological innovations in order to replace workers with machines. For individual capitalists and managers of enterprises, this is a rational thing to do. The result of labor-saving machinery is a reserve army of labor (not necessarily in the same country) that is brought in and out of the workforce and always keeps wages low. There will always be someone unemployed who will take a job at a lower wage than the incumbent.

But if wages are kept low and people are kept out of work, where will the demand for the products come from? Workers are always producing more for the capitalists than they earn, so the aggregate demand can never keep pace with the growing supply of products – this is the second contradiction of capitalism. In short, workers do not make enough money to provide sufficient demand for the goods they have produced. Across the whole system, the economy is driven to what is termed a "crisis of over-accumulation." Capitalists have more products than they can sell, or idle machinery that cannot be used to full capacity because there is insufficient market demand for the product. Idle capital and idle labor are found in the same place at the same time, with no apparent way of bringing them together for socially useful purposes. Wealth is accumulated, but on the basis of exploited workers who now cannot afford to buy all of the products they create. This becomes a "blockage point" for capitalism (Harvey, 2011).

Because of this structural tendency, the capitalist system is prone to crisis and instability. There are ways of forestalling it, such as selling to overseas markets, pressing down wages, or further investing in labor-saving machinery (we will discuss these later). But crisis is only delayed, not avoided; its tendencies are merely moved around. Therefore, capitalism contains within itself a recurring contradiction. It is not hard to find examples where wealth and surplus exist side by side with shortage and need. Indeed, booms and busts appear to be common cyclical occurrences. Over the course of the 20th century, a number of business cycles saw the wealth of national and regional economies grow or decline. We will see this phenomenon in the case of California in Section 3.6. Understanding the logic of the capitalist system puts those crises in a rather different perspective than the usual interpretations found in the field of Economics. Instead of being occasional hiccups, or unpredictable storms, that hit the economy from the outside, they are in fact inherent to the capitalist system itself.

How, then, has the capitalist system managed to get beyond crises in some cases, and to contain, absorb, or delay crises in ways that do not bring about its own downfall? Here we are thinking at the level of structures explained earlier rather than the level of particular capitalists or firms for whom crises may still be individually disastrous, leading to bankruptcy and unemployment. We can focus on four ways in which the capitalist system as a whole might restore the conditions for profitability (Harvey, 1982, 2010):

- *Devaluation*: This process involves the destruction of value in the system: money is devalued by inflation; labor is devalued by unemployment (for example, when an important industry leaves a place); and productive capacity is devalued, in fact literally destroyed, by wars and military encounters.

Recreating value gets the system moving, and capital circulating, once again, but not of course without significant political, social, and environmental costs.

- *Macro-economic management*: This involves devising ways of bringing together idle over-accumulated capital and idle labor. This might involve heavy government spending to create jobs and stimulate demand during periods of recession (e.g., President Obama's American Recovery and Reinvestment Act in 2009). It might also involve legislation that curbs excessive labor exploitation by establishing certain standards in working conditions and minimum wage levels. These strategies help in ensuring that demand in the economy is maintained, and have been conceptualized as part of a broader process of regulation (see Box 3.1).
- *Temporal displacement of capital*: This involves switching resources to meet future needs rather than current ones – for example, by investing in new public infrastructure (such as the New Deal strategy adopted by the Roosevelt administration in the U.S. during the 1930s), or by using over-accumulated capital as loans and thereby intensifying future production, and hence its crisis tendencies, but averting a crisis in the present.

KEY CONCEPT

Box 3.1 Regulation theory

Regulation theory was first developed in the 1970s and the 1980s by a group of French scholars, including Michel Aglietta, Robert Boyer, and Alain Lipietz. They used the word *regulation* in a broader sense than its usual meaning in English, where it is largely limited to a set of rules or procedures for governing action. In the French sense, regulation refers to the wider set of institutions, practices, norms, and habits that emerge to provide for periods of stability in the capitalist system. This mode of regulation involves both state and private sector actors, and when successful can foster a period of sustained capitalist growth and expansion. Essentially, the mode of regulation works by seeking compromises in the inherent tensions and contradictions that exist in the capitalist system. Historical periods of stability, known as regimes of accumulation do, however, eventually end in crisis and a new mode of regulation must be discovered. A regime of accumulation known as Fordism dominated North America, Europe, and Australia in the post-war decades (see Box 9.3), and featured a three-way balance between the interests of corporations, organized labor, and governments (although the precise features of the mode of regulation in each case varied). This regime crumbled in the 1970s, and many see neoliberalism as a new mode of regulation in which market-based, rather than state-mediated, solutions are sought for problems of economic management (see also Box 4.4).

- *Spatial displacement of capital*: This involves opening up new spaces for capitalist production, new markets, or new sources of raw materials. Rather than expanding the time horizon of the capitalist system using credit or loans, the spatial horizons of the system are expanded. This might mean the development of entirely new production sites in newly industrializing parts of the world or the re-creation or rejuvenation of old spaces.

This last form of crisis avoidance is particularly geographical in nature and of most interest to us here in our discussion of uneven development. It involves the ways in which capitalism needs space in order to function and the process through which the system values and then devalues different spaces according to its structural imperatives at a given point in time. In this sense, space is not just the container in which capitalism takes place. Rather, geography is fundamental and inherent to the successful operation of the capitalist system.

3.4   Inherent Uneven Geographies of Capitalism

Since the beginning of the Industrial Revolution, the capitalist global economy has been constantly in flux. With each era, the capitalist system has created production facilities, infrastructure, and even whole landscapes that have suited its needs at a given point in time. Whether it was the watermills of the 18th century, the cotton mills of the 19th century, the industrial suburbs of the mid-20th century, or the new manufacturing zones of the present (for example, in China in Figure 3.1), we can see how capitalism creates landscapes that match its changing requirements.

George Lin, with permission

Figure 3.1   A landscape of contemporary capitalism: an industrial park in Suzhou, China

Geographers Michael Storper and Richard Walker (1989, p. 141) call these spaces "territorial production complexes," with various kinds of complexes representing the outcome of a particular kind of capitalist imperative over 250 years of industrialization. They identify four spatial forms of such complexes that emerged historically:

- Regional complexes such as the manufacturing belts across the American Northeast, German cities along the Rhine and Ruhr river valleys, or the Osaka–Nagoya–Tokyo belt of southern Japan;
- Clusters of towns such as the textile region of North Carolina or the metal-working towns of the Connecticut Valley;
- City-satellite systems such as the textile towns scattered around Boston or Manchester in the 19th century, or the mining towns of the Mother Lode and Nevada centered on San Francisco after 1849 (or Calgary in Canada and Linfen in China today); and
- Large cities or metropolises such as greater London or Baltimore that contain a number of industries and their specialized districts.

While this list focuses on industrial areas, it is useful to add that there are other kinds of economic landscapes such as agricultural belts (e.g., the Great Plains in North America) and services-oriented landscapes (e.g., the Gold Coast in Australia). In general, capitalism's restless urge to expand leads to the production of new territorial complexes, and as more places have become incorporated into this capitalist imperative, industrialization has become pervasive on a planetary scale.

These territorial production complexes are not, however, static. As the system continues to grow and change, these landscapes become outdated, unprofitable, and inhibiting. They become impediments to future growth and must be devalued to make way for a new round of growth and exploitation. Going back to our satellite view described in Section 3.1, these landscapes must be "dimmed" before they can possibly be "bright" again. A fundamental feature of capitalism is therefore its constant search for a geographical solution for its crisis of over-accumulation – what might be called a *spatial fix* (Harvey, 1982).

This idea of a spatial fix has an interesting implication: capitalism's inherent tensions relate to more than just the processes of capital circulation and accumulation we have already discussed. There is a further set of internal tensions that are related to the spaces, and especially built environments, that capitalism creates at any given moment in time. However up-to-date and state-of-the-art a landscape of production might be when it is created, it is always destined to become redundant and outdated as the dynamic system moves forward. In short, the capitalist system is faced with a contradiction that is rooted in its own dynamism: as soon as it creates the landscape that will ensure profitability at a given moment, the process of making that landscape obsolete immediately begins.

A good metaphor for representing this inherently expansionary geographical process of capitalism is to be found in a game of playing cards, whereby different players represent different places or regions in the capitalist space economy

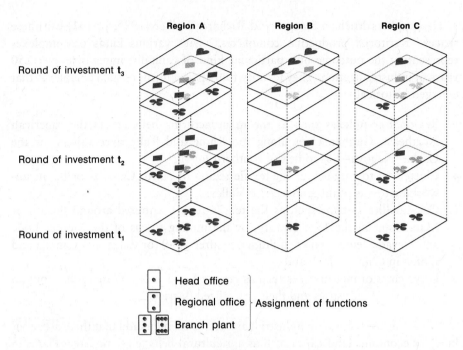

Figure 3.2 Spatial divisions of labor
Source: Gregory, D. (1989) Areal differentiation and human geography. In D. Gregory and
R. Walford (eds) Horizons in Human Geography. Totowa, N. J.: Barnes & Noble Books,
Figure 1.4.2.

(see Figure 3.2). This economic-geographical process is captured in a concept
known as the *spatial division of labor* (Massey, 1995). Each round of investment
is represented in a different suit, with different places being dealt a card of different
value (reflecting its position in the spatial division of labor). The characteristics
of a place (the hand of cards being held) can then be seen as the product of past
rounds of growth or decline, when the place was inserted into the spatial division
of labor in a particular way. This metaphor also illustrates how a place can rise
or decline over time in its importance. What it cannot capture, however, is the
way in which cards already in a hand can affect the new card that will be dealt,
which is, of course, a critical feature of the system we have described.

When a particular place becomes the "territorial production complex" for a
new round of growth and investment in the ongoing dynamism of the capitalist
system, it does not just develop the new transport infrastructure, technology,
institutions, housing, and workplaces that are appropriate to that epoch. It also
contributes to the emergence of an evolving set of social relations. Steel towns,
for example like Sheffield in the United Kingdom, Pittsburgh in the United States,
Hamilton in Canada, or Kitakyushu in Japan, developed working class cultures
that valued manual work, prized a collective working class identity through trade
union organizations, and were based on male-oriented cultural institutions, such
as working men's clubs and professional sports teams. Gender relations were,

at least in the past, based upon the assumption of a male breadwinner in each household and a role for women based on domestic work, and possibly the generation of a secondary income. There was, then, a distinctive set of social relations that emerged in association with a particular industrial sector: public institutions, gender relations, masculine and feminine identities, and class politics. The same could be said for mill towns, mining towns, export processing zones, high-tech districts, downtowns, and suburbs. Their economic and sociocultural relations do not determine each other, but they are inseparably interconnected (see more in Chapters 13 and 14).

A further implication of the emergence of social relations associated with particular industries is that future investment decisions may be made based upon these relations. When, for example, Japanese auto manufacturers started to locate production facilities in North America in the 1980s, they sought a workforce that was flexible and open to the particular forms of work organization favored by Japanese employers. This meant seeking people who were not steeped in big-city, heavily unionized workforces with long industrial legacies. One option was to locate in small rural towns – places close to highways and air transport infrastructure, but without any historical legacy of industrial development and thus unionization (Klier and Rubinstein, 2010). In other cases, though, industrial towns might be preferred where a period of decline had "broken" the union movement, or where labor had been devalued to the point that its collective ability to make demands was much diminished (see also Chapter 6). Hence, Nissan's manufacturing plant near Sunderland, a former mining and shipbuilding town in the northeast of England, established entirely new work practices, but did so in a town desperate for new investment and jobs.

We now have a picture of different territorial production complexes across a capitalist landscape that is swept by new waves of investment as the cycles of boom and bust occur. Each one creates particular social forms and lifestyles, and builds on past social forms and lifestyles. Each wave may also come with intense socio-spatial struggles that can be divided along lines that include class, ethnicity, and gender. In some contexts these struggles might be manifested as demands for better wages and working conditions, perhaps through trade unions. Or they might take the form of demands for social justice in the form of rights for women, ethnic and sexual minorities, and immigrants. More recently, campaigns for a living wage, and the anti-globalization and Occupy movements, have stirred debate on the social inequalities that capitalist development often brings.

3.5   Placing and Scaling Capitalism

We have seen that, in order to understand the development over time of a particular place or region, we need to understand not just the present nature of the economy, but also its historical and geographical background – the previous layering of rounds of investment. And once the social and cultural lives of people living in particular places are added into the analysis, we begin to see not just the systemic logics of uneven development under capitalism but also its human consequences, as ways of life are in turn created and dismantled over time. Just as capitalism must create physical forms, it must also engage in a process of creative destruction to break out of social formations as it grows and changes. New growth will often not arrive until old social norms have been broken down. Old industrial areas with strong working class traditions will not see new growth until after a significant period of decline that has broken the strength of unions and other embedded institutions and practices. And once old landscapes and spaces are broken down, they can at some future point in time be reinhabited by new productive forces. This suggests that a core geographical process in capitalism may take the form of a see-saw of uneven development, in which some places are sites of rapid investment and growth while others decline (Smith, 1984).

As implied in our imaginary view from space in Section 3.1, development in one area (e.g., industrialization in China) or region (e.g., Silicon Valley) is impossible without simultaneous underdevelopment in another area (e.g., decline in older manufacturing regions). But over time, the see-saw can swing back again so that older spaces are recreated as sites of new investment. The see-saw continues as the needs of the system to perpetuate growth and avoid crisis are satisfied. This process of capitalist restructuring helps us understand why we find continuous shifts of economic activity at different geographical scales from the global to the regional, the national, the urban, and the local. This section offers some further illustrations of these shifts at different geographical scales.

On a global scale, the shift of development from one continent to another has been dramatic in recent years. The most striking of these global shifts has been the rise of East and Southeast Asia as macro-regions of rapid growth due to substantial investment in the manufacturing industry. A first wave of development occurred in East Asia in Hong Kong, Singapore, South Korea, and Taiwan, and a second slightly later wave in Malaysia, Indonesia, and Thailand. This came to be known in the 1970s as the *new international division of labor* because labor-intensive manufacturing activities were increasingly shifting from developed countries to such rapidly industrializing economies (see also Box 10.4). Over the last two decades, it has been the rise of China as an industrial power that has captured popular attention, although India too is rapidly establishing a presence in high-technology manufacturing, services, design, and software programming (see Figure 3.3). Thus, at the global scale, we see the capitalist world system finding a spatial fix.

The same pattern of economic-geographical restructuring occurs at a regional scale as new territorial production complexes emerge in different parts of a country. As old industrial regions in Western Europe and North America succumbed to competition from newly industrialized economies (in Asia in particular), places

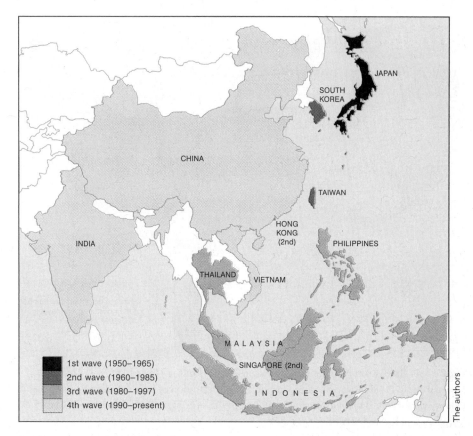

Figure 3.3   Waves of industrialization in East, Southeast, and South Asia, 1950–present

such as the Rust Belt of the United States, northern England, and the Ruhr Valley in Germany fell into decline. Their factories, infrastructure, and towns were seen as anachronistic relics of the past, as were their labor forces. Meanwhile, new growth zones were developed in the American South, West London and the Southeast of England, Paris in France, California in the United States, and Baden Württemburg in Germany. Figure 3.4 graphically illustrates this process in the United States. Over the course of economic crisis years in the 1970s, manufacturing employment dramatically declined in Northeast states while growth occurred simultaneously in the South and the West (Peet, 1983). A "weather map" of global economic activity in the 1980s, for example, would look like this:

> a series of highs building around the Pacific rim of much of East and Southeast Asia (from Japan to Hong Kong), as well as down the West Coast of the United States. It would have depicted most of Latin America stagnant but prone to violent political and economic upheavals and a series of deep depressions passing across the Ohio

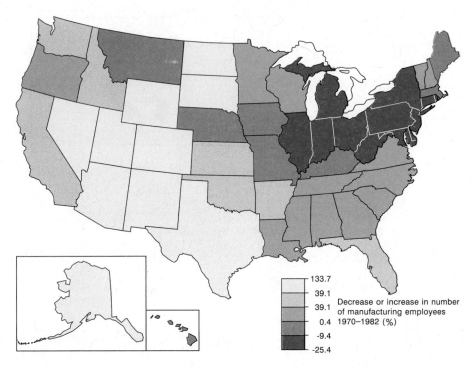

133.7
39.1
39.1   Decrease or increase in number
0.4   of manufacturing employees
         1970–1982 (%)
-9.4
-25.4

Figure 3.4   Industrial restructuring during the 1970s in the United States
Source: Peet, R. (1983) Relations of production and the relocation of United States manufacturing industry since 1960, *Economic Geography* 59(2), pp. 112–43, Figure 9. With permission Clark University and Richard Peet.

> Valley and Pennsylvania, through the British industrial heartlands, and across the Ruhr Valley of Germany. But the sun would be shining on Bavaria, Tuscany, Silicon Valley, and all manner of smaller industrial production districts. In the misty dawn of an East Asian renaissance, the heat island of China would just be visible (Harvey, 2011, p. 12).

Moving to the urban scale, neighborhoods and buildings can be erected as the vanguard of economic growth. In the 19th century, industrial towns in England grew up around the processing of natural resources such as cotton, tea, and rubber brought in by ship from distant colonies. In the early and mid-20th century, large-scale manufacturing of clothing, prepared foods, and electrical devices (including automobiles) formed the mainstay of urban economies in Western Europe and North America. In specific places, it was not just industrial landscapes that reflected these times. The small brick row houses of older industrial cities occupied by factory workers relatively close to their places of employment were the manifestation of a particular period of capitalist growth on the residential landscape. By the 1950s, industrial suburbs were being built to house the workforce, reflecting the growing use of private vehicles. Looking at the changing housing stock of any large contemporary city in an industrialized

The authors

Figure 3.5   Redundant industrial landscape in urban Philadelphia, U.S.

economy is, in fact, an interesting exercise in economic archaeology, as it reveals the built forms that were needed in any given period of economic growth: from the medieval alleyways of central London built in the earlier centuries, just wide enough for a horse and cart, to the car-clogged 20th-century freeways of Los Angeles and Toronto.

As the imperatives of capitalist growth changed over time, many of these landscapes became redundant (see Figure 3.5), only to be rediscovered at a later date when, thoroughly devalued, they could once again be used, except for very different purposes. In some industrial cities of Europe and North America, for example, garment factories made idle and derelict when production moved offshore in the 1960s and 1970s are now being renovated. They are prized as studio, office, or even living space by young professionals in growing industries such as graphic design, computer animation, and internet consulting. These previously poor and depressed neighborhoods may become the home of wealthy professionals, an urban regeneration process commonly known as gentrification. In this way, a devalued space of industrial capitalism has been revalued in a new era of post-industrial capitalism.

An example is provided by Liverpool in northwest England. Liverpool grew in the 19th century due to its importance as a seafaring center and as a point of transhipment for cargos from all over the world. Liverpool's docks then processed some 40% of the world's trade, making Liverpool a leading world city. Its prosperity reflected its appropriateness for a particular era of capitalist growth: mechanical cotton spinning had developed in nearby Manchester, and steamship technology linked the city with the world. From the 1960s onward, however, container ships began to increase massively in size, requiring larger and deeper

Figure 3.6   Galleries and apartments now occupy 19th-century industrial infrastructure in Liverpool, England

ports. Air travel took the place of ocean liners around the same time, and the products of the textile industries in Lancashire and the region around Liverpool were replaced with imports from abroad, particularly Hong Kong and Southeast Asian countries. Liverpool as a built environment designed to service capitalism at a given point in time became anachronistic. The city was not just outdated in relation to newer technologies of industrial production and transportation; its whole setup was an impediment to harnessing their growth potential in a changing global economy. For much of the second half of the 20th century, Liverpool was a city in economic decline. After it has been devalued, however, the infrastructure of the past can be revalued in another era. Hence, the derelict docklands of Liverpool are being refurbished to house new art galleries, tourist attractions, and apartments (see Figure 3.6). Today, similar to several older cities in northern England, Liverpool prides itself as being a newly regenerated business and cultural center (www.liverpoolvision.co.uk, accessed on March 8, 2012).

3.6   Going Beyond National Capitalism: A "Global California"?

Over the course of this chapter, we have demonstrated the way in which the internal logic of capitalist production needs an uneven economic landscape. The booms and busts of development are a see-saw process of devaluation and revaluation that help the capitalist system avert crisis. We often think of this uneven development as bounded within national territories. It is thus common to think of the American, British, Japanese, German, and lately Chinese economies as if each of these national forms of capitalism is distinct and insulated from each other. But there is a puzzle here that has been quietly with us since we began the chapter with the global shifts in economic activity. Why is it that some regions can go through the boom-and-bust cycle several times, whereas other regions in the same country find it hard to bounce back once they have gone into decline? If capitalism is such a dynamic system, surely all regions should be able to find the "right" spatial fix to earn a decent place in modern capitalism?

The answer to this question lies in the fact that regional economies are not bounded and self-contained, but are connected to other places in various ways. Regions in decline may revitalize if they are well embedded in extraregional and extranational networks that offer them new growth opportunities. We can use the example of California to show how external connections have brought about continued growth and prosperity. From its origin as an agricultural region, to the gold rush of the 1800s, to the present high-technology sector, California has been positioned to take advantage of new waves of growth in each era (Walker, 2008, 2010). We will focus on three key themes of this "global California":

- Labor migration and California's agriculture growth;
- Geopolitics and the rise of aerospace and defense industries in California; and
- Globalization and the role of California's information and communications technology (ICT) and motion pictures industries.

Interregional, and later international, migration played an extremely important role in the earliest development of capitalism in California based on extractive and agricultural activities. Prior to the gold rush of 1848 and the ceding of Alta California by Mexico to the United States in 1850, California was a sparsely populated territory north of the Baja Peninsula. The discovery of gold in northern California brought a frenzy of economic activity as prospectors and merchants arrived in droves and agricultural activities were expanded to feed a growing population centered around San Francisco. The completion of a transcontinental railroad in 1869 permitted Californian farmers to export their increasingly bountiful food crops to eastern states and the rest of the world. The westward flow migrants, the opening up of new agricultural land, the development of growing urban centers (San Francisco and Los Angeles), and the connections created by the railroad can be seen as the production of new spaces for American capitalism in the late 19th and early 20th centuries.

By 1870, San Francisco had 150,000 people and had become one of the ten largest cities in the United States. Meanwhile, Los Angeles was awakened by the discovery of oil: in the 1910s, over 20% of the world's oil production came from Southern California. In short, California (and the West more generally)

was providing one of the first spatial fixes for American capitalism, and the
new opportunities for profit that the system required. But just as California was
booming, devaluation was also going on. The workers that planted, harvested,
and processed California's crops were often migrant laborers, many from the
Dust Bowl states of Oklahoma, Texas, and Arkansas in the 1930s, others from
China or the Philippines. Their labor was often devalued to the point of barely
sustaining their survival, hence maximizing the surplus value being accumulated
by California's farmers and the agro-food corporations that dominated the state.
A new spatial fix was necessary.

Geopolitics provided such a spatial fix and sustained the capitalist imperative
of "global California." In the 1920s and the 1930s, aerospace emerged as a new
industry, and California became a major center for building and testing aircraft.
World War II sparked massive federal government expenditures in this sector, and
the state's manufacturing output tripled during the war years in the early 1940s.
When a new round of defense-related investment arrived, driven primarily by the
Cold War of the 1940s–1980s, California's existing aerospace corporations were
poised to move into missile and satellite production. Perhaps more than anywhere
else in the United States, California bore witness to the economic growth potential
(unevenly distributed) that follows the destructive devaluation of warfare in an
era dominated by geopolitical concerns (see Box 3.2).

FURTHER THINKING

Box 3.2   *Sustaining the global capitalist system:*
geopolitics and domination

As a system of production, capitalism is sustained through political-
economic arrangements premised on unequal exchange relationships
between core countries/regions and peripheries and the ways in which
value gets concentrated in some places and not others. Until around 50
years ago, the processes that determined how "developing" countries were
paid for their resources were shaped by the unequal power structures of
colonialism. By taking over territories through military force, European,
American, and Japanese colonialists asserted their ability to capture
resources and establish the prices that would be paid for them. In this
way, a much greater component of the value created was located in the
countries of the colonizers than in those of the colonized. In most of the
world, however, the formal colonialism that took hold throughout the 19th
century has dissipated, but power in the contemporary global economy is
still wielded by dominant states. It is not just capitalists who determine
what, and where, economic activities will take place, and on whose terms.

The wealthier nation-states of North America and Europe (in particular, the so-called G8 group of countries) have been so dominant, financially and militarily, that they have had ample opportunities to dictate the terms upon which less powerful countries will engage with them economically (see also Chapter 4).

This power might be used to force peripheries to open up their economies to the activities of core-based corporations, to expose their domestic markets to imported products, to allow outsiders to operate in their financial markets, to export their natural resources while importing higher value-added goods and services, to permit environmental degradation, to limit labor rights and working conditions, or to privatize common property resources such as genetic material and natural environments. Self-interest among local elites in such places will likely motivate them to collaborate and collude in this process. The implication is that we see crises of accumulation averted not just through the intensification of the capitalist production process, but also through the geographical expansion of capitalist processes into formerly communally or socially owned spheres – for example, common property resources, citizenship rights, or national assets. In the terms described in Chapter 2, we see the alternative economy being encroached upon by the capitalist economy. The possession of communal resources (e.g., oil and gas) is transferred from local communities to global traders who concentrate their corporate wealth elsewhere. This process, through which collective resources are appropriated by private interests, has been described as accumulation by dispossession (Harvey, 2003; Glassman, 2006). But we also see a global situation in which this process is advanced by one capitalist state over another, often through direct negotiations between governments (e.g., trade agreements) or through multilateral arrangements (e.g., the structural adjustment programs imposed by the International Monetary Fund, described in Box 4.6). This new form of political-economic domination has been dubbed the *New Imperialism* (Harvey, 2003).

Meanwhile, capitalist growth in California not only depended on the geopolitics of warfare, but also benefited enormously from California's dominant role in the globalization of culture and technology. The motion picture industry had emerged as a major sector in Los Angeles, which trumped New York in filmmaking by the 1920s. The success of this industry had much to do with its critical role in the globalization of American culture. Today, Los Angeles is closely associated with innovations in cultural industries, and Hollywood is synonymous with the production of blockbuster motion pictures (see Section 12.3). Since the 1960s,

California has also led the world in globalizing its technologies. This globalization of the California economy brought about another round of capitalist growth dynamics. By the 1960s, aerospace and other defense-related expenditures had laid the seedbed for an innovative electronics and computer industry in California. Silicon Valley, south of San Francisco, emerged as the iconic territorial production complex of contemporary high-tech capitalism (see more in Chapters 9 and 12). Hosting such powerful corporations as Hewlett-Packard, Intel, Apple, Google, and Facebook, the driving forces behind the personal computer and information technology revolution of recent decades, Silicon Valley remains the global epicenter of high-tech industries.

Taking a central position in the global networks of both high-tech and cultural industries, California has found a new spatial fix to help it through the nation-wide decline in industrial capitalism. California has thus successfully bucked the trend of deindustrialization that has occurred in the Rust Belt of the Northeast, and remains America's best hope in the face of global competition and industrial decline. Over the course of its development, California has shown the growth potential of devaluation, spatial fixes, and government involvement in the economy. Wealth creation requires the extraction of surplus value from labor. In California, a constant stream of out-of-state and international immigrants, low- and high-skilled, has made this possible. It has also demonstrated the role of pre-existing patterns of economic activity in attracting future rounds of investment – a process that has historically worked in California's favor. As Bardhan and Walker (2010, p. 32) note, "[u]ltimately, what sets California apart from other states [in the U.S.] is that it is more open to international trade and capital flows, being on the coast and boasting a large immigrant diaspora; it has a greater share of services; and it has a larger share of the 'innovation pie'. All these make its economy more prone to booms and busts." In short, California is a uniquely dynamic territorial production complex that thrives on its global connections.

3.7　Summary

We started this chapter with a dynamic image of the unevenness of capitalist development. Our focus on value and its creation and distribution led us to consider the fundamental structural processes underlying a capitalist economy: how value is created, how it is circulated, and the contradictions that exist in such a system. The result was an understanding that unevenness is not an accidental by-product of natural resource endowments, nor is it something that our economic system automatically levels out over time. Instead, spatial unevenness is quite fundamental to the workings of the capitalist system and something we should expect to continue. If unevenness did not exist, we might expect that the see-saw of capitalist development would create it. In other words, uneven development is both a cause and an outcome of capitalist growth.

The last three sections of this chapter examined the diverse geographies of capitalism arising from its systemic tendencies. There we showed different forms and scales of spatial fix in capitalism's relentless drive for profits and avoidance of accumulation crises. New territorial production complexes are continuously developed, and older regions and towns experience dramatic restructuring and transformations. These capitalist imperatives and shifts also play out at different spatial scales, from the global all the way to the local and even in the household, as the social formations of one era are redundant in the next.

As we work through other themes in this book, it will be worth bearing in mind the structural analysis that we have developed in this chapter. The following four chapters in Part Two will explore the making of the spatial economy. Chapters on the state, environment, labor, and finance will all, in different ways, highlight the actors and processes involved in the contest over the creation and distribution of value, between social groups and between places. In Part Three, we will examine how the capitalist space-economy is organized through commodity chains, the technological dynamism of capitalism, the imperatives of transnational corporations, and retail practices, along with their implications for spatial patterns of investment, wealth creation, and environmental change. In Part Four, chapters on clusters, gender, ethnicity, and consumer behavior will highlight place-based characteristics that capitalism as a system often incorporates and uses. Thus, while this chapter has established why capitalism fundamentally needs an uneven economic geography, much of this book will go on to examine precisely how distinct places and uneven spaces are actually created and used.

Notes on the references

- The single most important figure in the development of geographical approaches to studying the structure of capitalism has been David Harvey (1982, 2006, 2010). For an introduction to, and retrospective on, his monumental work, see Castree and Gregory (2006). See also Harvey's (2011) lecture on capitalism and uneven development published in *Economic Geography*.

- For state-of-the-art reflections on what was probably close to the peak of interest in Marxist theory in geography, see chapters by Peet and Thrift, Smith, and Lovering in the edited volumes by Peet and Thrift (1989). For more recent reviews of such a geographical political economy, see Sheppard (2011) and chapters by Mann, Glassman, Smith, and Yeung in Barnes et al. (2012).

- For case studies of regional economic restructuring, see Massey (1995), Allen et al. (1998), and Coe and Jones (2010) on the United Kingdom, Birch and Mykhnenko (2009) on continental Europe, and Bardhan and Walker (2010) and Walker (2010) on the United States.

- For a geographical review of theories related to international development, see Peet (2007).

Sample essay questions

- Explain the key differences between conventional accounts of uneven development and a geographical political economy approach.
- Is development inevitable for the "developing world"?
- Why is space so integral to the survival of capitalism as an economic system?
- Using a specific example, describe the ways in which a particular era of capitalist production became imprinted upon the landscape and social characteristics of a particular place.

Resources for further learning

- Several Marxist scholars maintain extensive and informative websites with excellent introductions to the field. See for example David Harvey's website: davidharvey.org; Erik Olin Wright's website at the University of Wisconsin: www.ssc.wisc.edu/~wright; and Bertell Ollman's site at New York University: www.nyu.edu/projects/ollman/index.php.
- International organizations such as the United Nations and World Bank have extensive websites containing studies of developing areas. See for example the UN Millennium Project www.unmillenniumproject.org, and the Data and Research page on Development www.worldbank.org.

References

Allen, J., Massey, D., and Cochrane, A. (1998). *Rethinking the Region*. London: Routledge.

Barnes, T. J., Peck, J., and Sheppard, E., eds. (2012). *The New Companion to Economic Geography*. Oxford: Wiley-Blackwell.

Bardhan, A., and Walker, R. (2010). *California, Pivot of the Great Recession*. Working Paper No. 20310, Institute for Research on Labor & Employment, University of California, Berkeley.

Birch, K., and Mykhnenko, V. (2009). Varieties of neoliberalism? Restructuring in large industrially dependent regions across Western and Eastern Europe. *Journal of Economic Geography*, **9**, 355–380.

Castree, N., and Gregory, D., eds. (2006). *David Harvey: A Critical Reader*. Oxford: Wiley-Blackwell.

Coe, N. M., and Jones, A., eds. (2010). *The Economic Geography of the U.K.* London: Sage.

Glassman, J. (2006). Primitive accumulation, accumulation by dispossession, accumulation by "extra-economic" means. *Progress in Human Geography*, 30, 608–625.

Harvey, D. (1982). *Limits to Capital*. Oxford: Wiley-Blackwell.

Harvey, D. (2003). *The New Imperialism*. Oxford: Oxford University Press.

Harvey, D. (2006). *Spaces of Global Capitalism: Towards A Theory of Uneven Geographical Development*. London: Verso.

Harvey, D. (2010). *The Enigma of Capital and the Crises of Capitalism*. London: Profile Books.

Harvey, D. (2011). Roepke lecture in economic geography – Crises, geographic disruptions and the uneven development of political responses. *Economic Geography*, 87, 1–22.

Klier, T., and Rubinstein, J. (2010). The changing geography of North American motor vehicle production. *Cambridge Journal of Regions, Economy and Society*, 3, 335–347.

Massey, D. (1995). *Spatial Divisions of Labour*. Second Edition. London: Macmillan.

Peet, R. (1983). Relations of production and the relocation of United States manufacturing industry since 1960. *Economic Geography*, 59, 112–143.

Peet, R. (2007). *Geography of Power: The Making of Global Economy Policy*. London: Zed Books.

Peet, R., and Thrift, N., eds. (1989). *New Models in Geography*. Two volumes. London: Unwin Hyman.

Smith, N. (1984). *Uneven Development*. Oxford: Blackwell.

Storper, M., and Walker, R. (1989). *The Capitalist Imperative: Territory, Technology and Industrial Growth*. Oxford: Blackwell.

Walker, R. (2008). At the crossroads: defining California through the global economy. In D. Igler and W. Deneven, eds. *A Companion to California History*. Hoboken, NJ: Wiley-Blackwell, pp. 75–96.

Walker, R. (2010). The Golden State adrift. *New Left Review*, 66, 5–30.

PART II

MAKING THE (SPATIAL) ECONOMY

CHAPTER 4

THE STATE
Who runs the economy?

Goals of this chapter

- To understand how the state shapes economic processes in multiple ways within its territory
- To recognize the different kinds of states within the global economy
- To appreciate the changing role of the state in an era of globalization
- To understand the notions of neoliberalism and neoliberalization
- To demonstrate why geographical scale matters in the reconfiguration of the state

4.1 Introduction

On October 3, 2008, as the financial crisis caused by bad debt in the U.S. housing market was unfolding with startling speed, the U.S. government signed into law the Troubled Asset Relief Program, commonly known as TARP. The program was designed to stabilize the financial system by purchasing assets and equity from struggling financial institutions. The sums involved were mind-boggling. A total of US$475 billion was allocated to TARP, of which US$388 billion was ultimately spent: US$245 billion on banks, US$80 billion on struggling auto manufacturers, US$48 billion on bailing out the giant insurer AIG, and US$15 billion on various credit markets (U.S. Treasury, 2010). The two massive banks Citigroup and Bank of America alone were allocated US$45 billion each under the scheme. A similarly massive bailout package was also enacted in the United Kingdom during 2008–09. A total of £117 billion (approximately US$185 billion) was spent on buying shares and providing loans to struggling banks, including the Royal Bank of Scotland (RBS) and Halifax Bank of Scotland (HBOS), as well as effectively nationalizing the Northern Rock bank to prevent its total collapse. The government also underwrote – that is, effectively insured – a massive £730 billion (US$1,150 billion) in bank assets and loans in order to keep capital moving through the financial system (NAO, 2009). The United Kingdom and the United States also engaged in rounds of "quantitative easing," or the pumping of tens of billions of dollars into the economy in order to stimulate growth. In November 2010, for example, the Federal Reserve (America's central bank) announced a US$600 billion injection of capital into the system. While the total costs to the taxpayer of these various schemes will take several years to assess, they undoubtedly achieved their short-term goal of propping up ailing national financial systems.

These huge bailouts are very revealing about the roles played by the state in the contemporary global system in several ways. First, they show the state performing perhaps its most fundamental role – as the ultimate guarantor of the economic system within its territory. The U.S. and U.K. governments intervened to prevent the widespread economic collapse that would have resulted if the financial system had ground to a halt. As we demonstrate in this chapter, however, this is just one of the many roles that states perform with respect to the economy. Second, while both administrations at the time – George W. Bush's Republican Party in the United States and Gordon Brown's Labour Government in the United Kingdom – were presumed to be broadly pro-market and neoliberal in their stance, there was also a political necessity behind fast and decisive action, since both parties would have to seek reelection by the citizens within their territory. States, then, are unavoidably both economic and political institutions, and it is the intersection and inseparability of these two domains that defines the nature of particular states. Put another way, states can only be understood as *political-economic* organizations. Third, the international institution charged with handling global economic crises – namely, the International Monetary Fund (IMF) – did not take a lead role in the design and delivery of these rescue packages as it had done in previous economic crises, such as the one that hit Asia in 1997–98. The speed and scale of the initial intervention required in this instance meant that it was beyond the reach of a multilateral institution and could only be provided by the states themselves. Fourth, the fact the U.S. and U.K. governments were able to take this action at short notice and to raise the necessary capital on global bond markets demonstrates another key theme of this chapter – namely, that states have different capacities to act within the global system. The subsequent IMF-brokered bailouts of Greece and Ireland in 2010, by contrast, concerned smaller economies that required external assistance.

In expanding on these initial observations, the chapter proceeds in five major stages. The next section (4.2) introduces and questions debates surrounding the supposed irrelevance of the state in an era of globalization. Section 4.3 demonstrates the wide range of ways in which states continue to exert influence over the economic activities that occur within their territory and across their borders, simultaneously performing the roles of ultimate guarantor, regulator, manager, business owner, investor, and basic service provider. These interventions in turn create uneven geographies, both in terms of how states "plug in" to the global economy and the distribution of economic activity within their borders. Section 4.4 profiles how different approaches towards these various roles, reflecting different political ideologies, intersect to produce tremendous diversity in state forms within the global economy. We use a four-fold typology of neoliberal, welfare, developmental, and authoritarian states to impose some organization on this diversity. In Section 4.5, we explore how state functions are being "rescaled" as the relationships between the state and both local and supranational levels of government are reconfigured. Finally, we look at how the state is also being "hollowed-out," with the result that private sector actors play an increasing role in delivering a wide variety of state functions (Section 4.6). For more on the basic terminology surrounding the state, see Box 4.1.

KEY CONCEPT

Box 4.1   Unpacking the state

Terminology concerning "the state" can be confusing, and so some clarity is useful. A state is an area of territory within which the population is governed by some kind of authority structure. A nation, on the other hand, is a significant grouping of people who have a shared culture and historical legacy. The two may or may not overlap to produce a nation-state where the political and cultural dimensions coincide. Many states have experienced traumatic journeys to *nation*hood characterized by internal contests and civil wars (e.g., the former Yugoslavia and several African states), while others are subject to competing claims of *sovereignty* from other states (e.g., Taiwan from China). Equally, there are a range of "nations without states" that continue to press their case for independence through separatist movements of different kinds (e.g., the Basques in Spain and France, the Palestinians in Israel, the Kurds across Iran, Iraq, and Turkey). For simplicity, however, and because our primary focus is on the exercise of authority with respect to the economy, we prefer to use the term "state" throughout the chapter. Importantly, linking back to our discussion in Chapter 1, the state is an unavoidably *territorial* organization.

We also need to have clarity in terms of the scalar language we use. While the state is fundamentally seen as a national-level structure, it is actually a *multi-scalar* organization that necessarily functions through different tiers of government. Many large countries have at least three tiers of government, operating at the national scale, the regional scale, and the local/city scale. China, for example, officially has four levels – the national, the provincial, the county, and the township – but also uses the terminology of the prefecture and the village. The regional level in particular can cause some confusion, as these areas are variously called states (e.g., India, the U.S.), provinces (e.g., Canada, China) and territories (e.g., Canada, India) as well as simply regions (e.g., the U.K., France). The degree of control exercised by these lower tiers of government varies depending on the context. In federal states such as the United States, Canada, Germany, and India, regional governments (e.g., California, Ontario, Bavaria, or Andhra Pradesh) have considerable freedom in policy terms. In more centralized states – for example, the United Kingdom, France, and Norway – regional tiers of governments are much more limited in their resources and decision-making capacities. As we shall see in this chapter, the relationship between the various tiers also needs to be seen in dynamic terms. Over the past 30 years in China, for example, certain provinces have been given increased freedom to shape their economic development.

4.2   The "Globalization Excuse" and the End of the State?

We begin by reviewing how, in some quarters at least, the state has increasingly become viewed as irrelevant to the operation of the global economy. This is commonly known as the *hyperglobalist* position, promoted through the popular accounts offered by business gurus. For example, an early contributor to these debates, Kenichi Ohmae, suggested the following in this book, *The End of the Nation State*:

> "...what we are witnessing is the cumulative effect of fundamental changes in the currents of economic activity around the globe. So powerful have these currents become that they have carved out entirely new channels for themselves – channels that owe nothing to the lines of demarcation on traditional political maps. Put simply, in terms of real flows of economic activity, nation states have *already* lost their role as meaningful units of participation in the global economy of today's borderless world." (Ohmae, 1995, p. 11, emphasis in original).

In this view, globalization is a force that integrates capital, technology, and people in a "borderless" world. Globalization, we are led to believe, thereby creates a single global market and a gigantic "global village" that cannot be controlled by individual states. In outlining this *new* global system, the hyperglobalists have created a powerful popular discourse surrounding the end of the state as an institution of political-economic governance. This discourse takes its power from mobilizing what can be thought of as the "globalization excuse" – that is, explaining the ongoing reconfiguration of state functions by invoking an all-powerful, external force – globalization. Globalization becomes the convenient explanation for whatever happens to the state, not as something that itself needs to be explained. Business leaders, politicians, and media commentators fall into this trap, repeatedly prefacing their pronouncements with statements such as: "Because of globalization, we have to . . . ."

The hyperglobalist accounts argue that power has shifted from states to the giant corporations and international financial institutions that orchestrate global flows of capital and technology. The state is deemed powerless to control its national economic affairs and the activities of the corporations within its territory (see Chapter 10). Instead, in deciding on where to invest, global corporations are seen to play states off against one another in order to obtain benefits in terms of political support, financial incentives, favorable environmental regulations, and market access. The successful outcomes achieved by some corporations in both advanced industrialized countries and poor developing countries have led to the universal claim by hyperglobalists that host states are no longer able to govern their economies effectively. Meanwhile, the successful global reach of these corporations is seen to offer supporting evidence for the state's declining role in domestic economic regulation. In short, the rise of corporate power and global finance are seen to explain the demise of the state, as if the two are in no way connected.

But is the state really withering away as alleged by the hyperglobalists? Not really. In this chapter, we argue for an alternative and more nuanced economic-geographical reading of the role of the state in today's global economy. More specifically, we argue that the role of the state in governing the modern economy can be understood in three ways. First, we reject the hyperglobalist position that separates the state from corporate and financial actors. Instead, firms and markets need to be seen as engaged with the state in a mutually dependent relationship. Firms need the state to function, while the state's political legitimacy will be challenged if it fails to deliver economic development through the activities of firms and markets within its borders. Territorial control, then, remains of paramount importance in the current era. Second, the hyperglobalist account gives the impression that all states are the same – clearly a gross oversimplification. Instead, we need to appreciate that there are many different varieties of states with differing abilities to enact control over their territories. While some states may succumb to global forces, hence in part at least seeming to corroborate the claims of the hyperglobalists, there are many others that have the influence to shape the "rules of the game" of the global economy. For much of the second half of the 20th century the nature of global capitalism reflected U.S. and West European economic dominance; in the 21st century, the debate rages as to how far the global economy now reflects the power of new centers of growth and profit such as Brazil, Russia, India and, in particular, China (the so-called "BRIC" economies). Third, and perhaps most important, we subscribe to a perspective that sees the state as always remaking itself in relation to wider economic processes. The state is not a static institution; rather it is a dynamic entity that is constantly reinventing itself in pursuit of economic growth and, in democratic contexts, electoral support. Overall, our central argument is that the territorial powers of the state are being *transformed* by globalization processes rather than necessarily eroded.

4.3   The State as the Architect of the National Economy

To begin our analysis of the intersections of state politics and the economy, we need to understand the ways in which states interact with firms and markets. In this section, we introduce the full range of economic functions performed by modern states and show how they work out very differently in different places. Here we focus on six important state functions that directly impinge on the national economy.

The State as Ultimate Guarantor

While markets may appear to be self-organizing, they can fail in the wake of large-scale upheavals such as financial crises and natural disasters. In these situations the state often steps in, becoming an institution of last resort and the ultimate guarantor of crucial economic instruments. This is the perhaps the most basic role for the state, within which we can identify four important elements:

- *Dealing with financial crises*: As the period 2007–09 showed (and was also seen in previous eras such as the Great Depression of the 1930s) market failure is particularly likely in the financial industry where the real or potential bankruptcy of major financial institutions can result in severe financial crises. While on some occasions states may choose to let struggling institutions fail – as seen, for example, in the collapse of the investment bank Lehman Brothers in late 2008 – in most cases the centrality of financial flows to the wider economy means that states intervene to stabilize and underwrite the system. As we saw earlier in the chapter, this is what happened in countries such as the United States and United Kingdom in 2008–09 in order to restore confidence in the financial system and stabilize the economy at large.
- *Guaranteeing national economic instruments*: The international economic credibility of a state in part depends on its ability to maintain the value of its

currency and government bonds. For example, U.S. Treasury Bills – a form of government-issued credit note – have strong credibility in the international financial community. Many institutional investors such as pension funds and insurance companies have purchased these Treasury Bills because of their attractive interest payments and the security of repayment upon maturity (see Chapter 7). By comparison, the bonds issued by smaller and less stable economies may be subject to a lot more uncertainty and are therefore less attractive to potential investors. In 2010, for example, both Greece and Ireland suffered severe financial crises in which they were unable to cover their large public sector debts by selling bonds. In terms of currencies, states ensure that their currency is the universal standard, or legal tender, accepted in their territory and try to maintain the relative value of that currency. The experiences of selected Latin American countries during the early-to-mid 1990s (e.g., Argentina and Mexico) and some Asian economies in the late 1990s (e.g., Indonesia, Thailand, and South Korea) are telling here. In both instances, the states in question were unable to guarantee their national currency with sufficient foreign reserves or gold, resulting in massive depreciation and subsequent financial crises as investors rapidly sought to sell their reserves of the currency.

- *Securing international economic treaties*: The state is also important in that no other institution has the political legitimacy to negotiate and sign international trade and investment agreements. States often engage in bilateral free trade agreements (FTAs) with one another in order to advance their common economic interests. As of March 2012, there were 319 active FTAs across the global economy, many with a regional focus (http://www.wto.org, accessed on March 6, 2012). States also engage in bilateral investment guarantee pacts in order to protect the commercial interests of their national firms in foreign territories. Smaller states such as Singapore have deployed these FTAs highly effectively in order to expand their connections with the global economy.

- *Property rights and the rule of the law*: States are also crucial in establishing and maintaining private property rights and upholding the rule of the law through a well-defined and enforced legal system. Property rights refer to the right of an individual or a corporate entity to own properties (e.g., land, building, machineries, ideas, designs, and so on) and derive income from these properties. This aspect of the state is a defining characteristic of modern capitalism (see Chapter 3) because without effective property rights, capitalists (individuals or firms) cannot capture any profits or rents derived from their properties and therefore do not have any economic incentive to invest. A state thus has numerous departments to deal with property rights issues such as land titling, patents and trademarks, and business registries, and enacts many laws to protect such rights.

The State as Regulator

States are also the primary regulators of the economic activities that take place within, and across, their borders. For those of a pro-market persuasion, the state should simply seek to enable and protect market-driven activities. At the same time, the state's primary source of legitimacy originates from its citizens, whom it is expected to protect from the potentially harmful effects of market and firm activities. The state therefore often engages in a wide variety of forms of regulation of economic activities, ranging from economic and environmental to social and ethical considerations:

- *Market regulation*: The state endeavors to uphold the "fairness" of the market mechanism. The extent to which the state actively regulates market behavior varies between different national economies. The United States, for example, is known for its strong preference for open market competition and an anti-monopoly or anti-trust stance. During the 20th century, many large American corporate empires were broken down into smaller business units in order to prevent excessive market power held by these corporate giants – for example, the dismantling of Rockefeller's Standard Oil in 1911, creating the precursors of today's ExxonMobil and Chevron, and the breakup of the former AT&T into many local "baby" Bell companies in 1984. Many states have established fair trade commissions and anti-monopoly units to ensure market competition in different industries. This anti-monopoly approach to regulating the market economy can be contrasted with the experience of other states that directly own and operate monopolies, mostly but not exclusively in developing countries. In these latter cases, as we shall see shortly, the state is often a direct owner and manager of business enterprises, replacing the market as the primary mechanism of economic governance.
- *Regulating economic flows*: The contemporary state also plays a very important role in regulating its borders. This function is particularly important in an era of accelerated globalization associated with massive cross-border flows of capital, commodities, people, and knowledge. In regulating capital flows, some states may be very restrictive in not allowing financial capital to enter and leave their countries without complying with lengthy regulatory procedures or paying hefty taxes. Malaysia's response to the Asian economic crisis of 1997–98, for example, was to impose strict controls on the flows of capital across its boundaries. Border regulation is of particular importance in relation to labor flows (see Figure 4.1). As globalization has intensified, so levels of international migration of various kinds (e.g., short-term/permanent, high skill/low skill) have grown, particular in terms of temporary migrant workers. Increasingly, major destination countries (e.g., the U.S., Canada, or Australia) are seeking to regulate the *types* of migrants that they allow in, using "points" systems, for example, to assess the skills, qualifications, and financial resources that potential migrants can bring into the economy. They are also facing the challenge of policing their extensive borders against a range

Martin Hess, with permission

Figure 4.1   The border crossing between China's Shenzhen and Hong Kong's Lo Wu

of clandestine and illegal immigration flows, as seen, for example, along the U.S.–Mexico border and the eastern and southern perimeters of the European Union.

The State as Manager of the National Economy

In managing their national economies, most states pursue a wide range of economic policies in order to sustain, shape, and promote economic development. These intersecting economic policies may be linked to trade, foreign direct investment, industry, and the labor market (Dicken, 2011). A critical issue here is the extent to which a particular state pursues these economic policies in a *strategic* manner in order to seek or advance its national competitive advantage.

- *Trade strategies*: States often actively manage trade in the interests of domestic producers. In most cases, the result is policies that seek to stimulate exports while being restrictive to imports. In terms of imports, while World Trade Organization (WTO) rules mean that tariff barriers have been dramatically reduced across the global economy, states may implement a variety of non-tariff barriers to curtail imports – including, for example, quotas, licensing regulations, labeling and safety requirements. For exports, the state may get involved in promoting exports through its various agencies, or manipulating the cost of exports through subsidies and exchange rate policies. For instance, the exchange rate of the Chinese renmimbi, particularly with the U.S. dollar, is a huge ongoing geopolitical issue given its effect on the relative cost of Chinese manufacturing exports.

- *Foreign direct investment (FDI) strategies*: Rising levels of cross-border investments by transnational corporations have been a key element of globalization processes over the last three decades. Most policies are concerned with increasing levels of inward foreign investment, although the recent economic crisis has seen a slight reversal of that trend (Dicken, 2011). In many cases, states combine tax incentives, the availability of prime land, and supporting industries and workforce characteristics to form an attractive package for foreign investors. States may also, however, seek to capture the gains from inward investment by insisting on certain levels of local purchasing and technology transfers, and/or by trying to limit the repatriation of profits. In certain sectors, states may try to restrict or ban FDI. France, for example, actively prevents foreign takeovers in a range of sectors including the cultural and defence industries (see also Box 4.2).
- *Industry strategies*: These policies are concerned with seeking to stimulate certain areas of economic activity, and usually involve some kind of financial incentive. They may be applied across the economy as a whole (e.g., the level of corporate tax) or applied selectively. Selective interventions may target certain sectors – either to prevent decline (e.g., autos, steel) or to reinforce growth (e.g., biotechnology) – certain kinds of firms (e.g., small startups) or particular geographical areas (e.g., depressed deindustrialized or inner city areas).
- *Labor market strategies*: Policies in this area have sought to promote enhanced labor market *flexibility* through processes of deregulation, and in some contexts such as the United Kingdom and the United States there have been shifts from welfare to *workfare* regimes in which citizens are expected to work in return for state benefits. This theme of state labor control strategies is developed more fully in Chapter 6.

The above policies often work out very differently in contrasting economies. Trade policy provides one example. The United States, a powerful advocate for free trade, does not necessarily apply the doctrine to its *imports* from around the world. In the past decade, the United States has imposed punitive tariffs on imports of steel from Europe, Japan, and many developing countries, and imports of textiles and garments from China. The United States justifies these tariffs on the basis of the strategic importance of domestic producers in these sectors and the unfair "cheap dumping" by exporting countries. It also continues to subsidize heavily domestic cotton production – a practice that is illegal under WTO rules – causing a trade dispute with Brazil and others in 2010. By contrast, the Asian newly industrialized economies (NIEs) explicitly engage in the strategic promotion of *exports*. For example, generous export subsidies and tax incentives have been offered to national champions such as Hyundai and Samsung from South Korea, and Acer and Tatung from Taiwan. Indeed, the Taiwanese state

CASE STUDY

Box 4.2   BHP Billiton's failed bid to buy PotashCorp

BHP Billiton is one of the world's largest mining groups, headquartered out of Melbourne, Australia, and London, United Kingdom, and with annual revenues of over US$40 billion. When, in August 2010, the company announced a US$39 billion hostile takeover bid for the Canadian firm PotashCorp, many might have expected it to become simply the latest example of a national player being bought up by a global giant. Instead, on November 3, 2010, the Canadian industry minister Tony Clement rejected the bid, saying that it was not of "net benefit to Canada." Under the rules of the Investment Canada Act, foreign takeovers must be able to demonstrate a net benefit for the country in terms of jobs, exports, production, and investment. This was just the second time in 25 years, however, that a takeover had been rejected on these grounds.

Scratching the surface of this apparently economic ruling quickly revealed the political forces at work behind it. PotashCorp is one of the world's leading producers of fertilizer – and its key ingredients of potash, phosphate, and nitrogen – and is headquartered in Saskatoon, Saskatchewan. The company dates back to 1975 when it was founded by the government of Saskatchewan before being privatized in 1990, and now accrues annual revenues of US$10 billion from operations across Canada, the United States, Brazil, and the Middle East. The potential takeover of PotashCorp was staunchly opposed by the Saskatchewan premier Brad Wall, who quickly lined up the support of four other provincial leaders and the three opposition parties in the national legislature, putting enormous pressure on the usually business-friendly minority Conservative government of Prime Minister Stephen Harper. At the heart of Wall's resistance was the fear of losing US$198 million a year in tax revenues, which BHP Billiton would have been able to avoid paying due to their other investments in the province. While the Harper administration had previously approved other takeovers of natural resource companies in the mid-2000s, protectionist arguments were being fueled by the backdrop of economic recession in 2010, and the potential BHP Billiton takeover was blocked. In summary, the PotashCorp case serves as a reminder not only of the continued ability of the state to police investment activity across its borders, but also that the decision-making processes that underpin its decisions are unavoidably political *and* economic in nature.

has pursued a wide range of strategies to develop its export-oriented IT industry since the early 1970s, including the establishment of the Hsinchu Science Park outside Taipei, direct investment in early key players in the sector including UMC (United Microelectronics Corporation) and TSMC (Taiwan Semiconductor Manufacturing Corporation), and steering developments in the industry through its agencies such as ERSO (Electronics Research Service Organization), ITRI (Industrial Technology Research Institute), and CETRA (China External Trade Development Council).

The State as Business Owner

With a few exceptions, states also directly engage in owning economic activities. This is an obvious and yet often overlooked aspect of state involvement in the economy; put another way, in some areas the state *is* the economy. It also challenges the notion that globalization is driven by *private* corporations, as some of these state-owned enterprises are actively globalizing as well. Different countries, however, experience different levels of state involvement in business enterprises. A useful distinction can be made here between *state-owned enterprises* (SOEs) and *government-linked corporations* (GLCs).

SOEs are public enterprises that are directly owned and managed by the state. CNOOC, the Chinese National Offshore Oil Corporation, provides a good example. One of China's largest oil companies, CNOOC was founded in 1982 and is headquartered in Beijing. In 2009, it had approximately 66,000 employees and an annual turnover of US$32 billion (http://en.cnooc.com.cn/, accessed December 10, 2010). The Chinese state not only owns a substantial 70% stake in CNOOC, but also directly appoints its top management. SOEs are commonly found in developing countries throughout the world, where the state has immediate developmental goals that can be achieved through direct state ownership and management of business enterprises. They are found in such diverse developing economies as Brazil (e.g., Petrobas), China (e.g., China Ocean Shipping), India (e.g., Indian Railways), Malaysia (e.g., Petronas), Mexico (e.g., CFE, an electricity monopoly) and Russia (e.g., Gazprom). SOEs are not only found in developing countries, however, as the examples of Statoil (Norway), Systembolaget (alcohol, Sweden), Enel (electricity, Italy) and Amtrak (U.S.) demonstrate. It is important to note that state ownership rarely means that the state owns 100% of a corporation – Enel and Gazprom have been part privatized, for example – but rather has a controlling stake. Interestingly, the global economic crisis from 2008 onward led to debates about whether we have seen a return to "state capitalism" in developed economies with leading financial institutions in countries such as the United States and United Kingdom effectively being renationalized as a result of government bailouts (e.g., Northern Rock and Royal Bank of Scotland in the U.K., and mortgage lenders Fannie Mae and Freddie Mac in the U.S.).

Government-linked corporations (GLCs), on the other hand, refer to business enterprises in which the state has a direct or indirect stake and yet leaves the management to professional managers. State intervention in these GLCs is often less significant in comparison to SOEs. The state may own some equity in these enterprises although they are mostly operated as profit-driven businesses. GLCs are commonly – but not exclusively – found among the largest transnational corporations from many developing economies such as Singapore (e.g., Singapore Airlines), Taiwan (e.g., Taiwan Semiconductor), Hong Kong (e.g., China Resources Enterprise), and Malaysia (e.g., the MUI Group). Among advanced industrialized economies, some states continue to have substantial stakes in their national firms. For example, the French and Italian states have significant ownership stakes in the automobile manufacturers Renault and Fiat, respectively. Indeed, the French state has the highest level of enterprise ownership in Europe, totalling US$400 billion and 5% of total employment in 2007. The distinction from SOEs is that the state does not necessarily own a controlling share and does not exercise direct managerial control. GLCs can be very active globalizers, as seen in the central role played by GLCs such as the Keppel Group, the Sembawang Group, and Singapore Technologies in the Singaporean government's regionalization drive – an initiative to promote outward investment across Southeast and East Asia – since the mid-1990s.

The State as Investor

In recent years, certain states have become major global investors through their own national funds, known as sovereign wealth funds (SWFs). As such, they have become important players in the global financial landscape (see Chapter 7). In general, SWFs are developed through the accumulation of capital via the following means:

- *Huge resource endowments in home countries*: Surplus national funds can be accumulated through the large-scale export of commodities derived from such endowments as oil fields (e.g., Norway and Saudi Arabia) and minerals (e.g., Russia).
- *Significant trade imbalances through exports*: Surplus foreign exchange reserves are turned into national funds for global investment (e.g., China, Hong Kong, and Singapore).
- *Longstanding conservative fiscal management*: Significant government budget surpluses over time are accumulated to form SWFs (e.g., Singapore).

Table 4.1 shows the world's 12 largest SWFs in late 2010, ranging from the Abu Dhabi Investment Authority's US$627 billion fund to the Qatar Investment Authority's US$85 billion fund. Four things stand out about this "sovereign fund capitalism" (Clark et al., 2010). First, with the exceptions of Norway, Hong Kong, and Singapore, the leading funds are based in rapidly developing countries where the pace of economic growth is allowing the accumulation of funds.

Table 4.1 World's twelve largest sovereign wealth funds in 2010 (US$ billion)

Country	Fund Name	Launch Year	Assets as of December 2010
UAE (Abu Dhabi)	Abu Dhabi Investment Authority (ADIA)	1976	627.0
Norway	Government Pension Fund-Global	1990	512.0
Saudi Arabia	SAMA Foreign holdings	n/a	439.1
China	SAFE Investment Company	1997	347.1
China	China Investment Corporation (CIC)	2007	332.4
Hong Kong, China	HKMA Investment Portfolio	1993	259.3
Singapore	Government of Singapore Investment Corporation (GIC)	1981	247.5
Kuwait	Kuwait Investment Authority	1953	202.8
China	National Social Security Fund	2000	146.5
Russia	National Welfare Fund (incl. Stabilization Fund)	2004	142.5
Singapore	Temasek Holdings	1974	133.0
Qatar	Qatar Investment Authority	2005	85.0

Source: http://www.swfinstitute.org, accessed on 14 February 2011.

Second, most of these countries are small economies, with the notable exceptions of China and Russia. The emergence of very substantial SWFs from these small economies indicates their intention to seek greater returns to their national wealth for sustaining future generations. The emergence of SWFs of such mammoth scale from small economies that are either resource-rich (e.g., Norway and Saudi Arabia) or resource-poor (e.g., the city-states of Hong Kong and Singapore) is a startling feature of the contemporary world economy.

Third, some of these SWFs have a long history of existence. However, as Clark et al. (2010, p. 2272) argue, "it was only during the first decade of the new millennium that governments around the world decided – seemingly en masse and independent of their "variety of capitalism" – that these special-purpose investment vehicles were crucial to achieving their policy objectives." Fourth, while most SWFs are managed professionally by national agencies, their close connections to ruling monarchs/regimes have raised significant concern among the broader international community. Critics argue that the investment objectives of SWFs are intertwined with state objectives. As Figure 4.2 illustrates, however, SWFs vary greatly in the extent to which they are pursuing strategic as opposed

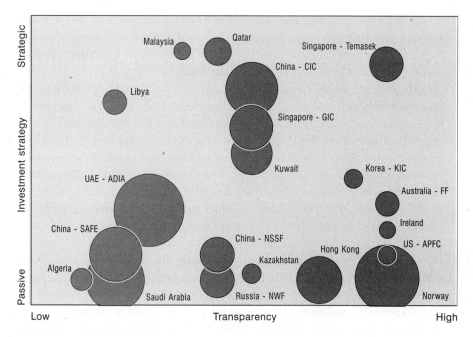

Figure 4.2   The global landscape of sovereign wealth funds, January 2011
Source: Adapted from http://wwsw.swfinstitute.org, accessed on 15 February 2011.
Note that the size of the bubble represents the size of the Sovereign Wealth Fund.

to economic objectives, and also in the extent to which there is transparency about those objectives. This mixing of financial with political imperatives has the potential to be an increasingly destabilizing and controversial aspect of state action within the global economy. Overall, the growing role of the state as investor is an important rejoinder to "the state is dead" narratives; here we have the state as an integral part of, not divorced from, the international financial system.

The State as Provider of Public Goods and Services

The state is also a direct provider of a wide range of so-called *public goods*. These are goods and services – for example, transport, education, health, infrastructure, and public housing – that are seen as too risky or unprofitable for individual private firms and/or fundamental to a population's well-being. Throughout the world, all states have at least to some extent taken up the role of providing public goods and services in lieu of the market and private firms, in many cases making the state the largest single employer in the country:

- *Transport services*: The state is involved in the provision of transportation at a variety of geographical scales ranging from the local to international connections. For example, the German federal state continues to own and operate Deutsche Bahn, one of the most successful train systems in the world. Most developing and some developed countries own and manage their national airlines (e.g., Singapore Airlines). This state-owned approach to public transport contrasts with the U.K. experience where British Airways was privatized in 1987 and British Rail in 1993.

- *Health and education services*: States are usually heavily involved in the provision of health and education services for their citizens. For example, at the tertiary level, most universities throughout the world today continue to depend on state-level funding. In some countries, private sector health and education provision is available alongside public sector provision for those who can afford it. In the United States, for example, the private college system funded by tuition fees and donations has thrived and produced some of the world's leading universities such as Harvard, Yale, and the University of Chicago. The state may also be important in providing research infrastructure and funding.

- *Infrastructure services*: The nation is often tasked with providing basic facilities and amenities such as roads, highways, airports, ports, and power supplies and communications networks. In developing countries, the capacity of the state to provide good public infrastructure is often critical to attracting inward investment. Some states also provide large public housing programs in order to improve the living standards of their citizens and/or to maintain the cost competitiveness of their labor force. The latter experience has been particularly relevant for Hong Kong and Singapore, two cities in Asia that link their highly successful public housing programs to export-oriented industrialization policies.

Taken together across all these areas of activity, state employment is clearly very significant. It can also, however, be geographically uneven in its importance. Put another way, public sector employment is often concentrated in particular places. Sometimes this is for obvious reasons, such as in capital cities; in other cases it may reflect a deliberate strategy to decentralize state employment and to try and promote economic growth in lagging regions. Levels of public sector employment vary significantly across the United Kingdom, for example, with increased levels of dependence on public sector jobs – above 40% of the total in some places–outside of the relatively prosperous Southeast and London where the rate tends to be below 20% of total employment.

To summarize, the state continues to exercise an enormous range of powers over the economic activities that occur both within and across its borders. The five categories of functions we have just reviewed are not separate, but rather interconnect and overlap in complex ways. State involvement in the domain of transport, for example, may straddle all five areas in many countries, encompassing basic guarantees of service delivery, strict safety and pricing regulations, intervention to promote growth in new transport technologies, state ownership of transport providers, and strong involvement in the management and provision of the underlying networks. More importantly, however, in overall terms these multiple functions intersect to produce enormous diversity in state–economy relationships within the global economy. Contrary to the hyperglobalist perspective, these multiple state–economy interactions create, over time, economies that operate in different ways. These can be thought of as different national varieties of capitalism or *national business systems* (see Box 4.3). In the next section we move on to make sense of this diversity by identifying some key categories of contemporary states.

KEY CONCEPT

Box 4.3   *National business systems*

How can we explain the fact that the nature of capitalism seems to vary between national contexts? To answer this question, we can turn to a concept developed by organizational sociologists – the *national business systems* approach (Whitley, 1999). Put simply, the approach argues that in different national economies, there are distinctive ways of doing business and that over time, these systems of business practices become *institutionalized* as ways of organizing economic activities. States policies are absolutely central to such processes. The national business system can be thought of as comprising three important elements:

- *Ownership patterns and corporate governance*: A diversity of corporate governance structures exists across different countries, ranging from public companies traded on stock exchanges, to state-owned enterprises, family businesses, cooperatives, and private companies. In some countries (e.g., the U.S. and the U.K.), the shareholders of most firms do not exercise direct control and prefer to leave their businesses in the hands of professional managers. In other countries (e.g., Scandinavia and South America), key shareholders, and sometimes their family members, get directly involved in management control. Elsewhere, the state may step in to develop dominant state-owned or government-linked enterprises (e.g., Venezuela, Saudi Arabia, and Singapore).
- *Business formation and management processes*: In countries with highly competitive markets (e.g., the U.S. and the U.K.), firms tend to be established by private entrepreneurs. The development and maturity of financial markets in these countries put tremendous pressure on top management to seek short-term returns and to pursue financially-driven corporate strategies. In contrast, in countries practicing "cooperative capitalism" (e.g., Germany and Japan), stable and enduring relationships between banks, firms, and the state tend to encourage strategic and long-term investments by corporate managers.
- *Work and employment relations*: The delegation of responsibility to skilled workers, and the demarcation of workplace functions, can vary

significantly between national business systems. In the United Kingdom, for example, the main divisions in the workplace exist between generalist top managers and managerial specialists, as well as between skilled workers and unskilled operatives. In Japan, it occurs between male, core employees and female, temporary workers. Diverse national business systems thus tend to foster different relationships between employers and their employees, and between different employees within the workplace.

In a variety of ways, then, distinctive national business systems can be identified, reflecting the ongoing interactions between firms and the gradually evolving institutional contexts in which they operate. We shall return to these issues from the perspective of the transnational corporations in Chapter 10.

4.4   Varieties of States

Present-day states have emerged from a rich variety of historical-geographical conditions, producing a wide range of *different* states rather than a homogeneous group. These states engage in a plethora of relationships with firms, markets, and international organizations. Whereas some states exert strong direct control over their domestic economies through strict economic regulation, other states have relegated such functions to other domains. In other words, the type of state is crucial in determining its capacity to control the economy and to resist and shape globalization processes.

In this section, we introduce four different types of states (drawing on Dicken, 2011; see Table 4.2 for a summary). To distinguish these varieties of states, we use two broad criteria:

- *Political governance systems*: These range from liberal democracy in neo-liberal states and social democracy in welfare states to firmer political control in developmental and authoritarian states.
- *The organization of economic institutions*: How firms, industries, state and non-state institutions relate to each other. In some contexts, the state will seek to shape and drive this web of relationships, while in others it will allow private sector organizations to take a leading role.

The diversity of state formations points to the highly *uneven* and *differentiated* relationships between these states and globalization tendencies. There is no single outcome of globalization in relation to these states. Instead, globalization processes occur in parallel with continued diversity.

In *neoliberal states*, government institutions seek to keep their distance from private firms and industries. The main role of the state in these economies is

Table 4.2 Varieties of states in the global economy

Type of States	Examples	Main Characteristics	Political Governance Systems	Organization of Economic Institutions
Neoliberal states	North America (e.g., Canada and the U.S.), the United Kingdom, Australia, New Zealand	Reliance on the market economy	Liberal democracy with multiple political parties	Strong role of capital markets and finance-driven investment regimes
Welfare states	Nordic countries (e.g., Sweden, Finland, and Norway) and most Western European countries (e.g., Germany and France)	Coexistence of substantial provisions of state benefits and the market economy	Socialist democracy with multiple political parties	Bank-centric financial systems and strong interdependency between capital, labor, and state
Developmental states	Japan and the newly industrialized economies (e.g., Brazil, Mexico, Singapore, Taiwan, and South Korea)	Relative autonomy of the capitalist state from corporate interests and voters	Soft authoritarianism dominated by a single large political party	Direct involvement of state in economy through industrial policies and strategic investments
Authoritarian states	Post-socialist countries in Eastern Europe, former Soviet Union, China, and Southeast Asia (e.g., Vietnam, Cambodia, and Laos)	Former communist states that have moved rapidly toward market-based economies	Strong authoritarianism dominated by a single political party	Coexistence of state-owned enterprises and market economy

to establish market rules via legislation and to enforce these rules through their regulatory capacity. In the United States, for example, competitive and antitrust or antimonopoly market rules have led to the development of a particular kind of investment regime that governs the market. In this investment regime, American firms tend to rely on capital markets for their investment needs, and their performance is measured according to the ability to maximize shareholder value. Hence, short-term investors exert a great deal of influence on corporate decision making, a regime that leads to particular kinds of flexible labor market practices (see Chapter 6). In this vein, New Zealand has experienced labor market restructuring that has occurred in tandem with massive privatization programs – both hallmarks of *neoliberalism* more generally (see Box 4.4). Overall, the economy in neoliberal states is underpinned by a well-developed financial market and a flexible labor market. The role of the state is to serve as the custodian of this system through complex sets of regulatory institutions and legislation.

KEY CONCEPT

Box 4.4   Neoliberalism

Neoliberalism is a political and economic ideology based on a strong belief in individual liberty, markets, and private enterprise. In stark contrast to the central economic management role played by the government in Keynesian welfare states, from this viewpoint the state's role in the economy should be minimized and limited to enforcing property rights and the operation of free markets and free trade. These core beliefs map onto a series of policy prescriptions encompassing:

- Fiscal discipline and the minimization of government budget deficits
- The reorientation of public expenditure from welfare, and redistribution toward the development of economic competitiveness
- The lowering of tax rates and enhancement of incentives for economic growth
- Financial liberalization to allow market control of interest rates and capital flows
- Trade liberalization and pursuit of reduction in both tariff and nontariff trade barriers
- The removal of barriers of foreign direct investment
- The privatization of state enterprises and increased private-sector involvement in public service delivery
- Widespread deregulation and the removal of barriers to competition

The development of neoliberalism can usefully be split into three phases (Peck and Tickell, 2002).

Its core ideas started to surface among politicians and academics during the economic crises of the 1970s in what can be thought of as a period of *proto-neoliberalism*. General Pinochet's Chile was arguably the first country to road-test a neoliberal reform program in the aftermath of the 1973 coup. The elections of Margaret Thatcher in the United Kingdom in 1979 and Ronald Reagan in the United States in 1980 heralded a second phase of *roll-back neoliberalism* in which state intervention in these economies was reduced through ongoing processes of privatization, deregulation, and marketization. By the early 1990s, neoliberalism had seemingly become normalized as policy-making "common sense" and a period of *roll-out neoliberalism* developed. This ongoing phase has been characterized by a creeping extension of neoliberal principles across all aspects of society and also its transfer around the globe through policy networks. The so-called Washington Consensus – the alignment of key officials at the U.S. Treasury, World Bank, and IMF to neoliberal principles – has been a powerful vehicle for the spread of these ideas through various forms of structural adjustment policies and "shock therapy" in response to economic crises around the world.

The work of economic geographers has been important in revealing three key aspects of the neoliberal project. First, it has explored both the origins of neoliberal thinking and the way it has traveled around the globe through different forms of *policy transfer*. Second, it has revealed that, as with globalization, neoliberalism is a complex matrix of ongoing and overlapping processes that are ever only partially effective and are always actively resisted. From this perspective, they are perhaps better thought of as *neoliberalization* rather than neoliberalism. Third, while it is possible to distill core neoliberal beliefs, in reality they come together in highly variegated and place-specific forms – or *actually-existing neoliberalisms* – that reflect the history and character of the preexisting state formation in which they have been initiated. For more, see England and Ward (2007), Peck (2010), and Springer (2010).

In *welfare states*, labor unions and state institutions play a more direct role in corporate governance and firm behavior. Private firms do not necessarily have a free hand in labor management; instead, there are stringent labor laws and other welfare provisions that shape the investment behavior of private firms. The role of capital markets in driving the national economy is less than in neoliberal states. The economies of both Germany and France, for example, are heavily funded by their national banks, many of which are state controlled, and less so by the capital markets (e.g., national stock exchanges) than in neoliberal states. As a result, German and French banks have substantial ownership stakes in, and management input into, large German and French firms. In this way, welfare states are able to control directly the national economy through their well-developed and regulated banking systems. Through state taxation, they are also able to provide a significant range of national welfare services for their citizens, ranging from unemployment benefits and medical insurance to retirement pensions and education.

In *developmental states*, the state is relatively autonomous from the influence of interest groups, business, and the population at large. This autonomy, often achieved through an element of political control, is necessary for the state to pursue interventionist policies favoring economic development. To achieve their economic development objectives, these states exercise economic control through developing elite state-sponsored economic agencies and strategic industrial policies (e.g., Japan's former Ministry of International Trade and Industry and South Korea's Economic Planning Board). These agencies are heavily engaged in consultation and coordination with the private sector, and these consultations become an essential part of the process of policy formulation and implementation. Through these elite agencies, the state decides on the "right" industries to nurture and the "best" firms to promote, creating a range of "national champion" firms, some of which are directly owned and managed by state institutions. The developmental state is also actively involved in regulating its domestic capital and labor markets in order to enhance the success of its industrial policies. In capital markets, the ministry of finance in most developmental states takes direct stakes in national banks and finances export-oriented industrialization programs through export subsidiaries and generous grants to its national champions. In labor markets, developmental states are often actively involved in subordinating the interests and rights of their workers. For example, in Taiwan and Hong Kong, labor unrests and strikes are managed through tough laws that curtail labor union activity.

In dynamic terms, over time many developmental states have progressed from what are termed *import-substitution* growth strategies to ones that are *export-oriented*. Import-substitution industrialization (ISI) strategies were commonly adopted by states in their early post-colonial period when nationalistic sentiments were particularly strong. Such strategies have been pursued, for example, with varying degrees of success, in many economies across East Asia and South America. Under the ISI strategy, the focus of economic development was on the domestic national economy, as evidenced by protectionist trade policies to reduce imports, tight exchange rate controls, the nationalization of large export-oriented firms, highly regulated labor markets, and strong subsidies for state-owned enterprises (SOEs) and domestically oriented private firms. Export-oriented industrialization (EOI) strategies have subsequently worked very well in the newly industrialized economies (NIEs) of East Asia (e.g., Hong Kong, Taiwan, Singapore, and South Korea) and Latin America (e.g., Brazil and Mexico). In contrast to ISI strategies, these programs are characterized by: devalued currencies to increase export competitiveness; state incentives to promote large export-oriented firms; the regulation of union activities to reduce labor militancy and to stabilize cheap labor supply; industrial policies favoring new industries with strong export propensities; and subsidies to export-oriented state-owned enterprises(SOEs) and private firms.

Authoritarian states combine a highly centralized political system with an increasingly open economic system. Many are former socialist states that have sought to liberalize their economies at the same time as maintaining strict political control. Their economies tend to be dominated by a mixture of SOEs and private firms (domestic and foreign), and the state continues to exercise a great deal of control of domestic economies through owning stakes in SOEs and strictly regulating private firms and industries. Authoritarian states are mostly found in Central and Eastern Europe and in Asia. While some of the post-socialist states in Europe have implemented both economic and political reforms, others – most notably Russia, but also former Soviet republics such as Belarus, Kazakhstan, Kyrgyzstan, and Uzbekistan – continue to control tightly political affairs. In Asia, China is a highly significant example of an authoritarian state. The opening of China to the global economy since 1978 has been accompanied by only slight advances in political freedom. The Chinese Communist Party remains the only legitimate political party in China, prompting a hybrid form of economic development characterized by strong communist control and expanding market freedom some refer to as "red capitalism" (Lin, 1997). In this system, the many state-owned enterprises are managed jointly by Party and firm managers. Vietnam has pursued a similar path since 1986 under its *doi moi* (renovation) program. Several states in North Africa and the Middle East (e.g., Saudi Arabia) also fall into the authoritarian category.

This four-fold typology allows us to appreciate the very varied institutional configurations that exist within different state forms and how these are influenced by political ideologies. However, we need to offer five caveats about conceptualizing state variety in this way. First, there will always be individual states that defy easy categorization – for example, Cuba and North Korea. And how should we think about the leftward shift in Latin American politics over the past ten to fifteen years that has seen a range of countries attempt to reverse neoliberalization trends – for example, in Venezuela's extensive nationalization program? Second, it is important to recognize that there is significant variation *within* each category. The U.S., U.K., and Australian states, while sharing some core neoliberal characteristics, are at the same time very different institutional configurations, reflecting their different histories and economic structures. This can be seen, for example, in the very different policies and attitudes toward universal healthcare in the United States and United Kingdom. Similarly, leading Asian developmental states such as Japan, South Korea, Taiwan, and Singapore have pursued highly distinctive growth paths. Third, in many cases apportioning all the activities of a state to a particular category may be simplistic – states will often adopt a range of policy positions across different domains. The New Labour administration that governed the United Kingdom from 1997 to 2010, for instance, adopted broadly neoliberal economic policies at the same time as it undertook a massive investment program in state education and health services. In political terms, this was seen as an attempt to develop a hybrid "third way" that could marry the benefits of markets with the social equity concerns of welfare states.

Fourth, it is important to recognize that different states have different capacities to implement policies and enact control over their territory. In some cases, a *weak* state may largely have succumbed to the power of oligarchs and corporate interests in what is essentially a dictatorship. *Dependent* states, meanwhile, may be so beholden to international institutions (e.g., the World Bank and the International Monetary Fund) or other powerful states (e.g., the U.S.) that they have little space for autonomous action. Most of these dependent economies are highly vulnerable to cyclical fluctuations in prices for the global commodities on which they heavily depend (e.g., coffee for Kenya and Ethiopia). In *failed* states, international institutions such as the United Nations may actually need to perform state roles on behalf of host governments if their authority disintegrates due to internal (e.g., civil wars or military coups) or external factors (e.g., international diplomatic or military pressure). The Caribbean state of Haiti, following the devastating 2010 earthquake, could be described in this way, as could Afghanistan in the post-U.S. invasion period. States may also choose to apply policies in a deliberately uneven way across their national space in order to enhance their position within the global system (see Box 4.5 on the notion of *graduated sovereignty*).

FURTHER THINKING

Box 4.5   Graduated sovereignty?

Thus far, we have considered the state as exercising influence and control in a uniform manner across its territory. In reality, however, state control is both socially and spatially uneven in its operation and effectiveness. In terms of social unevenness, certain portions of the population may be more or less excluded from state attempts to marshal the economy in the context of globalization. This may be voluntary on the part of some sections of the populace – for instance, in the case of those deliberately working in the illegal or black economy beyond the reaches, or at the margins of state regulation – or involuntarily – as in the case of temporary migrant workers that may not be granted the same range of rights and opportunities as full citizens. With regard to spatial unevenness, we can think of certain spaces in which either a different form of state control is in existence – for instance, in a tax-free export processing zone – or where the state simply cannot exert its controls – such as dangerous "no-go" areas in certain cities. Rather than states exerting unproblematic control over their society and territory, most in reality operate a form of *graduated sovereignty* (Ong, 2000) in which different zones of economic control overlap and coexist.

These ideas can be illustrated in the context of contemporary Malaysia, where it is possible to identify at least four zones of graduated sovereignty that are being created as the state has endeavored to integrate itself selectively into the global economy (here we draw on Ong, 2000). First, there are the spaces of illegal migrants from Indonesia, the Philippines, Bangladesh, and Myanmar, working as domestic servants, and on plantations and construction sites. Such workers have no rights to citizenship, and their short-term contracts are firmly enforced. Second, for Malaysia's diverse indigenous peoples, known in peninsular Malaysia as Orang Asli, differentiated government has meant displacement from traditional territories to increasingly marginal land. Third, in the case of manufacturing workers, industrial sites, and export processing zones (EPZs) are spaces exempted from a range of national labor and investment regulations, where labor unrest and potential strikes are quickly and firmly quashed, using force where necessary. Finally, the Multimedia Super Corridor (MSC), a 750-square-kilometer zone established in 1996 and stretching from Kuala Lumpur to the new international airport some 50 kilometers to the south, is an area designed to meet the needs of the global knowledge economy, offering a wide range of financial, technological, and labor market incentives.

Far from being a single territory governed in a uniform manner, therefore, contemporary Malaysia is in reality a fragmented space in which the experience of government depends on who and where you are. These ideas clearly have wider resonance: in all kinds of states it is possible to think about how different portions of the population are governed differentially and the graduated zones or spaces that are created as a result.

Fifth, our typology should not be taken to imply that the state is a *static* feature of the global economy; rather we view the state as a dynamic set of institutions. States can change their orientation in significant ways and then blur the boundaries of these categories. The post-socialist states of Central and Eastern Europe, for example, have moved from being centrally planned economies to authoritarian states with strong market elements in just two decades, while the so-called "Arab spring" of 2011 saw the overthrow of authoritarian regimes in Tunisia, Egypt, Libya, and Yemen. Equally, in the postwar period (1945–75) the states we currently understand as neoliberal were associated with a different state form – namely, the Keynesian welfare state. This state form combined a Keynesian approach to macroeconomics – typified by the state playing an active role in managing demand in the economy through its direct expenditures – with direct involvement in the provision of social services such as education, health, and housing (see also Chapter 2). Another element of this dynamism also relates to the *scalar* reconfiguration of the state – that is, the extent to which its territory is also influenced by sub- and supranational governmental organizations, a topic to which we turn in the next section.

4.5   Rescaling the State

In this section, we consider how the state is being reconfigured in scalar terms in an era of globalization. As noted in Chapter 1, the global space-economy can be conceived in terms of different geographical scales that range from the global and macro-regional to the national, regional, and local. States are actively reshaping their institutional structures of economic governance at both the *international* scales and *subnational* scales. In this *rescaling* process, the state relinquishes some of its control of national economic affairs to other governmental authorities at higher (international and macro-regional) or lower geographical scales (subnational, regional, and municipal).

Historically, the state executed its regulatory functions at the *national* scale. In other words, its policies and rules were applied throughout its territory, irre-spective of local conditions. Globalization tendencies and changing international relations are increasingly challenging this state-centric model of economic gover-nance. In particular, the rise of international and macro-regional organizations has increased the importance of international cooperation and the coordination of economic policies among states. This movement from national regimes of economic governance to authority at higher geographical scales can be described as a process of *upscaling* (Swyngedouw, 2004). These international and macro-regional organizations encompass *international organizations* with more or less global remit, and *macro-regional groupings* involving contiguous groupings of states. We will now consider each in turn.

Among the *international organizations*, the United Nations is the oldest and broadest in terms of its mission and scope. However, it has only limited means and resources to influence economic governance at the global scale. In this sphere, we must consider the highly powerful IMF, World Bank, and WTO and their influence upon the global financial system, development assistance, and trade, respectively. The IMF was founded in 1944, and had 187 member countries as of early 2012 (see Figure 4.3a). It is the central institution of the international monetary system, and is charged with promoting stability and efficiency within the global system from its Washington offices. The IMF is often called upon to resolve large financial deficits accrued by individual nation states. During the global economic crisis of 2008–10, it trebled its overall resources to US$750 billion and, as of mid-2010, had outstanding loan credits of US$54 billion with another US$146 billion allocated for countries to draw upon if necessary (www.imf.org, accessed December 10, 2010). The package offered to Greece

Figure 4.3 Institutions of global governance?
(a) The IMF, 700 19th Street, Washington, D.C. (b) The World Bank, 1818 H Street, Washington, D.C.

in May 2010, for example, was worth a potential US$26.4 billion over three years. However, the IMF not only lends to these states to help them through cash-flow crises (e.g., a lack of sufficient foreign currency), but also seeks to impose conditions on how these economies *should be* restructured. In other words, the IMF intervenes directly, through its rescue packages and associated conditions, in the state governance of their domestic economies. Consequently, many IMF-assisted states have historically been required to reduce their state budgets and liberalize control of the domestic economy in line with the IMF's neoliberal policy prescriptions (see Box 4.6).

FURTHER THINKING

Box 4.6   Shock therapy

As we saw in Chapter 3, a system capitalism has in-built crisis tendencies. Such crises offer political opportunities for dramatic intervention and for the spread of particular forms of economic management, particularly when states seek outside help from agencies such as the International Monetary Fund (IMF). Shock therapy is a term that has emerged to describe the imposition of a neoliberally-inspired package of policy measures at such points of crisis. The package usually includes the removal of price and currency controls, the withdrawal of state subsidies, trade liberalization, and the large-scale privatization of state assets. Such packages can be assembled and implemented extremely quickly, hence the term "shock" analogous to the short, sharp jolt offered by electrotherapy to the body. The reforms are often insisted upon by the IMF in return for the injections of capital required by struggling economies.

Before the 1980s, the role played by the IMF was mostly short term – for example, to ease temporary cash-flow problems. With the onset of the debt crisis in Latin America, however, many South American economies suffered from a sudden downturn in capital inflows, a massive devaluation of their national currencies, and hyperinflation. One of the first and most well-known shock therapy packages was implemented in Bolivia in 1985 to stem hyperinflation. Decree 21060 contained over 200 separate laws, including provisions that allowed the peso to float against the dollar, cut two-thirds of the employees of the state oil and tin companies, ended price controls, eliminated subsidies to the public sector, and liberalized trade rules. Inflation fell rapidly, with the package being lauded a success, but the medium-term impacts on Bolivia's economy were negative. In their early days, these IMF packages were known more neutrally as *Structural Adjustment Programs* (SAP) and were also implemented in other Latin American countries such as Argentina, Chile, and Peru in response to debt crises.

Since the 1980s, shock therapy has been implemented around the world in response to economic crises, most notably during the early 1990s in

Central and Eastern Europe (e.g., Poland and Russia), and in East Asia in response to the 1997–98 Asian economic crisis (e.g., Indonesia, South Korea, and Thailand). Commentators have become increasingly critical of such schemes, however, due to both the huge social costs of rapid economic restructuring and the uneven post-therapy performance of the "treated" economies. In all such debates, however, it is important to remember that while the implementation of shock therapy reflects the increased power of international institutions such as the IMF, it also requires the support of domestic political coalitions. For more on shock therapy, see Klein's (2007) outstanding account.

The Washington-based World Bank was also founded in 1944, has 187 members, and provides development assistance, in terms of both knowledge and finance, to over 100 developing countries on an ongoing basis (see Figure 4.3b). This assistance covers a wide variety of activities including basic health and education provision, social development and poverty reduction, public service provision, environmental protection, private business development, and macro-economic reforms. In 2008–09, it granted US$58.8 billion in development assistance through a mixture of loans and grants, with Africa and South Asia being the biggest recipients (www.worldbank.org, accessed December 10, 2010). Some states have benefited from the World Bank's assistance and engaged in a virtuous circle of successful development pathways (e.g., postwar Germany, Japan, and the Asian NIEs). Other weaker and dependent states have benefited less from the Bank's economic advice and development assistance (e.g., Chile and Peru). This uneven outcome of development interventions by the Bank (and the IMF) has created discontent in certain developing economies, where some portray the World Bank and IMF as *causes* of the development gaps in today's global economy.

The WTO was created in 1995 as the successor to the General Agreement on Tariffs and Trade (GATT) established in 1947. As an organization it is responsible for the regulation of global trade, and operates a multilateral rules-based system derived by negotiation between its member states. By early 2011 the WTO had 153 members, accounting for 97% of global trade (www.wto.org, accessed March 6, 2012). In addition to administering trade agreements and providing a negotiating forum, it handles trade disputes, monitors national trade policies, and provides technical assistance and training for developing countries. For critics, the rise of the WTO and its enforcement of global trading rules has actually increased uneven development at the global scale. While developing countries find it hard to engage in protectionist trade policies because of WTO rules, some of the richest states in the world – most notably the United States and some within the European Union – continue to flout WTO rules by protecting their domestic producers in politically sensitive sectors such as steel, clothing, and agriculture. Trade liberalization under the auspices of the WTO therefore does not necessarily reduce its uneven impacts, as different countries benefit differentially from free trade. The very slow progress of the Doha round of negotiations initiated in 2001 (and still ongoing in 2012) is reflective of these tensions, with agricultural subsidies and tariffs proving to be a major sticking point.

While the triumvirate of the IMF, World Bank, and WTO date back to the mid-20th century, a new grouping of states has become increasingly influential in recent times – namely, the G20. The G20 has its roots in a rich country grouping, the G7, formed in 1976 to bring together Canada, France, Germany, Italy, Japan, the United Kingdom, and the United States. In turn, the G8 was formed in 1997 with the addition of Russia, while the G20 was created in 1999 in response to the international financial crises of the late 1990s. The grouping includes the G8, the European Union, and the major developing markets including Argentina, Brazil, China, India, South Africa, and South Korea (see Figure 4.4). Together, the countries account for 90% of world gross national product, 80% of world trade, and two-thirds of world population. The organization is constituted by regular meetings of the finance ministers and central bank governors of its members, although the economic crisis of 2008–10 saw the initiation of annual leaders' summits as well (see Figure 4.5). The G20 describes its mandate as, "the premier forum for international economic development that promotes open and constructive discussion between industrial and emerging-market countries on key issues related to global economic stability" (www.g20.org, accessed February 15, 2011). G20 summit meetings have increasingly become the focus of anti-globalization sentiments and protests. The June 2010 leaders' summit in Toronto, Canada, for example, saw considerable protests from a wide range

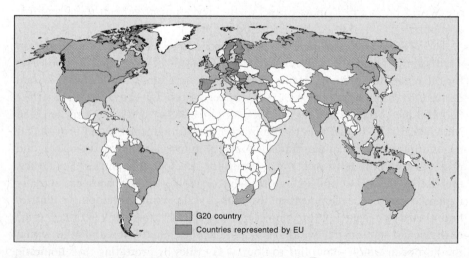

Figure 4.4 Map of the G20 countries

Figure 4.5   The second G20 leaders' summit, London, April 2009

of civil society organizations. Going forward, it seems likely that the G20, or an expanded version of it, will be the crucial forum for forging international economic cooperation.

Macro-regional groupings, on the other hand, have emerged primarily for member states to engage in economic integration and, to a limited extent, political integration. As such, regional economic blocs are a significant addition to the regulatory architecture of the global economic system. The initial stimulus for such formations comes from a desire to reduce barriers to trade and therefore enhance levels of intra-regional trade. Following Dicken (2011), we can identify four key types of bloc (see Table 4.3 for a range of examples). As we move down the list, the level of economic and political integration increases:

- *The free-trade area*: Trade restrictions between member states are removed, but states retain their individual trading arrangements with nonmembers – for example, the North American Free Trade Agreement (NAFTA) between Canada, the United States, and Mexico since 1994.
- *The customs union*: Members operate a free-trade agreement between themselves and have a common trade policy for nonmembers – for example, the MERCOSUR customs union between Argentina, Brazil, Paraguay, and Uruguay since 1991 (Venezuela has since joined).
- *The Common Market*: This has the characteristics of a customs union, but in addition, allows the free movement of factors of production (e.g., capital and labor) between members – for example, Caribbean Community (CARICOM) since 1973.
- *The economic union*: Promotes harmonization and supranational control of economic policies. Only the European Union (EU) comes anywhere close to this form as evidenced by the adoption of the euro as the sole currency of

Table 4.3   Major regional economic blocs in the global economy

Regional Group	Membership	Date	Type
EU (European Union)	Austria, Belgium, Bulgaria, Cyprus, Czech Republic, Denmark, Estonia, France, Finland, Germany, Greece, Hungary, Ireland, Italy, Latvia, Lithuania, Luxembourg, Malta, Netherlands, Poland, Portugal, Romania, Slovakia, Slovenia, Spain, Sweden, United Kingdom	1957 (European Common Market) 1992 (European Union)	Economic Union
NAFTA (North American Free Trade Agreement)	Canada, Mexico, United States	1994	Free Trade Area
EFTA (European Free Trade Association)	Iceland, Norway, Liechtenstein, Switzerland	1960	Free Trade Area
MERCOSUR (Southern Cone Common Market)	Argentina, Brazil, Paraguay, Uruguay, Venezuela	1991	Customs Union
ANCOM (Andean Common Market)	Bolivia, Colombia, Ecuador, Peru, Venezuela	1969 (revived 1990)	Customs Union
CARICOM (Caribbean Community)	Antigua and Barbuda, The Bahamas, Barbados, Belize, Dominica, Grenada, Guyana, Haiti, Jamaica, Montserrat, St. Kitts and Nevis, Saint Lucia, St. Vincent and the Grenadines, Suriname, Trinidad and Tobago	1973	Common Market
AFTA (ASEAN Free Trade Agreement)	Brunei Darussalam, Cambodia, Indonesia, Laos, Malaysia, Myanmar, Philippines, Singapore, Thailand, Vietnam	1967 (ASEAN) 1992 (AFTA) 2010 (ASEAN + China, ASEAN + India, ASEAN + Australia and New Zealand)	Free Trade Areas

Source: Adapted from Dicken, 2011, Table 6.2.

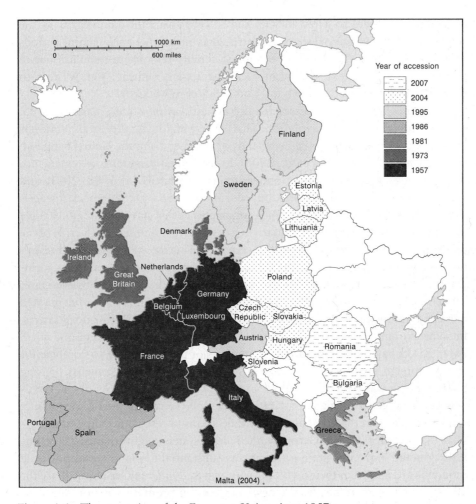

Figure 4.6   The expansion of the European Union since 1957

several member states in 2002. The EU started out as the European Common Market in 1957, but now encompasses 27 European countries in an advanced form of regional integration (see Figure 4.6).

Three further points should be made about these groupings. First, numerically, the vast majority of agreements fall under the first two headings (i.e., free-trade areas and customs unions). Second, economic blocs may develop over time and move down this list, as has been the case with what is now the European Union. Third, it is important to recognize that all these regional economic forms are initiated by, and derive legitimation from, their member states. However, as states increasingly participate in such regional initiatives, some claim that they are ceding power over their domestic economies. This claim is in part based on the experience of EU member states that have had to adjust their domestic budgetary expenditure in order to avoid deficits and thereby maintain the integrity of the single European currency system. Countries may also experience painful periods of adjustment as certain kinds of economic activity relocate to take account of the new regional context (e.g., manufacturing activity shifting from western to central/eastern Europe, and from the United States to Mexico).

Meanwhile, rescaling processes are also at work *within* states, manifested in a tendency toward regional devolution. This process of *downscaling* of the state has been particularly evident in the heartlands of neoliberalism, the United Kingdom and the United States (Jones et al., 2005). In the United Kingdom, the emergence of institutions such as Training and Enterprise Councils (in 1988–2001, and subsequently Learning and Skills Councils) and Regional Development Agencies (RDAs, since 1998) in the management of investment incentives, human resource development, and regeneration initiatives has somewhat diminished the role of the national state in domestic economic governance. This reconfiguration and rescaling of economic governance in the United Kingdom in favor of local states forms was an integral part of an ongoing neoliberal project first initiated by Margaret Thatcher in the 1980s. Over time, local and regional authorities have been granted increased autonomy over economic decision making and investment initiatives. In some extreme examples (e.g., the Welsh Development Agency and one-north East), RDAs have committed massive amounts of public funding to compete for specific foreign investors (South Korea's LG and Germany's Siemens). During 2010, funding was removed for both Learning Councils and RDAs by the U.K.'s new coalition government as part of a wider drive to reduce public spending, a stark reminder that these remain *political* decisions to restructure the state.

In the United States and Germany, the federal system of governance has historically granted individual state governments substantial autonomy and power in economic affairs within their own states (known as *Länder* in Germany). Since the 1990s, however, there has been renewed interest in promoting major American cities as the "growth poles" of the U.S. economy. This phenomenon is broadly known as *urban entrepreneurialism*, reflecting the way in which metropolitan governments and city mayors are becoming more entrepreneurial in their economic policy initiatives, particularly toward attracting inward investment. The process has led to stern inter-urban competition for trade and investment, a phenomenon widely observed throughout Western Europe as well.

While the United States and the United Kingdom are well-known examples of devolution and downscaling, these processes also have echoes in states that are not necessarily neoliberal in their political-economic outlook. For example, the gradual opening of China to international economic activities since 1978 has been accompanied by a significant process of decentralization of economic decision making from the central communist state to local and provincial authorities. This decentralization process has subsequently produced a dramatic surge in economic activity, most notably in the coastal provinces. The local and regional governments in these provinces (e.g., Guangdong and Lower Yangtze) are extremely aggressive in their efforts to attract foreign investment. Overall, then, when considering the changing nature of the contemporary state, it is important to consider the extent to which certain state functions have been rescaled to other levels of authority.

4.6   Hollowing-Out the State?

The ongoing reconfiguration of contemporary states is not only about scalar restructuring, however. Many states are also simultaneously experiencing processes whereby private actors are playing an increasing role in managing the economy. These processes constitute more than a simple "chipping away at the edges" (Pinch, 1997) of the state apparatus through rationalization and privatization processes. Instead, they reflect a *hollowing-out* that has seen state functions taken over by either public-private partnerships or private forms of regulation, thereby changing the nature of how state functions are delivered. This rise in nonstate forms of economic regulation is often described as being a shift from *government* to *governance*, reflecting a move away from control through directly elected and accountable representatives. This process sees the devolution of state power to nonelected but state-sanctioned QUANGOS (Quasi-Autonomous Non-Governmental Organizations), independent civil society organizations (also known as NGOs) and sometimes even private entities. It often intersects with the rescaling of economic governance described above. The downscaling of certain state functions to local and regional tiers of government in many contexts, for example, has occurred in tandem with an increased role of private sector representatives in deciding how public economic development funds are allocated and spent.

Let us consider some examples of these hollowing-out processes. The privatization of former state-owned enterprises (SOEs) does not just represent a transfer of ownership from the public to the private sector. It also entails a transfer of economic management and governance rights from the public sector to the hands of new private shareholders of these former SOEs. In developing countries throughout the world (e.g., India, Brazil, and Thailand), the partial retreat of the state from ownership of firms in various industries points to the rise of industry-specific private regulation. In the telecommunications industry, for example, the privatization of former state-owned telecommunications providers has led to new forms of service standardization and pricing strategies that are regulated by competitive market forces rather than state directives. The entry of foreign firms into these newly privatized industries has further strengthened the role of transnational corporations in shaping the country-specific operation of these industries. In other words, the regulation and governance of telecommunications industries in many developing countries are as much shaped by the largest private telecommunications firms as by communications ministries.

In the realm of *regulatory regimes*, we are also witnessing a certain degree of hollowing-out of the state as national legal systems increasingly intersect with private mechanisms. This phenomenon is perhaps most obvious in the world of global finance (see also Chapter 7). The regulation of the financial health of national firms used to be held in the hands of states and their designated institutions, usually the central banks or ministries of finance. Today, however, the importance of private coordinating and evaluative agencies such as credit rating agencies (e.g., Moody's and Standard & Poor's), institutional investors (e.g., Goldman Sachs and JP Morgan) and pension funds (e.g., the California Public Employees Retirement System and the Universities Superannuation Scheme in the U.K.), and accountancy firms (e.g., PriceWaterhouseCoopers) has grown rapidly in an era of global finance. The role of the state in ensuring order and transparency in their national financial markets has been partially taken over by these private institutions. It is no exaggeration to say that many of the largest firms throughout developed and developing countries are more worried about their ratings with Standard & Poor's and the buy/sell advice by Goldman Sachs than their standing with home country state authorities. This is because favorable credit ratings and "buy" advice by Goldman Sachs can significantly enhance the share prices of these large firms and assist their search for capital to fund corporate growth. Likewise, states are increasingly concerned about maintaining the ratings of their government bonds: AAA is the strongest, signifying that there is virtually no risk of default. Such ratings are crucial in a postrecession context where massive public sector deficits are financed through bond sales. Iceland, Ireland, Portugal, and Spain, for example, had their national sovereign debt ratings lowered by these agencies, leading to panic in the financial markets.

In this regard, the business media also play a significant role in shaping how business practices are governed. Business magazines such as *Forbes*, *Fortune*, and *Business Week* and business newspapers such as *The Wall Street Journal* and *The Financial Times* often play a de facto monitoring role in the business world. Corporate scandals reported in these magazines and newspapers, true or otherwise, can have very damaging effects on a firm. In fact, the state often gets involved in investigating the wrongdoings of private firms *after* the media breaks the news (e.g., the role of the media in the 2001 collapse of Enron in the U.S.). *The Financial Times* and other media in the city of London were implicated, for example, in the pessimistic corporate discourses on the financial crisis that swept Asia in 1997 and 1998. Likewise, during the crisis that unfolded in 2007–08, revelations about the state of the financial sector by the BBC's Business Editor Robert Peston were purported to be a key factor in undermining confidence in the British banking system, causing one observer to ask, tongue-in-cheek, "has Robert Peston caused a recession?" (Fenton-O'Creevy, 2008).

In addition to these private agents, there are also many "quasi-private" institutions that operate at the *international* scale to regulate specific economic activities. These international institutions range from industry associations and watchdogs to environmental agencies and private foundations. What these institutions have in common is that they are increasingly taking over regulatory and coordination functions previously held by states. For example, in terms of global standards in accounting, the International Accounting Standards Committee (IASC) was established in 1973, becoming the International Accounting Standards Board (IASB) in 2001. As an independent private-sector body, the IASC aims to create a set of global accounting standards that can be applied and implemented throughout the world. By the late 1990s, all stock exchanges in the world had adopted the reporting standards recommended by the IASC. This achievement should be viewed in a context where many state authorities (e.g., Japan and Germany) had previously found it very hard to implement a common standard among their domestic firms, let alone uniform international standards recommended by the IASC. In other words, one may argue that the IASC is much more effective in governing global accounting practices than individual states. We will look more at the emerging private governance regimes surrounding global commodity chains in Chapter 8.

4.7   Summary

In this chapter we have shown that it is far too simplistic to claim the irreversible demise of the state in managing national economies. As we have seen, states continue to shape profoundly the economic activity within, and across, their borders in a wide range of ways: from the assurance of basic laws and property rights, through the provision of basic infrastructure and education, to direct ownership of companies and a range of financial and tax incentives. At the same time, states are not all alike: there is a tremendous range of different state forms in today's world, from neoliberal variants through to welfare and developmental variants wherein states play a much stronger role in economic development processes. These different states in turn have widely differing abilities both to control their economies and to exert influence on international institutions. Rich nations such as the United States, Japan, and Germany, for example, can wield substantial power in these international organizations through their voting rights, financial contributions, and the appointment of key personnel.

Overall, however, the state continues to be an extremely important actor in the global economy. Ongoing processes of rescaling and hollowing-out, while significant, have not diminished the state's significance as perhaps the single most important shaper of political-economic activity worldwide. Importantly, the state itself is always implicated or involved in directing these processes through its conscious decisions whether to engage with international organizations or to devolve power and authority to local states. In short, states remain critical institutions through which international, regional, and local economic issues are evaluated and resolved. Together with the transnational corporations that we will describe in Chapter 10, they are architects rather than passive observers of globalization.

Notes on the references

- Dicken et al. (1997), Yeung (1998), Kelly (1999), Amin (2002), and Dicken (2004) offer some of the best geographical analyses of globalization.
- O'Neill (1997) provides an excellent perspective on the state in economic geography. Dicken (2011) offers a wealth of useful material on contemporary states and their interaction with transnational corporations, while Glassman (1999) shows how the state is rapidly internationalizing to project its power.
- For more on neoliberalism and its geographies, see Peck and Tickell (2002), Brenner et al. (2010), Peck (2010), and Springer (2010).
- Brenner et al. (2003), Swyngedouw (2004), and Keil and Mahon (2009) offer nuanced accounts of the contemporary rescaling processes affecting the state.

Sample essay questions

- Why is the end-of-the-state thesis flawed?
- How do states differ in their approach to economic governance?
- How does neoliberalism influence state behavior?
- What does the rescaling of governance mean in geographical terms?
- To what extent is the state still capable of managing its national economy?

Resources for further learning

- http://www.globalpolicy.org/nations-a-states.html: The website of the Global Policy Forum provides useful insights into the history and formation of modern states.
- http://www.imf.org/external/index.htm, http://www.wto.org/ and http://www.worldbank.org: The websites of three of the world's most powerful financial and economic institutions offer a wide range of information on how these international organizations can shape a whole variety of political-economic processes in specific nation states.
- http://www.g20.org/ and http://www.oecd.org/home/: The G20 and OECD are other important groupings involved in the governance of the global economy.
- http://europa.eu/index_en.htm: The European Union portal contains a whole host of information on the genesis and working of the world's largest macro-regional organization.
- http://www.aseansec.org/, http://www.apec.org/ and http://www.nafta-sec-alena.org/: ASEAN, APEC and NAFTA are other good examples of regional economic groupings.

- http://www.direct.gov.uk/en/index.htm: The web portal for the public services operated by the U.K. government.
- http://www.bis.gov.uk/: Website of the U.K.'s Department for Business, Innovation and Skills.
- http://www.ca.gov/: The website of the state of California reveals the range of state powers in a federal context.
- The 2011 movie *Too Big to Fail* is based on Andrew Ross Sorkin's book of the same name, and depicts the dealings between the U.S. government and major banks during the crisis of 2008. See also Charles Ferguson's 2010 film *Inside Job*.

References

Amin, A. (2002). Spatialities of globalisation, *Environment and Planning A*, 34, 385–399.

Brenner, N., Jessop, B., Jones, M., and MacLeod, G., eds. (2003). *State/Space*. Oxford: Blackwell.

Brenner, N., Peck, J., and Theodore, N. (2010). Variegated neoliberalization: geographies, modalities, pathways. *Global Networks*, 10, 182–222.

Clark, G. L., Monk, A., Dixon, A., Pauly, L. W., Faulconbridge, J., Yeung, H. W-C., and Behrendt, S. (2010). Symposium: sovereign fund capitalism. *Environment and Planning A*, 42, 2271–2291.

Dicken, P. (2004). Geographers and "globalization": (yet) another missed boat? *Transactions of the Institute of British Geographers*, 29, 5–26.

Dicken, P. (2011). *Global Shift: Mapping the Changing Contours of the Global Economy*, Sixth Edition. London: Sage.

Dicken, P., Peck, J., and Tickell, A. (1997). Unpacking the global. In R. Lee and J. Wills, eds., *Geographies of Economies*. London: Arnold, pp.158–66.

England, K., and Ward, K., eds. (2007). *Neo-liberalization: States, Networks, Peoples*. Oxford: Blackwell.

Fenton-O'Creevy, M. (2008). Has Robert Peston caused a recession? Social amplification, performativity and risks in financial markets. http://www.open2.net/blogs/money/index.php/2008/10/17/robert-peston?blog=5 (accessed on December 10, 2010).

Glassman, J. (1999). State power beyond the "territorial trap": the internationalization of the state. *Political Geography*, 18, 669–696.

Jones, M. R., Goodwin, M., and Jones, R., eds. (2005). Special issue on devolution and economic governance. *Regional Studies*, 39, 397–553.

Keil, R., and Mahon, R., eds. (2009). *Leviathan Undone: Towards a Political Economy of Scale?* Vancouver: UBC Press.

Kelly, P. F. (1999). The geographies and politics of globalization. *Progress in Human Geography*, 23, 379–400.

Klein, N. (2007). *The Shock Doctrine: The Rise of Disaster Capitalism*. London: Penguin.

Lin, G. C. S. (1997). *Red Capitalism in South China: Growth and Development of the Pearl River Delta*. Vancouver: UBC Press.

NAO. (2009). *Maintaining Financial Stability across the U.K.'s Banking System*. London: National Audit Office.

O'Neill, P. (1997). Bringing the qualitative state into economic geography. In R. Lee and J. Wills (eds.), *Geographies of Economies*. London: Arnold, pp. 290–301.

Ohmae, K. (1995). *The End of the Nation State: The Rise of Regional Economies*. London: HarperCollins.

Ong, A. (2000). Graduated sovereignty in South-East Asia. *Theory, Culture and Society*, 17, 55–75.

Peck, J. (2010). *Constructions of Neoliberal Reason*. Oxford: Oxford University Press.

Peck, J., and Tickell, A. (2002). Neoliberalising space. *Antipode*, 34, 380–404.

Pinch, S. (1997). *Worlds of Welfare*. London: Routledge.

Springer, S. (2010). Neoliberalism and geography: expansions, variegations, formations. *Geography Compass*, 4, 1025–1038.

Swyngedouw, E. (2004). Globalisation or 'glocalisation'? Networks, territories and rescaling. *Cambridge Review of International Affairs*, 17, 25–48.

U.S. Treasury. (2010). *Troubled Asset Relief Program: Two Year Retrospective*. Washington, D.C.: U.S. Treasury.

Yeung, H. W-C. (1998). Capital, state and space: contesting the borderless world. *Transactions of the Institute of British Geographers*, 23, 291–309.

Whitley, R. (1999). *Divergent Capitalisms: The Social Structuring and Change of Business Systems*. New York: Oxford University Press.

CHAPTER 5

ENVIRONMENT/ECONOMY
Can nature be a commodity?

Goals of this chapter

- To understand how resources drawn from the natural environment become incorporated into economic processes and get assigned an economic value
- To appreciate the changing ways in which ownership over nature is established
- To assess the use of markets as a mechanism of environmental protection
- To appreciate how the line between human/economic systems and natural systems is blurred

5.1 Introduction

On April 20, 2010, an explosion rocked BP's Deepwater Horizon oil rig in the Gulf of Mexico, killing 11 workers. The rig, located 50 miles off the Louisiana coast, was extracting oil from geological formations around 13,000 feet below the sea bed and then drawing it up through 5,000 feet of water. A few days after the explosion the rig sank, causing the pipe that carried the oil to buckle and break. Safety shut-off mechanisms failed to place an emergency cap on the well, and so thousands of gallons of oil started to gush into the Gulf of Mexico (see Figure 5.1). Over the next three months, about 5 million barrels of oil (almost 800 million liters) entered the waters of the Gulf. This became the largest accidental oil spill ever recorded. By the middle of May, the contamination started to reach the shoreline of Louisiana, and within a few weeks, oil and tar balls were being reported in Alabama, Mississippi, and Florida. Repeated attempts to cap the well failed, and it was not until mid-September 2010 that the U.S. federal government certified that the oil had been permanently plugged.

Figure 5.1   The Deepwater Horizon oil rig, April 2010

The consequences of a pollution episode on this epic scale have been hard to determine. The immediately noticeable effects were on the coastal environments and communities where the oil reached land. Beaches and wetlands became contaminated, while fishing, shrimping, tourism, and other local livelihoods were heavily affected. The worst predictions did not come true though, because ocean currents kept some of the oil away from the coastline, and warm temperatures aided in the effort to disperse chemically the pollutant. Nevertheless, the social costs of the spill were substantial. Almost 500,000 individuals and businesses applied to the Gulf Coast Compensation Fund, which was established with $20 billion from BP and administered by the U.S. federal government. By February 2012, over $6 billion had been paid out, while longer-term compensation claims for damage, and future lost income, are still ongoing. In a stunning indictment of how environmental impacts are accounted for in economic terms, one study suggested that American GDP might actually gain from the disaster. This bizarre calculation was plausible because, although fishing and tourism jobs were affected and total oil production was reduced, the enormous cost and employment associated with the clean-up operation may have had a net positive effect on economic output (Di Leo, 2010).

The costs of the disaster that were not associated with people's homes and businesses are much harder to evaluate. Thousands of birds were found dead due to oil contamination, along with dolphins, turtles, and other marine life. Doubts exist about the long-term ecological effects of the chemical dispersants used to fight the oil slick. Studies are also suggesting that a large quantity of oil remains submerged in the Gulf of Mexico and may have continuing impacts on aquatic ecosystems. Furthermore, all of this damage has occurred in a marine environment that is already highly stressed and vulnerable. Each year the Mississippi River carries huge quantities of waste products from agricultural and urban areas across the central and eastern United States, including fertilizers, pesticides, treated sewage, and urban runoff. The result is a huge expanse of algal growth and bacteria in the spring and summer that reduces the dissolved oxygen in the Gulf waters, making them uninhabitable for most marine life. The "dead zone" stretches over 8,000 square miles, roughly the same area as the Deepwater Horizon oil contamination.

Despite the dangers involved, drilling for oil and extracting other resources from the earth remain central to both our economic life and the ways in which we relate to nature. This chapter examines how nature has become a part of our economic system, and the example of oil extraction in the Gulf of Mexico illustrates a few key points. The first point is that the natural environment is a fundamental part of the production system. As we adopt increasingly urban lifestyles dependent on manufactured goods, it is often easy to forget that ultimately all of our needs (from housing and food, to computers and cars) are made through processes that transform naturally occurring materials into other forms. Nature is therefore the fundamental basis for the economy, not something outside of it. The existence of facilities such as the Deepwater Horizon rig, and the purpose they serve, reminds us that this transformation of nature into a resource is something that we depend upon every time we fill up a car with fuel. The tragic loss of life on the Deepwater Horizon rig also reminds us that, however technologically advanced the process might be, our transformation of nature is ultimately always carried out through acts of human labor.

A second point is that nature's transformation into something useful is a process that also requires social institutions and structures to make it happen. In particular, it requires the establishment of ownership and extraction over a piece of the natural world: in this case a reserve of oil tucked into the earth's crust, thousands of feet below the ocean floor. It also requires that an economic value be assigned to that resource in order for the massive engineering work of extraction to be worthwhile. It is only when that value reaches a certain level, and when the necessary technology is available, that a natural oil deposit becomes an exploitable resource.

A third point is that our use of the planet's resources and degradation of its environments are difficult things to incorporate into economic ways of thinking. While the costs of an oil spill can be quantified when they affect livelihoods or when an expensive clean-up is required, it is much less easy to put a price on the otherwise unvalued plants, birds, and marine wildlife that are damaged. It is also very difficult to establish a cost for the oil that is not yet spilled but instead goes on to be used as fuel or for manufacturing processes. The burning of hydrocarbon fuels is known to be a major contribution to human-induced climate change, but the effects are so diffused, long term, and uncertain that it is perhaps easier to evaluate the costs of the oil we lose (into the ocean) than the oil we use (at the fuel pump)!

In this chapter we explore these issues in four main sections. The first (Section 5.2) asks how natural resources are conventionally counted (or not counted) in economic thinking. While the market mechanism is most often relied upon to determine the price of a natural resource, there are also creative, albeit limited, ways of evaluating the costs of resource extraction and environmental damage. We then take a step back from these techniques of price setting and ask instead how natural resources are incorporated into economic processes in the first place. Section 5.3 examines the various ways in which natural resources can be subject to this process of *commodification*. This involves understanding the conditions necessary for nature to have both ownership and value attached to it. A geographical approach is especially useful here because it demands that we appreciate how the physical qualities of nature affect its *commodification* and how commodification processes vary across space. In Section 5.4 we explore the ways in which the *impacts* of the capitalist system on the natural environment can also become incorporated into processes of commodification. In particular, we will look at the ways in which pollution has become a tradable commodity and how a wide range of economic activities have emerged to meet increasing demand for renewable energy and resource efficiency. Finally, in Section 5.5, we consider how our human bodies are themselves becoming increasingly commodified, thereby blurring the traditional distinction between humans and nature.

5.2   How Is Nature Counted in Economic Thought?

In this section we present two ways of analyzing the worth of natural resources in contemporary economics. The first is based on the establishment of a price for a commodity based on supply and demand in the marketplace. The second is through an effort to establish the wider value of a natural environment (and thus the cost of disturbing it through resource extraction), by giving it a monetary value that is missing in the market mechanism.

In conventional economics, the valuation of nature can be deceptively straightforward. It is established in the form of a price set for units of a given natural material such as wheat, gold, or lumber. The actual mechanism for determining a price is in a market exchange where buyers and sellers come together. In some cases such as a local cattle auction, the price reached for a head of cattle will be locally specific. But for many natural raw materials such as metals or oil, a global price is established at large commodity exchanges located in major cities. Such exchanges now exist in many countries around the world, but the largest and most important ones operate in just a few major centers: the Chicago Board of Trade and China's Dalian Commodity Exchange for agricultural products; the London Metal Exchange for metals such as tin, aluminum, and copper; and the New York Mercantile Exchange for energy products such as oil and natural gas. Some trading at these exchanges are for immediate purchases or "spot trades," but most are for "futures," whereby future deliveries of a commodity are purchased at a specified current price.

Conventional economic analysis views these exchanges as highly efficient mechanisms for establishing the value of natural raw materials. These exchanges allow producers and traders of natural substances to find a buyer for their commodities even before they are extracted from the earth. Such exchanges also allow an equilibrium price to be reached so that supply and demand can be matched. In establishing an equilibrium price for a commodity, it is the volume of the commodity entering the market, along with the volume of demand, that together determine the price. The volume of supply can, in some cases, respond relatively quickly to market prices. If, for example, the price of tuna in the supermarket increases, then more fishing boats will quickly seek to capitalize on the windfall (and the increasing supply will bring the price back down again). But if a cartel of producers restricts supply, as in the case of the Organization of Petroleum Exporting Countries (OPEC), then prices can be pushed higher. Alternatively, a change in demand for a product can move its price. The increasing demand for artificial sweeteners in the second half of the 20th century, for example, led to a long-term decline in the world price of sugar, devastating many Caribbean economies. More recently, the industrial expansion and increasing wealth of China in the late 1990s and the 2000s significantly pushed up the global prices of many raw material commodities. Between 2000 and 2010, for example, Chinese consumption of copper almost tripled, accounting for 40% of all global production. Over the same period the price of copper on global markets saw a four-fold increase, from around $2,000 per ton to over $8,000 per ton (Pleven, 2011; London Metals Exchange, 2012).

While commodity exchanges "discover" the price for a natural raw material, the price in question is the outcome of a range of factors. The ways in which these factors come together are the product of daily analysis and calculated guesswork by economists and traders in the various commodity exchanges around the world. What is absent in this calculation, however, is any evaluation of broader environmental impacts (in the present or the future) of extracting and using a resource. The market is, in other words, entirely blind to the environmental effects of using nature as a resource, unless these effects are factored into its price such as by adding the cost of conforming to government environmental regulations. As we saw in the case of oil, the costs of its use or its accidental release as a pollutant are seldom part of any cost calculation. The economic valuation of a commodity as the outcome of a supply-and-demand equation therefore results in an equilibrium price that obscures many of the environmental costs of the product.

There are, however, techniques available to calculate some of these costs, largely from the field of environmental economics that emerged in the 1970s. They involve the quantification of a wide range of impacts associated with environmental degradation. In the case of an oil spill, such as the one that occurred in the Gulf of Mexico, some costs are straightforward to assess in economic terms. As noted earlier, many direct costs can be estimated in the case of the Deepwater Horizon disaster: the costs of containing a spill, cleaning up pollutants, compensating individuals and businesses affected, accounting for lost oil production, and the costs of litigation and damages awarded (e.g., for loss of life or injury). What these costs do not tell us, however, are the effects of damage to ecosystem components that have no direct and immediate economic uses today. The field of environmental economics provides some tools for assessing the value of nature in these more indirect ways (see Box 5.1).

However sophisticated they become, economic techniques for assigning a value to nature, both when it is left undamaged and when it is degraded, have several features in common. First, they all attempt to convert environmental damage into economic costs. In this way, they attempt to translate the value of the nonhuman into very human terms – money. This approach means that dolphins or sea turtles are valued, but only to the extent that they can be expressed as a measurable economic benefit, and not for their intrinsic worth. Second, despite the appearance of a comprehensive accounting of nature's value, environmental economics still leaves many uncertainties because it can be based only upon our current evaluation of ecosystem services. The future value and importance of an ecosystem to both human society and the wider biosphere may not be fully appreciated or even comprehensible today.

Economic thinking, then, is concerned with assigning a current monetary value to nature, allowing natural "things" to be integrated into a common framework of analysis. A geographical approach, by contrast, allows us to step back from the determination of monetary value and instead to think about *how* nature becomes a resource to be traded in the first place – that is, how it becomes commodified.

5.3 Incorporating Nature: Commodification and Ownership

In this section we move away from an approach to nature that is based on assessing its value and instead consider the conditions under which nature is incorporated into economic processes. We have noted that economies are based upon the transformation of nature. At its most basic level, an economic activity involves the extraction, transformation, and trading of substances that we, as human societies, have found on earth. In most cases, this means the geological deposits, life forms, waters, and atmosphere of the planet. Whether in relatively basic forms such as fresh food, water, and fuel, or transformed through production processes of varying complexity into clothes, vehicles, computers, machinery, and so on, all commodities are derived from the natural environment's resource base.

In this sense, it is useful to think of a *system* or *cycle* of material flows and balances (see Figure 5.2). This allows us to trace the flows of physical and chemical inputs from extraction through production, manufacturing, and

FURTHER THINKING

Box 5.1   *Evaluating nature through the ecosystem services approach*

The ecosystem services approach involves assigning a value to the services and benefits provided by an intact and unharmed natural environment. These services might take on a variety of forms:

- Provisioning services: the material goods provided by an ecosystem, which might be summarized as food, feed, fuel, and fiber.
- Regulating services: the role of an environment in maintaining larger ecological systems.
- Cultural services: the recreational, spiritual, and aesthetic services of an ecosystem.
- Supporting services: the ways in which an environment provides support to other environmental systems, for example as a sink for pollutants.

In each of these ways, a healthy ecosystem provides services that are of value to human society. Once they are identified, the next step in the calculation is to estimate a monetary value for each, which can be done using a variety of techniques:

- *Revealed preference* establishes a monetary value through observed economic behavior. If, for example, recreational fishing licenses are sold for access to Gulf waters, then these along with travel costs and other expenses might be used to infer this component of the cultural services provided by the Gulf ecosystem.
- *Stated preference* techniques ask survey respondents to place a value on certain ecosystem services. For example, a respondent might be asked what he or she would be willing to pay in order to protect or restore an environmental amenity that has meaning to him or her, for example a clean beach or the sighting of dolphins.
- *Cost-based methods* look at the cost of restoring an environment after it has been damaged, or the replacement cost of providing a similar amenity in an alternative location. For example, if losses to a particular species in the Gulf of Mexico, such as lobsters, were compensated by restocking with lobsters cultivated elsewhere, then this would be a restoration cost.

By combining these various techniques, environmental economists can establish a value for natural assets even when they are not traded in a commodity market. (For more on the ecosystem services approach to environmental economics and its application to the Gulf of Mexico oil spill in particular, see National Academy of Sciences, 2011).

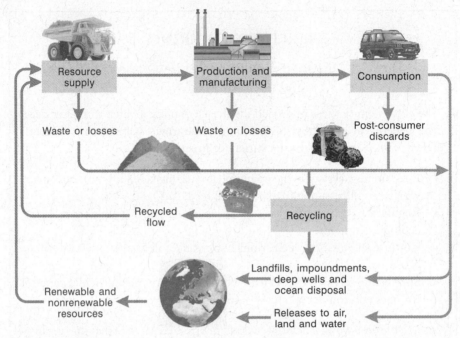

Figure 5.2   The economy as a system of material flows
Source: Adaptation courtesy of the U.S. Geological Survey.

utilization to recycling or disposal. Such an analysis reiterates how economic activity places two sets of demands on the natural environment. First, it acts as provider of inputs to the production process in the form of natural resources, but also at every other stage in the process as materials and fuels are consumed in transportation, retailing, and other stages of economic activity. The scale of resource use by human societies is quite staggering. For example, in 2007, economic activities in the 27 countries of the European Union used over 8.3 billion kilograms of material, including 1.7 billion kilograms of plant and animal biomass, 1.9 billion kilograms of fossil fuels, and 0.3 billion kilograms of metal ores. This amounted to about 16,500 kilograms of physical material per person per year (or 45 kilograms per person per day). And all of it is ultimately derived from nature.

A second demand upon the environment is to act as a receiver of the outputs of production processes in the form of waste disposal (e.g., landfills and ocean disposal) and pollution (i.e., releases onto land or into the air/water). In 2008, the European Union generated 2.6 billion tons of material waste or 5,237 kilograms per person per year. More than half of it came from mining, construction, and demolition. In the same year, Europeans generated just over 5 billion tons of greenhouse gas emissions, or 10,000 kilograms per person (all data from EU, 2012). It is a sobering thought, then, that most of us in high-income countries

Figure 5.3   The Rosia Poieni Copper Mine in Romania

collectively use resources and generate waste materials equivalent to our own body weight every single day. Figure 5.3 gives a sense of the massive quantity of material moved in one mining operation – in this case, the Roşia Poieni copper mine in Romania. The mine has about 60% of the European Union's copper reserves and involves the processing of 9 million tons of ore per year.

In this section we will focus on the first of these two demands upon the environment by exploring the ways in which the "natural world" becomes converted into usable, ownable, and tradable commodities – that is, how nature becomes commodified and thus a part of economic processes.

Creating a Commodity

In some instances, the ways in which nature is harnessed and incorporated for economic purposes is fairly obvious: a mine, a quarry, a farm, and a dam are clearly all sites where natural materials are converted into commodities. But it is important to note the ways in which nature becomes a resource. The key transformation that occurs in the process of commodification is the *valuing* of natural substances. This might be value based on the uses that they have – for example, as food, shelter, clothing, or inputs into making something else. It might also be based on the value that they have in exchange for other commodities. While the extraction of natural resources for their use value is still practiced, for example in subsistence agriculture, in most cases our engagement with the natural world is mediated by exchange relationships, for example in supermarkets, building supply stores, clothing shops, or water, gas, and electricity supply systems.

In order for the economic exchange of natural materials to occur, they must be assigned a monetary value. But the specific elements of the natural world that will be valued are variable over time and across space. There are two principal variables that have determined the history and geography of nature's commodification:

- *Technological and scientific knowledge*: New knowledge can reveal an element of nature that was not previously valued as a commodity. This can happen in two ways. First, new technologies might create a demand for a substance that previously did not exist. Uranium, for example, was not a commodity one hundred years ago, but is now a prized economic resource for use in nuclear power generation, medical technologies, and weapons manufacturing. Only with demand for the substance does it become a commodity to which a value is assigned. Second, technology may "create" a commodity through new forms of extraction, which make it possible to use something that was always there but never before usable. Sunlight, for example, has always been freely available, but only with the development of photoelectric cells has it been possible to harness the sun's heat and light to generate electricity that can then be stored and traded.
- *Economic circumstances*: Changing economic conditions can also turn a natural substance into a commodity, sometimes very quickly. Knowledge about the usefulness of a substance might exist, and the technology to obtain it might be available, but it might be just too expensive to extract it as a tradable commodity. When circumstances change, for example other supplies of the substance run dry or demand increases, a part of nature can quickly become an important resource. One example is found in the oil sands located mostly in the province of Alberta, Western Canada. Scientists and mining engineers had been aware of these oil-bearing sand deposits for some time, but the high cost of extraction (double that of conventional oil wells) had effectively rendered them without any status as a commodity. It has only been in recent years that the global price of crude oil has risen to heights that make the oil sands worth owning and exploiting. In this way, largely desolate landscapes have suddenly become booming sites of economic activity and environmental degradation (see Figure 5.4).

The correct technological and economic circumstances are therefore required before a piece of nature becomes an economically valued resource. It is worth noting, however, that once nature becomes a resource it does still retain some *material* properties and these mean that nature is not entirely subsumed by human attempts to tame it. For this reason, different resources have very different characteristics as commodities. One obvious dimension of this point is that many natural resources occur in certain places and not others. There is some flexibility and human agency in this process. For example, agricultural crops or forestry can be cultivated wherever appropriate soil and climate conditions exist. Such cultivation can be even further afield if artificial environments, such as greenhouses, are created. But in many cases, the location of a commodity, such as a coal seam underground, a fishing ground at sea, or a climate appropriate to

C. Major, used with permission

Figure 5.4   The oil sands of Alberta, Canada

tropical fruits, cannot easily be moved or modified. This means that extractive industries have a logic to their geographical distribution across space that is rather different from the manufacturing or service sectors. In essence, resource extraction activities must generally locate in situ or where resources are to be found.

A second argument can be made in relation to the physical nature of a commodity, especially when it is a plant crop. The geography of its production, and the social system that supports it, will be shaped not just by the environmental characteristics needed for its cultivation such as climate, soil, water supply, and so on, but also by the nature of that cultivation process itself. Tea, for example, grows on bushes and the leaves can be plucked all year round in areas such as Karnataka state in southern India. As a result, the crop tends to be grown on plantations with a permanent onsite labor force that often lives in homes owned by the plantation. Coffee, on the other hand, is a seasonal crop that requires harvesting labor for a short time of the year. In southern India this is from November to February. As a result, seasonal migrant labor is used. Where coffee is grown by smallholders, its cultivation is combined with farming other crops (Neilson and Pritchard, 2009). In each case, then, the nature of the crop dictates a different geography of settlement, labor mobility, and organization of production as a piece of nature is converted into a commodity.

A third feature of natural resources is that, once extracted, each commodity has distinctive characteristics that determine the ways in which it can be used. Milk, for example, can be produced anywhere that a herd of cows can be fed, either through open grazing or in controlled feedlots. But its physical properties require that it be transported to a consumer as quickly as possible in refrigerated containers. Likewise, there is no getting around the fact that water is a bulky and heavy commodity and one that is prone to leaking and evaporating. Water supply is therefore best drawn from sources relatively close to its consumers.

Figure 5.5   Grain wagons operated by the federal and provincial governments in Canada

On the other hand, although diamond deposits are concentrated in very few places around the world, once processed, diamonds are eminently transportable and virtually indestructible. Thus the concentration of diamond dealers in a few cities, such as Antwerp (Belgium) and Tel Aviv (Israel), is perfectly feasible. Each commodity, then, has its own post-extraction geography.

A fourth and final characteristic of natural resource extraction, especially on the scale now demanded by urban-industrial societies, is that it requires significant levels of investment, infrastructure, and logistical support. As a result, governments have often been central to the emergence of resource-based economies. For mining, agricultural, or forestry activities to develop, for example, it is usually governments that provide the infrastructure that enables them to happen in far-flung places. In developed countries, schools, hospitals, roads, railways, sewers, policing, and so on are all facilities usually provided for by agencies of the state. The state's facilitation of commodity production may go even further. Figure 5.5, for example, shows grain wagons owned and operated by the government of Canada and by the province of Alberta.

When all of these features of commodity creation and commodity transformation are put together, they can lead to very distinctive patterns of development and decline across space. Furthermore, each commodity, as we have noted, has its own peculiar history and geography.

Establishing Ownership

We have seen that natural resources can enter the economic sphere in various ways and can create a variety of economic geographies. If, however, nature is to be sold in the form of resources, then a further requirement is that it has to belong to someone who will do the selling. This will, in most instances, be an individual, a group of people, a firm, or a state.

In very simple terms, we can identify four possible models through which ownership is established over a natural resource commodity:

- *Communal access*: Such rights allow individuals to use resources but not to own them. The rights of shifting cultivators to access and use forest resources would be one example. In reality, though, most land ownership is governed by states, which might then permit communal use rights. One important exception, however, is the fishery resources of the open seas. Some legal frameworks do apply to fishing in open waters, for example a limited international ban on whaling. In general, any fishing boat operators can appropriate marine resources, as they cannot be said to belong to anyone while they remain in the sea.
- *State ownership and state exploitation*: In this model, the benefits of extraction are shared by all citizens as government revenue. The development of hydroelectricity or water supply projects by public utilities is one example. As we will see later, the role of states in natural resource extraction has been changing, but in diverse ways around the world.
- *State ownership and private exploitation*: Here, corporations pay some kind of licensing fee to extract a publicly owned natural asset – for example, when governments grant timber licenses to logging firms entitling them to extract wood from a defined territory, or when oil or mining companies purchase exploration rights. In both cases, royalties are then usually paid to governments based on the amount of the resource that is extracted.
- *Private ownership and private exploitation*: This is a common economic model and it occurs when, for example, a farmer owns the land that he/she cultivates and is entitled to reap all of the profits that result from the operation.

In reality, these categories of ownership are overlapping and increasingly blurred. There have been, however, three widespread trends around the world. The first involves the expansion of the concept of private ownership into contexts where it was not previously applied, thereby erasing communal or open access resources and replacing them with privatized access. The second trend involves the increasing importance of some state-linked corporations as key players in global resource industries, especially in the developing world. The third trend is the opposite of the second and involves the privatization of formerly state-controlled resource industries or markets. We will consider each of these in turn.

Expanding Private Ownership

In some cases, ownership of nature has long been established in deeply rooted legal traditions. In Western Europe, for example, the ownership of agricultural land has been entrenched in legal structures of private property for several centuries.

For other resources, or in other places, however, the definition of who owns a resource is more uncertain and contested. In countries with minority indigenous populations, their claims to territory, often based on very different legal traditions, have usually been neglected by majority or dominant populations. In Canada and Australia, for example, indigenous groups have unresolved claims to land and mineral resources across much of the territories now administered by those nation states. Such land claims are only slowly being settled by contemporary governments and native groups. In many parts of the world, marine resources provide a particular problem, precisely because of the characteristics of the natural processes being harnessed. Who, for example, will own migratory species of fish, such as salmon, when they traverse the U.S.–Canada border? And who will exploit offshore oil reserves when they lie in the South China Sea in between China, Malaysia, the Philippines, and Vietnam? Clearly, ownership and tenure rights are not always straightforward in the commodification of nature.

We do, however, see an intensifying trend toward frameworks for defining ownership over resources. In some contexts, this integration of land and mineral resources into a system of ownership is relatively recent. This is especially true, for example, where groups that live in tropical forested environments have traditionally employed a system of shifting cultivation. Shifting cultivation involves the controlled burning of a patch of forest and then planting crops until the soil loses its fertility (see Figure 5.6). The crop cultivation is then moved to a new patch while the old patch is allowed to regenerate its forest cover and soil nutrients over a number of years. Where such a system has been used, ownership of land is not individually assigned. While a forested landscape might be the traditional territory of a village or a particular cultural group, it is not the property of an individual.

K. Barney, used with permission

Figure 5.6   Shifting cultivation of glutinous rice in Khammouane Province, Laos

In recent years, however, governments have often tried to replace this system of communal "use rights" with permanent titles to land. In the small Southeast Asian country of Laos, for example, the government's Land and Forest Allocation program has been assigning ownership rights in rural villages since 2001. Unfortunately, imposing the concept of land ownership is often inconsistent with the way rural people actually use natural resources. In a case described by Barney (2009), one-third of a village's forest land was granted to a Japanese pulp and paper firm for a eucalyptus plantation to supply the rapidly growing Chinese market for paper products. This severely restricted the space available for shifting cultivation, thereby placing private plantation land uses in direct conflict with local livelihood practices.

This Laotian example represents a case where shared resources, often used to produce food for subsistence rather than for the market, are taken into private ownership or use rights. Natural assets that were collectively used are therefore made available to individuals or firms for market-oriented production and profit-making. This is a process that was fundamental to the early development of capitalism in Europe, when common grazing lands were enclosed to create private farms and estates. It is a process that has been conceptualized as "primitive accumulation" or "accumulation by dispossession" (see Box 5.2).

State-Owned Resource Companies

A second trend in recent years has been the expanding role of large government-linked corporations in seeking control of natural resources. In several emerging economies around the world, such enterprises dominate natural resource sectors. In Russia, for example, where state-owned enterprises from the communist era were privatized in the 1990s, a process of amalgamation has more recently recreated massive conglomerates with control over key resources and close links with political leaders. Gazprom, for example, is the world's largest natural gas producer and is linked to Russia's political leadership in complex ways. Oil extraction in the Middle East is also carried out almost exclusively under the auspices of state-owned companies. Indeed, all of the world's ten largest oil and gas corporations, measured in terms of the reserves they have access to, are government-linked companies located in Iran, Saudi Arabia, Venezuela, Kuwait, Russia, Qatar, UAE, Iraq, Turkey, and Libya.

State-linked companies are also expanding globally and are seeking to secure access to resources wherever they are to be found. China has been particularly active in using development aid, foreign policy, and market power to gain access to resources from around the world to fuel its rapid economic growth. This has led Chinese government-linked enterprises into deals with various countries in the Middle East, Africa, and elsewhere. The Chinese typically offer loans and funding for infrastructure in exchange for guaranteed supplies of oil, natural gas, and other resource commodities (*The Economist*, 2012). A particular

FURTHER THINKING

Box 5.2 *Primitive accumulation and accumulation by dispossession*

In Chapter 3 we discussed the characteristics of a capitalist system. Its key feature is that one group has ownership of the means of production and the profits deriving from production, and a second group has rights over its labor, which it sells to the first group for wages. Ever since scholars have identified capitalism in this way a key question has been how it came about. In early writing, for example that of Karl Marx in the mid-19th century, this question related to Europe and how capitalism emerged from a feudal system of production. Marx understood this transition to occur through a process he called "primitive accumulation." Essentially it referred to the ways in which people became separated from the means of production and became waged laborers instead. This happened in a variety of ways – for example, by enclosing common property where anyone could hunt, gather, graze animals, or cultivate crops, and turning it into private property such as estates, plantations farms, and so on.

Primitive accumulation is not, however, a process that has been completed, and even today capitalism exists in competition or coexistence with other possible ways of organizing production. People are still being drawn into the process of selling their labor instead of producing what they need for themselves, as we saw in the case of Lao farmers whose land was allocated to a eucalyptus plantation by a Japanese firm. There, and in many other cases, the class relationship within which people work is being transformed.

Recently, the concept of primitive accumulation has been expanded to include the erosion of a wider range of collective rights. David Harvey's (2003) work has been especially influential. Harvey notes that rights to collectively funded provisions such as state pensions, social housing, education, and healthcare have been cut away in many countries. Similarly, communal spaces such as wilderness areas, urban public spaces, and other public assets are increasingly being drawn into capitalist market relations rather than being subject to collective regulation and sharing. This process of privatizing the "commons," – that is, causing a loss to many and rich benefits for just a few – is what Harvey calls "accumulation by dispossession."

concern has been the role of state-linked companies in gaining access to land in areas where land markets and titling are not well established. Some have called this a "global land grab," as described in Box 5.3.

CASE STUDY

Box 5.3 A global land grab?

As concerns over food and energy supplies have increased, some countries have used a variety of means to gain access to resources overseas. A particularly sensitive aspect of this process has involved foreign government-linked enterprises gaining access to land holdings in order to secure food supplies. Some have characterized this as a "global land grab." It has mainly happened in the developing world, where governments have been keen to see new investment and where land tenure is often ill-defined and insecure. One estimate suggests that between 2000 and 2011 around 80 million hectares have been subjected to deals worldwide. This is a vast amount of land, equivalent to more than the total area of farmland in Britain, France, Germany, and Italy combined. Africa in particular has been a focus of attention, accounting for almost two-thirds of that total (*The Economist*, May 5, 2011, The Surge in Land Deals). Investors may be private corporations with government backing, government-linked corporations, or the investment arms of sovereign wealth funds (explained in Section 4.3). Some reports suggest that Chinese government-linked enterprises have been big players in the African land market, but many others are involved too. Specific examples include the Saudi Arabian company Hadco, which acquired 25,000 hectares of cropland in Sudan with a majority investment from the Saudi government's Industrial Development Fund. The Abu Dhabi Fund for Development is similarly financing the development of 28,000 hectares of farmland in Sudan to grow various animal feeds and food crops for export to the United Arab Emirates (Cotula et al., 2009). Those concerned with development and food security have roundly criticized these deals, pointing out that they provide little rent or employment for the local economy, while undermining local food security.

Privatization of Collectively Controlled Resources

While the early 2000s saw the rapid emergence of state-linked resource corporations in the developing world, the trend in the developed world was the opposite. Resources once the exclusive preserve of state-linked enterprises became privatized. In Britain, for example, until the late 1980s, coal was mined by the National Coal Board, natural gas was supplied by British Gas, steel was made by the British Steel Corporation, and water was supplied by regional water boards. All were owned by the government, and all were sold off as private companies in the late 1980s.

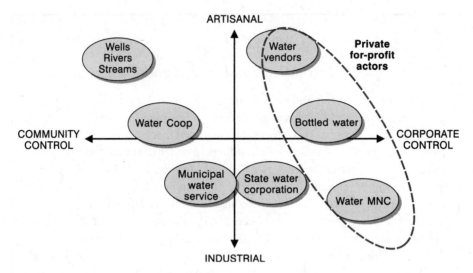

Figure 5.7   Modes of urban water supply provision
Source: Redrawn from Bakker, K. (2003) An uncooperative commodity: privatising water in England and Wales, John Wiley & Sons, Inc.

Water is a particularly interesting case. There are a variety of ways in which people might be supplied with fresh drinking water in urban areas, ranging from individual access to wells and water courses, to private corporate ownership of large-scale water supply systems. Figure 5.7 depicts the various options along two axes of community versus corporate control and small-scale artisanal versus large-scale industrial systems. Over the last two decades, water has been privatized in many contexts around the world. In other words there has been a movement toward the bottom-right portion of Figure 5.7.

Water is, however, in many ways an "uncooperative" commodity when it comes to private ownership (Bakker, 2004, 2010). It is very difficult to extend complete private ownership over a substance that is so essential to human health, so ubiquitous, so bulky and heavy, and so dependent on massive infrastructure. Nevertheless, in a number of poorer countries, development aid and loan packages from international financial institutions have been conditional upon the privatization of public utilities. It is estimated that over 500 million people worldwide are now served by private water companies. At the same time, such private ownership has become increasingly concentrated in just a few hands. The global top five water firms, Suez-Lyonnaise des Eaux, Vivendi, and SAUR International in France, Thames Water in Britain, and Bechtel from the United States, are currently responsible for over 80% of global private water service delivery (Harris, 2009).

One further characteristic of water that makes privatization problematic is that it is so central to a country's ecosystems. It is virtually impossible for a utility company to reflect truly the various environmental consequences of extracting water from the environment and the disposing of wastewater and sewage. Water is too integral to life to be fully accounted for in this way. This point highlights a key contradiction that always exists when nature is commodified. As soon as a commodity is extracted from the natural system of which it is a part, and is packaged, owned, transported, and used in some way, a cost is incurred by the earth's natural system that can never be fully accounted or even understood in economic calculations.

This section has highlighted some of the preconditions necessary for the commodification of nature, including appropriate levels of technology, economic demand, and systems of ownership. We have also seen that the systems through which ownership is established are changing, but in geographically uneven ways. Furthermore, it is clear that when discussing commodification we also need to be aware of the physical features of the resource in question. The natural characteristics of a commodity remain important even when it has been harnessed for economic purposes.

5.4   Valuing Nature: The Commodification of Environmental Protection

In the previous section we considered the ways in which nature becomes harnessed for economic purposes. We now turn to the side- and after-effects of economic processes, and consider how protecting the environment from pollution and degradation can also become commodified. In a general sense, being "sustainable" or "green" is used as a selling point for many products, and in that sense environmental protection is being integrated as part of the value of a commodity. This added value is often incorporated through the certification of products as organic, sustainable, or local. But in more specific ways, the protection of natural environments from the waste, outputs, and effects of economic production is becoming a commodity with a tradable value. In this section we examine the marketization of pollution permits as one form of commodifying environmental protection, and the emergence of a "Green Economy" including energy- and resource-efficient technologies as another.

Commodifying Pollution

Perhaps the best example of environmental protection being commodified is found in the development of markets on a global scale to trade permits for air pollution. One of the first pollution markets was created in the United States in the 1990s. The U.S. Environmental Protection Agency's (EPA) Acid Rain Program was started in 1995 to reduce sulphur dioxide emissions from coal-burning electricity generation plants. Acid rain is caused by a chemical reaction in the atmosphere involving the by-products of fossil fuel burning such as sulphur dioxide and nitrogen oxides. The consequences of acid rain include damage to forests and other plants, corrosion of buildings, and effects on human health. The EPA's Acid Rain Program was a "cap-and-trade" system, which means that an absolute maximum was placed on emissions from all polluters and the right to pollute could then be traded between them. The unit being sold in this market was a permit to release a ton of sulphur dioxide, thereby giving each producer an incentive to lower its costs by reducing its emissions and therefore its need for permits.

As with all such market-based schemes, the theory is that those producers who can most easily reduce their emissions will be the first to do so. The market will therefore "find" the most efficient means of reducing pollution produced across all emitters. The first phase of the Acid Rain Program (1995–2000) reduced sulphur dioxide emissions by 40% and was judged to be a success. Since 2000, the scheme has been expanded to include all forms of fossil fuel-fired power generation and to encompass nitrogen oxide emissions. The program did, however, have certain advantages. It was initially focused on just one nation (the U.S.), one pollutant (sulphur dioxide), and one sector (coal-fired power plants); it had reliable baseline measurements of emissions; and it could implement continuous monitoring of pollution sources. Almost all of these advantages were missing in future cap-and-trade systems for which the Acid Rain Program provided a model.

The most extensive cap-and-trade system now in existence is designed to reduce the pollutants that cause global warming and climate change, the so-called greenhouse gases. In 1997, the signatories to the United Nations Framework Convention on Climate Change met in Kyoto, Japan, and developed an agreement that sought to reduce emissions of greenhouse gases. The Kyoto Protocol came into force in February 2005. By 2010 a total of 191 countries had ratified it. The notable exception was the United States, which accounted for about one-sixth of global emissions in 2005. Ratification of the protocol committed these countries to reducing greenhouse gas emissions by a set percentage from their 1990 levels by 2012. While some countries such as most of the European Union states met these obligations, many significant polluters did not, including Australia, New Zealand, Spain, and Canada. Canada even announced its withdrawal from the protocol in December 2011. The shape of the post-Kyoto framework for reducing emissions of greenhouse gases remains a subject of intense political negotiation.

Under the Kyoto Protocol, each country established a cap on its carbon dioxide emissions, and governments then allocated emissions permits that could be traded. A variety of markets emerged, including several individual U.S. states since 2008 and an Australian system since 2003. The most highly developed market, however, is the European Union's Emissions Trading System (ETS). The ETS initially covered 12,000 big industrial polluters across EU member countries. A country's total allowable emissions were divided up among different sectors: power generation, steel, glass, cement, pulp and paper, and ceramics. Firms emitting more than their permitted level could buy other companies' unused allocation, or pay fines, while those who were reducing their pollution could sell unused permits. In theory, if there was extensive over-pollution, the price of permits would rise, therefore inducing firms to find ways of cutting emissions levels.

Aside from purchasing emissions permits from other polluters, firms could also offset their emissions by investing in projects that reduced greenhouse gases in other ways. If, for example, a developing country decided to conserve a forested area as a "sink," or store, of carbon dioxide rather than exploiting it as an economic resource, then this conservation project could be converted into, and sold as, a certain number of emissions permits. This system was known as the Clean Development Mechanism (CDM). In some cases, such carbon offsets are also used to compensate for emissions by polluters who are not part of the emissions trading scheme. For example, an airline might offer consumers the opportunity to purchase carbon credits to offset the carbon footprint of their flight.

The ETS and other carbon trading schemes have achieved some successes: they have created an international governance regime through which emissions reductions can be fostered; they have established a price for carbon emissions that has now to be factored into the costs of production; they have attempted to allocate responsibility for emissions reductions to those who could do so efficiently; and they have created a means through which carbon sinks such as forests can be valued as a resource without being exploited. Furthermore, the carbon trading market is growing rapidly. It expanded from US\$11 billion worth of transactions in 2005 to over US\$140 billion in 2010. Some estimate that it could reach over US\$2 trillion by 2020 to become the world's largest commodity market.

Nevertheless, carbon trading has faced numerous criticisms. Some of these criticisms relate to the specific ways in which the ETS was implemented in Europe, but they also address some fundamental problems in creating such a market for pollutants:

- *The reliability of baseline measures.* While carbon dioxide is relatively straight-forward to measure, the sources of other more potent greenhouse gases such as methane are less well understood and so baseline measurements are uncertain. It is also difficult to establish the equivalence of different gases in terms of their effects on global warming.
- *The difficulties of continuous monitoring.* Climate change is caused by multiple gases and they come from many different sources. It is impossible to monitor all sources continuously, and so verification of what is being traded in a market for emissions permits is difficult. Furthermore, many countries do not have extensive monitoring capabilities even for those gases that can be effectively measured.
- *The allocation of permits when markets are first created.* Establishing a brand new market requires decisions to be made about how many pollution permits will be assigned to individual firms before the trading begins. In this process it is difficult to avoid "rewarding" large polluters with the corresponding initial allocation of such permits. This was a major criticism of the European ETS

scheme. At a larger scale, there is also an argument that carbon trading does not recognize past responsibility for climate change. High-income countries of the developed world have caused global warming but developing countries are consigned to low-carbon futures. The debate between the United States and China is the most heated in this instance.

- *Ensuring "additionality."* Where carbon credits are granted because of a conservation project under the Clean Development Mechanism, it is not always clear that the market has actually preserved something that would not have been preserved otherwise. It is also uncertain how climate change is affected by the geographical distribution of greenhouse gases. For example, it is uncertain whether a forest conservation project in Indonesia really does compensate for a coal-fired power station in Indiana, in the United States.

- *Avoiding "leakage."* The market for carbon emissions has many possibilities for leakage, displacement, and substitution. Carbon emissions avoided in one place may be spatially displaced to another location. There is also an issue of temporal leakage. If a forest conservation project receives certification as a carbon offset credit, it is difficult to ensure that it will remain in place for a long period of time.

- *Conflicts of interest in market institutions.* Critics have claimed that the firms involved in regulating the market, for example those verifying emissions reductions or certifying carbon offset projects, have had vested interests in the market or have been involved in managing the projects that they certify. Carbon markets may also behave like financial markets, with speculation and price swings caused by market players and not the actual users of permits themselves (Bond, 2011).

- *Market volatility.* The price of carbon emission permits has not kept pressure on polluters to reduce their emissions. For example, the downturn in economic activity in Europe in 2008–12 reduced emissions significantly, because firms were reducing capacity or closing down. This caused a huge drop in the price of permits, and the market therefore exerted no pressure on remaining polluters to reduce their emissions. From a high point of around 30 euros (US$47) per ton of carbon in 2008, the price or permits by early 2012 was less than 10 euros.

- *Carbon offsets have unintended side-effects.* Allowing forest conservation and other projects around the world to be valued as carbon offsets has some unintended consequences. In particular, the rights of forest users such as shifting cultivators may not be recognized if an area is designated for pre-servation.

All of these criticisms illustrate the difficulties in commodifying nature. Carbon markets attempt to create a market out of something that has many properties that are resistant to marketization. Critical assessments have also gone beyond these specific problems with the operation of carbon markets. It could be argued that making a market out of the atmosphere is simply another example of capitalism's drive to find new ways of generating possibilities for profit. Arguably, a far more effective but less popular and less profitable strategy would be to place a tax on every ton of greenhouse gases emitted (Solomon and Heiman, 2010).

The Rise of a Green Economy?

While carbon trading appears poised to become a major commodity market in its own right, there has also been much discussion about the possibility of environmental protection forming an entire sector of the economy. The so-called "green economy" represents those activities that are increasingly needed in order to address a low-carbon, sustainable future. What might be included in such a sector is not always very clear, but generally it is taken to include activities such as the following:

- Renewable energy generation activities from wind, water or sunlight, and biofuels.
- Energy-efficient products such as battery technologies, electric vehicles, energy-saving building materials, lighting systems, mass transit, and water-efficient systems.
- Agricultural and resource products such as organic food and farming, and sustainable forest products.
- Greenhouse gas reduction activities and such as carbon storage, pollution reduction technologies, and air and water purification technologies.
- Recycling and waste management, and recycled or reused products.
- Professional services in training and compliance certification.

Depending on how widely the definition of the green economy is cast, it is already a very significant sector in some countries. One report has suggested that this sector accounted for 2.7 million jobs in the United States in 2010, while the United Nations Environment Program estimates that worldwide there might be over 20 million jobs in wind-, solar-, and biofuel-related activities alone by 2030 (Muro et al., 2011; Davies and Mullin, 2011).

Proponents argue that as the world faces up to the necessity of a resource-scarce and low-carbon future, there will be a wave of demand for these types of products and services and those countries that have developed the necessary technologies and capacities will reap the rewards. For this reason, some governments have been keen to promote the green economy, seeing it as a win-win proposition offering environmental protection, economic growth, and possibly more secure energy supplies as well. In the United States, for example, the federal government's anti-recession spending package, the American Recovery and Investment Act of 2009, allocated US$60–80 billion over several years to clean technology and energy efficiency initiatives such as advanced batteries and alternative fuel vehicles. China has been even more aggressive in promoting green technologies, spending over US$50 billion in 2010 on low-carbon energy technologies such as solar and wind power, high-speed rail systems, and hybrid or electric vehicles. Other countries have not made these kinds of financial commitments, but have sought to reorient their regulatory frameworks to encourage the green economy. In Ireland, for example, the government has proposed streamlining planning approvals for green projects, greening public sector procurement practices, and developing financial services specifically focused on green enterprises (Davies and Mullin, 2011).

Questions still surround the promises of a green economy. For example, some argue that the high-technology jobs created may not match the skills of those who have lost jobs in other sectors. There are also risks involved in shifting to low-carbon energy sources. In the case of rechargeable batteries, global supplies of lithium are heavily concentrated in just a few places, especially Bolivia, Chile, and Argentina, meaning that supplies are limited and costs have remained high. The political climate in these dominant supplier countries may also create uncertainty. In the case of nuclear power, the radiation leakage from the Fukushima Daiichi nuclear power station in Japan in 2011 has made many countries far more nervous about switching to greater reliance on nuclear generation. More generally, some doubt that we can ensure a livable environment into the future by putting our faith in new technologies, markets, and private-sector innovation. A more radical transformation of lifestyles and production processes may be necessary. Whatever the answers to these questions, it seems likely that there will be growing demand for products and services that reduce environmental impacts and use energy and resources more efficiently. In that sense, the commodification of environmental protection through the emergence of a green economy is well under way.

5.5   Human Nature: The Body as Commodity

We have so far considered the ways in which nature becomes commodified as a resource, and the ways in which the conservation of environmental integrity can also be commodified through pollution permits. In both cases we have seen that the commodification of nature is a problematic process and the interface between human society and nature is not easily reducible to a market value. Within each of these examples, however, nature is consistently treated as separate and distinct from human society. Whenever we ask how nature is impacted, valued, and sold as a commodity, we are setting up this binary between humanity on the one hand and nature on the other.

In this section we begin to undermine the certainty that lies behind the distinction between humanity and nature. There are a number of ways to do this. One would be to examine how nature is something that has been *produced* by human activity, for example through the genetic modification or selective breeding of organisms, or the transformation of landscapes. The argument that nature, as we experience it, is now the product of hundreds of years of intensive human economic activity is quite compelling. Another way of blurring the boundary between humans and nature would be to see ourselves as *part* of nature. Despite, or perhaps because of, our technological sophistication, frequent outbreaks of food safety scares and new communicable diseases should remind us that we are subject to the natural world, which literally enters us every day. A specific approach to overcoming the human/nature division in this way has been conceptualized in Actor-Network Theory (see Box 5.4).

FURTHER THINKING

Box 5.4   Actor-Network Theory (ANT)

One way in which some scholars have sought to tackle the binary distinction between humans and nature is to think of the natural world as a player in social processes such as production just as much as the human participants involved. This means that we have to take seriously the physical properties of natural materials, and we can view both humans and natural substances as having "agency." Obviously the agency of nonhuman plants, animals, and minerals is not conscious, but thinking in this way allows their physical properties to be given appropriate emphasis.

These ideas have been articulated in Actor-Network Theory (ANT), which has become influential in the field of Geography since the mid-1990s. ANT argues that we should collapse the artificial binary distinction between human and nonhuman that pervades our ways of thinking, and instead focus on the sprawling networks of both animate and nonanimate actors through which social and economic life is played out. Actors acquire their agency through their relationship to other things in a mixed network or *heterogeneous association* of both human and nonhuman entities. By tracing the relations between humans and the stuff of nature, without presuming that either takes priority in social processes, ANT provides a way of overcoming the human/nature binary.

In Economic Geography, ANT has served to bring in nature as an actor, but it has also had further implications. Methodologically, ANT has emphasized the importance of tracing relations and practices rather than assuming how actors will behave. ANT also challenges notions of space, place, and scale. Rather than seeing the world as organized into hierarchical scales, ANT sees social systems as operating through combinations of network connections. Such relational and network thinking has been very influential in Economic Geography. (For more on ANT see Johannesson and Bærenholdt, 2009).

An even more intriguing argument against the separation of humanity and nature is to be found in evidence that the human body itself is being increasingly treated as an economic resource. In other words, the elements of nature that comprise the human body are themselves also open to commodification. There is, of course, a long history of treating bodies as commodities. Under slavery, African bodies were treated as assets that could be owned and traded; blood transfusion goes back hundreds of years; and body parts and corpses have long been sold as raw materials for medical training and research. It is, however, only in the last few decades that biotechnologies have advanced sufficiently to create tradable commodities out of genetic material, bodily tissues, and organs. These technologies include DNA sequencing, organ transplantation, genetic modification and cloning, tissue culturing, cryogenic storage of bodily tissues, and *in vitro* fertilization. In a sense, this reiterates the point made in Section 5.3 that nature only becomes a commodity once the technological circumstances enable it.

The commodification of the human body raises some profound questions about where the line between humanity and nature should be drawn, when a part of a person becomes a tradable "thing," and what aspects of their bodily self a person has ownership over. We can address these questions by examining two forms of bodily commodification. The first relates to the ownership and commodification of the information that our bodies contain in the form of genetic material. The second concerns the commodification of actual organs, tissues, and other materials.

The information contained in a person's genetic material comprises a unique aspect of their individual body and yet when copied it is generally deemed to be no longer their own property. The development of genetic sequencing has allowed scientists to create a complete profile of a human genome and individual parts of it – for example, those associated with a particular disease. The question then arises whether that information belongs to the person whose genes are being analyzed, or to the person or company that owns the technology to conduct the sequencing? In general, legal judgments in the United States have upheld the property rights of those doing the genetic sequencing rather than those from whom the DNA is extracted (Parry, 2012). Furthermore, this model of ownership has been globalized through trade agreements concerning intellectual property rights such as the Trade-Related Intellectual Property Rights agreement, or TRIPS, administered by the World Trade Organization.

The second example of bodily commodification concerns the extraction and use of physical material from a living body – for example, a kidney, part of a liver or lung, bone marrow, stem cells, skin, blood, eggs, or sperm. While blood, semen, and eggs can be legally sold in many places by the individual who produced them, the sale of major organs is prohibited in most countries, with a few exceptions such as Iran. Nevertheless, numerous cases of payments being made for organ donations have been documented, including cases of "transplant tourism" where individuals (many from India and Bangladesh) are brought to cities such as

CASE STUDY

Box 5.5   Selling cells

In 1951, Henrietta Lacks was dying of cervical cancer in a segregated ward for "colored" patients at the Johns Hopkins Hospital in Baltimore. Just before she died, a sample of her cancerous cell tissue was taken and sent without her consent to Dr. George Gey in the tissue culture lab at Johns Hopkins. Gey had been trying to cultivate human tissue for use in medical research but had been unsuccessful to that point. The cells from Henrietta Lacks' cervix were the first ones in medical history to be kept alive and reproduced. After many replications they became an "immortal cell line" that has been used widely in medical research ever since, known as HeLa cells. They were used to develop the polio vaccine in the 1950s and have contributed to numerous drug discoveries and to research for more than 60,000 scientific papers. While the cells were taken from Lacks without permission, and then shared freely by Gey with medical researchers around the world, the cell line is now bought and sold as a raw material for medical research, generating billions in profits for the pharmaceutical industry.

The story of Henrietta Lacks has resonated in recent years because her race, class, and gender meant that the medical establishment did not give a second thought to her rights over the cells that were taken. But, in fact, most individuals still have very little control over what is done with their unique DNA-coded cells when they are removed from their bodies, even though now they would be presented with a consent form to sign. In another case, in the 1980s, a patient named John Moore sued a doctor who had filed for a patent on his cells. The cells had been used to create a commercialized cell line estimated to be worth $3 billion. In a landmark decision in 1990, the California Supreme Court found in favor of the doctor and declared that Moore had no right to any of the profits generated by his cells. These examples illustrate the commodification of human tissues, the complexities of ownership over such commodities, and the way that something given as a donation may later be commercialized. (For more on the story of Henrietta Lacks' cell line, see Rebecca Skloot's 2009 bestseller, *The Immortal Life of Henrietta Lacks*.)

New York where one of their kidneys is extracted. While the recipient might be paying US$150,000, the donor will receive less than one-tenth of that amount (Parry, 2012). Such practices are generally seen as unacceptable, not just because the monetary benefits of the transaction are so unevenly divided, but because donors are often forced into selling a part of themselves because of poverty.

With a huge shortfall of kidney donations relative to the need for transplants in many countries, this black market in organs is likely to continue. It raises some interesting questions about the contradictory ways in which bodily commodification is viewed. While some bodily materials can be sold, others cannot. Thus there is inconsistency in what is seen as sacrosanct and beyond commodification, with some bodily tissues open for sale while others are not. Another inconsistency is that while the selling of organs as commodities is illegal, it clearly still happens with relative impunity through a variety of channels. Finally, it is worth noting that certain bodily material may shift in and out of commodity form over time. For example, a cell culture or bone marrow may be donated without compensation but can, later on in its "career," become subject to ownership, sale, and therefore commodification. The case of the HeLa cell line forms a fascinating example (see Box 5.5).

These brief examples illustrate how the boundary between humanity and nature is a complicated one. Not only are we very much a part of nature, but our material beings are convertible into commodified resources in a variety of ways. The conceptual distinction that we usually draw between humans and the natural world, then, with humans imposing economic processes on external nature, is therefore more uncertain than it first appears.

5.6   Summary

In this chapter we started out with approaches rooted in Economics and identified some of their shortcomings. The analysis of commodity pricing, for example, is purely an exercise, albeit often a very sophisticated one, in establishing the relationship between supply and demand. While environmental economists do have tools to assign value more carefully to both pristine nature and the economic consequences of environmental change, we saw that these analyses tend to be limited to those aspects that can be quantified in monetary terms.

The rest of the chapter moved away from asking how nature is valued and instead explored how elements of the environment are incorporated into economic processes through commodification. A geographical approach that emphasizes the contexts and institutional bases of commodification allowed us to examine how both resource inputs, and the pollutants that result from it, are turned into economic processes. In the case of inputs, we saw how the material features of a natural resource affect the geographies of its economic exploitation, how establishing ownership is a key precondition for its commodification, and how patterns of ownership over natural resources are changing around the world. In the case of pollutants and specifically greenhouse gas emissions, we saw how a market in environmental protection might be established and some of the problems that arise. More broadly, we saw that environmental protection is also commodified in the form of a green economy creating the technologies and products needed for a low-carbon and resource-efficient future.

Most of this chapter was underpinned by an implicit separation between the world of nature on the one hand and the world of humanity/economy/society on the other hand. The final section sought to blur this distinction by showing that human bodies are themselves subject to commodification.

Notes on references

- For overviews of environmental economic geography, see Bridge (2010) and Bakker (2012).
- The National Academy of Sciences (2011) report on evaluating the costs of the Deepwater Horizon oil spill provides a good applied introduction to the ecological services approach in environmental economics.
- Webber (2008) provides a clear and accessible account of primitive accumulation in contemporary China, while Harvey's (2003) original statement on "accumulation by dispossession" has been applied in many contexts.
- Bakker (2010) provides a clear discussion of the global trend toward urban water privatization.
- For a detailed report on land deals in Sub-Saharan Africa and a balanced assessment of the "global land grab" debate, see Cotula et al. (2009).
- Solomon and Heiman (2010) provide an accessible overview of the problems of carbon markets. On the use of carbon sinks for offset credits, see Gutierrez (2011). For an update on the operation of global carbon markets, see World Bank (2011, updated annually).

Sample essay questions

- Is assigning a price a sufficient way of valuing nature?
- How do the physical properties of a resource affect the geographies of its commodification and use?
- Discuss the advantages and disadvantages of assigning ownership over natural environments.
- Do market mechanisms, such as those created for carbon emissions trading, provide an effective means of reducing pollution?

Resources for further learning

- The websites of various commodity exchanges explain their history and operation in some detail. See, for example, http://www.nybot.com (New York Board of Trade), http://www.cbot.com (Chicago Board of Trade), and http://www.lme.co.uk (London Metals Exchange).

- http://unfccc.int: The website gives full information on the history and details of the Kyoto Protocol.
- http://ec.europa.eu/environment/climat/emission.htm: This EU site provides further information on the Emission Trading Scheme.
- The United Nations Environment Program Green Economy Initiative has numerous resources about low-carbon and resource-efficient technologies and sectors: http://www.unep.org/greeneconomy/.

References

Bakker, K. (2003). Archipelagos and networks: urbanization and water privatization in the South. *The Geographical Journal*, **169**, pp. 328–341.

Bakker, K. (2004). *An Uncooperative Commodity: Privatising water in England and Wales*. Oxford: Oxford University Press.

Bakker, K. (2010). *Privatizing Water: Governance Failure and the World's Urban Water Crisis*. Ithaca: Cornell University Press.

Bakker, K. (2012). The "matter of nature" in Economic Geography. In T. Barnes, J. Peck, and E. Sheppard, eds., *The Wiley-Blackwell Companion to Economic Geography* pp. 104–117.

Barney, K. (2009). Laos and the making of a "relational" resource frontier. *The Geographical Journal*, **175**, 146–159.

Bond, P. (2011). Emissions trading, new enclosures and eco-social contestation antipode, *Antipode*, **44**: 684–701.

Bridge, G. (2010). The economy of nature: from political ecology to the social construction of nature. In R. Lee, A. Leyshon, L. McDowell, and P. Sunley, eds., *Compendium of Economic Geography*. London: Sage, pp. 121–134.

Cotula, L., Vermeulen, S., Leonard, R., and Keeley, J. (2009). *Land Grab or Development Opportunity? Agricultural Investment and International Land Deals in Africa*. London/Rome: IIED/FAO/IFAD.

Davies, A., and Mullin, S. (2011). Greening the economy: interrogating sustainability innovations beyond the mainstream. *Journal of Economic Geography*, **11**, 793–816.

Di Leo, L. (2010, June 15). Oil spill may end up lifting GDP slightly. *Wall Street Journal*.

Economist, The. (2012, January 21). The world in their hands: State capitalism looks outward as well as inward.

EU (European Union). (2012). Eurostat Database. http://epp.eurostat.ec.europa.eu/portal/page/portal/environment/data/database (accessed February 22, 2012).

Gutierrez, M. (2011). Making markets out of thin air: a case of capital involution. *Antipode*, **43**, 639–661.

Harris, L. M. (2009). Gender and emergent water governance: comparative overview of neoliberalized natures and gender dimensions of privatization, devolution and marketization. *Gender, Place & Culture: A Journal of Feminist Geography*, **16**, 387–408.

Harvey, D. (2003). *The New Imperialism*. Oxford: Oxford University Press.

Johannesson, G. T., and Bærenholdt, J. O. (2009). Actor-network theory/network geographies. In Thrift and Kitchin, eds., *International Encyclopedia of Human Geography*. Amsterdam: Elsevier Vol. 1, pp. 15–19.

London Metals Exchange. (2012). LME Copper Price Graph, http://www.lme.com/copper_graphs.asp (accessed February 25, 2012).

Muro, M., Rothwell, J., Saha, D., with Battelle Technology Partnership Practice. (2011). *Sizing the Clean Economy: A National and Regional Green Jobs Assessment*. Washington, D.C.: Metropolitan Policy Program, Brookings Institution.

National Academy of Sciences. (2011). *Approaches for Ecosystem Services Valuation for the Gulf of Mexico After the Deepwater Horizon Oil Spill: Interim Report*. Washington. Available at http://www.nap.edu/openbook.php?record_id=13141&page=R1.

Neilson, J., and Pritchard, B. (2009). *Value Chain Struggles: Institutions and Governance in the Plantation Districts of South India*. Oxford: Wiley-Blackwell.

Parry, B. (2012). Economies of bodily commodification. In T. Barnes, J. Peck, and E. Sheppard, eds., *The Wiley-Blackwell Companion to Economic Geography*, pp. 213–225.

Pleven, L. (2011, December 14). *As China goes, so go commodities. Wall Street Journal*.

Solomon, B., and Heiman, M. (2010). Integrity of the emerging global markets in greenhouse gases. *Annals of the Association of American Geographers*, 100, 973–982.

United States Geological Survey (2002). *Circular 1221: Material flows in the economy*. Denver, Colorado.

Webber, M. (2008). The places of primitive accumulation in rural China. *Economic Geography*, 84, 395–421.

World Bank. (2011). *State and Trends of the Carbon Market*. Washington, D.C.: World Bank.

CHAPTER 6

LABOR POWER
Can workers shape economic geographies?

Goals of this chapter

- To question whether capital's mobility always gives it bargaining power over labor
- To understand how labor markets are socially regulated and embedded in specific places
- To appreciate how workers can sometimes shape the economic system
- To examine the consequences of migration for labor
- To reflect on the possibilities for alternative or noncapitalist labor geographies

6.1   Introduction

Until 2008 the offices, seminar rooms, and lecture halls of Queen Mary College in the University of London were cleaned by a private contractor called KGB Office Cleaning Services. Many of the cleaners were employed for just two hours a day between 6:00AM and 8:00AM, at the national minimum wage. From January 2008, however, the college's practice of contracting out its cleaning services was ended and cleaning staff were taken on as full employees of the college. As a result, they received more hours, higher rates of pay, and better benefits. Moreover, Queen Mary became the first university in the country to declare itself a "living wage" campus. This means that all employees would be paid not just the national minimum wage, but a wage that was deemed to be sufficient for survival in an expensive city such as London. At a time when subcontracting such services was and still is the norm, this was a dramatic change in the way a university treated its lower-level workers. The results at Queen Mary were, on the whole, very successful. Cleaners were happier with their new terms of employment, other staff saw improvements in the standard of cleanliness, there was almost no increase in total cleaning costs, and the university was able to boast that it had taken the moral high ground (Wills, 2009).

The impetus for this change came from a number of sources: a local citizens coalition was doing research on low-paid work in London in 2005 and 2006; faculty members in the Department of Geography at Queen Mary were collaborating in this research; and students assisted by conducting research on their college's cleaning staff in particular. The wider coalition of activists in the living wage campaign involved local religious institutions, trade unionists, the student union, faculty members, local politicians, and of course the cleaners themselves. Together they changed the nature of employment relations at Queen Mary and improved the well-being of families in the surrounding low-income neighborhoods where most of the cleaners live. They achieved this success despite initial resistance from the college authorities, and despite the absence of a trade union to negotiate formally on their behalf.

The Queen Mary case is not the only example of cleaners, and low-paid workers in general, pressing for better working conditions. The Justice for Janitors (JfJ) campaign in the United States has been a highly successful movement for improving working conditions among cleaners and security guards and has received much attention (Savage, 2006). Beyond the work of cleaners, living wage campaigns also have a long history, starting in Baltimore in the 1980s. In short, there are now many examples of low-paid workers in the urban service sector, and their allies, actively campaigning for better working conditions (Merrifield, 2000).

What can we learn from these examples? The first point to make is that the cleaners at Queen Mary had geography in their favor. Many workers live with the threat that their jobs might be done elsewhere by even lower paid and unseen labor. For many decades now, factory jobs manufacturing clothing, toys, electronic equipment, and so on have been moving from the early-industrializing countries of Europe and North America to late-industrializing countries with lower labor costs. Increasingly, we have also seen office jobs being moved "offshore" to office parks in Mumbai, Manila, and elsewhere. But in the case of the cleaners at Queen Mary, there was no practical way for the college to move the jobs to where they could be done more cheaply. The physical space of the college was in East London, and so the cleaning jobs had to be there too. The question of the relative mobility of capital or employers and labor or employees will be the focus of Section 6.2.

A second point to note in the Queen Mary example is that the cleaning work was outsourced, until 2008, to a private contractor. The cleaners were employees of the contractor, not the college. This illustrates the way in which labor markets are increasingly being organized. The subcontracting of service work is a widespread phenomenon, even in public-sector settings such as universities and hospitals, and reflects the fact that governments around the world have sought to cut costs and introduce competitive processes in many of their supporting functions. More generally, many governments have moved toward deregulating their labor markets so that employers can hire and fire staff more easily, providing very little job security for many people. The result is a trend in many places toward the increasing use of temporary, part-time, and casual labor. The larger point, however, is that labor markets are not simply the product of a direct employer–employee negotiation, but rather they have various institutions shaping them, including but not limited to governments. This is a point that we will develop in Section 6.3, taking particular note of the geographical variations in labor markets that result from the local institutions that regulate them.

The third point is that even the most marginalized workers may, in the right circumstances, actively shape their workplaces, their relation to their employers, and the wider economic system of which they are a part. Labor is not, therefore, always a passive victim of corporate capitalism, nor is it just an input to the production process as is machinery or raw materials. The active role played by labor in shaping the geography of the economy will be a key point developed in this chapter. In some cases this active role might take the form of trade unions representing workers collectively. But the Queen Mary example also reminds us that workers may be aided in redefining their relationship with their employer through a variety of actors. The coalition of community, religious, student, and other groups was critical in forcing change at the college and supporting workers who were not represented by formal labor organizations. Coalitions such as these have become more and more common as trade unions change their strategies for organizing workers, and work closely with other groups in the community. In Section 6.4 we consider the various ways in which labor becomes an increasingly active force in shaping economic processes.

A final point is illustrated by an unfortunate aspect of the Queen Mary story. When the cleaning work was taken back by the college, only about half of the existing cleaners applied to continue in their existing jobs. The others disappeared. In many cases this was because they were migrants working without official papers or permission. When the college started to hire its own cleaners it requested various forms of documentation that many workers could not provide. This is symptomatic of the increasing role played by migrant workers, both legal and illegal, in many economies around the world. Where employers do not relocate in order to find lower-cost workers, they may instead try to recruit migrant labor that is, for a variety of reasons, less expensive. We will examine the implications of labor migration in Section 6.5.

Most of this chapter will focus on waged labor and the relationship in which workers sell their labor power to employers. This is the core labor process that exists in capitalist societies. But we will end the chapter with a reminder that many different relationships may also coexist. Section 6.6 will briefly outline some of the alternative forms of work that exist, for example, in cooperatives, households, or labor exchange schemes.

6.2 Is Labor at the Mercy of Globally Mobile Capital?

When workers demand better pay or working conditions, a popular argument is that if they make themselves too expensive or demanding, then jobs and investment will simply go elsewhere. This argument implies that the *relative* geographic mobility of capital compared to that of labor gives the former tremendous bargaining power when negotiating with the latter. In particular, the persistent threat that corporations may relocate production becomes a powerful tool for employers when determining wages and benefits, contracts, and investment strategies with employees. In this way, the *potential* reorganization of the activities of transnational corporations on a global scale (to be discussed further in Chapter 10) becomes a critical tool for firms negotiating with workers and local interest groups. Workers, in turn, become increasingly pressured into having to defend their interests in particular places, and in the process may enter into direct wage competition with workers in other places in a bid to secure jobs. This interplace competition fosters a progressive ratcheting-down of labor terms and conditions as mobile capital seeks out the best rates of return, a process sometimes known as the "race to the bottom."

This competition can occur at the international and/or subnational scales. The United Kingdom and Spain may compete, for example, to host a new Japanese car manufacturing plant to serve the European market. Locations within those countries may also be drawn into competition with one another in order to attract much-needed investment. Regional and local governments may become embroiled in this competitive dynamic, offering up large sums of public money in the form of incentives in a bid to secure investments. To give one example, in 2008, state and local levels of government offered Volkswagen an estimated US$577 million in incentives to locate an auto assembly plant in Chattanooga, Tennessee. This amounted to about US$290,000 for each of the 2,000 jobs that would be created in the factory (although many more jobs would be created indirectly) (Platzer and Harrison, 2009).

The general thrust of the "capital mobility" argument is persuasive. The geographic mobility of capital *has* certainly increased over the last few decades, due to a number of overlapping factors. Deregulation, as seen in the lowering of barriers to international trade and foreign direct investment, increasingly enables the transfer of both finished goods and inputs for the production process, such as materials, technologies, people, and capital (see Chapter 4). In some cases governments may also provide ready-made factory or office space for foreign investors, and, in making investment easy, they may also make departure and relocation easy as well. Dramatic improvements in production, transport, and communications technologies have facilitated increasingly complex geographic arrangements of production and the rapid global circulation of financial capital.

Technological changes have also, it should be noted, allowed automation in many production processes so that jobs are displaced by machinery that is sourced globally. This long-run trend is very clear in agriculture, mining, and manufacturing, all of which now employ far fewer people than a few decades ago.

While production capital is mobile, it is also true that workers are attached to particular places in a variety of ways, which makes them much less mobile than the corporations that employ them:

- *Labor reproduction*: The reproduction of labor – the ways in which workers are sustained from one day to the next and socialized from one generation to the next – is necessarily local. Social reproduction is carried out by various institutions of everyday life, such as the family, religious congregations, schools, clubs, and so on. These institutions are usually deeply rooted geographically, and thus keep people committed to particular places.
- *Place attachments*: In addition to social institutions, places also foster identities and cultures. They become sites of familiarity, routine, affection, and friendships. These emotional ties to place can be hard to break.
- *Households*: Individuals do not make decisions about where to live in isolation from the people around them. It is increasingly the norm (and necessity for some) for household incomes to be derived from the work of several family members, so it is not always possible for one person to relocate to where better opportunities might lie.
- *Regulation*: Many people are limited in their geographical mobility because governments regulate where they can live and work. Most obviously, mobility across national borders is restricted, but in some contexts migration within national borders is also limited (e.g., through China's *hukou* household registration system).
- *Local cultures of work and credential recognition*: People may be limited in their potential to move across space because of linguistic and cultural differences, and because their credentials and skills might not be formally recognized in a different context. For example, in some federally governed countries such as the United States and Canada, a lawyer would have to pass a bar exam in each state/province in which he or she wished to practice law.

In all of these ways, then, we can see that labor is limited in its spatial mobility and so there is some truth to the idea that mobile corporate capital has the upper hand over immobile place-based labor.

Inevitably, however, the world is more complicated than is suggested by the simple portrayal of powerful mobile capital on the one hand and labor rendered powerless by its immobility on the other. How might we construct a counter-argument? One important point is that employers are not nearly as mobile as this view suggests. The single largest employer in almost every territory around the world is the government or the public sector. While governments can move their employees around, for example by locating some office functions in remote areas to create employment, such jobs are generally not in danger of leaving. It is also important to remember that many jobs cannot move: a mining job has to be located at a mineral deposit, a cleaning job has to be where the office is located, and farm work has to be in the fields. So the idea that capital is highly mobile needs to be treated with some caution. Even firms engaged in activities that *could* be done elsewhere, such as making cars, software, advertising, and so on, are in reality rooted in places in all sorts of important ways, including their dependence on a source of labor that is appropriate to their needs.

This place-based rootedness of capital will be discussed elsewhere (e.g., Chapter 12) so it is not an argument that we will pursue here. In this chapter we will instead explore several ideas that suggest labor might have more power than we usually realize to shape the geography of economic activity. First, the relative power and mobility of capital and labor are not "natural" outcomes of a capitalist system, but are shaped in a variety of ways by the institutions of government that regulate the labor market. As a result, labor market regulation is quite uneven and varies between countries and even between subnational jurisdictions. But more important for our argument here, labor markets are "socially regulated" in ways that can be contested. Second, even labor that is rooted in place is not always powerless in the face of corporate power. In many instances labor may play a more active role in shaping the economic landscape than is often assumed. This may involve individual workers, trade unions, or broader coalitions. Finally, the suggestion that labor is not mobile has its limits, especially in an age where labor migration is increasingly common. In some cases the movement of labor has had important impacts on economic development processes, but we will also note that the use of migrant labor is quite often a way of undermining the power of workers.

6.3   Geographies of Labor: Who Shapes Labor Markets?

We noted in Chapter 2 that markets rarely exist in the pure form of competitive buyers and sellers with full freedom and knowledge to maximize their gains. In practice, markets are embedded in the social institutions that shape them. More than any other example, this applies to the market for labor. In countless ways, the labor market is influenced by institutions that exist beyond the scale of the relationship between employer and employee. For this reason, labor markets display enormous geographical variability. In this section we will highlight how labor markets are regulated and how this varies across space.

The most obvious institutional intervention in the labor market is through labor laws enacted by governments. These laws shape everything from health and safety in the workplace, to pension schemes, to employment contracts, to the rules governing the collective organization of labor in trade unions. In most cases these laws are created and enforced at a national level, but some may also be shaped by such subnational units as states or provinces or by supranational agreements. In Canada, for example, the regulation of labor markets is the responsibility of subnational provincial governments rather than the federal government. Supranational involvement in labor laws, on the other hand, can be found in Europe where the European Union creates directives that must be implemented through national laws in member states. One example is the creation of European Works Councils. This directive, established in 1994 and strengthened in 2009 and 2011, requires companies to create channels of communication for their workers across national borders when they have substantial operations in several European states (Wills, 2001). Another example is found in EU rules concerning the employment of workers through temporary employment agencies.

A more specific form of local labor market regulation is through control over access to trades and professions. Most trades and professions require practitioners to have some kind of accreditation in order to be licensed. For electricians, nurses, teachers, and every other form of licensed trade or profession there are regulatory bodies that are given a mandate by governments to ensure that those employed in these fields have the appropriate training. In some cases just one regulatory body exists. For example, the State Bar of California is the only organization mandated to authorize lawyers to practice in the state. But accountants often have many regulatory bodies for chartered, financial, or management accountants. As accreditation and regulatory bodies, these are not the same as trade unions, but they do still represent the interests of their members in certain respects. In particular, they act as gatekeepers to the practice of trades and professions, thereby closely controlling the market supply of such labor. Only those accredited by such organizations, which usually operate at arm's length from the government, can legitimately practice trades and professions (Girard and Bauder, 2007).

Another form of intervention into labor markets is in the form of public sector employment. It is often overlooked that governments are enormous employers and not just regulators of employment. Of the world's largest employers shown in Table 6.1, all but three giant corporations – Wal-mart, McDonald's, and Hon Hai – are government agencies. This means that simply through determining how their own employees are treated, governments can play a major role in shaping the standards applied throughout the labor market. If governments provide attractive pay and benefits, then private-sector employers will have to do the same in order to attract talent. Conversely, if government employees are poorly treated, then the overall standard is also likely to be lower.

The examples discussed so far relate to fairly obvious forms of government control over the labor market. But the supply and characteristics of labor are also shaped indirectly in other important ways. Families are ultimately the source of a new generation of labor. Of course family needs are usually the main reason

Table 6.1 The world's biggest employers, 2010

Employer	Government-Owned?	Number of Employees
U.S. Department of Defense	Yes	3.2 million
Chinese People's Liberation Army (1)	Yes	2.3 million
Wal-mart	No	2.1 million
McDonald's (2)	No	1.7 million
China National Petroleum Corporation	Yes	1.7 million
State Grid Corporation of China	Yes	1.6 million
National Health Service (U.K.)	Yes	1.4 million
Indian Railways	Yes	1.4 million
China Post Group	Yes	0.9 million
Hon Hai Precision Industry	No	0.8 million

Notes: (1) data from 2008; (2) includes all franchise employees
Source: Data available on http://www.bbc.co.uk/news/magazine-17429786
http://www.businessinsider.com/the-10-biggest-employers-in-the-world-2011-9?op=1,
http://www.dailymail.co.uk/news/article-2038401/Defence-Department-McDonalds-
Walmart-worlds-biggest-employers-list.html, and http://www.economist.com/blogs/
dailychart/2011/08/biggest-employers

why people work in the first place. The supply of labor is thus dependent, in part, on decisions made by parents a generation ago, not by considerations of current supply and demand. The family is also critical in shaping the socialization of individuals into the types of adults, with different skills and abilities, that they will eventually become. This role is not, of course, simply determined within the family itself; it is a product of culture, government policy, religion, educational and training institutions, and other influences. Labor markets will be affected by the birth rate that may be influenced by government policies (e.g., China's one-child policy), the nature of the educational system that in some contexts may be heavily geared toward science and technology (e.g., Singapore), and the culturally accepted roles of men and women that may be shaped in some contexts by religious teachings. Perhaps less obviously, other social institutions, such as prisons, may also play a part in shaping the labor market (see Box 6.1 on Chicago). In all of these ways, the characteristics of the labor market are determined by its embeddedness in geographically-specific social institutions.

While governments claim sovereignty over their territories, there are many instances where they choose to implement their authority to regulate labor markets in a very selective manner. In domestic spaces of the home, for example, many governments do not inspect or regulate the working conditions of people employed as domestic workers. This lack of attention to decent working conditions is a complaint of many organizations advocating for the rights of domestic workers around the world. Another example is the use of industrial estates and export processing zones as havens for investors to locate their activities, while

CASE STUDY

Box 6.1   Chicago's prisons as labor market institutions

Prisons are often thought to be a holding tank for those deemed a danger to society, but in fact they often represent a revolving door for those who survive at the economic margins. Research in Chicago has examined how penitentiaries shape the urban labor market in important ways (Peck and Theodore, 2008). In 2010, the state of Illinois had around 47,000 inmates, and released about 35,000 prisoners each year. The recidivism rate – the likelihood that prisoners would return to jail – was around 50%! In general, then, there was a constant cycling of convicts, especially young African-American men, in and out of the correctional system. During their periods out of jail, these young men are returning to the same deprived neighbourhoods from which they had been arrested and taken. Once back there, work is very hard to find, and their criminal record immediately disqualifies them from a wide range of jobs, benefits, and services. With poor employment prospects and racial stigmatization, ex-convicts find few options available to them. But the consequences of their unemployability are much broader. With so many ex-offenders potentially looking for work in certain neighbourhoods of Chicago, the labor market is flooded with people who have few options, which drives down the pay and conditions that anyone can expect (ex-convict or not). Prisons have therefore become "boundary institutions" in the labor market, shaping the employability of a certain group in the population, and by extension undermining the quality of work that is available to others as well.

enjoying some immunity from government interference. One scholar has described these zones as examples of "graduated sovereignty" to suggest the way in which governments may withdraw from their regulatory responsibilities, including the oversight of work conditions and labor laws (Ong, 2000; see Box 4.5).

Labor markets are also shaped by market intermediaries – that is, those who play a go-between role in linking supply/workers and demand/employers of labor. An increasingly important group of players are temporary staffing agencies that supply employers with short-term contract-based staff members. By using temporary staffing agencies, employers avoid a direct contractual relationship with the people working for them, and can hire and fire such workers at will. In many countries around the work the temporary staffing industry has grown dramatically, in tandem with a growing prevalence of short-term, part-time, low-paid, and precarious terms of employment (see Box 6.2 on the temporary staffing industry).

CASE STUDY

Box 6.2   *The temporary staffing industry*

Temporary staffing agencies represent labor market institutions that supply client companies with nonpermanent workers. Essentially, they sell the labor of their workers to client firms, and make a profit by taking a share of the workers' wages. This may be an appealing arrangement for many employers. It allows them to make rapid changes in the size of their workforce, to reduce the cost of hiring, to avoid the risks and responsibilities of a standard employment relationship such as pensions, benefits, holidays, and so on, and in some cases it may also reduce wage costs after paying the fees of the temp agency. Since the 1970s, temporary staffing has developed into a significant component of many national labor markets. One estimate suggests that the number of people working through temp agencies worldwide has more than doubled, from 4.5 million in 1997 to 9.7 million in 2007. In recent years, major temporary staffing firms have expanded globally, with the two largest players being Adecco, based in Switzerland with global revenues of over US$26 billion, and Manpower, based in the United States with revenues of over US$17 billion.

Although the numbers of temp employees may be a small fraction of the total workforce, the existence of such employment relations may be a factor in undermining standards in the labor market as whole. In particular, standard employment contracts that included permanent, full-time work with employee benefits have increasingly given way to insecure work that may be part-time, on contract, outsourced, fixed-term, and home-based. The temporary staffing industry is therefore an example of a labor market institution that acts as an intermediary between employer and employee, but it also *shapes* the labor market as a whole in important ways. Its presence in the labor market both reflects and reinforces changing standards of "normal" practices, rights, and responsibilities in labor relations. (For a review of research on the temporary staffing industry, see Coe et al., 2010.)

So far, we have laid out a range of institutions and agencies that shape the operation of labor markets in various countries. Clearly, the labor market is far from being a simple negotiation between buyers and sellers of a commodity in order to establish a price. Instead, we can see that the conditions under which people work are rooted in the cultural, social, and political institutions of their specific context. This means that labor markets exhibit a great deal of geographic variability and unevenness, with different places constructing the relationship between employers and employees in quite different ways. For this reason, economic geographers have often examined the *local* labor market in order to highlight these variations (Peck and Theodore, 2010).

At a large scale, we can classify styles of labor market regulation into some very broad patterns. In particular, we can differentiate between a model that minimizes regulation and one that prioritizes government involvement in labor market processes. Broadly speaking these correspond to a model typified by the U.S. economy, and one found in the German economy. Table 6.2 lays out the features of these two different models. These models have quite different ways of dealing with trade unions, training, unemployment benefits, job security, and other features of the labor market. While the table treats these models as distinct, in reality there is considerable variation within the groups shown, and they could more accurately be thought of as a continuum of different styles. Although American and Canadian contexts are seen as exemplifying the "liberal market" model, these two countries are also distinctly different. For example, union membership in Canada in 2009 accounted for about 27% of the labor force, but only 12% in the United States. Similarly, although the United Kingdom is different from its European neighbors in many ways, it is nevertheless subject to the same EU directives on labor market regulation, such as the European Works Councils and standards for temporary employees mentioned earlier. Nevertheless, Table 6.2 provides some of the contrasting features that help us to differentiate labor market regulation around the world.

Within the broad parameters set by labor market institutions in different places, it is also important to remember that firms themselves will have different approaches to labor relations. Even within the same local labor market, different workplaces and different industries will inevitably adopt diverse practices, and these will change over time. In some sectors, new production technologies will change the way in which employees are managed. In the case of call centers, for example, digital technology allows employees' work to be allocated, monitored, and analyzed in great detail. Alternatively, a firm may adopt a variety of organizational tools in order to manage its workforce, including shift systems, use of overtime, changing job descriptions, and the employment of part-time, short-term, and temporary workers (see Box 6.2 above). Firms may also have varied cultures in terms of their attitudes toward union organization. In some contexts, employers attempt to constrain the activities of unions within the workplace. They may simply ban unions completely in the workplace, establish one-union rather than multi-union workplaces, insist on workplace bargaining and dispute resolution as opposed to collective bargaining at larger spatial scales, or establish worker–management councils within the workplace to try and negate the need for union representation. The term *lean production* has been used to describe the nature of the production system in contexts where most or all of these various mechanisms coincide (see Box 6.3).

Notwithstanding the variability of strategies across firms, it is possible to see patterns in how local labor markets are configured across space. In particular, the

Table 6.2   Different national labor conditions: two ideal types

	Liberal Market Economies	Coordinated Market Economies
Classic example	United States	Germany
Other examples	United Kingdom, Canada, Australia, and New Zealand	Japan, Austria, Switzerland, Italy, Belgium, the Netherlands, Denmark, Sweden, and South Korea
Education and training	• Weak systems of vocational training; limited company involvement • Strong post-compulsory/ higher education • Substantial doctoral programs in basic sciences and engineering, but with weak links to employers and industry technologies	• Strong systems of vocational education and training; significant involvement of industry organizations and unions • Limited post-compulsory/higher education • Substantial doctoral programs in basic sciences and engineering, with close links to large employers
Employment practices	• Short-term employment relationships • Market-led employment adjustment through external labor market • Individualized and competitive employment relations	• Medium-term employment relationships • High standards of employment protection • Macroeconomic stabilization of employment levels
Industrial relations	• Company-based, uncoordinated wage bargaining • Limited workplace roles for unions	• Formal or informal coordination of wages across key industries • Employer associations and unions play major roles in wage determination • Employee-elected bodies and representative bodies play key roles in company decision making

Sources: Based on Peck (1996, pp. 246–7) and Peck and Theodore (2007, p. 746).

KEY CONCEPT

Box 6.3   *Lean production*

A form of production system often termed lean production involves a variety of strategies to manage the workforce. The essence of lean production, which originated in Japan in the postwar period and has subsequently spread widely, is that it combines economies of scale with economies of scope derived from various practices of *flexibility*. Three types of flexibility are encapsulated in lean production. First, functional flexibility concerns increased levels of job rotation and multi-skilling. Second, numerical flexibility reflects a growing use of contracting-out and part-time and casual labor. Third, temporal flexibility describes the shift to new forms of shift patterns such as 24-hour working and overtime. These changing working patterns are often combined with the very latest productive technologies. Proponents emphasize that these processes enhance product and process innovation, whereas critics argue that flexibility is most often implemented to facilitate cost reductions. But despite seemingly positive descriptors such as "quality management" and "team-working," these changes are not simply about worker empowerment, but are also about the micro-management of workers' time and activities, a system that might be described as "management-by-stress" (Moody, 1997, p. 87). New forms of groupings designed to increase worker input to the production process such as teams and job circles can actually serve to bypass unions or avoid them altogether. In some situations workers and their representatives are forced to develop proposals to reduce costs and improve quality in competition with other plants. The usual outcome of lean production is a combination of job losses, and increased work and time pressure on the primary workforce. For more, see Moody (1997).

ways in which labor is disciplined and controlled is often distinctively arranged in specific places. This has led some researchers to conceptualize "local labor control regimes" as explained in Box 6.4. An example of such localized regimes can be found in Southeast Asia. Across the region, several countries have localities where rapid industrialization has occurred in recent decades. Here we will discuss Cavite and Laguna (the Philippines), Penang (Malaysia), and Batam (Indonesia). These places share several important features: high levels of female employment in light manufacturing and assembly work in the electronics and garments industries; substantial inward foreign direct investment from North America, Western Europe, and leading East and Southeast Asian economies; strong state provision of locational and export incentives; and very low levels of unionization.

KEY CONCEPT

Box 6.4   *Local labor control regimes*

The local labor control regime (LLCR) is a useful concept for understanding the inherently placed-based nature of economic activity. The development of LLCRs is driven by one of the basic contradictions of capitalism that exists between the potential spatial *mobility* of many firms and their associated bargaining power, and the need for firms to extract profits from concrete investments in particular localities, which requires a certain measure of *stability*. An LLCR is the place- and time-specific set of mechanisms such as social relations, norms, rules, and habits that coordinate the links between production, work, consumption, and reproduction in particular localities (Jonas, 1996). Importantly, this formulation extends beyond the workplace to incorporate the domains of consumption such as housing, recreation, and household consumption, and reproduction such as education, training, healthcare, and welfare. It naturally follows that the full range of worker, household, firm, civil society, and state institutions are, or at least may be, involved in shaping the LLCR. In part, a LLCR constitutes the unique, place-specific relations between firms, workers, unions, and regulatory institutions that enable workers to be integrated into a production system.

At the same time, a LLCR is shaped by multiscalar processes, and there is a broad range of extralocal worker, employer, and regulatory connections that will influence its character. Put another way, every local regime is 'nested' within labor control regimes operating at larger scales that will influence, but not determine, the nature of employer–worker relations in that particular locality. Of particular importance is the way in which the regime is integrated into production systems organized at the national and international scales that actively seek to take advantage of the differences between various regimes in terms of labor costs, skills, unionization rates, and so on. The global capitalist economy is characterized by a bewildering variety of different LLCRs. Even within the same country and the same industry, the differences may be profound. In the U.S. computer industry, the highly flexible innovative labor regime of Silicon Valley, California, is reputedly very different from the large-firm dominated regime found in the Route 128 area near Boston, Massachusetts. A corollary of this great variability is that the problems and possibilities facing workers will differ dramatically between different places. After all, it is only certain kinds of businesses that benefit from interlocality flows of investment and aggressive spatial restructuring. Others serve to benefit far more from becoming progressively involved in the LLCR on a long-term basis.

We might expect to find that labor control is achieved in much the same way, but in reality we find considerable variation (Kelly, 2002).

The varied strategies of labor control across these three locations can be seen as operating at several different scales. First, at the level of the *individual*, employers try to establish individual relations with workers rather than allowing collective representation through, for example, unions. Through the extensive use of dormitories and hostels, employers attempt to provide a tightly controlled and regulated nonwork environment in which workers may be offered counseling and a range of religious, educational, and sporting activities. Second, the *workplace* is a site of labor control. Where collective organization is allowed, labor relations are usually confined to the factory level through the use of in-house unions and labor-management councils. This is to prevent independent trade unions from operating across different workplaces and firms, which may facilitate the comparison of wages and working conditions. Third, the *industrial estate* is another scale of labor containment. Estate management companies provide services to tenants that go way beyond infrastructural provision, including tight security, establishing and maintaining worker databases identifying non-union employees, undertaking employer–worker mediation, and fostering links and contacts to local politicians. In this way, estate management companies actively seek to maintain their jurisdictions as union-free spaces. Fourth, *national regulatory policies* can serve to constrain worker organization through trade union legislation. This may not be through legislation that bans unionization outright, but more likely through highly bureaucratic and sector-specific registration processes that present an effective barrier to worker organization. Fifth, different *spaces of migration* can be identified that allow flexibility for employers and promote labor cooperation. Employing workers either from other regions of the country or from abroad can provide employers with a flexible workforce that is often disinclined to change jobs or prevented from changing jobs through the conditions of their work visa and/or insulation from the distractions of family or social life. Such workers may also be highly motivated by the desire to send money home to their families.

The precise configuration of these different strategies, however, is quite distinct in each of the three localities (see Figure 6.1). In Penang (Malaysia), the use of domestic and foreign migrant workers is common, as is their accommodation in carefully controlled hostel environments. In-house unions and firm-level industrial relations combine to keep disputes at the factory level, while national legislation provides a strong barrier to worker organization at larger scales. In Batam (Indonesia), the housing of domestic migrants in hostels, supported by a militaristic approach to preventing organization, has provided an effective system of control (Figure 6.2). In Cavite and Laguna (Philippines), rather than migrant labor, hostel spaces, or national legislation, it is the enclaved nature of industrial estates as carefully controlled spaces combined with strategic links into local politics that facilitate labor control. There is clearly geographic variability in the relative importance of different strategies of labor control in each of the

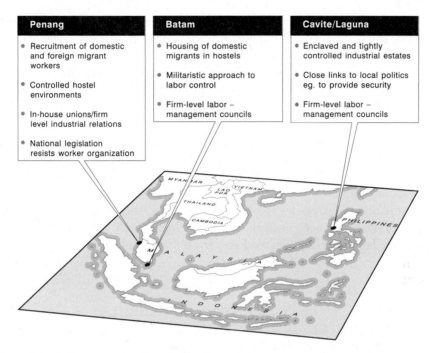

Figure 6.1   Labor control regimes in Southeast Asia

Figure 6.2   Worker dormitories in Batam, Indonesia

localities. Moving beyond Southeast Asia, we can therefore expect that there would be equally dramatic variability in labor control strategies elsewhere in the developing world.

We have seen in this section that labor markets are far from being determined by the "pure" economic processes of supply and demand. Instead, they are rooted in social institutions of many kinds. As a result, they display a high degree of geographic variability. There is also, however, a larger point that concerns our argument in this chapter about whether workers can play a role in shaping the economic system of which they are a part. A part of the answer to this question has to be "yes" because labor markets are, as we have seen, so deeply shaped by social institutions and not just by the objective price-setting mechanisms of supply and demand. Once we recognize this fact, the labor market begins to look a lot more open to change through democratic institutions. If we are collectively shaping the labor market, then workers can collectively change it.

6.4   Labor Geographies:
Workers as Agents of Change

We have established that labor markets are shaped by social institutions and are therefore open to change. But there are also more direct ways in which workers, either individually or collectively, are able to intervene in the economic system. This has been done in a variety of ways and has led to the emergence of a specific field, Labor Geography, which seeks to understand how workers exert *agency* in the face of the economic structures around them (Coe and Jordhus-Lier, 2010). In this section, we will discuss three examples of labor agency: small-scale forms of resistance in the workplace; coalitions between labor organizations and community groups in particular places; and international coalitions in support of workers' causes.

Micro-politics: Action within the Workplace

Some forms of labor agency manifest themselves at the scale of the individual and/or groups of workers within a workplace and are designed to exert pressure or express displeasure toward an employer. All employer–worker relations inherently embody tensions concerning a wide range of issues including worker autonomy, training and skills development, wages, working hours, promotion opportunities, job security, job status, nonwage benefits, and workplace facilities. In contesting these issues, workers have their own resistance strategies, which can be used to either support or protest employer strategies. These strategies can vary along three key dimensions. First, they may either proactively initiate debate with employers or reactively respond to employer actions. Second, they may be either official or unofficial, depending on whether or not they operate within approved industrial relations frameworks and laws. Third, they may be either physical (e.g., striking or working to rule) or nonphysical (e.g., using persuasion and bargaining). In many cases, and mirroring the interconnecting labor control strategies implemented by states and firms, workers will use a combination of resistance strategies at the same time.

The coping strategies employed by women working in the Jamaican information-processing industry provide an interesting geographical example (Mullings, 1999). The industry exports a range of services associated with the collection, transmission, storage, and processing of information, using information and communications technologies, to mainly U.S. clients. Often working out of the export processing zones of Kingston and Montego Bay, the activities undertaken by these Jamaican subcontractors range from simple data entry work for magazine subscriptions to software localization. Growth in the sector, as in other Caribbean states, has been driven by the labor cost savings of up to 90% that can be secured by U.S. firms using data entry clerks, secretaries, and telephone operators located in Jamaica. Some 90% of the workforce is female with the highest concentrations in the data entry rather than management strata, and the average age is only 25. In short, the Jamaican data processing industry mobilizes a workforce of nonunionized, low-wage female workers. As in the Southeast Asian examples recounted earlier, labor control is enacted through a range of practices including premises designed to obscure workers from the public gaze and prevent interaction, strict control on toilet and drink breaks, and close electronic monitoring of workers' productivity and accuracy.

The worker–manager relation in this context is clearly one of constant tension. While few workers have overtly resisted work conditions, many have engaged acts of everyday resistance that may hamper the ability of firms to meet contractual deadlines for their clients. Examples include absenteeism, working to rule, rejecting overtime when offered, and "finger dragging" (i.e., deliberately slowing the work rate when basic targets have been met). These are activities that can actually reduce the wages that workers receive, and may indeed jeopardize their employment in the industry. The viability of these strategies and the psychological boost that workers may receive from them can only be understood in terms of the role of the extended family in the Jamaican context. Through place-specific forms of household organization, geography clearly matters in shaping worker actions. The basic wages earned in the information processing industry are not enough to meet the income needs of a family. Hence, many families survive by pooling a variety of resources: remittances received from overseas family members, formal wages, and by exchanging unpaid services within and between family groups. It is these collective household economies that cushion the effects of the low wages and harsh discipline of the information-processing industry, and make viable a series of workplace resistance and coping strategies. While these forms of resistance may at first seem ineffective, astute employers will realize in such circumstances that the productivity of their workers would be much greater if the sources of irritation were removed – in which case the action has been successful.

Defending Place: Worker Actions in Situ

Despite the apparent global mobility of capital, workers still have considerable potential to shape the economic development of particular localities and the geography of production in industrial sectors. The role of labor unions in the Canadian auto manufacturing sector provides a good example. The Canadian province of Ontario has the highest output among all auto manufacturing jurisdictions in North America, with the sector directly accounting for about 100,000 jobs and 15% of Canada's exports across all sectors (Stanford, 2010). While Toyota and Honda have major production facilities that are not unionized, the "Big Three" Detroit-based manufacturers Ford, General Motors, and Chrysler and their suppliers all have unionized plants in a corridor from Greater Toronto to Windsor (see Figure 6.3). Organized labor has shaped the geography of this industrial cluster in a variety of ways.

First, when unions negotiate collective bargaining agreements with employers they will obviously influence workplace employment conditions. In the case of large auto assemblers, they may also negotiate a firm's geographical investment strategy. For example, in negotiating contracts with major auto manufacturers in southern Ontario, the Canadian Auto Workers union (CAW) has consistently sought to protect local jobs by including future investment in their collective agreements. In the 1990s, when the Big Three Detroit-based auto manufacturers

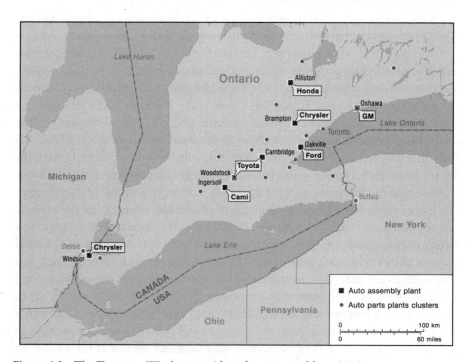

Figure 6.3   The Toronto–Windsor corridor of auto assemblers, 2012

The authors

Figure 6.4   The Ford plant in Oakville, Ontario, flying the flags of Canada, Ford, and the CAW

were profitable, this was done in a quite aggressive manner. By the 2000s, however, with Ford, Chrysler, and GM suffering losses, it was more often a case of the union offering concessions to encourage the firms to maintain production and reinvest locally. When GM was deciding where to locate production for its new Camaro vehicle, for example, it was the union local in Oshawa, Ontario, that supplied concessions in their collective agreement to bring the work to their facility (Siemiatycki, 2012). Whether through actively ensuring future investment and production levels through a collective agreement or by conceding elements of a collective agreement to attract investment, unions have played a key role in shaping the geography of production among the Big Three in Ontario. It is telling that outside Ford's major assembly plant in Oakville, Ontario, the flags that fly are those of Canada, the company, and the Canadian Auto Workers union (see Figure 6.4).

A second area of influence for organized labor may be in the development of an innovative cluster of firms in the same industry. From Oshawa to Windsor, southern Ontario represents a major cluster of auto assembly firms and parts suppliers. Research has show that unions have played a key role in the sharing of information in such clusters and therefore the spread of innovative ideas and practices (Rutherford and Holmes, 2007). This has not necessarily been through a process of collaboration with managers, but instead through the transfer of information on an informal day-to-day basis and during collective bargaining processes. In this way, the union assists in creating advantages for firms located in a geographical cluster. For example, when bargaining with one employer the union might point out the existence of specific practices or strategies at another firm – information that managers would not otherwise have access to. A firm that relocated to a nonunionized and lower-cost location would lose out on the benefits of this kind of information flow.

Finally, unions have played a key role in pressuring both national and local governments to take an active role in supporting the auto industry. In some cases this might be through community unionism, in which labor organizations join forces with other local movements in order to protect jobs and defend workers' rights. In other cases, it may be through cross-class alliances in which unions collaborate with local firms, employer organizations, and local governments to advocate for action to support the local industry, perhaps by lobbying higher levels of government.

One example of this process is found in Windsor, an auto manufacturing center in southwest Ontario adjacent to Detroit. In 2007, as Ford proposed closing its engine manufacturing plant in Windsor, a coalition of local organizations launched a "Manufacturing Matters" campaign to raise awareness of the auto sector's importance in the local economy and to pressure governments to do more to support it (Ross, 2011). The campaign was led by the Canadian Auto Workers union, but ultimately included a wide range of local players: elected officials, public sector unions, teachers and professors, community-based organizations, and local business owners. In May 2007, 38,000 people attended a rally of "community unity" in Windsor. The argument developed by the campaign was not against the Ford Motor Company or other employers. Instead, it emphasized the centrality of manufacturing jobs to the community as a whole, and the role of governments in creating trade policy, currency valuation, and industry support in order to foster the industry's growth in a competitive environment. Although clearly not the only reason, such community mobilization and advocacy by the auto workers' union undoubtedly played a part in the $14 billion of auto industry support provided by the federal government of Canada and the provincial government of Ontario in 2009 (Stanford, 2010).

Upscaling Worker Action: Organizing across Localities

The final kind of geographical strategy we will consider in this section consists of organizing across localities, or upscaling worker activism. This upscaling can take place at different spatial scales. It might be between workers in different work-places, different towns or cities, distant parts of the same region or country, or transnationally between workers in different countries. The important point is that it involves workers collaborating with others outside of their own employment situation in order to create solidarity and to tackle employers at a larger scale. The purpose is usually either to confront a single employer at the scale at which corporate decisions are being made, or to unite workers across many employers to ensure that undercutting wages and benefits is not the basis upon which firms are competing. The advantage of such a strategy is that it prevents different groups of workers being played off against each other and it enables more resources to be tapped when national and international organizations are involved.

Upscaling might be organized formally through local, national, or international trade union structures, or it might be relatively organic and grassroots in nature, connecting across localities outside of formal worker organisations. In many countries, the formation of national trade unions, for example on occupational or sectoral grounds, has traditionally been the favored means of coordinating worker representation and action. Confusingly, national trade unions in the United States are usually called "international" unions. As globalization dynamics have gained pace in recent decades, however, and national labor agreements and unionization rates have come under pressure from both governments and corporations, increased emphasis has been placed on worker internationalism – that is, worker activism coordinated at the international scale to try and tackle the global reach of firms and capital.

For an example of upscaling by labor organizers we can turn again to the example of the Justice for Janitors Campaign that was introduced at the start of this chapter. Fewer and fewer buildings are cleaned by people actually employed by the company, school, or hospital that uses the space. Furthermore, the cleaning industry in many parts of the world has become increasingly centralized in the hands of a few major corporations who bid for contracts. For example, Integrated Service Solutions is a company headquartered in Denmark and owned by American investors that has cleaning and other operations in over 50 countries. With more than 500,000 employees in 2011 it is one of the world's biggest private employers (ISS, 2011).

Organizing cleaners, security guards, and others working in the building services sector has always been a challenge for trade unions. Having migrated from all over the world, employees often work at night in isolation and they may have language barriers and feel that they have little in common with each other. They may also be part-time and temporary employees with other jobs to go to after they finish their shift. These difficulties have been addressed in imaginative ways in the last few decades. The Justice for Janitors (JfJ) campaign is one of the most celebrated cases. JfJ was started in California in the 1980s by the Service Employees International Union (SEIU). Faced with a declining union membership among office cleaners and healthcare workers in Los Angeles, the union adopted a strategy that did not just involve a traditional effort to collect new memberships through the local branch of the union. Instead it established a creative media campaign to raise awareness of the hard and essential work done by janitors and the poor pay and working conditions they experienced. The union also reached out to immigrant communities that were not well represented among union organizers. This was therefore a form of alliance building and "community unionism" that we noted earlier.

The next move was to organize all janitors in a given area, thus moving up from the scale of the workplace to the scale of the city. In this way, various cleaning companies could not outbid each other on local contracts by undercutting wages. The results in Los Angeles were impressive. Between the mid-1980s and the mid-1990s, unionization among downtown LA office cleaners increased from around 10% to around 90% (Savage, 2006). The next move by the SEIU was to organize all of the local janitors' unions into a statewide branch of the union, thereby creating a powerful force to negotiate with the increasingly large companies employing cleaners. In recent years, the SEIU has advocated for a nationwide consolidation of unions in the United States along occupational lines. In this way, larger unions would concentrate in strategic know-how, financial resources, and bargaining power. This move, however, has been controversial and has led to the splintering of the main national trade union coalition, the AFL-CIO, in 2005. Several of the country's largest unions, including the SEIU, the Teamsters, and UNITE HERE formed a new federation called the Change to Win Coalition.

Despite controversies concerning what the SEIU's Justice for Janitors model might mean for the national level union in the United States, the JfJ model has spread globally (Aguiar and Ryan, 2009). With branches already existing in Canada, from 2004 the SEIU assigned staff to Australia, Poland, the United Kingdom, India, France, Switzerland, Germany, the Netherlands, South America, and South Africa. The purpose was to develop alliances and relationships with local unions working with janitors, security guards, and other workers in these countries. In Australia, for example, the SEIU invested considerable resources in working with the Liquor, Hospitality and Miscellaneous Workers' Union (LHMU) to create a campaign called "Clean Start – Fair Deal for Cleaners" starting in 2006 (Ryan and Herod, 2006). One year later, the LHMU claimed that 70% of cleaners in downtown offices across Australia were covered by the "Clean Start Responsible Contracting Principles" that the union had established (Aguiar and Ryan, 2009).

Through the kinds of upscaling practices adopted by the SEIU and the Justice for Janitors campaign, real improvements have been possible in the working conditions of cleaners. Collectively, workers have strengthened their negotiating power with their employers and have changed the scale at which these negotiations will take place. There are, however, pitfalls in this strategy. First, local solidarity is still a key part of mobilizing workers, and strategies devised and coordinated from far away will not always resonate with workers in different contexts. Second, maintaining solidarity among workers across long distances is often difficult, especially when workers in one country might see themselves as competing with those in another country. Third, transnational organizing networks that span the globe need to stay aware that it is still largely governments that set the regulatory conditions for labor relations within their jurisdictions. Global workers' movements may therefore not be the most effective way to influence policy at the national scale.

While Justice for Janitors represents an upscaled organization coordinated by a union, there are many other examples of labor-based movements that are not trade unions in the traditional sense. Although they may not represent people directly in negotiating their employment conditions, their work is explicitly aimed at improving the welfare of workers. One example is the Clean Clothes Campaign (CCC). Established in 1989 in the Netherlands, the CCC advocates to improve the working conditions of workers in the global garments and sportswear industries. The targets of its activities include companies making such products, but also consumers and governments. In some instances the organization provides direct support to workers who are demanding better working conditions. The CCC is an alliance of more than 200 diverse organizations including women's rights, consumer advocacy, and poverty reduction activists, as well as trade unions. Although primarily rooted in 15 European countries, the CCC cooperates with similar campaign in the United States, Canada, and Australia. Most of its campaigns target conditions in garment and sportswear factories in the developing world. Campaigns may target specific companies accused of labor abuses, seek to raise consciousness about the labor "behind the label" in general, or even champion the cases of individual workers involved in disputes (Hale, 2005). An important part of the strategy is to mobilize consumers to boycott or pressurize manufacturers, retailers, and designers. This is not, then, a traditional contestation between workers and employers, but is instead a transnational mobilization in support of workers rights.

This chapter began by asking whether place-bound labor is rendered powerless by mobile capital. We have shown that the social regulation of labor markets, and collective action by labor and its allies, can in fact exert considerable influence. Whether it is through small-scale action in the workplace, defending existing jobs in a given locality, or upscaling actions to become transnational movements, labor plays a significant part in shaping the geography of economic activity. While this certainly does not mean that workers have power that is equal to their employers, we can see that at the very least, labor has to be seen as an *active* agent in the production process. One last piece in the argument remains, however. While the contest between capital and labor often assumes mobile capital and immobile labor, the global workforce is increasingly a mobile one. It is therefore important to examine not just how workers' organizations move across borders, but also how workers themselves become international migrants.

6.5 Migrant Labor

In Section 6.2 we discussed several reasons why labor is relatively immobile. While this is generally true, it is now the case that more people are moving than ever before. In 2010, the number of people living outside their country of birth was estimated to be about 213 million or 3% of the world's population (UN, 2009). In addition, internal migrations within nations, for example from rural to urban areas and between regions, are measured in hundreds of millions. In China alone, it is estimated that internal migrants number around 150 million (Chan, 2011). What might this kind of mobility mean for the power of labor? In this section we will suggest that for some it represents empowerment, but for the majority of migrants the experience is quite the opposite.

Mobile Elites

Elite labor migration has expanded dramatically over the last 20 years as part of the broader dynamics of globalization. A small minority of very wealthy individuals may hold assets such as homes and businesses in several countries and may structure their lives in order to avoid higher tax rates in certain places. Global mobility for such individuals is very easy and they may exert considerable power over economic development policies. In the United Kingdom, for example, tax structures and the regulation of the financial sector have been carefully configured to the needs of very wealthy and highly mobile individuals, in the hope that they will remain based in London (Beaverstock and Hall, 2012). More commonly, however, elite migration is the result of organized overseas secondment and assignment schemes within transnational corporations.

A form of global elite mobility that is of particular interest involves highly skilled professionals and entrepreneurs working in the information technology sector. Innovation and economic growth in clusters of high-tech activity such as Silicon Valley, Shanghai, and Taiwan is dependent on attracting the best brains from around the world. Migrants from India and China play a particularly important role. It is estimated, for example, that between 1995 and 2005, over half or 52.4% of Silicon Valley startup firms had at least one immigrant as a key founder. Indian immigrants accounted for 15.5% of all startups, and immigrants from China and Taiwan another 12.8% (Wadhwa et al., 2007).

The economic power of highly skilled migrants also provides benefits for the places they have come from. Increasingly, skilled workers who have gained experience in the United States are moving back to their home countries and fostering entrepreneurial ventures there. Research at the height of the dot.com boom in the late 1990s showed that over one-third of scientists and engineers in Silicon Valley's high-technology workforce were foreign born, with the vast majority being of Asian descent (Saxenian, 1999). By 1998, there were 2,001 firms led by an ethnic Chinese manager and 774 by an ethnic Indian manager, together accounting for 24% of the total number of firms. The continued success and dynamism of Silicon Valley is thus increasingly based on trans-Pacific flows of highly skilled migrants, particularly those from Taiwan, India, and China, but also in much lesser numbers from Vietnam, the Philippines, Japan, and Korea. These migrants contribute both directly to the economy, as engineers and entrepreneurs, and indirectly, as traders and middlemen linking California to the technologically advanced areas of Asia.

There has also been an ongoing globalization of Silicon Valley's ethnic networks. Many of the new firms set up by immigrant entrepreneurs are global actors from the outset, with the ability to use ethnic network connections to access Asian sources of capital, development skills, and markets. Such links are especially apparent between the technology communities of Silicon Valley and Taiwan's predominant technology region, Hsinchu (Saxenian, 2006). The key actors are a cohort of U.S.-educated Taiwanese engineers who have the experience and language skills to operate effectively in both regions. This community is constituted by three international migration flows. First, there is the continued flow of Taiwanese migrants to the United States, and Silicon Valley in particular, seeking educational and economic opportunities. Second, there are now increasing numbers of returnees to Taiwan, who have become an integral part of the technology industries of Hsinchu. Many immigrants have returned to Taiwan to start businesses there while keeping close links with former colleagues and classmates in Silicon Valley. Return migration grew appreciably in the 1980s at about 200 engineers and scientists per year on average and accelerated in the 1990s at 1000 per year in mid-decade. Third, there is a growing population of circulatory migrants or "astronauts" that work in *both* regions. With families based on either side of the Pacific – most commonly in California due to lifestyle factors – these engineers travel between the two areas as often as once or twice a month. They include Taiwanese "angel" investors and venture capitalists, as well as managers and engineers from firms with operations in both regions. The bridge between Silicon Valley and Hsinchu is thus constituted by a combination of permanent, temporary, and transient highly skilled international migrants. Similar examples exist in other locations, including India and mainland China, based on the same kinds of networks that bring together migration flows and business networks (Saxenian, 2006).

Temporary Migrant Workers

While IT workers and other professionals benefit from their relatively privileged positions, a far greater number of people are engaged in temporary labor migration, with much less stability and security in the places where they work. Temporary migrant workers, and the government programs that recruit them, have in recent years become a key component of the labor market in many settings.

Some economies are now effectively dependent on the recruitment of large numbers of foreign workers. In Singapore, the number of foreign workers grew from 612,200 or 29% of the total workforce in 2000 to 1.04 million or 34% of the total workforce in 2009 (MoM 2009). In South Africa, the workforce in farms and mines is drawn largely from neighboring states. In South Africa's gold mines, for example, 38% of the 267,894 workers in 2006 were foreign, taken from Botswana, Lesotho, Mozambique, and Swaziland (Crush and Williams, 2010). Perhaps the most extreme examples are found in oil-rich economies in the Middle

East where foreign workers make up almost the entire workforce in some cases. One estimate suggests that 99% of Dubai's private sector workforce comprises "immediately deportable non-citizens" (Davis, 2006). Even in countries that do accept permanent immigrants, such as Canada, temporary foreign workers are being increasingly used, for example, as agricultural labor, domestic childcare workers, or low-paid service sector employees.

The phenomenon of temporary foreign workers is important because such workers have fewer rights than citizens of the countries in which they are working. Such workers might face a threat of deportation, nonrenewal, or blacklisting from future contracts; visa conditions that permit work only for a specified employer; and other forms of surveillance and control. All of these leave workers with a quite different relationship to their employers than those with full rights of residency and citizenship. Ultimately the host government takes little responsibility for them. If they are unable to work, or no longer needed, then they are simply sent home. In Singapore, for example, when the country experienced a recession in 1998, many foreign workers were sent home, but unemployment among local residents reached only 3.5% (MoM, 2009). Migrant workers, therefore, provide a buffer for the resident population.

While in many cases temporary foreign workers are employed legally, examples of undocumented or illegal migrant workers are also common. In 2010, about 11 million people or 5% of the U.S. workforce was estimated to be unauthorized immigrants (Passel and Cohn, 2011). In Southeast Asia, cross-border migration in significant numbers is found in several settings. Large numbers of Lao migrants are employed in farming operations in the northeast of Thailand, while Burmese migrants are found in the northwest of the country. In East Malaysian plantations growing palm oil and rubber, large numbers of Filipino and Indonesian workers are found. In such circumstances, workers have even fewer rights as they are effectively outside of any protection that could be provided by labor laws.

The two contrasting examples, of transnational professionals and temporary foreign workers, both highlight the ways in which labor is increasingly mobile. This suggests a need for caution when characterizing capital as mobile and labor as stationary, as we did at the start of this chapter. However, mobility is clearly not always the same as empowerment. While mobile professionals and entrepreneurs often enjoy a privileged transnational lifestyle, the far more numerous legions of temporary foreign workers live with few protections in their employment relations and the constant threat of deportation.

6.6 Beyond Capital versus Labor: Toward Alternative Ways of Working?

So far, this chapter has examined the role of workers within the capitalist system. However, it is possible to think of a wide range of noncapitalist or *alternative* economies within which the nature of work might be different from the dominant capitalist system. As we saw earlier in the case of the Jamaican information-processing sector, economic geographers need to recognize the diverse range of ways of working that together sustain livelihoods in communities across the globe. In the context of workers, it is possible to identify two alternative kinds of employment spaces that are being carved out within contemporary capitalist economies (Leyshon et al., 2003):

- Alternative *formal* employment spaces: The emergence of "third-sector" community-owned and run enterprises – for example, health, youth, or arts centers – is intended to fill the gaps in welfare provision between the state and private sectors. Box 9.5 profiles a well-known example of a workers cooperative, the Mondragon Cooperative Corporation in Spain.
- Alternative *informal* employment spaces: These encompass a variety of informal nonmarket work activities, which may be paid work hidden from the state such as a builder paid cash-in-hand to evade taxes, unpaid work undertaken for one's own household such as childcare and cleaning, or unpaid work for another household such as work exchanged within family networks.

The importance of these alternative ways of working will clearly vary both socially and spatially. While in many developed societies they exist in addition to, and toward the margins of, mainstream capitalist activity, in developing world cities they may constitute the chief forms of daily economic exchange for large sections of the population. For example, estimates suggest that between 40% and 60% of total employment within the cities of Latin America, Africa, and Asia is informal in nature.

We should not assume that these other forms of economy are unproblematic alternatives that might simply replace capitalist labor relations if more people were engaged in them. Rather, the challenge for economic geographers is two-fold. First, it is necessary to explore geographical variations in the prevalence and significance of these activities, and the ways in which these ways of working connect to, and interact with, the dominant capitalist system. Second, we need to consider how such activities might offer a potentially important starting place for thinking about how capitalist labor relations might be resisted, challenged, and even changed, even if change is gradual and incremental. In reality, however, these initiatives quite often reveal as much about the difficulties of disconnecting from capitalist dynamics as they do about the potential for change. The Mondragon case in Box 6.5, for example, clearly reveals the problems inherent in combining a successful cooperative venture with the imperatives of the global marketplace. Equally, some third-sector schemes can be read as state-driven initiatives designed to conceal gaps in formal welfare provision. Alternative economic geographies are seemingly always open to incorporation by capitalist activity. Equally, they may simply supplement the mainstream while remaining only notionally separate from it, or may support it without really providing a challenge to its dominance. While alternative economies need to be read critically then, this in no way detracts

CASE STUDY

Box 6.5   The Mondragon Cooperative Corporation

The Mondragon Cooperative Corporation (MCC) was formed in the Basque area of Spain in 1955. Unhappy with the close relations between large businesses and the Spanish government in the aftermath of Spanish civil war, Father José María Arizmendiarrieta set-up the first cooperative in order to establish a degree of economic and political independence for the area. Since its creation the cooperative has expanded greatly, and now encompasses over 250 separate companies and bodies, divided into three industrial groups and a series of research and training centers (including a university). By 2010 the MCC employed about 85,000 people, with 39% of its workforce in the Basque region, 42% in the rest of Spain, and 19% abroad. It had assets of €33 billion and total sales of €14.8 billion. The cooperative accounts for about 3% of all economic activity in the Basque region. Underpinning the whole endeavor are principles of democratic participation in its business organization, the creation of jobs, the human and professional development of its workers, and the development of the social environment.

From its formation until the early 1990s, the MCC was able to use these principles to inform a model of growth that was widely lauded. Since then, however, it has gone through a period of expansion and restructuring driven by the logics of global competition that has challenged its founding principles, thereby illustrating the difficulty of instituting truly noncapitalist forms of working. Traditional labor-intensive production methods have been steadily replaced by the greater use of technology and offshore production and markets. In addition, the MCC has widened its growth strategy to include joint ventures, often with noncooperatives. For example, the cooperative now has 6,000 employees in China. By 2010, the MCC had 27 overseas plants and exports of €3.6 billion. As the cooperative has expanded overseas, only about 50% of its employees are now coop members, leading to a two-tier system of participation. Growing involvement in global markets has thus diluted the implementation of the cooperative's ideals. Managers now find themselves wrestling with depressingly familiar dilemmas, such as whether to keep production in the Basque country, where pay rates are higher, and risk job losses and firm closures, or establish joint ventures in regions where wages are low. For more see http://www.mondragon.mcc.es/ing/index.asp, and Ramesh (2011).

from the fact that there *are* different ways of living and working, and they present alternatives to the usual arrangements of formal waged labor.

6.7   Summary

We started this chapter by asking whether capital, in the form of employers, always has the upper hand over workers because of the looming threat to take jobs elsewhere. In other words, is the "threat to exit" always a potent weapon for mobile capital to control and discipline relatively immobile labor? We answered this question in three ways. First, we argued that the relative power and mobility of capital and labor are not natural outcomes of a capitalist system, but are shaped in a variety of ways by governments and other institutions that regulate labor markets. As a result, labor markets are quite uneven, varying between countries and even between subnational jurisdictions. More importantly, the implication is that local labor markets are socially regulated and therefore contestable. This means that there are opportunities for labor to shape collectively its relationship with capital.

Our second argument was that labor has proven capable of resisting actively the power of capital in many instances. This may involve individual workers, trade unions, or broader coalitions of social movements. Quite often, such acts of resistance take advantage of the geographies of employment. Finally, we examined how the portrayal of labor as immobile is limited, especially in an age where labor migration is increasingly common. In some cases such mobility is empowering, as high-tech workers have returned and become entrepreneurs in their home countries. But in other cases, temporary foreign workers are now found in the worst kinds of employment in many countries around the world. In the last section of the chapter we saw how waged labor represents just one of many forms of employment relationship.

Notes on the references

- Castree et al. (2004) provide a student-friendly and explicitly geographical approach to the study of labor.
- McGrath-Champ et al. (2010) and Bergene et al. (2010) provide collections of case studies in the field of labor geography.
- Herod (2001) charts the shift from "geographies of labor" to "labor geographies" and offers a range of compelling case studies of the latter.
- Examples of different ways of working are described and evaluated in Leyshon et al. (2003).

Sample essay questions

- In what ways are business- and state-imposed mechanisms of labor control interrelated?
- How and why do the mechanisms of labor control employed vary between different places?
- How can labor work with others in the local community to assert its agency?
- What are the main geographical strategies through which workers can assert their agency?
- What are the possibilities and limits of workers organizing themselves transnationally?

Resources for further learning

- http://www.ilo.org: The website of the International Labor Organization, a United Nations agency, has a wealth of information of labor and working conditions worldwide.
- http://www.icftu.org: The International Confederation of Free Trade Unions (ICFTU) is one of the most important international union federations (see also the World Federation of Trade Unions: http://www.wftucentral.org).
- http://labornotes.org: The home of Labor Notes, a regular report on labor activism worldwide.
- http://www.unglobalcompact.org: The website of the Global Compact, a UN initiative to bring together companies, labor, and civil society groups to promote universal principles in the areas of labor, human rights, the environment, and corruption.
- http://www.etui.org: The research arm of the European Trade Union Confederation (ETUC), which offers a range of reports and information on labor issues in Europe.
- http://www.labourstart.org: This is a trade union website that contains up-to-date stories about worker struggles worldwide.

References

Aguiar, L., and Ryan, S. (2009). The geographies of the Justice for Janitors. *Geoforum*, 40, 949–958.

Beaverstock, J., and Hall, S. (2012.) Competing for talent: global mobility, immigration and the City of London's labour market. *Cambridge Journal of Regions, Economy and Society*, 5, 271–288.

Bergene, A. C., Endresen, S. B., and Knutsen, H. M., eds. (2010). *Missing Links in Labour Geography*. Farnham, U.K.: Ashgate.

Castree, N., Coe, N., Ward, K., and Samers, M. (2004). *Spaces of Work: Global Capitalism and the Geographies of Labor*. London: Sage.

Chan, K. W. (2011). Internal labor migration in China: trends, geography and policies. In United Nations Population Division, *Population Distribution, Urbanization, Internal Migration and Development: An International Perspective*. New York: United Nations, pp. 81–102.

Coe, N., Jones, K., and Ward, K. (2010). The business of temporary staffing: a developing research agenda. *Geography Compass*, 4, 1055–1068.

Coe, N., and Jordhus-Lier, D. (2010). Constrained agency? Re-evaluating the geographies of labour. *Progress in Human Geography*, 35, 211–33.

Crush, J., and Williams, V. (2010). Labor migration trends and policies in Southern Africa. *SAMP Policy Brief*, 23, South African Migration Programme, Kingston, Ontario and Cape Town, South Africa.

Davis, M. (2006, Sept/Oct). Fear and money in Dubai. *New Left Review*, 41, 47–68.

Girard, E., and Bauder, H. (2007). Assimilation and exclusion of foreign trained engineers in Canada: inside a professional regulatory organization. *Antipode*, 39, 35–53.

Government of Ontario. (2007). *Ontario's Auto Industry*. Toronto: Government of Ontario, Ontario Investment and Trade Services.

Hale, A. (2005). Organizing and networking in support of garment workers: why we researched subcontracting chain. In A. Hale and J. Wills, eds., *Threads of Labour: Garment Industry Supply Chains from the Workers' Perspective*. Oxford: Wiley-Blackwell, pp. 40–68.

Herod, A. (2001). *Labor Geographies*. New York: Guilford.

ISS (Integrated Service Solutions). (2011). ISS Annual Report 2011. http://inv.issworld.com/annuals.cfm (accessed on April 10, 2012).

Jonas, A. E. G. (1996). Local labour control regimes: uneven development and the social regulation of production. *Regional Studies*, 30, 323–338.

Kelly, P. F. (2002). Spaces of labour control: comparative perspectives from Southeast Asia. *Transactions of the Institute of British Geographers*, 27, 395–411.

Leyshon, A., Lee, R., and Williams, C. C., eds. (2003). *Alternative Economic Spaces*. London: Sage.

McGrath-Champ, S., Herod, A., and Rainnie, A., eds. (2010). *Handbook of Employment and Society: Working Space*. London: Edward Elgar.

Merrifield, A. (2000). The urbanization of labor: living-wage activism in the American city. *Social Text* 18, 31–54.

MoM (Ministry of Manpower). (2009). *Report on the labor force in Singapore*. Singapore: Government of Singapore.

Moody, K. (1997). *Workers in a Lean World*. London: Verso.

Mullings, B. (1999). Sides of the same coin? Coping and resistance among Jamaican data-entry operators. *Annals of the American Geographers*, 89, 290–311.

Ong, A. (2000). Graduated sovereignty in South-East Asia. *Theory, Culture & Society*, 17, 55–75.

Passel, J., and Cohn, D. (2011). *Unauthorized Immigrant Population: National and State Trends, 2010*. Pew Hispanic Center. Available at http://pewhispanic.org/reports/report.php?ReportID=133.

Peck, J., and Theodore, N. (2007). Variegated capitalism. *Progress in Human Geography*, 31, 731–772.

Peck, J., and Theodore, N. (2008). Carceral Chicago: making the ex-offender employability crisis. *International Journal of Urban and Regional Research*, 32, 251–281.

Peck J., and Theodore, N. (2010). Labor markets from the bottom up. In S. McGrath-Champ, A. Herod, and A. Rainnie, eds., *Handbook of Employment and Society: Working Space*. Cheltenham: Edward Elgar, pp. 87–105.

Peck, J. (1996). *Work-place*. New York: Guilford Press.

Platzer, M. D., and Harrison, G. J. (2009). *The U.S. Automotive Industry: National and State Trends in Manufacturing Employment*. Federal Publications. Paper 666. http://digitalcommons.ilr.cornell.edu/key_workplace/666.

Ramesh, R. (2011, March 30). Basque country's thriving big society. *The Guardian*. http://www.guardian.co.uk/world/2011/mar/30/basque-country-big-society-spain.

Ross, S. (2011). Social unionism in hard times: union-community coalition politics in the CAW Windsor's Manufacturing Matters campaign. *Labour/Le Travail*, **68**, 79–116.

Rutherford, T., and Holmes, J. (2007). "We simply have to do that stuff for our survival": labour, firm innovation and cluster governance in the Canadian automotive parts industry. *Antipode*, **39**, 194–221.

Ryan, S., and Herod, A. (2006). Restructuring the architecture of state regulation in the Australian and Aotearoa/New Zealand cleaning industries and the growth of precarious employment. *Antipode*, **38**, 486–507.

Savage, L. (2006). Justice for janitors: scales of organizing and representing workers. *Antipode*, **38**, 645–66.

Saxenian, A. (1999). *Silicon Valley's new immigrant entrepreneurs*. Public Policy Institute of California, San Francisco, California.

Saxenian, A. (2006). *The New Argonauts: Regional Advantage in a Global Economy*. Cambridge: Harvard University Press.

Siemiatycki, E. (2012). Forced to concede: permanent restructuring and labour's place in the North American auto industry. *Antipode*, **44**, 453–473.

Stanford, J. (2010). The geography of auto globalization and the politics of auto bailouts. *Cambridge Journal of Regions, Economy and Society*, **3**, 383–405.

UN (United Nations). (2009). *International Migration 2009*. http://www.un.org/esa/population/publications/2009Migration_Chart/2009IttMig_chart.htm (accessed November 18, 2010.)

Wadhwa, V., Saxenian, A., Rissing, B., and Gereffi, G. (2007). *America's New Immigrant Entrepreneurs*. Master of Engineering Management Program, Duke University; School of Information, U.C. Berkeley.

Wills, J. (2001). Uneven geographies of capital and labour: the lessons of European works councils. *Antipode*, **33**, 484–509.

Wills, J. (2009). *The Business Case for the Living Wage: The Story of the Cleaning Service at Queen Mary, University of London*. Department of Geography, Queen Mary University of London. http://www.geog.qmul.ac.uk/docs/staff/8041.pdf.

CHAPTER 7

MAKING MONEY
Why has finance become so powerful?

Goals of this chapter

- To appreciate the changing role of banking in financing modern production
- To understand banking deregulation and the global reach of finance
- To demonstrate why geography matters in the globalization of financial markets
- To comprehend how financial capital has entered into everyday life through the processes of financialization

7.1 Introduction

On a snowy day in November 2006, the school board in Whitefish Bay, a village of about 13,508 people in Milwaukee County in the U.S. state of Wisconsin, made a major financial decision that it was to regret two years later. It followed the recommendation of a trusted local investment banker and borrowed substantial funds from a German-Irish bank named Depfa, at low but variable interest rates, to invest in high-yield financial instruments (see Figure 7.1). Together with boards from four other nearby districts, it invested some $200 million ($165 million of which was borrowed) in the deal with limited understanding of the financial instruments that we now know as collateralized debt obligations (CDOs; see Box 7.1 for a glossary of key financial market terms). According to *The New York Times* report on November 2, 2008, what seemed like a good plan for financing the teachers' retirement benefits with returns from this investment led to the complete annihilation of the board's investment. The $200 million was actually put into the purchase of three instruments managed and sold by the Royal

Wisconsin school boards and New York transportation officials relied on the same Dublin-based bank to help them invest pension funds, without paying close enough attention to the enormous risks involved. Here is an example of how the global financial crisis spread from the United States across the Atlantic and then back again.

①

In November 2006, five Wisconsin school districts borrowed from Ireland's Depfa Bank to invest in presumably high-grade bonds managed by the Royal Bank of Canada. The districts were hoping to expand their retirement fund.

②

The schools' money was actually used as insurance on $20 billion in corporate bonds – a promise to pay bondholders if corporations failed to pay their debts. The investment was registered in the Cayman Islands and managed by RBC, ACA and UBS.

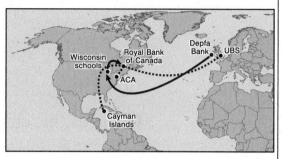

③

The investment went sour in early 2008, and the schools would likely not repay Depfa. That and other woes caused a crisis at the bank. Its parent in Germany, Hypo Real Estate, received a $75 billion bailout from the German government in October 2008.

④

Depfa had also guaranteed dozens of bonds issued by entities like the Metropolitan Transport Authority in New York. As Depfa's troubles escalated, those guarantees became more expensive to its clients such as state agencies.

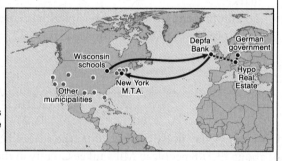

Figure 7.1 Global finance gone mad in 2007–2008, from Wisconsin and Toronto to Dublin, the Cayman Islands, and Germany

Source: Adapted from www.nytimes.com, www.npr.org, and online.wsi.com, on January 3, 2012.

KEY CONCEPT

Box 7.1   A glossary of common financial terms

Global financial services facilitate circulation for all parts of the economic system, but they are also commodities/services in themselves. In this global financial market, the following financial instruments and institutions are common:

Asset-based securities (ABS): These are securitized debts backed by the value of the underlying assets (e.g., properties) and repayment from these debts (e.g., mortgages). A particular form of ABS is mortgage-backed securities (MBS) that are issued as investment products to investors against the value of the income streams from mortgages.

Capital markets: These are markets for the selling and buying of equity securities (stock exchanges) or debt securities (bond markets).

Collateralized debt obligations (CDO): A specific type of security formed through the bundling together of different kinds of debts and loans to form structured financial products tradable in international markets. Buyers of CDOs are unlikely to be fully aware of the underlying debts and their associated risks.

Credit default swaps (CDS): Insurance contracts developed by lenders to manage the default risks that are associated with their lending (i.e., assets). These default risks are securitized and traded as CDS among financial institutions and other investors.

Derivatives: These are complex securities whose value is derived from the value of underlying "things" such as commodities (e.g., coffee or gold), assets (e.g., land and buildings), rates (e.g., interest and exchange rates), debts (e.g., bonds), indices (e.g., the Dow Jones or the FTSE), economic aggregates (e.g., GDP or inflation), and, more extremely, probabilities (e.g., likelihood of a snowstorm or terrorist attack).

Hedge funds: These are private funds that invest in a broad range of financial products and hedge their risks through investment in (1) shares, bonds and commodities that they own and (2) risky securities and derivatives that they borrow from a third party, with the obligation of returning the same securities on a mutually agreed future date.

Private equities: These are securities not publicly traded on stock exchanges and are issued by private firms that pool together capital from interested parties.

Securities: Financial instruments negotiable between the seller and the buyer whose value is determined by the market mechanism. They include

> equities (e.g., stocks and shares) and debts (e.g., bonds) issued by firms and state agencies.
>
> *Securitization*: A process through which debts and loans are packaged into, and sold via, financial securities at prices determined by their underlying risks (e.g., defaults) and returns (e.g., payments to these debts and loans).

Bank of Canada, which had business links with the local banker's employer based in St. Louis, USA. These instruments were a complicated form of insurance guaranteeing about $20 billion worth of corporate borrowing (bonds). The entire $200 million would be gone if just 6% of these borrowings defaulted. If there was no default, the school boards would receive about $1.8 million a year from the Royal Bank of Canada after paying off their own debt to Depfa.

These events took place in a period when all sorts of public, quasi-public, and private agencies were trying to cash in on the boom in global finance. Many were novice investors, blinded by promises from investment bankers of big profits and quick returns. When the crisis hit, the school board investors immediately became victims of this "pain from global gamble," a term coined by *The New York Times* report. In a matter of months, the entire $200 million investment was wiped out as a repayment to creditors of the corporate borrowings insured with the Royal Bank of Canada. Meanwhile, school boards in Whitefish Bay and four other districts told their Dublin-based creditor, Depfa, that they were no longer able to service their $165 million loan. While the lessons we should learn from this tragic story have both moral and financial dimensions, this chapter focuses on the geographical aspects of the extent to which money and finance make the world go around and literally connect all our fortunes. For instance, how has money become such a global commodity and how has this form of *financial capital* – investment funds originally meant for productive purposes such as building infrastructure and making goods and services – been turned upside down to the extent that it has become a "productive" force in its own right? Why has the U.S. dollar, a national form of currency, become a global currency tradable and deposited almost everywhere? What did the U.S. debt crisis in August 2011 have to do with debt defaults in the euro zone and the rise of East Asian economies, particularly China?

By unpacking the historical and geographical evolution of the global financial system, this chapter aims to draw attention to both the broader structural processes and the varied everyday geographies of global finance, known as "financial capital" throughout this chapter. These geographies help us appreciate better the diverse connections laid bare by the global financial crisis that developed in 2007, unfolded fully in 2008, and continued in different forms into the 2010s. By tracing capital's global reach, we can understand more fully the complex linkages in a globalizing financial sector that bring together the changing fortunes of school boards, investment banks, regulators, and even national governments. Our choice of global finance is a deliberate one because it offers unique economic-geographical insights into the inner workings of contemporary capitalism described in Chapter 3.

Next, Section 7.2 will examine the common understanding of the rise of "place-less capital" and the end of geography, showing how this popular understanding of global finance underestimates the geographical linkages and variability inherent in financial systems. Section 7.3 will focus on the role of finance as a productive input into economic processes. Here we will discuss the varied geographies of different national financial systems and their integration with industrial production. Section 7.4 will consider the financial sector as an industry in its own right, producing financial products such as pension funds, unit trusts, and derivatives. We will also focus on the dominant places within the financial sector, namely global cities and offshore financial centers. Finally, Section 7.5 will show how the global economy is becoming increasingly "financialized," with the circulation of finance increasingly being divorced from its productive allocation and use.

7.2   Is Global Finance Placeless?

Since the late 1980s, popular writings on finance and the future of financial markets have been full of rosy predictions of seamless global financial integration and uninterrupted flows of global capital across different national jurisdictions. For example, the economist Richard O'Brien opened his famous 1992 book *Global Financial Integration: The End of Geography* with the following:

> The end of geography, as a concept applied to international finan-
> cial relationships, refers to a state of economic development where
> geographical location no longer matters in finance, or matters much
> less than hitherto. In this state, *financial market regulators* no longer
> hold full sway over their regulatory territory … For *financial firms*,
> this means that the choice of geographical location can be greatly
> widened provided that an appropriate investment in information and
> computer systems is made… *Stock exchanges* can no longer expect
> to monopolize trading in the shares of companies in their country
> or region, nor can trading be confined to specific cities of exchanges.
> Stock markets are now increasingly based on computer and telephone
> networks, not on trading floors. Indeed, markets almost have no fixed
> abode. For *the consumer of financial services*, the end of geography
> means that a wider range of services will be offered, outside the
> traditional services offered by local banks … Money, being fungible,
> will continue to try to avoid, and will largely succeed in escaping, the
> confines of the existing geography (pp. 1–2; emphasis in original).

This passage is symptomatic of the "irrational exuberance" – the term used by former U.S. Federal Reserve Chairman Alan Greenspan in December 1996 – surrounding global financial integration that prevailed up to the 2008 financial crisis, and which saw the constraints of geography being inevitably overcome by technological innovations and other forms of societal change. O'Brien's vision suggested that financial firms only had to invest in computer technology to open up a wide variety of potential locations. Equally, stock exchanges would become far less tied to particular cities, and consumers would be presented with a proliferation of financial products and services. In short, finance would become increasingly *placeless*. This observation is not at all surprising if we take into account the common perception of finance's hypermobility.

Thinking about today's global financial system, we can see merit in some elements of O'Brien's account. The workings of the financial industry have indeed been profoundly transformed by the impacts of technologies such as computers, networks, satellites, and even smartphones. Money has become largely electronic, replacing tangible forms such as coins, notes, checks, and credit notes. Travelers can withdraw local currencies from ATMs in most destinations through their VISA PLUS/Electron cards. Smartphones can be used for financial transactions in lieu of physical payments. As a result, money can now be transferred almost instantaneously at a global scale. Overlapping financial markets in different time zones allow financial firms to engage in 24-hour trading. These connections between different markets at the international scale have enabled financial institutions to increase their international lending activities and to respond extremely rapidly to exchange rate changes in international currency markets. Finally, new technology has undoubtedly altered how financial firms interact with their clients. For example, as individuals we increasingly interact with our banks through internet or phone banking rather than bricks-and-mortar branches, and we can use the search capability of the internet to compare and contrast different financial products and services (see, for example, www.gocompare.com, accessed on January 3, 2012). All sorts of credit facilities are offered to us through many different distribution channels, including commercial television advertisements, with the result that access to finance seems to be widely available.

This "end of geography" conception of finance, as an essentially top-down and hypermobile phenomenon, has much to do with the almost exclusive focus of academics and the media on global finance. Capital markets define everything we see happening in the financial realm, and finance seems to come out of nowhere! Its "real" socioeconomic role in our everyday routines seems to be entirely divorced from these global accounts. And yet, the core geographical element of O'Brien's argument, the end of geography, has clearly not come to pass. The overall geography of the global financial industry is curiously resistant to change. The vast majority of financial assets today remain confined to investments in, and institutions of, their home jurisdictions. In other words, the "home bias" effect is very strong even in global finance.

Moreover, a small number of leading cities – most notably London, New York, Hong Kong, Singapore, Shanghai, Tokyo, Chicago, Zurich, and San Francisco – continue to control the lion's share of the international financial transactions in the global economy. In the March 2011 issue of the *Global Financial Centres Index*, these cities were ranked as the top nine global financial centers (www.longfinance.net, accessed on January 3, 2012). The top three, London, New York, and Hong Kong, accounted for approximately 70% of equity trading. The *Index* also noted that these three "are likely to remain powerful financial centres for the foreseeable future [and] the relationships between London, New York and Hong Kong are mutually supportive" (p. 6).

In 2010, London, New York, and Chicago accounted for a combined 50% share of the world's $4 trillion global foreign exchange trading and 70% of all foreign exchange derivatives transactions undertaken in the United States and the European Union (www.dbresearch.com, accessed on January 3, 2012). As the world's leading financial center and home to 240 foreign banks, London alone accounted for 37% of the world's daily foreign exchange trading and 50% of Europe's investment banking, managed four-fifth of the assets held by Europe's hedge funds, employed 591,789 people in financial and professional services, and produced a gross value added of £68.7 billion for the U.K. economy (www.thecityuk.com, accessed on January 3, 2012). Indeed by some measures, the concentration of control in these leading centers has actually increased at the same time as the use of information technologies has been expanding rapidly.

How can we explain this apparent paradox of accessibility to finance becoming more global and yet financial centers in specific places consolidating their dominance? Why has the end of geography forecast by O'Brien not become more apparent in reality? What has happened to money and finance as sources of capital in modern capitalism? What has changed since the late 20th century, when we saw the transformation of *money* from an everyday tool into *global finance*? To understand these global shifts, we need to start in the next section by looking at the origins of modern money and banking.

7.3 Financing Production: The Evolution of Banking

This section focuses on the role of finance as a productive input into economic processes. In particular, we examine the role of the modern banking system as a financial intermediary between savers (creditors) and investors (debtors). In general, banks pool together our money/savings as capital and use that capital to finance productive activity such as commodity production, the manufacturing of goods, and the provision of services. This mode of money and finance operated successfully long before the emergence of global finance and its associated capital markets, securitization, private equities, hedge funds, and the like (see Box 7.1).

We will explore these latter developments in the next section. Two important issues of ownership and accountability arise for our consideration in this section: who owns money and how does it get transformed into working capital, and what/who accounts for this transformation? These questions clearly raise issues about the *agency* of finance. To understand these twin issues, we need to probe into the modern financial system by first asking what does finance really do in relation to the banking system? We then examine the geographical evolution of national financial systems in advanced industrialized economies, focusing in particular on the United Kingdom and the United States. Finally, we showcase some relatively recent changes in the organization of the financial sector to pave the way for our analysis of the rise of global finance in Section 7.4.

The Role of Finance in Production

In modern times, most people think of finance as "things that banks do." The real question here then is what might these "things" be? In theory, the banking system is about the pooling of financial resources among those with surplus funds (savers) to be lent out to those in need of such funds (borrowers) for investment in productive activity. Banks lend a lot more than they have in deposits and this is called a reserve ratio. In this saving–investment relationship, a creditor receives compensation in the form of interest on his/her deposit held temporarily by the bank and a borrower pays a cost for using the funds pooled from many creditors depositing with the same bank. In general, the interest payable by the borrower is significantly higher than the interest received by the creditor. The "middle agent" or intermediary – the bank – will retain the difference in saving–lending interest rates as compensation for its service of rendering financial intermediation and taking on the risk of lending. In this system of exchange of funds, all three parties – the depositor, the bank, and the borrower – receive returns in relation to their inputs into the overall capitalist system. Geographically, these three parties need not be in the same area/region for such exchange to take place. This unique form of financial geography allows for capitalism to grow in different regions and countries in tandem with the spatial reach of the banking system. In short, modern banking has oiled the wheels of capitalism and influenced its structural outcomes (see Chapter 3).

This possibility for exchanging funds through borrowing and lending via the banking system has enabled two forms of finance: consumer and corporate. In *consumer finance*, credit is extended to individuals for their acquisition of tangible assets (e.g., mortgages) or services (e.g., education loans). More will be said about consumer banking and finance in Section 7.5. In *corporate finance*, investment in existing or new productive activities and places becomes viable and the economy continues to grow. Entrepreneurs with innovative ideas in different places can turn these ideas into real products and/or services, even if they and their local social ties cannot generate sufficient funds of their own to do so (see also the Facebook example in Chapter 1). They can borrow working capital from banks and put these funds into their productive activity in the hope of making sufficient profits in future that exceed the repayment of these loans and interest charges. This productive activity includes everything from agriculture (e.g., acquiring land and farming equipment) and manufacturing (e.g., building factories and purchasing inputs and equipment) to services (e.g., IT innovations, retail and wholesale businesses, or nonspeculative property development).

In short, capitalists can borrow financial capital to fund their profit-making productive activities. The banking system will allocate financing to those economic endeavors with the highest rate of return and therefore to the most efficient uses of financial capital. It will also reallocate financial capital from one part of the economy to another, prompting the emergence of new sectors, regions, and economies (see Chapter 3). The greater the efficiency of this capital exchange system via banks and other related financial intermediaries, the higher the productivity gains to the overall economy will be, as surplus funds will be deployed rationally – hypothetically at least – by economic agents in the economy.

In this capitalist system, money serves primarily as a means of payment, a unit of account, and a store of value. A critical transformation occurred when the importance of commodity money (e.g., notes and coins) decreased with the rise of *money of account* associated with the development of modern banking systems. From a means of payment (commodity money), money has now become a unit of value that can be stored, transmitted, exchanged, and even traded through an evolving banking system in contemporary capitalism. In other words, money can be turned into financial capital, through the modern banking system, to fund production. The vast majority of monetary payments do not take place with physical notes and coins but, rather, via the financial system. For example, by 2011, the total stock of notes and coin in the United Kingdom was only £58 billion, whereas the total stock of money was £2.2 trillion. This means that more than 97% of all the money in the economy is not in the form of notes and coins (www.thecityuk.com, accessed on January 4, 2012). The next two subsections trace the geography of these developments in banking and money in the 20th century and its reorganization during the late 20th century.

The Changing Geography of National Banking Systems

Banks are not the same everywhere because each banking system develops according to different national and institutional contexts. Several key factors stand out in shaping the spatial evolution of the financial system (Dow, 1999, pp. 40–42):

- *Rationale*: Some financial systems were developed through private initiatives such as in Scotland or through state financial requirements such as in England. Others were a part of the federal policy of maintaining regional balance among different states such as in the United States.

- *The presence of significant stocks of wealth*: Some international financial centers such as London, New York, and Zurich have benefited from the long-standing residence of wealthy people and institutions that in turn have sustained their preeminence in global finance.
- *The role of the state*: Some state institutions had to develop their financial systems in order to raise finance for wars (e.g., the Bank of England) or to overcome political constraints on growth (e.g., the U.S.). In the U.S. federal system, there were inherent limits to the out-of-state expansion of city-based banks such as those in New York. A prime case is Citibank, which needed to grow beyond the city of New York (described in detail in Box 7.2).

CASE STUDY

Box 7.2 *Why did Citi expand beyond the United States?*

During much of the 20th century the U.S. banking system placed significant limits on the growth of its domestic banks. Some of these banks responded through the rapid internationalization of their banking operations. One prime example here is Citibank. Founded in 1812 as the City Bank of New York and already one of the largest U.S. banks by 1868, it could not grow further within the United States due to the tightly controlled intrastate and interstate bank branching regulations. The limited domestic access to the full banking services market compelled Citibank to internationalize much earlier than most global banks as we know them today. Its international footprint started as early as 1897 when it became the first major U.S. bank to establish a foreign department. Its first overseas office was established in Buenos Aires, Argentina, in 1914 and some of its offices in London, Shanghai and Calcutta were opened earlier at the turn of the 20th century by one of its wholly-owned subsidiaries. By 1930, the bank had 100 branches in 23 countries outside the United States. To overcome further federal regulatory constraints in the postwar era, Citibank came up with an innovative ownership solution in the late 1960s. In 1976, the bank renamed itself Citicorp, which was not a bank under the U.S. law. This nonbank ultimate ownership enabled Citibank to operate as a credit card bank in all 50 U.S. states. In 1998, it merged with the insurance giant Travelers Group to become Citigroup Inc. To date, it remains the world's largest financial services network, spanning more than 160 countries with approximately 16,000 offices, 260,000 staff, and over 200 million customer accounts around the world (www.citigroup.com, accessed on January 4, 2012). Internationalization allows formerly national banks to diversify into new products and markets and to outsource business functions in order to achieve cost and market competitiveness.

As a critical input into production and consumption, financial capital and its providers – banks – were intended to serve their communities of customers. The different mix of the above three factors led to the development of different national financial systems with varied geographies; some were characterized by a regional and bank-oriented system, others by a national and market-oriented system. We can compare, for example, the historical evolution of the national banking systems in the United Kingdom and the United States (Martin, 1999). In the United Kingdom, London did not emerge as the national center of finance until the early 20th century. Prior to that, a complex system of local and county banking developed in tandem with industrialization during the 18th and 19th centuries. In short, the geography of banking closely aligned with the geography of the Industrial Revolution. This regional and bank-oriented system evolved alongside the growth of commercial banks in London and reached its peak in about 1810 when there were more than 900 of these small banks in the United Kingdom. With the arrival of shareholder banking, the growth of the British Empire, and subsequent waves of mergers and acquisitions, the U.K. banking system became increasingly concentrated and centralized in the hands of national banks, most of which were headquartered in London. By 1914, there were only 66 independent banks in the United Kingdom, 20 of which had more than 100 branches. Large national banks such as the Midland (now acquired by HSBC), Lloyds, and Barclays were headquartered in London and each had over 500 branches.

By contrast, the U.S. banking system did not become nationally oriented and remained local and state-based for a long period of time. It was thus very fragmented and decentralized. The westward development of the country during the 19th century encouraged the growth of local banks and related institutions that capitalized on the relatively large distances between different cities and their isolation from New York, the older and most established commercial and financial center of the United States. The development of regulation by each individual state in the first half of the 20th century further exacerbated this trend in the United States toward a regionally-based banking system. Various federal and state banking laws (e.g., the McFadden Act of 1927 and the Glass-Steagall Act of 1933) controlled tightly intrastate bank branching and strictly prohibited banking institutions from crossing over to other states in terms of establishing branches or owning subsidiaries. This control has only been partially relaxed since the mid-1970s. Even in the 1990s, the establishment of new offices within a U.S. state by an out-of-state bank was still largely prohibited.

A similarly decentralized and regionally/state-based system could also be found in certain European countries such as Germany, France, and Italy. For example, the Italian banking system up until the mid-1980s was also characterized by the presence of many small local banks operating in restricted territorial areas and a few national banks, an outcome of a very restrictive regulatory regime in terms of the geographical mobility of banks and their operating areas. There was a strong coupling of local banks with small- and medium-sized firms in Italy's famous industrial districts (see Chapter 12). These European financial systems are thus best understood as decentralized systems whose regulation was not well integrated at the national level. They were driven primarily by bottom-up institutions such as local or regional banks.

These varied geographies of banking systems have important implications for understanding the spatially differentiated role of finance in supporting production and consumption. To summarize, a local and regional banking system has three implications for the local community:

- It is more likely to serve the local economy and business community.
- There is more creation, retention, and capture of locally originated capital.
- The system is less vulnerable to external financial crisis, but more susceptible to changing local economic conditions.

In contrast, a nationally-oriented banking system is often less forgiving when it comes to serving some local communities because of:

- More cautious and restrictive lending practices in less prosperous local economies and business communities.
- A greater likelihood for operations in poorer areas to be cross-subsidized during upswings in business cycles.
- A greater likelihood of rationalization of operations in economically depressed areas during downswings in the business cycle, thus effectively increasing the financial exclusion of communities in these areas.

Reorganization of the National Banking Sector

Since the early 1970s and especially the late 1980s, national deregulation, the IT revolution, and the increasing global reach of finance have led to significant reorganization within many banking sectors. One dimension of this reorganization merits particular attention because it presages the rise of finance as *the* dominant force in global capitalism. Massive consolidation of banks has occurred on the back of the post-Fordist global economy in which service sectors have come to the forefront of generating economic value and wealth (see also Chapter 9). Many national banking systems have undergone major consolidation since the 1980s in the belief that while size does not necessarily lead to economies of scale, it will likely help in other respects such as hedging between markets, influencing state policy, and making the consolidated bank less vulnerable to financial market downturns. This wave of banking consolidation has profoundly reshaped the existing geography of banking and finance through the closure of bank branches and the rise of financial exclusion.

Partly because of its nationally oriented banking system, banking consolidation was particularly pronounced in the United Kingdom. It is estimated that between 1986 and 1996, for instance, bank branches decreased by one-quarter from some 14,008 to 10,334 (Martin, 1999). Over this period of reorganization, large British banks became highly profitable. Midland Bank's profit soared from £434 million to £1.2 billion on the back of closing over 500 branches. Its rival Barclays Bank doubled its profit to £2.3 billion while closing over 800 branches. In the United States, a similar process of branch closing by savings associations occurred during the same period. Low-income neighbourhoods in Los Angeles, Chicago, New York, Boston, and other U.S. cities witnessed a pattern of selective branch closures. In Los Angeles, three of the four largest retail banks operated a hub-and-spoke system through which designated branches specialized in certain kinds of debt and investment products and took referrals from other branches. Since then, bank branches have functioned more as sales outlets than as fully fledged transactions-handling operations offering the entire spectrum of financial services. A great deal of customer segmentation has occurred in relation to this reorganization. In the Beverly Hills branch, the Bank of America offers various "private banking facilities" to service their high-net-worth customers (Pollard, 1999). In San Francisco, the bank extends its service to Chinatown in order to tap into increasing affluent Asian migrants.

At the national level, this trend toward banking consolidation in the United States has continued unabated during the 2000s. Figure 7.2 shows that the market share of the top four American banks in terms of U.S. deposits rose rapidly from about 16% in 1998 to almost 37% in 2009. Many of these large banks in the United States and the United Kingdom were seriously burdened by the 2008 financial crisis and further consolidation occurred. In both countries, the consolidation process is often very spatially selective, leading to the exclusion

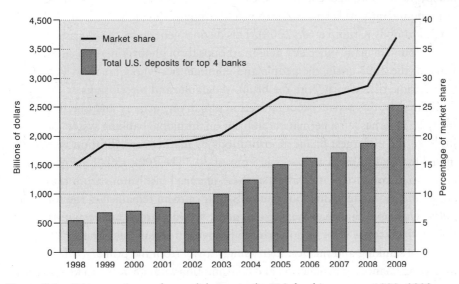

Figure 7.2 Concentration and consolidation in the U.S. banking sector, 1998–2009
Source: Data from www.fdic.gov, accessed on January 3, 2012.

of low-income inner-city areas and rural communities from the mainstream bank-led financial system. The simultaneous availability of a large number of vacant properties and overcrowded high-density housing in American cities, for example, can be explained by the unwillingness of these banks to finance "high-risk" areas for (re)development and mortgage purchase. The role of financial and related institutions in shaping place development is thus very apparent through their discriminatory lending and credit allocation decisions that in turn can reinforce existing class and ethnic segregation patterns in the city. Known as "financial exclusion," this highly spatially differentiated access to financial services by different social groups and its resultant impact on the production of urban poverty has occurred in both developed countries and developing countries alike.

In the period leading up to the global financial meltdown that started in 2008, this exclusionary practice of "redlining" by mainstream lenders (i.e., the major banks) was replaced by highly differentiated and racialized patterns of so-called subprime lending by independent mortgage companies that were encouraged by the broader deregulation of financial markets, the remarkable glut of global capital chasing after borrowers, very lax underwriting standards, and aggressive or even predatory business practices (see more in Section 7.5). In this new wave of stratified and unequal "greenlining" by subprime lenders from the late 1990s onward, low-income and racially marginalized neighborhoods were offered mortgage-related financial products in the name of equity extraction and quick profits. Studies have shown that 44% of African-Americans, as compared to 30% of Caucasian-Americans, borrowed from less heavily regulated subprime lenders such as independent mortgage companies. Subprime loans are five times more likely in predominantly African-American neighborhoods than in Caucasian-American neighborhoods (Crump et al., 2008). This distinct geography of subprime lending can be found in many inner-city neighborhoods throughout American cities. In the wake of the 2008 crisis, the people and communities who were most vulnerable to subprime risks clearly bore the brunt of defaults and foreclosures.

Ultimately, this broader reorganization of banking activity and functions since the late 1980s has left a permanent imprint on the global banking system of today. In 1999, the 20 largest financial institutions by market capitalization were dominated by Anglo-American banks (15 out of 20). The 2008 global financial meltdown so disproportionately affected these financial institutions from the United States and the United Kingdom that only four of them remained in the top 20 by 2009. This rapid change in the fortunes of the world's largest financial institutions (mostly banks) has a lot to do with the fundamental shift since the 1990s in the nature of banking and finance, increasingly away from the primary role of intermediating finance for production and consumption to a new role as a key platform for investment services and financial profits. The next section will examine the origin and key elements of this transition in banking and finance in late capitalism.

7.4   The Rise of Global Finance

As the global financial system evolves over time and space, we find banks and other financial institutions increasingly keeping their financial resources within the financial system rather than financing investment in production and related economic activity. This process of "churning" occurs when credit creation is not driven primarily by the need to finance the "real" economy, but rather by perceived or actual gains from investment in the kinds of financial instruments described in Box 7.1. In other words, banks have now been turned into institutions of financial investment – a kind of value-added service increasingly overshadowing their traditional business of handling savings and lending. Since the 1970s, banks have turned their attention to credit extension that goes well beyond the financial needs of borrowers in the productive sector (i.e., what we are terming here the "real" economy). Banks can now gain more from lending to investors of all sorts, including those who speculate on various financial instruments. More particularly, banks are actively engaged in *securitization* by issuing securities with underlying asset values rather than just engaging in direct lending to borrowers. Many commercial banks have also set up investment banking arms to profit from the rapidly escalating possibilities of the financial securities and derivatives available in capital markets. As a result, banks are no longer merely financial intermediaries, but are also active investment agents in their own right.

Meanwhile, the financial services industry has broadened to include not just banks and bank-related institutions. Importantly, financial services are now provided by a wide range of non-bank institutions such as securities and commodities trading firms, fund management companies, and hedge fund firms. These non-bank institutions pool vast sums of financial capital in the form of pension funds, life insurance, mutual funds, unit trusts, and private equities for investment at the individual customer level and, more frequently, at the institutional level. These non-bank investments can be put into conventional bricks-and-mortar tangible assets such as properties and businesses. Increasingly, however, these institutional investors are engaged in securities that are issued on stocks, shares, currencies, and commodities, and their derivative financial instruments such as swaps, options, futures, and collateralized debt obligations. Since the 1980s, this rapid proliferation of financial services and investment instruments worldwide has led to the emergence of what we now know as global finance. It is today possible for an individual or institutional investor to invest, via the above securities and related financial instruments, in virtually any asset, real or otherwise, located in territories that are connected into the global financial market.

Through this phenomenon of global securitization and financial trading, the geography of finance is no longer confined to localities; it has become much more global in scale and scope with two profound consequences (Martin, 2011). First, local financial transactions and assets are increasingly plugged in to global financial markets. This creates a much greater degree of dependence for local financial outcomes on the behavior and decisions of actors and institutions located elsewhere than ever existed before. Second, events within these global financial markets have much greater intended and unintended impacts on the conditions and processes at work in local financial systems in particular places. As clearly evident in the global financial meltdown that unfolded in 2008, local housing and mortgage bubbles were deeply linked into, and destabilized, global financial markets with serious consequences for local people and communities.

This shift from finance for production to global finance within two to three decades merits further discussion. In particular, the emergence of global finance is closely associated with rapidly escalating global capital flows during the past two decades. Since the mid-1980s, global capital flows as a percentage of world GDP have grown rapidly. From a low of 3% in 1983, these flows peaked at almost 14% of world GDP in 2004. As shown in Figure 7.3, the role of banks and private flows has decreased during the past three decades. Instead, portfolio flows of equities and debts have become the largest source of global capital flows into advanced economies, underscoring the critical importance of financial securities traded in capital markets. The pattern of cross-border financial holdings also indicates much intensified linkages in the financial sector across different countries and regions. By 2000, finance had arguably become truly global. As such, it was no longer tied to the productive needs of domestic economies as outlined in

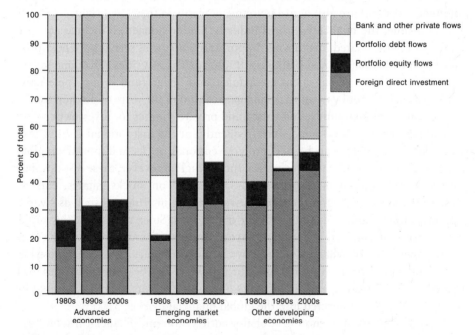

Figure 7.3   Relative importance of different types of global capital inflows since the 1980s (percent of total)
Source: Data from the World Economic Outlook database on www.imf.org, accessed on January 3, 2012.

the previous section. In the next two subsections, we explain the rise of global finance in relation to wider conditions such as changing regulatory frameworks and interplace competition.

Deregulation as a Driver of Global Finance

Prior to the 1960s, a global financial system did not really exist as such, as national banking systems were closely regulated and largely self-contained within specific territories. Governments sought to maintain autonomy over monetary policy, control their exchange rates, and closely limit foreign investment (and ownership). Hence financial capital did not flow freely across borders, and global financial institutions did not really exist. There was a system in place for trade, but financial systems were primarily national and relationships between them were governed by supranational organizations such as the Bank for International Settlements (BIS) and the International Monetary Fund (IMF) (see Chapter 4). These international institutions served as the lenders of last resort for central banks in different nation-states; they were the ultimate guarantors of nationally-oriented banking and finance systems. A limited form of internationalization of finance came about when national banks initially expanded activities abroad in order to service transnational corporations producing overseas from the 1960s onward. In other words, financial capital started to follow the internationalization of production, but it remained tightly regulated by home governments.

This national system of financial regulation based on such international organizations as the BIS and the IMF described in Chapter 4 began to break down in the early 1970s due to intense competition between different financial centers and their embedded regulatory frameworks: "from the late 1950s to the early 1970s financial globalisation was stimulated by the uneven geography of market regulation, and competition between the world's major financial centres was largely shaped by the regulatory contours between them" (Martin, 1994, p. 263). As shown in Table 7.1, the regulatory context for the international financial system has evolved through several stages over the last 150 years. The first major challenge to the post-war nationally-based system of financial regulation was the development of "stateless monies" through the eurodollar markets. These eurodollar markets refer to U.S. dollars held outside the United States and therefore outside of U.S. regulatory control. In the late 1950s and the early 1960s, the eurodollar markets were particularly appealing to the Soviet Union and China. Because of the Cold War, these two large communist states were unable to obtain U.S. dollars through trade with the United States. The availability of U.S. dollars through the eurodollar markets was vital in resolving their need to buy goods and services in international markets.

London was the major center of these eurodollar markets up until the 1970s when the British government tightened its regulation of foreign banks operating eurodollar deposits and businesses. Subsequently, these banks moved their

Table 7.1 The changing regimes of financial regulation in the global economy

Period	Regime of Regulation	Characteristics
1870s–1920s	International Gold Standard	Value of a national currency guaranteed by the amount of gold held by the country's central bank.
1945–1973	The Bretton Woods system	Central banks of countries other than the U.S. maintained fixed exchange rates between their currencies and the U.S. dollar. This was the first international monetary order established under the auspice of the U.S.
Mid-1970s to mid-1980s	Successive international deregulation	Collapse of the Bretton Woods system. New technologies, financial innovation, and the need for risk-managing financial instruments created the environment in which the New York Stock Exchange was deregulated in 1975 and successive British governments implemented deregulation from 1978 onward.
Late 1980s to late 2000s	New re-regulation and monetary integration	Establishment of the Basel Banking Accord: a set of minimal capital requirements for banks to ensure liquidity and stability in the international financial system. Increased systemic risk from exposure to derivatives and fallout from associated scandals, as well as the East Asian crisis, leading to demands for global re-regulation. Completion of European monetary integration through the creation of a single currency (euro) and a common monetary policy space among the member states of the European Union (European Central Bank).
2008 onward	Re-establishing the global financial order?	Massive systemic risks led to the collapse of collateralized debt obligations (CDOs) and other complex financial derivatives and instruments. Nationalization of major banks in many developed countries, including the U.S., U.K., and Germany. Major global imbalances in debts and bond-holding led to potential currency wars and sovereign debt crises.

Source: Adapted from Budd (1999, Table 1, p. 119).

eurodollar operations to other less-regulated tax havens such as Hong Kong, Singapore, the Bahamas, and the Cayman Islands. This emergence of an offshore currency market essentially meant a global, rather than *inter*national, financial system could emerge. During the early 1970s, massive oil price increases saw the huge accumulation of wealth in oil-rich states and the recirculation of so-called petrodollars globally, largely through European banks in the eurodollar markets. Again, money disentangled from its national origins became a big part of the global economy, culminating in serious challenges to the U.S.-dollar-based Bretton Woods system of international financial system. The existence of these eurodollar markets meant that there was already some de facto deregulation in the international financial system.

Table 7.1 illustrates how, since the mid-1970s, other factors created pressures that ultimately forced regulators to relax controls in response to processes that were increasingly beyond their control anyway. Competitive international deregulation became the dominant trend, with financial centers and authorities mimicking each other in lowering their control of, and constraints on, financial sector activity. In particular, significant changes occurred in this period to allow foreign banks to operate in the United States and facilitated U.S. banks expanding overseas. The "Big Bang" of 1986 in the United Kingdom also allowed British banks to do what other institutions had previously done (e.g., trade in securities and engage in investment banking) and allowed foreign banks to operate in the United Kingdom. Similar kinds of "follow-thy-neighbor" changes occurred elsewhere driven by the logic that domestic financial sectors would otherwise lose business to rival financial centers. This deregulation accelerated rapidly from the 1980s onward:

> During the course of the 1980s and 1990s, a tidal wave of deregulation (and re-regulation) swept across the globe, beginning in the United States and the U.K. but quickly spreading to other developed and developing countries. Financial markets were dramatically redrawn as nations became locked in a process of 'competitive deregulation', in a 'race to the bottom' to free money and finance from the regulatory structures built up during the post-war decades (Martin, 1999, pp. 12–13).

To summarize, the role of the state in facilitating this process of deregulation of global capital was critical. In addition, the state in the United States and the United Kingdom was experiencing an ideological change from a welfare regime to a neoliberal market regime (see Chapter 4). The rise of the so-called Reaganomics and Thatcherism during the 1980s led to the massive privatization of public goods, from transport to communications to utilities. This major liberalization and marketization of public utilities and services resulted in the widespread securitization and public offering of assets formerly belonging to these government institutions and enterprises. These securities were subsequently sold to national financial institutions such as pension funds and other institutional investors. The wealth and assets of citizens of these countries became more entangled with these financial market instruments. The deregulation of financial markets was a necessary step in this move toward a neoliberal market regime of economic governance. However, as finance increasingly became freed from its national context, did its global reach entail the demise of existing financial centers? Interestingly, the next subsection shows that these financial centers have not only remained dominant in an era of global finance, but also that their intensified competition with each other plays a crucial role in driving the globalization of finance.

Putting Global Finance in Its Place

In global finance, the leading international financial centers are almost always global cities in their own right (see Box 7.3). Finance continues to be highly concentrated in these global cities despite deregulation on a global scale. These financial centers have a peculiar mix of actors (e.g., bankers and financiers), institutions (e.g., capital markets), and supporting services (e.g., accounting and legal firms) that facilitate their central role in spearheading global finance. In other words, these global cities are the nerve centers of global finance; they compete fiercely against each other for finance and investment. Finance involves a large number of individuals and institutions whose spatial orientations are more often than not local. They need places in which they reside, locate, accumulate capital, and make savings or investment decisions. In reality, financial services are among the most spatially concentrated of all economic activities. While the evolution of the banking industry described in Section 7.3 has made physical proximity to borrowers, customers, and clients ever less necessary, locating near other financial and business service institutions has become more important because of patterns of geographical agglomeration in the financial sector.

As we will see in more detail in Chapter 10, within the production system for a particular product or service, there are some operations, such as corporate decision making, which need to be located in agglomerations, while other activities such as production, sales and marketing, and customer support can be more spatially disparate. In other words, the spatial decentralization of economic activity facilitated by space-shrinking technologies requires the very kinds of control and coordination that can only be achieved through centralizing certain important corporate functions (e.g., finance) in key global cities (see Figure 7.4). More specifically, this dispersion creates different challenges of integration and coordination for firms. Information has to be acquired about what is going on across the production system as a whole, and social interaction is required to allow important interpersonal relations such as trust to develop. Financial centers, then, should not be considered on their own and purely in terms of the local relationships they contain, but as nodes in global networks, where certain kinds of local relationships are used to facilitate and support other kinds of nonlocal (and possibly global) relationships. In the terms explained in greater detail in Chapter 12, there are lots of traded and untraded interdependencies within these

KEY CONCEPT

Box 7.3   Global cities

There have been many attempts over the last 20 years to try and capture the essence of the global urban hierarchy and its interactions with the changing geography of the world economy. A leading example is Sassen's (2001) detailed research on the so-called global cities of New York, London, and Tokyo. Her thesis is that globalization has created a new strategic role for major cities with control and organization functions. There are four stages to her basic argument: (1) cities are key command points in the world economy; (2) this creates a demand for financial and business services that are leading sectors in advanced economies; (3) cities become key sites for production and innovation in these sectors; and (4) cities in turn constitute the main markets for business and financial services. In this view, global cities – of which New York, London, and Tokyo are leading examples – are not just the site of corporate headquarters, but full-blown global service centers with a wide array of innovative support activities such as legal, banking, insurance, accountancy, management consultancy, advertising, and software firms. While other scholars have used different measures to construct global urban hierarchies (e.g., number of Fortune 500 firm headquarters, scale of the financial sector), most studies agree, that there is a leading group of cities – most notably New York, London, Tokyo, Paris, Zurich, and Frankfurt – that play a disproportionate role in controlling and coordinating the global economy. Three caveats about this approach should be noted. First, beyond this very top tier, the ranking can alter significantly, depending on what kind of measure is being used. Second, any ranking of cities must be seen in dynamic terms: the recent rise of cities such as Beijing and Shanghai, for example, will profoundly reshape existing patterns. Third, an ongoing challenge for economic geographers is to reveal the nature of the connections between key cities, rather than their static attributes. It may well be that the notion of a hierarchy is outdated in today's interconnected world (Taylor, 2004).

international financial centers. These interdependencies can be briefly summarized as follows:

- The location of some of the world largest transnational corporations that are served by these financial institutions.
- Strong interfirm cooperative relationships that require flexible and changing teams of staff in advanced producer services such as law, accountancy, insurance, banking, and consulting.

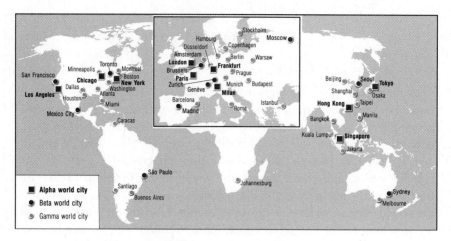

Figure 7.4   The global network of financial centers and global cities
Sources: Adapted from Dicken, P. (2011), Global Shift: Mapping the Changing Contours of the
World Economy, Sixth Edition. Copyright Guilford Press. Reprinted with permission of The
Guilford Press.

- The constant search for new business, which means that maintaining and
  expanding personal relationships through informal social activities near to
  their workplaces is vital. These relationships are embedded in thriving and
  dense networks of social institutions such as business associations, financial
  specialists, and so on.
- Acting as a crucible for rapid dissemination of market data and, more
  important, for these masses of information to be digested, assessed, and
  capitalized upon.
- Offering unparalleled concentrations of financial expertise: top graduates in
  the United States and the United Kingdom head for Wall Street in New
  York and the City of London, respectively. Spatial concentration also enables
  specialization in complex financial instruments. The greater the complexity
  and opacity of some financial instruments such as hedge funds (see Box 7.1),
  the more the need for the concentrations of specialized expertise. There is also
  the cumulative effect of reputation for customers and attractiveness to new
  employees.
- High levels of innovation due to concentrated expertise and the sharing of the
  fixed costs of market institutions (e.g., expensive security systems) and third-
  party service providers (e.g., document transportation and legal/accounting
  advice), thus creating scale economies.

   Now that we have demonstrated why geography matters in global finance in
theory, it may be useful to look at an example of a global financial center – New
York. As is well known, Wall Street, a simple street in New York City, plays a
dominant role in global financial markets. Wall Street is in the heart of lower
Manhattan, America's leading financial district and, together with Broad Street,
defines the boundary of an area that hosts the world's most powerful financial

institutions. These institutions range from the New York Stock Exchange (NYSE) and NASDAQ to the headquarters of many largest financial groups such as Goldman Sachs and Merrill Lynch (now part of Bank of America).

The NYSE on Wall Street is particularly significant after its 2007 merger with Euronext N.V. – an integrated platform of the exchanges of Brussels, Paris, and Amsterdam – which resulted in "the world's leading and most liquid equities exchange group" with more than 8,000 listed issues of stocks and shares for trading from more than 55 countries (www.nyse.com, accessed on January 4, 2012). Its stock market indices, such as Dow Jones Industrial, literally dictate the global price movements of stocks in numerous other stock exchanges, while its daily trade of stocks represents more than 80% of the value of all listed U.S. firms and over 70% of the largest 100 global corporations. Before the merger, its average daily value of trade was $46.1 billion in 2004, a figure already exceeding the annual GDP of many poor countries. After the merger in 2007, it grew substantially. In 2011, NYSE's equity exchanges had a total global market capitalization of $14.9 trillion and transacted an average daily trading value of approximately $57.7 billion. The NASDAQ on Broadway is home to two of the three largest U.S. companies, three of the top five of *Fortune* magazine's World's Most Admired Brands and seven of the *Fortune*'s top ten Fastest Growing Companies. Across its global exchanges, NASDAQ lists 3,500 companies in 46 countries, representing $6.9 trillion in total market value (www.nasdaqomx.com, accessed on January 4, 2012).

There is no doubt that New York, particularly Wall Street, has served as a key node in the global financial system and has come to dominate the global financial economy since the last decades of the 20th century. Wall Street institutions are extremely well connected into financial circuits operating at the global scale. Its global dominance owes much to the historical growth and role of New York City in the expanding hierarchy of urban centers in the United States, and the rise of American power and global finance in the post-Cold War era of financial globalization (see more in Table 7.1). Equally, the contemporary flows of people, capital, and knowledge into, and out of, New York City have continued to ensure its integral role in the global financial system. It is therefore not surprising that a social movement, known popularly as Occupy Wall Street, started in Manhattan's Liberty Square on September 17, 2011 as a bottom-up response to counter the excessive power and greed of Wall Street banks and firms in the new millennium (see Figure 7.5). The movement has since spread to over 100 cities in the United States and over 1,500 cities worldwide (occupywallst.org, accessed on January 3, 2012).

With the deregulation and re-regulation of the international financial system described in this section, a new breed of international financial centers, particularly those in Asia, has emerged in recent decades (see Figure 7.4). Up until the mid-2000s, Hong Kong, Singapore, and Tokyo were the only three Asian centers in the top twenty international financial centers. By 2011, Asia was much more significant, with eight centers in the top twenty (against six North American centers and five European ones). One good example is China's world city,

Figure 7.5 The Occupy Wall Street movement in New York City

Shanghai. In comparison with New York, Shanghai is a relatively new entrant to the world financial markets. From a relatively low ranking in the mid-2000s, Shanghai rose very rapidly to share its market position with Tokyo, an established global city. As a rapidly growing national financial center, however, it still lags well behind New York and London in terms of market openness, market size, and the variety and sophistication of financial products available (see Table 7.2). In

Table 7.2 Shanghai's position in global financial markets, 2008–2009 (in US$ billion)

Indices	Shanghai	Hong Kong	Tokyo	New York	London
Stock market capitalization, March 2009	1,863	1,307	2,611	7,946	1,676
Capital raised in stock market in 2008 (IPO plus secondary issues)	28	52	11	151	125
Outstanding domestic bonds at end 2008	2,210	50	11,077	24,622	1,223
Daily average turnover of foreign exchange market, April 2007	9	175	238	664	1,359

Source: Data from www.hangseng.com, accessed on March 3, 2011.

April 2009, the Chinese government unveiled a grand plan to transform Shanghai into an international financial center by 2020. In other cases, new financial centers such as the Cayman Islands and British Virgin Islands have arisen in response to the desire of financial institutions to evade national and international regulatory constraints and the propensity of wealthy individuals to engage in speculative/illicit financial transactions. Almost overnight in the late 1970s, for example, these former tax havens were transformed into offshore financial centers (see Box 7.4).

7.5   Circulating Capital: Financialization

This chapter has so far demonstrated how, in recent decades, the productive role of finance has been increasingly replaced by one of circulating investments that are detached from the "real" economy. While global finance operates through dominant financial centers filled with sophisticated financial elites, professional organizations, and modern technologies, its increasing complexity and velocity of circulation have gone far beyond the imagination and comprehensibility of most, if not all, of these individuals and/or institutions. Once the brainchild of the nation-state to fund the development of the economy, global finance has now become a primary driver of state policies and the behavior of all sorts of economic agents, from individuals to firms and even school boards! It is no longer amenable to control and regulation by any single territory or institution. After four decades of unabated growth and transformation since the 1970s, global finance now increasingly shapes the global economy in its own image.

In this penultimate section, we explore how this strange condition wherein global finance seemingly dominates all economic and political considerations has come about. We argue that it has much to do with a complex and all pervasive process now known as *financialization* in which all sorts of assets and things are transformed into financial instruments for trading among individuals and firms in the international capital markets. Buyers of complex financial instruments are no longer connected to, or even aware of, the underlying assets of their investment. This increasing disconnection between assets and investments through financialization has created enormous opportunities for profit creation and appropriation via the capital markets. Capital accumulation becomes increasingly possible through financial channels. Financialization thus enables the ever growing power and influence of financial markets, financial institutions, and financial elites at a global scale because of the immense interpenetration of domestic and international circuits of financial capital. Through financialization, fixed properties such as housing are financialized into structured investment vehicles such as mortgage-backed securities (MBS) that can be easily traded among global investors and financial institutions including pension funds, asset management firms, and

CASE STUDY

Box 7.4   *The Cayman Islands as an Offshore Financial Center (OFC)*

The Cayman Islands, a British Overseas Territory, is a good example of the place-selectivity and competition within global finance (Roberts, 1995; Hudson, 1998). As financial institutions started to move offshore in the 1960s, the Cayman Islands was a preferred destination due to a no-tax policy. In 1972, it had more than 3,000 registered companies and 300 trust companies. By 1993, some 25,000 companies were registered on the Islands, of which more than 55% were purely offshore entities. More than 80% of the 532 banks were "shell" branches with nothing more than a nameplate or a filing cabinet in another bank/company. In the aftermath of the 2008 global financial crisis, the Cayman Islands suffered a significant drop in revenues and real estate values (www.caymannewsservice.com, accessed on January 4, 2012). As the capital of the world's hedge fund industry, with assets of $2.3 trillion based on the Islands, its government suffered from huge deficits and was forced to borrow up to US$435 million from banks and others to keep itself afloat.

The dynamics of the Cayman Islands' development since the 1960s have been driven by a combination of different economic and political contexts. In economic terms, the Islands' attraction to foreign banks is clearly related to the activities of transnational corporations, many of whom have established financial departments on the Cayman Islands to serve as profit centers benefiting from tax minimization and transfer pricing (how internal transactions within business are accounted for). In this way, foreign banks are attracted to the Islands because they can cater to the needs for financial services of these transnational corporations. These foreign banks (e.g., Citibank) also offer customized packages of services (e.g., "global custody," the handling of global financial portfolios for clients) and other international private banking services to clients that have registered operations on the Islands. In political terms, the Cayman Islands experiences significant inter-island competition from other nearby OFCs such as Bermuda, The Bahamas, and Panama. While each occupies a special niche in the market, they also gain market share from the common practice of risk diversification among bank clients who do not want to base all their finances in one jurisdiction. The Cayman Islands underwent a major political struggle over national development options during the 1960s and found the OFC route to be the most viable and easy-to-implement growth strategy. In other words, its expansion as an OFC has been driven both by changing international economic circumstances and by the limited domestic options for other modes of economic development.

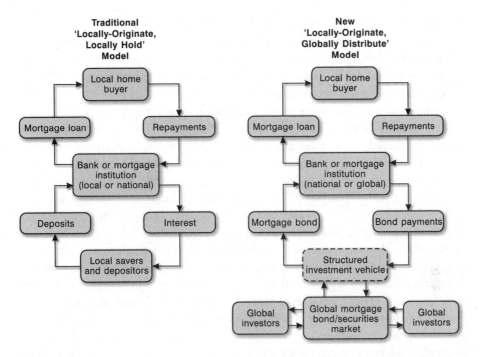

Figure 7.6   Global finance and the shifting relationship with local mortgage lending
Source: Adapted from Martin, R. (2011), 'The local geographies of the financial crisis: from the housing bubble to economic recession and beyond', *Journal of Economic Geography*, 11(4), 587–618, Figure 2.

investment banks (see Figure 7.6; see also the definitions in Box 7.1). Utilizing our retirement savings, pension fund managers buy into these MBS in order to profit from claims on future repayments by mortgagors and debtors. Payouts generated from these investments are used to finance the retirement incomes that in turn support our everyday consumption. This same process of financialization can be applied to many other assets and derivatives such as commodities (e.g., coffee and gold) and even carbon emissions.

At the same time, our consumption behavior is increasingly embedded in financialization. As more unsecured credit is extended to consumers through personal loans, credit cards, hire purchase schemes, buy-now-and-pay-later, and cash-back purchases, our borrowings have been aggregated to form securitized debt instruments and derivatives that can be traded in the capital markets. This process of financializing personal debt has greatly increased the incentives among credit-extending institutions such as credit card companies, retail banks, and saving societies to bankroll a dramatic expansion in household borrowing across the advanced industrialized societies, particularly the United States and the United Kingdom. Since the 1980s, credit card companies have pioneered the selling of their cardholders' debts as securitized debts to investors in the international capital markets. Financialization has led to the rapid emergence of collateralized debt obligations (CDOs) and other complex financial derivatives and instruments that are associated with massive systemic risks and the 2008 collapse of global financial markets. As related in the previous section, it is important to note that financialization has been fostered not only by the activities of financial corporations, but also by deregulatory government policies that have been largely in the favor of financial sector interests.

As the economy is becoming increasingly financialized, the circulation of capital becomes largely divorced from its productive allocation and use and thus can be easily repackaged by, and circulated among, individuals, institutions, and even nation-states. When finance evolves into capital flows for their own sake, there are reasons for us to begin to worry. Those engaged in the financial sector make money not just through the efficient allocation of funds to profitable ventures in what we might call the "real" economy, but also through betting on market directions and buying and selling the various instruments described in Box 7.1. Today, these financial instruments are collectively worth much more than "real" production, and the processes increase the vulnerability of the wider economy to financial crisis. In his widely read analysis of global capitalism over two decades ago, renowned geographer David Harvey (1989, p. 194) warned us that "I am therefore tempted to see the flexibility achieved in production, labour markets and consumption more as an outcome of the search for financial solutions to the crisis-tendencies of capitalism, rather than the other way round. This would imply that the financial system has achieved a degree of autonomy from real production unprecedented in capitalism's history, carrying capitalism into an era of equally unprecedented dangers."

How does financialization actually work? One particularly powerful mechanism has been the massive securitization of the world economy since the 1980s, which has opened up numerous opportunities for individuals and financial institutions to engage in a better form of saving – investment! This claim that saving should take the form of financial market investment has gone largely unquestioned since the early 1990s. This growth in investment opportunities has undoubtedly fueled the substantial conversion of individual savings into institutional investments of all forms including shares and stocks, unit trusts, mutual funds, insurance and pension funds, index funds, bonds, and, more recently, derivatives of debt obligations (see Box 7.1). On the other side of the financialization equation is the massive accumulation of consumer and corporate debts. By the turn of the new millennium, large-scale personal and household borrowing had become routine and mundane in the United States, the United Kingdom, and Canada, and reached a historical peak in 2009. This process of financialization helps us understand the rise of subprime finance and its pivotal role in the global financial crisis that started to unfold with the collapse of Lehman Brothers, a household name in investment banking, in September 2008 (see Box 7.5).

In response to the global financial crisis that unraveled from 2008 onward, various nation-states have taken concerted efforts to rein in the debilitating effects of national financial crises. First, massive rescue packages in the form of trillions

CASE STUDY

Box 7.5   Subprime and the crisis of global finance in 2008/2009

Subprime finance was deeply implicated in the 2008/2009 global finan-cial crisis. All sorts of credit facilities were extended to borrowers who previously would not have been granted such facilities. These subprime borrowers usually had low, irregular, or unverifiable income and tended to have poor credit histories and scores. They were lured into borrowing by a wide range of dubious lending practices undertaken by creditors, brokers, or even home-improvement contractors who were often not subject to fed-eral banking supervision. These practices involved engaging in deception or fraud, manipulating the borrower through aggressive sales tactics, or taking unfair advantage of a borrower's lack of understanding about loan terms.

Through securitization of these subprime mortgages, banks were able to separate out their risky lending business (e.g., credit card and mortgage lending) by trading these asset-backed securities (ABS) in the capital mar-kets, effectively eliminating the need for matching funding from savings and deposits for this kind of growth in assets (lending). Through this process of securitization at the global scale, lenders could profit from a "local" subprime loan even if the borrower defaulted within months. This was because once such a loan was packaged into an ABS and sold in global capital markets, the lender would have already recouped the loan amount plus profits and passed the future risk and liability to investors elsewhere who had bought the ABS. Meanwhile, the capital markets trading ABS were intimately linked with the millions of mortgagors, credit card holders, and instalment plan borrowers whose interest and principal payments on secu-rities issued by their lenders fed directly into what Langley (2008, p. 154) termed the "wheels of the mortgage-backed securities market." Martin (2011, p. 595) calls this a shift from a "locally-originate and locally-hold" model of mortgage provision to a securitized "locally-originate and globally-distribute" model in an era of global finance (see Figure 7.6).

In late 2007 and throughout 2008, an unlikely scenario occurred when a crisis of confidence hit the subprime market due to concerns about housing oversupply, leading to falling home prices that in turn reduced the mortgage-servicing ability of subprime borrowers. When a large number of these subprime borrowers could not repay their lenders, the latter's creditors, such as mainstream commercial banks and investment banks, got into trouble because these banks and their clients had bought into financial instruments derived from these subprime mortgages that became literally worthless. The CDO market thus collapsed and dragged down with it the entire global financial market.

of dollars of "new" money have been launched by central banks in developed and developing countries alike in order to kick-start "real" economies heavily weighed down by the financial crisis. The long-term net effects of these efforts on the changing geography and political economy of contemporary finance are still unclear. Second, major economic powers such as the G7 countries have made attempts to (re)order the global financial system through re-regulation, ranging from regulations on short selling and credit default swaps to corporate governance and capital requirements in financial institutions. These attempts, however, face significant resistance by the movers and shakers in global financial centers. Third, the escalating debt crisis in the United States and the European Union in 2011–2012, partly an outcome of the financial crisis three years earlier, has substantially increased volatility in global financial markets and turned the "real" economy upside down again. These uncertainties in economic outlook led, once again, to serious instability in financial markets, as seen, for example, in dramatic slumps in stock market values.

7.6   Summary

This chapter has critically examined how money and finance have experienced different transformations over time and space and how today's world of global finance came into being. Using the catastrophic loss of pension funds among five school districts in the American Midwest, we began the chapter with the promise and perils of global finance. To a certain extent, this promise is predicated on the popular discourse of finance as "placeless capital" that can reach every corner of the global system. We have argued against this "end of geography" reading of finance precisely because we understand that the *actually existing* geographies of finance clearly favor dominant financial centers and places.

To expand on this economic-geographical reading of money and finance, we have investigated the processes through which finance, mediated through the modern banking system, became a productive force in the real economy. In this developmental phase of modern capitalism, finance and banking remained fairly constrained within the territorial boundaries of nation-states. There was thus no such thing as global finance as we know it today. Commercial and merchant banks served mostly clients from their home countries and regions. This began to change in the 1970s when competitive deregulation of the financial sector and new technologies enabled capital to become an economic force in its own right and increasingly divorced it from the real economy. New sources of financial capital were found in emerging markets and developing economies. This global shift in the financial sector and its changing role within the wider economy led to the emergence of global finance. Still, we have argued that geography matters in putting global finance "in its place." We have shown how financial centers such as Wall Street and the Cayman Islands are pivotal in the orchestration of global finance to an extent that dwarfs their territorial and population size. In short, finance is global only because of the existence of these dominant places that serve to mediate massive global financial flows.

As financial flows through global financial centers have continued to expand since the 1990s, we have witnessed the growing domination of the world economy by global finance. These intensified processes of financialization have fundamentally reshaped the saving and borrowing behavior of individuals and firms and turned virtually everything, whether tangible or intangible, into potentially tradable financial instruments. The financial economy has grown to be a system that outstrips the importance of the real economy, whereby money and finance are now widely viewed as the primary source of investment returns and economic growth. This financialization has, not surprisingly, turned the potential promise of global finance (democratization) into a peril best illustrated by the global financial crisis that unfolded in late 2008. The severe destruction of value and wealth, the massive growth of debts, and the unprecedented number of housing market foreclosures that ensued are reflective of the centrality of the financial system within today's global economy.

To further understand these severe impacts of financialization and financial capital/crisis, we need to construct a nuanced view of how the real economy is organized. This is the central task of the next part of this book, which offers detailed analysis of how consumption and production are enabled and organized through commodity chains, technological innovations, transnational corporate activity, and retailing.

Notes on the references

- Martin (1999) and Clark and Wojcik (2007) are major collections for understanding different geographical perspectives on money and finance.
- Clark (2005, 2006) offers the most compelling argument for why the geography of finance matters.
- Dymski and Li (2004), and Leyshon et al. (2008) are excellent references on the development of money and banking. See Yeung (2009) and Lai (2011) for two recent geographical studies of financial market development in China.
- For geographical case studies of global financial centers, see Roberts (1995), Hudson (1998), and Clark (2002).
- Leyshon and Thrift (2007), French et al. (2009), Engelen et al. (2010), and Pike and Pollard (2010) provide useful geographical accounts of financialization of everyday life. For other work in the social sciences, see Krippner (2005), Dore (2008), Langley (2008), and Aalbers (2009).
- For some of the insightful geographical studies of the subprime crisis since 2008, see Crump et al. (2008), Walks (2010), and Martin (2011).

Sample essay questions

- Why is financial capital becoming increasingly global in its nature and operation?
- What are the factors that led to the spatial restructuring of national banking systems?
- Why do some international financial centers continue to be so influential and powerful in an era of global finance?
- How does financialization affect everyday social life?
- How did the 2008 financial crisis arising from bad mortgage loans in the United States become global?

Resources for further learning

- www.longfinance.net: The Long Finance website contains a comprehensive analysis of global financial centers, including the *Global Financial Centers Index*.
- http://www.nytimes.com/2008/11/02/business/02global.html?_r=1&ref=the_ reckoning: The original report on the financial woes of Wisconsin school boards during the late 2008 global financial crisis.
- www.gfmag.com: An online magazine about global finance that provides a wide range of useful reports and databases on banks and finance.
- www.fdic.gov: The Federal Deposit Insurance Corporation (FDIC) website provides comprehensive financial information about every FDIC-insured institution in the United States.
- www2.isda.org: The official website of the International Swaps and Derivatives Association provides a comprehensive guide to one of the most complex financial markets in the world.
- www.nyse.com: The New York Stock Exchange website offers detailed information on various financial instruments and market participants in today's global financial markets.
- www.thecityuk.com: The City of London website contains useful data and information on its role as *the* global financial center.
- www.citigroup.com: The Citigroup global corporate website gives a good indication of what it means to be a global bank.

References

Aalbers, M. B., ed. (2009). Symposium on the sociology and geography of mortgage markets: reflections on the financial crisis. *International Journal of Urban and Regional Research*, 33, 281–443.

Budd, L. (1999). Globalisation and the crisis of territorial embeddedness of international financial markets. In R. Martin, ed., *Money and the Space Economy*. Chichester, U.K.: John Wiley, pp. 115–37.

Clark, G. L. (2002). London in the European financial services industry: locational advantage and product complementarities. *Journal of Economic Geography*, **2**, 433–53.

Clark, G. L. (2005). Money flows like mercury: the geography of global finance. *Geografiska Annaler*, **87B**, 99–112.

Clark, G. L. (2006). Setting the agenda: the geography of global finance. In S. Bagchi-Sen and H. Lawton Smith, eds., *Economic Geography: Past, Present and Future*. London: Routledge, pp. 83–93.

Clark, G. L., and Wojcik, D. (2007). *The Geography of Finance: Corporate Governance in a Global Marketplace*. Oxford: Oxford University Press.

Crump, J. R., Newman, K., Belsky, E. S., Ashton, P., Kaplan, D. H., Hammel, D. J., and Wyly, E. (2008). Cities destroyed (again) for cash: forum on the U.S. foreclosure crisis. *Urban Geography*, **29**, 745–84.

Dicken, P. (2011). *Global Shift: Mapping the Changing Contours of the World Economy*, Sixth Edition. London: Sage.

Dow, S. C. (1999), 'The stages of banking development and the spatial evolution of financial systems', in Ron Martin (ed.), *Money and the Space Economy*, Chichester, U.K.: John Wiley, pp. 31–48.

Dore, R. (2008). Financialization of the global economy. *Industrial and Corporate Change*, **17**, 1097–1113.

Dymski, G., and Li, W. (2004). Financial globalization and cross-border co-movements of money and population: foreign bank offices in Los Angeles. *Environment and Planning A*, **36**, 213–40.

Engelen, E., Konings, M., and Fernandez, R. (2010). Geographies of financialization in disarray: the Dutch case in comparative perspective. *Economic Geography*, **86**, 53–73.

French, S., Leyshon, A., and Thrift, N. (2009). A very geographical crisis: The making and breaking of the sub-prime crisis. *Cambridge Journal of Regions, Economy & Society*, **2**, 287–302.

Harvey, D. (1989). *The Condition of Postmodernity: An Enquiry into the Origins of Cultural Change*. Oxford: Blackwell.

Hudson, A. C. (1998) Placing trust, trusting place: on the social construction of offshore financial centers. *Political Geography*, **17**, 915–37.

Krippner, G. (2005). The financialization of the American economy. *Socio-Economic Review*, **3**, 173–208.

Lai, K. P. Y. (2011). Marketization through contestation: reconfiguring China's financial markets through knowledge networks. *Journal of Economic Geography*, **11**, 87–117.

Langley, P. (2008). *The Everyday Life of Global Finance: Saving and Borrowing in Anglo-America*. Oxford: Oxford University Press.

Leyshon, A., French, S., and Signoretta, P. (2008). Financial exclusion and the geography of bank and building society branch closure in Britain. *Transactions of the Institute of British Geographers*, **33**, 447–65.

Leyshon, A., and Thrift, N. (2007). The capitalization of almost everything: the future of finance and capitalism. *Theory, Culture and Society*, **24**, 97–115.

Martin, R. (1994). Stateless monies, global financial integration and national economic autonomy: the end of geography? In S. Corbridge, R. Martin, and N. Thrift, eds., *Money, Power and Space*, pp. 253–78. Oxford: Blackwell.

Martin, R., ed. (1999). *Money and the Space Economy*. Chichester, U.K.: John Wiley.

Martin, R. (2011). The local geographies of the financial crisis: from the housing bubble to economic recession and beyond. *Journal of Economic Geography*, **11**, 587–618.

O'Brien, R. (1992), *Global Financial Integration: The End of Geography*, New York: Council on Foreign Relations Press.

Pike, A., and Pollard, J. (2010). Economic geographies of financialization. *Economic Geography*, 86, 29–51.

Pollard, J. (1999), 'Globalisation, regulation and the changing organisation of retail banking in the United States and Britain', in Ron Martin (ed.), *Money and the Space Economy*, Chichester, U.K.: John Wiley, pp. 49–70.

Roberts, S. M. (1995). Small place, big money: the Cayman Islands and the international financial system. *Economic Geography*, 71, 237–56.

Sassen, S. (2001), *The Global City: New York, London, Tokyo*, Second Edition, Princeton, NJ: Princeton University Press.

Taylor, P. J. (2004), *World City Network: A Global Urban Analysis*, London: Routledge.Walks, R. A. (2010). Bailing out the wealthy: responses to the financial crisis, Ponzi neoliberalism, and the city. *Human Geography*, 3, 54–84.

Yeung, G. (2009). Hybrid property, path dependence, market segmentation and financial exclusion: the case of the banking industry in China. *Transactions of the Institute of British Geographers*, 34, 177–94.

PART III

ORGANIZING ECONOMIC SPACE

CHAPTER 8

COMMODITY CHAINS
Where does your breakfast come from?

Goals of this chapter

- To demonstrate how different commodities commonly used or consumed in our everyday life are brought to us through the capitalist system
- To introduce commodity chains and their basic components as core concepts to explain these production conditions
- To appreciate the differentiation of commodity chains in terms of their structure and geography
- To recognize the possibilities for, and limitations of, different ways of organizing and regulating commodity chains

8.1   Introduction

In November 2010, McDonald's launched a new commercial focusing on its breakfast offerings. In the "Where does breakfast come from?" advertisement (vimeo.com/16516135, accessed on March 8, 2012), a little girl asks her father this apparently innocent question during a car ride on their way to eat breakfast. Avoiding answering, the father passes the question on to the mother who, in a moment of panic, dreams up a fairy tale about a faraway kingdom known as the Kingdom of Breakfast. While the mother describes a group of giants in this kingdom who take a long journey every morning to make deliveries to breakfast wizards, the video sequence shows a beautiful farm, golden-ripe wheat, healthy hens, and two young farmers loading cartons of fresh eggs onto their old Chrysler truck and making their way to McDonald's. The breakfast "wizard" turns into a young McDonald's associate who uses his magic wand (spatula) to make everything "hot" and "tasty." As the family enjoys a breakfast of scrambled eggs, beef patties, buns, coffee, and pancakes with maple syrup and butter, the final sequence of the video features the little girl asking her parents the even more vexing question of where *she* came from!

The intriguing thing about this McDonald's commercial is not just that it asks the same question as this chapter, but more importantly its indication that the little girl's parents, while very familiar with McDonald's meals, seem to know nothing at all about the origins of their breakfast and so resort to telling a fictional tale. However, in their ignorance they are surely not alone. Even on McDonald's own U.S. website (www.mcdonalds.com, accessed on March 8, 2012), there is only very limited information about the origins of their breakfast items. All egg products are sourced from "world class" supply chain companies such as Cargill Kitchen Solutions. McDonald's dairy products are typically supplied by global food companies such as Kraft or Dean Foods. The coffee comes from Gaviña Gourmet Coffee, which sources the beans from Central and South America. Lopez Foods supplies its beef patties, made mostly from U.S. beef and a small proportion sourced from Australia and New Zealand.

But where do these large food suppliers source their raw materials that are eventually transformed "magically" into a McDonald's hot breakfast? Many *perishable* inputs may be sourced locally or within the same country, but other inputs are not. Many will have traveled tens of thousands of miles to reach McDonald's. Knowing *where* your breakfast comes from can make an important difference in terms of both health and ethical concerns. (Chapter 15 will further discuss these consumption concerns.) Some breakfast items are produced by small firms, while others are distributed globally by transnational corporations: for example, Smucker's strawberry jam (by J.M. Smucker Co.), Lurpak butter (by the Danish Dairy Board), Skippy peanut butter (by Unilever), Corn Flakes® cereal (by Kellogg's), Minute Maid orange juice (by Coca-Cola), and Pike Place™ Roast (by Starbucks). If these brands and their geographical origins appear complicated and concealed, then the challenge is even greater when it comes to the generic own-label brands sold by retailers such as Wal-Mart or Tesco.

How can we conceptualize the complex journeys taken by such simple commodities as our breakfast foodstuffs across the global economy? In this chapter, we develop the notion of *commodity chains* as a way of understanding the connections and interdependencies between different workers, consumers, firms, and institutions involved in commodity stories. In following the travels of different commodities, our focus is on the geographies and governance of connections within the global economy and the normative dimensions of those interdependencies such as standards and regulations. There are four main sections in the chapter. First, we consider how capitalism as a system hides the connections or social relations inherent to a particular commodity, and reveal the implications of this concealment (Section 8.2). Second, we explore in detail the nature of commodity chains, revealing how they vary in terms of their structure, geography, coordination, and institutional context (Section 8.3). Third, we look at the regulation of commodity chains through standards initiatives of various kinds that seek to improve various aspects of production conditions along the chain (Section 8.4). Fourth, we explore the potential interconnections between different commodity chains and, in particular, how waste products may be used as starting points for new commodity chains (Section 8.5). Where individual commodity chains begin and end is perhaps not as obvious as it first appears.

8.2   Capitalism, Commodities, and Consumers

As explained in Chapter 3, capitalism can be thought of as a *commodity* exchange system. A commodity is simply something useful that enters the market and is available for purchase at a price. However, commodities are much more than just material things, such as books or food. In the contemporary world, more and more areas of our everyday life have become caught up in processes of *commodification* (see also Chapter 5 on nature). Domains as varied as culture (e.g., music, museums, and galleries), religion (e.g., celebrity preachers), knowledge (e.g., MBAs and intellectual property rights), the environment (e.g., carbon credits), war (e.g., private armies and ammunitions) and the human body (e.g., trade in human organs or genetic materials) have become commoditized and thus subject to the vagaries of the market mechanism.

While commodities are central to the capitalist system, at the same time they may serve to hide important dimensions of how they are produced and brought to us. The *exchange value* of a commodity – the price – is often indicative of how the commodity was created: the cost of the human labor and skills/knowledge that went into its production, the costs of machinery, buildings, electricity, trucks and so on that were required, and the profits extracted at various points in the process. And yet the simple price tag itself reveals nothing of the production process the commodity has undergone and the necessary social relations that connect this production to the commodity's eventual customer or user. As a result, consumers in the capitalist system are largely ignorant of the geographical origins and histories of the commodities that they consume. The purchase of a commodity through monetary relations (see also Chapter 7) serves to *disconnect* producers and consumers, encouraging an abdication of responsibility on the part of consumers for the terms and conditions under which the commodity was made. The consumer can simply benefit from the *use value* of whatever they have purchased – that is, the usefulness of a particular product to an individual. This poses profound challenges to both conscientious consumers who actively want to know the history (and geographical origin) of the commodities they consume, and economic geographers who want to understand connections and interdependencies within the global economy. In reality, even just buying a McDonald's Big Breakfast or drinking coffee in a Starbucks café makes the consumer complicit, albeit unknowingly in many cases, in complex webs of connections across the globe (see Box 8.1).

CASE STUDY

Box 8.1   Coffee, Cafés, and Connections

Founded in Seattle, Washington's famous Pike Place Market in 1971, Starbucks has become the largest chain of coffeehouses worldwide. Still headquartered in Seattle, Starbucks organizes its operations on the basis of distinct geographical segments: U.S. and international. Within the United States, it has many roasting and distribution facilities in different states. Internationally, it has coffee buying operations in Lausanne, Switzerland, roasting and distribution activities in the Netherlands and the United Kingdom, regional support centers in Toronto (Canada), Miami (Americas), Amsterdam (Europe), Hong Kong (Asia Pacific), and Shanghai (Greater China), a farmer support center in San Jose (Costa Rica), and thousands of retail stores worldwide. As of January 2011, Starbucks had some 11,181 owned or licensed stores in the United States, and 5,525 stores across more than 50 countries, and was serving millions of customers each day. The company offers a range of over 30 coffees and teas in addition to wide variety of snacks and other beverages. Through its marketing and store information strategies, Starbucks endeavors to create a "knowledgeable" culture of coffee drinking in its cafés. The corporate website, for example, has an extensive Coffee Education area, several pages of which describe and contrast the coffees from Latin America, Africa, and Southeast Asia (under the heading *The Flavors in Your Cup*), and it used to market its coffee with the motto, "Geography Is a Flavor". Elsewhere on the website, individual coffees are described in evocative tones such as "Kenya: Beloved, like the elephant that symbolizes it. Bold, like the sun coming up over the Great Rift Valley." The strategy is clearly to turn coffee drinking from a routine activity into a more meaningful consumption process involving certain kinds of knowledge about coffee as a commodity with a particular history and geography.

However, it is possible to offer a more critical reading of this sophisticated marketing strategy. The information on offer in Starbucks presents a highly partial interpretation of coffee and its production process. As a commodity, coffee also has many other less palatable stories to tell. For instance, the structures of domination and exploitation inherent in today's global coffee industry – and indeed their colonial origins – are entirely overlooked. The story of the global coffee industry over much of the last two decades has been one of rising production and falling prices, and as a result, increasingly marginal working and living conditions for millions of farmers and farm workers in a range of poor tropical countries, many of which are highly dependent on coffee exports (which make up over 50% of Ethiopia's total exports, for example). Over the same time period, the leading coffee roasters (such as Nestle and Kraft) and retailers (e.g., Starbucks) that dominate the global industry have been able to maintain healthy profit margins on their coffee products. In short, Starbucks offers a deliberately selective and romanticized reading of the global coffee industry in its literature and store displays. For more, see Oxfam (2003), Daviron and Ponte (2005), and www.starbucks.com.

Moreover, the *images* we receive about commodities in our everyday life may actively serve to further conceal or even misrepresent the origins and social relations of those commodities. Advertising – a significant economic sector its own right as we shall see in Chapter 11 – is extremely important here. The fairy tale in McDonald's "Where does breakfast come from?" commercial is a typical example of concealing the real origins of our breakfast items. Through the creation of various images, advertisers seek to establish time- and place-specific meanings for particular goods and services that may be a far cry from the realities of their production. Think, for example, of advertisements for gold jewelry in developed-country markets. Through skillfully combining pictures and words, these advertisements tend to emphasize certain *personalized* values and emotions that are associated with the products: love, passion, romance, commitment, and so on (for an engaging attempt to destabilize and subvert advertisements and their central messages, see www.adbusters.org).

But a more critical reading might ask: What is missed out in this representation of gold as a commodity? A gold necklace bought from a jeweler in a rich world city (Figure 8.1) may be the end point of a series of links that connects consumers to high-security global logistics firms, ring manufacturers in Italy, gold traders in Zurich, black male migrant miners from Lesotho working in appalling conditions in South Africa's gold mines, and women left behind in Lesotho working long hours for negligible pay in the textile industry. In this way, "the gold windows of Tiffany's in New York are linked to the gold widows in Lesotho" (Hartwick, 1998, p. 433). The harsh reality of the gold industry is that notions of love and commitment stand rather incongruously alongside the geopolitical legacies of the South African apartheid regime, oppressive slave-like working conditions in mines, and abandoned women working in unregulated factories. These geopolitical, racialized, and gendered relations are all part and parcel of the very commodity displayed on the windows of leading jewelry retailers.

Curiously, in some cases, places of origin are constructed in certain ways to make the products more appealing and positive. Many upmarket consumers are willing to pay more for products made in places well known to them: Swiss watches, Italian clothing, French wines, Belgian chocolates, German cars, and

The authors

Figure 8.1 The jewelry shop window – the start or the end of a complex commodity chain?

Japanese digital cameras. Product packaging often displays images and labels that give a caricatured view of their places of origin. For instance coffee beans are regularly packaged and labeled to emphasize their "exotic" tropical origins (e.g., Starbuck's Ethiopia Sidamo™), effacing the poverty levels of these severely underdeveloped countries that try to tussle with leading coffee brands for the ownership of their national names (Figure 8.2). Equally, Italian clothing manufacturers are increasingly reliant on cheap temporary workers from China to churn out their "Made in Italy" fashion products. Advertising, then, acts as a powerful force further accentuating the disconnection of producers from consumers.

By now, the significance of the question in the chapter's title – where does your breakfast come from? – should be clearer. It alludes to the fact that even the most mundane and everyday acts of consumption tie us into these webs of connections. Commodities, then, need to be thought of as much more than just their immediate market and use values. Instead, every commodity should be seen as "a bundle of social relations" (Watts, 1999, p. 307), or, put another way, as representative of the whole system of connections between different groups of people that have enabled the consumer to make a purchase. In this way, the working conditions and gender relations that underlie commodity production – and that may be unacceptable to certain consumers – can be revealed, challenged, and, eventually, improved (see also Chapter 15). In the contemporary era, this is increasingly about revealing interdependencies at the *global* scale, even in the case of relatively perishable foodstuffs.

Figure 8.2   Geography is a flavor...

8.3   Linking Producers and Consumers: The Commodity Chain Approach

How then do we bring together all the diverse actors involved in the global travels of orange juice or a T-shirt? The notion of the *commodity* or *production chain* is central to conceptualizing such systems. Figure 8.3 outlines a much simplified commodity chain for our daily breakfast, illustrating the transformation of initial raw materials into final outputs in the form of consumable foodstuffs. These outputs are then brought to us through such services as logistics and retailing. In more complex ways, this transformation includes core activities (e.g., production, marketing, delivery, and services) and support activities (e.g., merchandising, technology, finance, human resources, and overall infrastructure). The commodity

Figure 8.3   The basic commodity chain of our breakfast
Source: From left to right: © ZoneCreative/iStockphoto; © IP Galanternik D.U./iStockphoto; © Doug Berry/iStockphoto; © xyno/iStockphoto; © Alexey Avdeev/iStockphoto

chain, then, is not simply about manufacturing processes: many of the inputs to the chain and many of the final commodities produced will take the form of intangible services. In producing a tangible product (such as a mobile phone) or a service (consumer or merchant banking), the various corporate activities and interfirm relationships are linked together in a chain-like fashion, with each stage adding value to the process of production of the goods or services in question.

Coffee – a recurrent item in many breakfasts and thus also in this chapter – offers a good example of such a commodity chain. Further to the geographies of coffee beans explained in Box 8.1, Figure 8.4 shows how the market value (price traded) of a bag of coffee beans is derived and the costs and profits encountered by various producers and intermediaries. It is clear that very substantial differences in the value derived from this commodity chain exist between a coffee bean farmer who sells to a middle buyer for $0.14 per kg of raw beans, and a retailer who sells to end consumers for $26.40 per kg of roasted beans. More specifically, roasters capture about 30% of the value added in the coffee commodity chain, while retailers (22%) and international traders – such as Neumann, Volcafe, Cargill, and others – also do well (8%). The result is that at least 60% of the economic value derived from coffee goes to developed-country firms, although some estimates put the figure as high as 80%. How do we make sense of these enormous differences in value creation and capture along the commodity chain? Who holds the most power in this chain and how is it governed? What kinds of standards are imposed to ensure a certain degree of conformity and consistency in these commodities?

Understanding the sequence and range of actors involved in a particular commodity chain, sometimes known as its *input-output structure*, is simply the first step toward developing a good understanding of commodities and their production processes. However, there are three further important dimensions to all commodity chains that we will now consider in turn: their *geography*; the way in which they are coordinated and controlled, that is, their *governance*; and the way in which local, national, and international conditions and policies shape the various elements in the chain – their *institutional frameworks*.

Geographical Structures

In very simple terms, the geography of a commodity chain can range from being *concentrated* in one particular place to being widely *dispersed* across a wide range of localities. As the earlier discussion of breakfast commodities makes vividly clear, it is hard to identify a commodity chain in the contemporary global economy that is not global to at least some degree, even if it is just seen in the sourcing of one or two inputs or in a limited export market for the final good/service. Many bring together an extensive range of international connections. *Global commodity chains*, as they have come to be called, are one of the primary organizational features of the world economy (Gereffi, 1994). This geographical reach is important not only because it determines precisely which actors are connected together

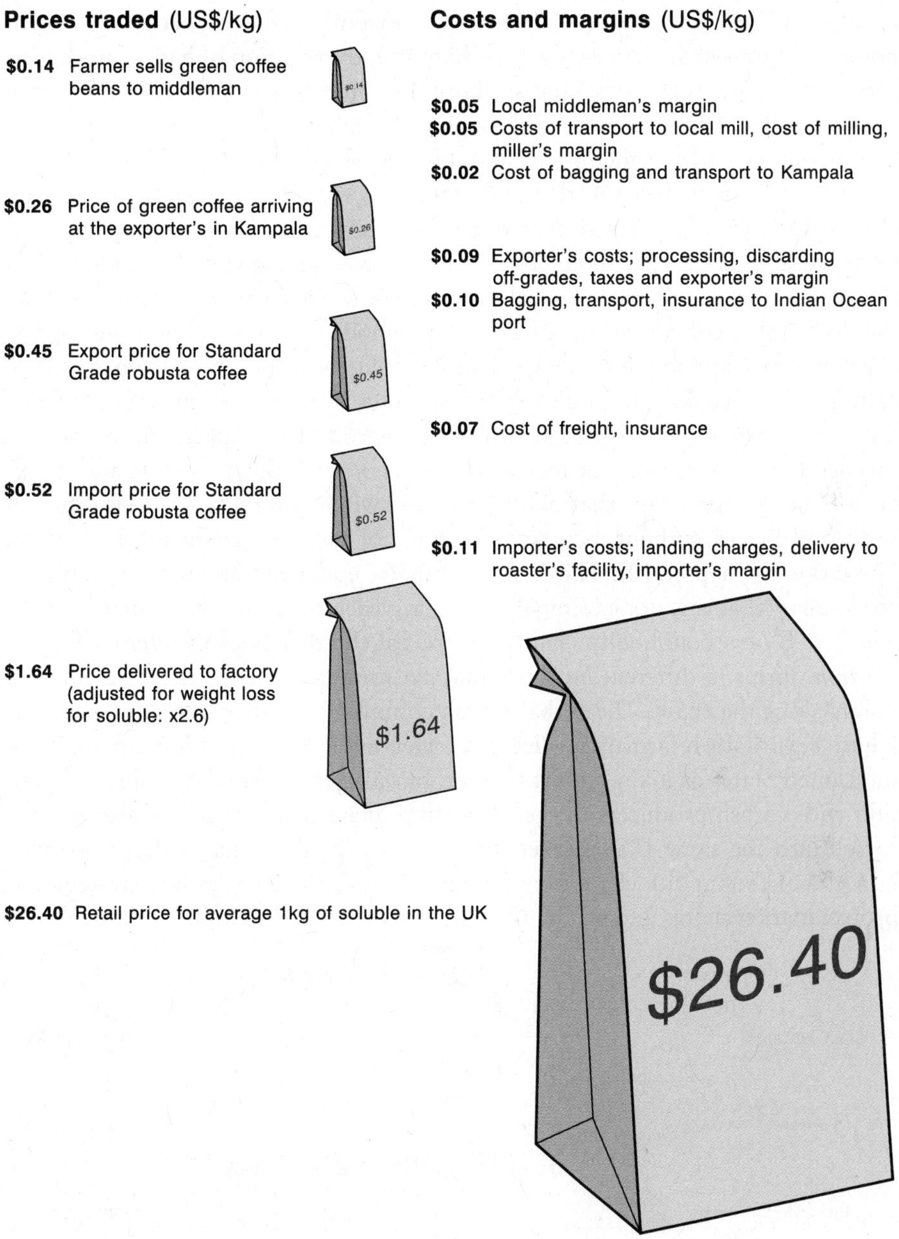

Prices traded (US$/kg)

$0.14 Farmer sells green coffee
          beans to middleman

$0.05 Local middleman's margin
$0.05 Costs of transport to local mill, cost of milling,
          miller's margin
$0.02 Cost of bagging and transport to Kampala

$0.26 Price of green coffee arriving
          at the exporter's in Kampala

$0.09 Exporter's costs; processing, discarding
          off-grades, taxes and exporter's margin
$0.10 Bagging, transport, insurance to Indian Ocean
          port

$0.45 Export price for Standard
          Grade robusta coffee

$0.07 Cost of freight, insurance

$0.52 Import price for Standard
          Grade robusta coffee

$0.11 Importer's costs; landing charges, delivery to
          roaster's facility, importer's margin

$1.64 Price delivered to factory
          (adjusted for weight loss
          for soluble: x2.6)

$26.40 Retail price for average 1kg of soluble in the UK

Costs and margins (US$/kg)

Figure 8.4   The coffee commodity chain: who gains most?
Source: Adapted from Oxfam, 2003, p. 24, with the permission of Oxfam GB, Oxfam House, John
Smith Drive, Cowley, Oxford OX4 2JY UK. www.oxfam.org.uk.
Oxfam GB does not necessarily endorse any text or activities that accompany the materials, nor has
it approved the adapted text.

across the global economy, but also in revealing the unequal geographical distribution of value, and associated economic development benefits, between different points along the chain. As we saw in Chapter 3, the location of high-value-added activities (design, marketing, and so on) in key cities is particularly important in the spatial inequality engendered through these commodity chains.

We can make five further points about the geographical structures of commodity chains. First, in general, the *geographical complexity* of global commodity chains is increasing, enabled by a range of developments in transport, communication, and process technologies. We can now source our daily goods from a much greater range of geographical origins than was the case previously. Second, the geographic configurations of global commodity chains are becoming *more dynamic* and liable to rapid change. This flexibility is derived both from the use of certain space-shrinking technologies (see Chapter 9) and from new organizational forms that enable the fast spatial switching of productive capacity. In particular, this flexibility comes from the increased use of *external* subcontracting and strategic alliance relationships that allow firms to switch contracts between different firms and places without incurring the costs of moving production themselves (see more in Chapter 10). Third, and relatedly, understanding the geography of commodity chains is not as simple as locating each stage in a particular place or country. Global commodity chains also reveal the dynamics of *interplace competition*. Firms in different localities may be vying for market share at different points along the chain. The global catfish commodity chain is revealing here (see Figure 8.5). Catfish farmers in Vietnam's Mekong delta are not just connected to the United States as a key export market, but also through relations of competition with catfish producers in the Mississippi delta region who are also seeking to sell into the same U.S. market. In other words, different localities involved in a global commodity chain may engage in competitive upgrading strategies to protect market shares and profitability (see Box 8.2). In developing countries such

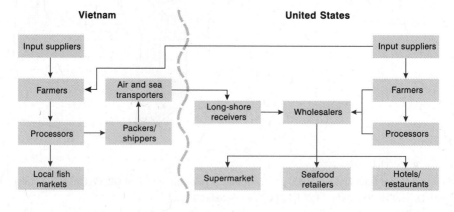

Figure 8.5   The catfish commodity chain
Source: Adapted from Duval-Diop, D. M. and Grimes, J. R. (2005) Tales from two deltas: Catfish fillets, high-value foods, and globalization, Economic Geography, 81, 177–200, Figure 5. This material is reproduced with permission of John Wiley & Sons, Inc.

KEY CONCEPT

Box 8.2   Upgrading strategies in global commodity chains

It is important to view the value structures of commodity chains in dynamic terms. Upgrading refers to the potential for firms, or groups of firms, to improve their relative position within the system as a whole. It is useful to distinguish between four different types of upgrading (Humphrey and Schmitz, 2004):

- *Process upgrading*: Improving the efficiency of the production system by either reorganizing the production process or introducing superior technologies. For example, a car manufacturer might introduce robot technology to speed up its assembly lines.
- *Product upgrading*: Moving into making more sophisticated products or services. For example, a basic food processing firm might start making prepared frozen meals, or a financial firm might offer new kinds of insurance products.
- *Functional upgrading*: Acquiring new roles in the chain (and/or abandoning existing functions) in order to increase the overall skill content and level of "value added" of the activities undertaken. An electronics manufacturer, for example, might move from simple assembly to original equipment manufacturing (OEM) to own-design manufacture (ODM) to own-brand manufacturing (OBM) (see more in Chapter 10).
- *Intersectoral upgrading*: Using the knowledge derived from a particular chain to move into different sectors. For example, a firm might use its experience of making televisions to enable it to move into computer monitor making.

At the level of the individual firm, successful upgrading strategies can transform fortunes. The Turkish clothing firm Erak, for example, has developed from a clothing manufacturer supplying Germany's Hugo Boss into a global clothing brand and retailer through its Mavi jeans products. More broadly, successful upgrading lay behind the emergence of the newly industrializing economies of Asia and Latin America from the 1960s onward (see also Chapter 4). The Taiwanese electronics industry, for example, has benefited from all four kinds of upgrading processes to develop from a base for foreign-owned electronics assembly into one of the world's leading centers for designing and producing new computer technologies in the global economy. For many other developing countries, facilitating upgrading across a wide range of sectors remains a key policy concern, as they seek to gain a greater share of the spoils of global commodity chains for clusters of local firms.

as Brazil, China, and India, localities that are home to firms successfully pursuing such upgrading strategies are liable to experience rapid economic growth and positive development outcomes.

Fourth, it is important to re-emphasize that global commodity chains are not just a feature of agricultural and manufacturing sectors, but are also apparent in many *service sectors*. For example, many service firms now find it advantageous to conduct routine data processing and software programming functions in overseas sites – India, the Philippines, Mauritius, Jamaica, and Trinidad and Tobago are prime examples – where there is relatively low-cost labor. Fifth, and finally, we need to connect these ideas about the geographical extensiveness and complexity of global commodity chains with the arguments about the geographical *clustering* of economic activity (see Chapter 12 for more on this). Some kinds of interactions within commodity chains will take place within the same locality, due to, for example, the intensity of transactions or the importance of place-specific knowledge to the activity in question. For these "nodes" in the larger mosaic of economic landscapes, commodity chains are often the organizational forms that connect firms and other economic agents in these clusters with their counterparts elsewhere in the global economy. These inter-cluster connections can be critical to the upgrading and further development of local production and innovation capacities.

Understanding the input-output and geographical structures of a commodity chain is undoubtedly important, but still leaves questions of control and power relations unanswered. Think of McDonald's or Wal-Mart. *Who* controls the organizational structure and nature of their global commodity chains? *Who* decides where inputs are purchased from, and where final goods and services are sold? *Who* shapes the restless geographies of commodity chains? This brings us on to the important issue of governance.

Governance Processes

So far we have seen how commodity chains are constituted by a mix of intra-firm and inter-firm linkages, and a combination of near and distant connections. In most cases, however, the chain will have a primary coordinator or a lead firm driving the system as a whole. In Figure 8.6, we distinguish between chains that are *producer-driven* and those that are *buyer-driven* (Gereffi, 1994).

Producer-driven chains are commonly found in industries where large industrial transnational corporations (TNCs), the *producers* of specific products and/or services, play the central role in controlling their global production (see also Chapter 10). This situation is found in many capital- and technology-intensive industries such as aircraft, automobile, computer, semiconductor, pharmaceuticals, and machinery manufacturing. These chains are notable for the degree of control exercised by the administrative headquarters operations of TNCs. Producers dominate such chains not only in terms of their earnings and profitability, but also through their ability to develop new products and markets, and to exert

Producer-driven commodity chains

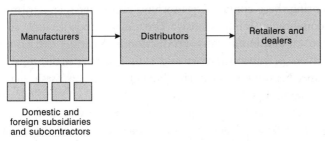

Buyer-driven commodity chains

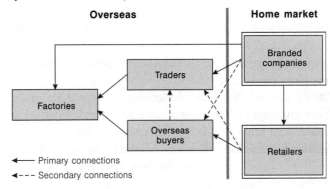

Figure 8.6   Producer-driven and buyer-driven commodity chains
Source: Adapted from Gereffi, G. (1994) "The organisation of buyer-driven global commodity chains: how US retailers shape overseas production networks" in G. Gereffi and M. Korzeniewicz (eds.) Commodity Chains and Global Development, Westport, Praeger, 95–122, Figure 5.1.

control "upstream" over raw material and component suppliers and "downstream" over the distributors and retailers of their products. Profits are secured through the scale and volume of production in combination with the producers' ability to lead technological and know-how developments. The automobile industry provides an excellent example of this kind of producer-driven chain. Leading producers or assemblers such as Toyota and Volkswagen coordinate production systems involving literally thousands of subsidiaries and tiers of subcontractor firms dotted around the world, as well as extensive global systems of distributors and dealers.

Buyer-driven chains, on the other hand, tend to be found in industries where large retailers (such as Wal-Mart, Tesco, Carrefour, and IKEA) and brand-name merchandisers (such as Adidas, Nike, and The Gap) play the central role in establishing and controlling the global production systems of their commodities, usually located in export-oriented developing world countries (see also Figure 8.6). These retailers and brand-name merchandisers are collectively known as *buyers* in this conceptual framework because they source their manufactured commodities from suppliers (i.e., producers) all over the world. It is thus important to note that buyers in these chains are *not* final consumers, but retailers, merchandisers, and wholesalers who bring these commodities to final consumers. This form of commodity chain is common in labor-intensive consumer goods sectors, such as clothing, footwear, toys, and handicrafts. Production is generally carried out through multi-tiered layers of subcontractors and suppliers that make finished goods subject to the specifications of powerful buyers. Profits in these chains are derived from the bringing together of design, sales, marketing, and financial expertise, allowing the retailers and merchandisers to connect overseas factories with the main consumer markets. Hence control is enacted through the ability of these large buyers to shape mass consumption patterns through strong brand names and meet this demand through global sourcing strategies from their suppliers/producers (e.g., the example of the coffee value chain and Starbucks in Box 8.1). These commodity chains are characterized by buyers who have the power and capacity to dictate the production terms and even standards of their suppliers/producers.

We can use the case of clothing to appreciate more fully the role of these powerful buyers in *driving* commodity chains. In doing so, their buying behavior has crucial implications for our understanding of production (mal)practices and working conditions in many of their suppliers. Armed with direct access to millions of consumers, large buyers such as Wal-Mart and Nike can achieve huge economies of scale in their global sourcing of clothing manufactured in developing countries. Price competition in the lower end of this industry is particularly intense due to its relatively low capital and technology requirements and thus low barriers to entry. This tendency toward a "race to the bottom" among low-cost suppliers creates an environment for breeding working conditions that are unacceptably difficult or dangerous, commonly known as *sweatshops*. While the history of such sweatshops can be traced back to the Industrial Revolution in the 19th century, their global manifestations today are a consequence of the emergence of these buyer-driven global commodity chains. During the 1990s, Wal-Mart and Nike came under intense criticism from activists and the mainstream media for their links to overseas suppliers in Africa and East and Southeast Asia where sweatshop conditions could be found. Understanding the geographical and organizational dynamics of these global commodity chains enables us to identify better these buyer–supplier relationships and their impact on not just the price and choice for final consumers, but also the "distant strangers" whose labor makes affordable the commodities on the shelves of global retail giants. As many of these sweatshops tend to employ young female workers and are located in developing countries, the clothing global commodity chain offers an extremely valuable window for us to understand gender relations along the entire commodity chain, from designer clothing targeting young and affluent female consumers to mass production employing tens of thousands of female workers.

Table 8.1 Characteristics of producer-driven and buyer-driven chains

| | Form of Economic Governance | |
	Producer-Driven	*Buyer*-Driven
Controlling type of capital	Industrial	Commercial
Capital/technology intensity	High	Low
Labor characteristics	Skilled/high wage	Unskilled/low wage
Controlling firm	Manufacturer	Retailer
Production integration	Vertical/bureaucratic	Horizontal/networked
Control	Internalized/hierarchical	Externalized/market
Contracting/outsourcing	Moderate and increasing	High
Suppliers provide	Components	Finished goods
Examples	Automobiles, computers, aircraft, electrical machinery	Clothing, footwear, toys, consumer electronics

Source: Adapted from Kessler and Applebaum (1998).

To sum up briefly, the distinction between producer- and buyer-driven commodity chains is a useful first step toward understanding commodity chain governance. We summarize and contrast the basic characteristics of producer- and buyer-driven commodity chains in Table 8.1. In reality, however, governance is far more variable both within and between different economic sectors. Equally, a chain may not necessarily be coordinated by either a large manufacturer (producer) or a retailer (buyer). Box 8.3 describes a particular kind of intermediary in coordinating global commodity chains – the Japanese *sogo shosha*. As neither manufacturers nor retailers, but well positioned in global networks of manufacturers and retailers, these very large trading companies possess strong organizational and logistical capabilities to coordinate even the most sophisticated global commodity chains. In the internet era, there is also evidence that new forms of *infomediaries* with extensive access to online consumer information (such as Google and AOL) may play an increasingly important role in mediating global production activity.

Other logistical and supply chain service providers are also playing increasingly important roles in facilitating the global reach of commodity chains. One good example is Hong Kong-based Li & Fung Group, which offers a powerful combination of intermediary and logistics services to its clients. In 2010, the group generated $15.9 billion in revenues from activities in 40 countries employing 27,000 staff and coordinating a sourcing network of over 15,000 suppliers (www.lifung.com, accessed March 8, 2012). While Li & Fung is a broadly based trading, distribution, and retail group, the lion's share of activity is accounted for by its export trading arm, Li & Fung Trading, which manages the entire supply chain for global brands (e.g., Disney, Levi Strauss, and Reebok) and

CASE STUDY

Box 8.3   Trading giants – the Japanese Sogo Shosha

The Japanese *sogo shosha* provide a fascinating example of the importance of logistics and distribution companies in the global economy. *Sogo shosha* translates directly from the Japanese as "general trading company," but this does not really do justice to the range of functions they perform. The *sogo shosha* have long played an important role in the Japanese economy. During the 1960s, they were the first Japanese firms to venture overseas, acting as global marketing and intelligence-gathering networks and thereby facilitating subsequent trade and Japanese foreign direct investment. The seven leading *sogo shosha* – Marubeni, Mitsui, Mitsubishi, Itochu, Nissho-Iwai Nichimen, Sumitomo, and Tomen – are now massive commercial, financial, and industrial conglomerates. They each operate huge networks of subsidiaries and affiliates across the global economy. In 2008, Mitsubishi registered profits of US$4.9 billion on total sales of US$240 billion across 500 affiliates, ranging from an aluminum smelter in Mozambique to the Kentucky Fried Chicken franchise in Japan. In the same year, Mitsui ran 154 offices across 67 countries, had 98 overseas trading subsidiaries, and employed almost 40,000 staff. In 2009, the combined annual sales of these trading companies reached US$835 billion, about a sixth of Japan's GDP (www.bloomberg.com, accessed on March 8, 2012). Each of these firms handles tens of thousands of different products. More specifically, they perform four specific functions (Dicken, 2011):

- *Trading intermediation*: Matching buyers and sellers in long-term contractual relationships
- *Financial intermediation*: Serving as a buffer between buyers and suppliers
- *Information gathering*: Gathering and synthesizing information of market conditions around the world
- *Organization and coordination of complex business systems*: As seen in the case of large infrastructure projects

The role of the *sogo shosha* has changed in several important ways over the past two decades. First, their share of Japanese trade has fallen substantially: in 1991 they accounted for almost 60% of Japan's imports and 50% of exports, while by 2002 these figures had dropped to 22% and 12%, respectively. In part these changes reflect how Japan's manufacturing companies have established their own marketing and sales networks overseas. Second, the nature of the business undertaken by the *soga shosha* has changed, with 79% of foreign affiliates being involved in service activities – including finance, insurance, transportation, and project management – in contrast to the historic focus on manufacturing. Some of them (e.g., Mitsubishi) now use their market knowledge to take control of an entire supply chain and to improve their pricing power (www.economist.com, accessed on March 8, 2012). Third, the *soga shosha* have been accumulating capital throughout the 2000s in order to buy into major assets abroad, particularly in the United States. In 2008 alone, Itochu, the fourth largest *soga shosha* with annual sales of US$130 billion, completed ten deals to the tune of a total of US$2 billion. Its investment campaign aimed to increase to US$6 billion by 2010, a pathway followed closely by other *soga shosha* (www.strategy-business.com, accessed on March 8, 2012). For instance, Mitsui paid US$1.2 billion for Mexican gas power plants in December 2009, and its investment in the Marcellus Shale gas deposits in the United States had reached US$5.4 billion by 2010.

retailers (e.g., Wal-Mart), with a particular specialty in garments. The company controls and coordinates front-end operations such as design and production planning, materials sourcing, and manufacturing, and the back-end tasks of quality control, product testing, and logistics. It is able to draw upon a network of some 15,000 suppliers in meeting the precise needs of its transnational clients although, importantly, it does not own any production capacity itself. Customers are able to place orders and track the production and shipping of products through dedicated online sites. In some cases, Li & Fung manages the entire sourcing operations of retailers. In March 2009, for example, it signed a deal to become the main global apparel and accessories sourcing agent for all Liz Claiborne brands (including Lucky Brand and Mexx). The company bought Liz Claiborne's sourcing operations for around $80 million, with Liz Claiborne paying a commission fee on all products purchased through the Li & Fung trading network. Beyond the large customers, Li & Fung's strength also derives from its ability to match small and medium-sized Western retailers with small and medium-sized manufacturers across Asia while benefiting itself from scale economies derived from its huge buying power.

In this regard, it is useful to think of *relational* forms of governance that fall in between the producer- and buyer-driven models (Gereffi et al., 2005). These can be thought of as close inter-firm relationships that develop on a relatively even footing. The example of the trade in fresh vegetables such as peas and beans between Kenya and the United Kingdom is useful here (Dolan and Humphrey, 2004). Up until the mid-1980s, trade was handled through a series of market relationships between Kenyan farmers and exporters, and U.K.-based importers and traders, wholesalers, and retailers. However, as U.K. supermarkets such as Tesco, Sainsbury's, and Asda have grown in size and expanded their market share, they have started to control and coordinate more directly the commodity chain. In order to attract customers, the supermarkets have introduced new varieties of fresh goods, placed heavy emphasis on quality, provided year-round supply, and increased the processing of products so they require little or no preparation before cooking. As a result, instead of buying vegetables through wholesale markets, they have developed closer, non-market-based relationships with U.K. importers. By the mid-1990s, a relational governance system had emerged in which the supermarkets work directly with a limited number of U.K. importers with whom they have established long-term relationships. The importers have moved beyond a trading role, offering a range of services including processing and handling, monitoring quality, finding new sources of supply, and supporting African producers and assessing their performance. The governance of a commodity chain should always be seen in dynamic terms, however. Over time, the previously buyer-driven relationships between importers and Kenyan exporters are also becoming more relational as the capabilities of the exporters grow and they are able to take on more processing of the produce.

Institutional Contexts

Commodity chain governance, then, is a highly dynamic affair. Its nature does not depend only on the sector or industry in question, but also on the precise array of places that are connected together by the chain. This is because every point on the chain is connected to, and shaped by, the *institutional context* in which it is situated or embedded. In reality, the intersections between commodity chains and their institutional contexts are many and varied. These institutional contexts may be related to international rules and agreements, host country government regulations and preferences, industry-wide standardization and requirements, or even third-party monitoring activity.

We need to discern between different institutional contexts in two ways to make sense of this complexity. First, we can distinguish between *formal* and *informal* institutional frameworks. The former relates to the rules and regulations that determine how economic activity is undertaken in particular places (e.g., trade policy, tax policy, incentive schemes, health and safety/environmental regulations), while the latter describes less tangible, place-specific *ways-of-doing business* that relate to the entrepreneurial and political cultures of particular places. Second, it is useful to think about how institutional context is important at different *spatial scales*. At the subnational scale, local and regional governments may implement a range of policies to try and promote certain kinds of economic development in the locality (such as tax holidays for firms that undertake more research and development, and minimum wage legislation to protect local workers). At the national scale, nation states still wield a huge range of policy measures to try and promote, and steer, economic growth within their boundaries (see Chapter 4 for more).

At the macro-regional scale, a variety of regional blocs have considerable influence on trade and investment flows within their jurisdiction (see the case of the North America Free Trade Agreement in Chapter 10). And at the global scale, institutions such as the World Trade Organization (WTO) and the International Monetary Fund (IMF) shape the rules of the game for global trading and financial relationships. Even a relatively simple global commodity chain, then, will crosscut and connect a wide range of multi-scalar institutional contexts because it involves not just firms and economic actors, but also other non-firm actors involved in the regulation, coordination, and control of the chains. A banana commodity chain linking Ecuador and France, for example, may be affected by corporate strategies of French and American banana importers, Ecuadorian and French economic policies, the rules and regulations of the Andean Common Market and the European Union, WTO rules and regulations, in addition to any more localized policy initiatives within the two countries. In St. Lucia, a small Caribbean island state with a population of about 170,000, two-thirds of its 10,000 banana farmers lost their revenue when its traditional export market, the United Kingdom, had to surrender its preferential trade arrangement with St. Lucia in August 2005. This took place because U.S.-controlled banana companies such as Chiquita filed complaints with the WTO against the British trade preference for its former colony and the WTO ruled in Chiquita's favor. Since then, St. Lucia's banana industry has suffered a terminal decline due to severe competition from other lower cost banana producers in Latin America.

We can further illustrate the profound impacts that changing institutional contexts can have on commodity chains by returning to the example of the coffee industry. Over 90% of the world's coffee is grown in some 60 countries in the developing world (Central and Latin America, Africa, and East and Southeast Asia), while the vast majority of the coffee (worth $70 billion in retail sales each year) is consumed in developed countries (see Figure 8.7). The commodity provides a livelihood for some 25 million coffee-farming families around the world. The coffee commodity chain that straddles these two groups of countries is depicted in Figure 8.8 (Ponte, 2002). The changing institutional context for the coffee chain can be considered at both the global and national scales. At the *global* scale, since 1962 the international trading of coffee has been governed by a successive series of seven International Coffee Agreements (ICAs), most recently concluded in 2007 (and coming into force in 2011) and managed by the International Coffee Organization (ICO), the main intergovernmental organization set up in London in 1963 under the auspices of the United Nations and constituted by representatives of a wide range of coffee importing and exporting countries (http://www.ico.org, accessed on March 8, 2012).

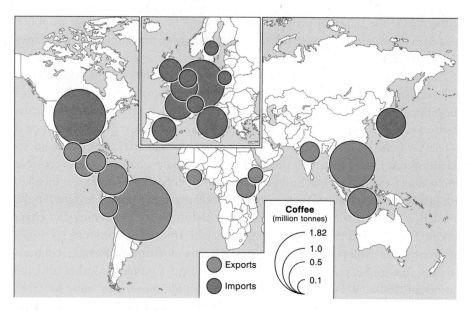

Figure 8.7   The top twelve coffee-producing (exports) and consuming (imports) countries, 2009.
Source: Adapted from Data from www.ico.org, accessed September 21, 2010.

Between 1962 and 1989, the first four ICAs combined price control bands and export quotas to support an international coffee trading system that was widely credited with raising and stabilizing coffee prices, which had fluctuated tremendously during the 1950s due to wars and severe weather conditions. At the *national* scale, many exporting countries established coffee marketing boards. These were government institutions that controlled markets and monitored quality within producing countries and acted as a link to exporters and international traders (Figure 8.8). For the millions of individual farmers and growers, these national coffee boards provided an important buffer between themselves and international markets. On July 4, 1989, however, the quota and controls provisions in the 1983 ICA were suspended in the face of rising production levels and low-cost competition from nonmember exporting countries, most notably Vietnam. Coffee prices thus were at record lows during the 1990–92 period. For example, exports of Robusta coffee grown in the central highlands of Vietnam expanded dramatically during the 1990s – rising from just 100,000 tons in 1990 to 1 million tons by 2005 and remaining at or around that level ever since – creating an oversupply in the global market and putting severe downward pressure on prices.

The ending of the ICA regime has dramatically altered the balance of power in the coffee commodity chain, as the now liberalized, market-based coffee trade regime has led to lower and more volatile coffee prices. Since the early 1990s, bargaining power has been concentrated in the hands of consuming-country firms, and in particular a small group of brand-name roasters and instant coffee

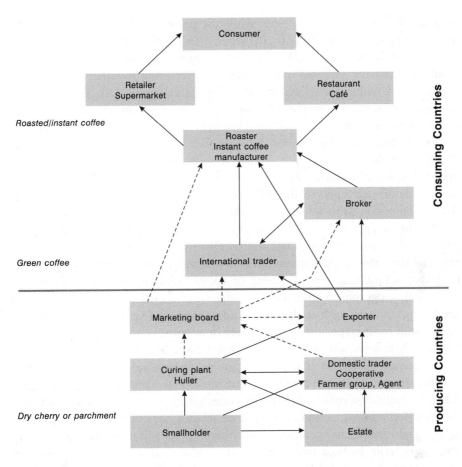

Figure 8.8   The coffee commodity chain: the changing institutional framework
Source: Adapted from Ponte, S. (2002) The 'latte revolution'? Regulation, markets and consumption in the global coffee chain, World Development, 30, 1099–1122, Figure 1.

manufacturers. The domination by these actors is in large part a simple story of concentrated market power and economies of scale: the top four international coffee traders account for 40% of the global trade; the top five roasters handle about 50% of the world's coffee; and the top 30 global grocery retailers account for about one-third of global coffee sales. The ICO has now lost the power to enforce quota and price controls provisions in pre-1989 ICAs and has become merely a forum for promoting *market-enhancing* coffee policies and consumption. Its latest 2007 ICA states, for example, that ICO members recognize the market distorting measures that may hinder the increase in coffee consumption.

Meanwhile, large coffee roasters and manufacturing firms are increasingly applying stringent quality standards that have implications all the way down the chain to the farmers. It has been estimated that the percentage of income from the coffee chain that is retained in the developed markets has gone up from

55% to 78% in the post-1989 era, while the proportion of income accrued by growers has fallen from 20% to 13% (see also Figure 8.4). At the same time, the national coffee marketing boards in the exporting countries have either been eliminated or have retreated into a restricted overseeing role that has left them marginalized within the global commodity chain (Figure 8.8). Buyer–grower relationships are now essentially "arms-length" *market* connections, with prices being set by international commodity markets such as the New York Board of Trade (Arabica) and the London International Financial Futures and Options Exchange (Robusta). As a result of these changes to the interlinked international and national institutional contexts, millions of coffee smallholders and farmers worldwide are left open to the full brunt of the global coffee market and its fluctuating prices. At its worst, this can lead to a situation where farmers receive less for their crop than it actually costs them to produce it. The example of coffee clearly shows how changing institutional frameworks can significantly affect all three of the other basic dimensions of a commodity chain, namely the input-output structure (e.g., the bypassing of coffee marketing boards in exporting countries), geographical structure (e.g., the rapid growth of production in new producing countries such as Vietnam), and governance (e.g., the accumulation of power with roasting/processing firms from developing countries).

Overall, this section of the chapter has shown how commodity chains are organizational platforms that link distant producers and consumers together, within certain institutional contexts, across the global economy. The precise form taken by individual commodity chains varies greatly – both within and between different sectors of the economy – in terms of their structure, geography, governance, and institutional context. Understanding this variability and complexity empowers us in at least two ways. First, it is a vital step toward identifying the different *winners* and *losers* in the globalization of commodity chains. Such an understanding is crucial not just because it is about corporate entities. Even more importantly, it is also about the people who work for these firms and the conditions of their workings. Second, a fuller appreciation of global commodity chains can enable us to *change* how commodity chains function and how they are regulated/governed. We now move on to consider the potential for different ways of trying to change, or re-regulate, commodity chains in ways that negate some of the less desirable aspects of commodity production, circulation, and consumption.

8.4   Re-regulating Commodity Chains: The World of Standards

There are many approaches through which different players may seek to alter the prevailing ways in which a commodity chain operates. For example, in Chapter 6 we examined how certain groups of workers have the necessary *agency* to challenge established ways of working in their industry through engaging in different forms of *production* politics, while in Chapter 15 we will explore the potential for different forms of *consumption* politics – interventions initiated at the consumption end of commodity chains through a desire to improve conditions at various points up the chain. Here, we focus on how global commodity chains are increasingly being re-regulated through efforts to impose consistent *standards* on the ways commodities are produced and distributed within the global economy (Nadvi, 2008).

As Table 8.2 illustrates, however, the world of standards is a broad and varied one. First, they can be applied to different aspects of the commodity chain – for example, health, quality, environment, or labor conditions. Second, the precise

Table 8.2   The world of standards

Attribute of Standard	Variability
Field of application	<ul><li>Quality assurance</li><li>Environmental</li><li>Health and safety</li><li>Labor</li><li>Social/economic</li><li>Ethical</li></ul>
Form	<ul><li>Codes of conduct</li><li>Label</li><li>Standard</li></ul>
Coverage	<ul><li>Firm/commodity chain specific</li><li>Sector specific</li><li>Generic</li></ul>
Key drivers	<ul><li>International business</li><li>International NGOs</li><li>International trade unions</li><li>International organizations</li></ul>
Certification process	<ul><li>First, second, or third party</li><li>Private-sector auditors</li><li>NGOs</li><li>Government</li></ul>
Regulatory implications	<ul><li>Legally mandatory</li><li>Market competition requirement</li><li>Voluntary</li></ul>
Geographical scale	<ul><li>Regional (e.g., a U.S. state)</li><li>National</li><li>Macro-regional (e.g., the EU)</li><li>Global (e.g., the UN Global Compact)</li></ul>

Source: Adapted from Nadvi and Waltring (2004).

form they take may be a code of conduct, a label on a finished product, a tightly specified technical standard, a set of voluntary initiatives, or some combination of all of these. Third, the standard may apply to a particular chain (e.g., fresh tomatoes or dolphin-safe tuna), a sector (e.g., fresh fruit and vegetables), or be generic (e.g., a safety standard to sell an electrical good in a particular national market). Fourth, while consumer campaigns are clearly the domain of end-consumers (and sometimes labor unions), standards may be developed by firms, NGOs, trade unions, or international organizations (such as the UN Global Compact), and usually involve a combination of some, or all, of these institutions. Fifth, the certification or accreditation of standards (i.e., the assessing of whether they have been met) may be undertaken by a variety of different parties (both public and private, for profit and not for profit). Sixth, the regulatory impacts will vary from the voluntary (e.g., seeking "Fairtrade" status for a product) to the mandatory (e.g., safety standards for plastic toys, organic certifications for food, and environmental certification for forestry products). Finally, standards are innately *geographical* in two different ways: in terms of the territory within which they apply to the consumption of particular products or commodities, and in the way in which they have implications further down a chain that may connect many disparate territories (see Box 8.4 on the environmental certification of dolphin-safe tuna that connects actors in California and Thailand, among others).

CASE STUDY

Box 8.4 Environmental certification of the dolphin-safe tuna commodity chain

In the early 1990s, the Earth Island Institute (EII), a California-based non-governmental organization, initiated a highly successful consumer-driven global environmental campaign. The EII scheme has played a key role in defining, monitoring, and regulating the use of the label/term "dolphin-safe" in the tuna-packing industry (Baird and Quastel, 2011). At the time of the initiative, much of the industry was concentrated in Thailand, with 25 different tuna-canning firms in operation. Thailand's Unicord, the world's largest tuna canner at the time, was canning about 500 metric tons of tuna daily and employing more than 7,000 people. The main market for canned tuna was the United States, followed by other developed countries such as the United Kingdom, Italy, Germany, and France. As described in Figure 8.9, the project is a certification scheme premised on perceived consumer demand, product labels, U.S. government labeling laws and UN resolutions, and strong advocacy in the media. In other words, the initiative is predicated on a complex interwoven set of environmental discourses in the name of marine mammal protection (e.g., the U.S. Marine Mammal Protection Act 1972) and nongovernmental and private regulatory efforts by EII and private-sector adopters. The scheme's success can be measured in terms of the significant decline in dolphin deaths and injuries associated with the harvesting of yellow-fin tuna in the eastern tropical Pacific and other tuna caught by drift-netting in the western Pacific and Indian Oceans. From an estimated number of 252,000 dolphin deaths associated with tuna fishing in 1973, the number declined to 8,258 in 1984 and about 1,000 in 2008 (www.earthisland.org, accessed on March 8, 2012).

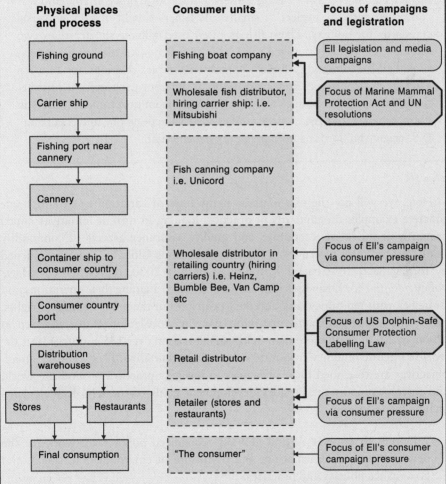

Figure 8.9 The regulation of the dolphin-safe tuna packaging industry by nongovernmental organizations
Source: Adapted from Baird, I. and Quastel, N. (2011), "Dolphin-safe tuna from California to Thailand: localisms in environmental certification of global commodity networks", *Annals of the Association of American Geographers*, 101(2), 337–55, Figure 2.

> The geography of this label regulation is particularly fascinating. As EII reached out to Thailand in the early 1990s, the monitoring process was widely adopted by all of Thailand's canneries primarily because EII had threatened to ruin their reputations for violating dolphin-safe standards and EII's system would bring legitimacy to their new labeling practices. By 1993, significant differences between the Thai and U.S. branches of EII were developed in relation to their framing of the certification standards for dolphin-safe tuna. The U.S. head office preferred a single universal standard to regulate the labeling of dolphin-safe tuna worldwide that would include banning *all* fishers using gillnets. But the local office in Thailand was concerned with the neglect of small-scale fishers caught in EII's universal standards for gillnets. These fishers often used gillnets shorter than two kilometers and rarely targeted tuna. This case demonstrates that the regulation of global commodity chains depends on complex relationships between a whole host of actors and institutions in different geographical jurisdictions. Today, EII continues to face challenges from government leaders in Latin America and Asia who have branded its actions of "universalizing" U.S.-origin standards as fundamentally neocolonial.

Here, we will use the first of these many lines of variation to explore some concrete examples of standards in practice, looking in turn at initiatives aimed primarily at the labor, economic, and quality assurance aspects of commodity chains. At the most global level, the United Nations Global Compact, launched by former Secretary-General Kofi Annan in July 2000, represents a *strategic policy initiative* for businesses that are committed to aligning their operations and strategies with ten universally accepted principles in the areas of human rights, labor, environment, and anti-corruption (www.unglobalcompact.org, accessed on March 8, 2012). With more than 8,500 signatories in over 135 countries, it is the world's largest voluntary corporate sustainability initiative. The Compact's main objectives are to embed these principles in business practices around the world and thereby support the broader development goals of the UN. Key elements include the following:

- Adopting an established and globally recognized policy framework for the development, implementation, and disclosure of environmental, social, and governance policies and practices.
- Sharing best and emerging practices to advance practical solutions and strategies to common challenges.
- Advancing sustainability solutions in partnership with a range of stakeholders, including UN agencies, governments, civil society, labor, and other nonbusiness interests.

- Linking business units and subsidiaries across the value chain with the Global Compact's local networks around the world, many of these in developing and emerging markets.
- Accessing the UN's extensive knowledge of and experience with sustainability and development issues.
- Utilizing UN Global Compact management tools and resources, and the opportunity to engage in specialized workstreams in the environmental, social, and governance realms.

In contrast to the generic focus on good business practice in the UN Global Compact, the fair trade movement is more centrally concerned with the *economic returns* that primary producers secure for their commodities. One of the leading examples is the U.K. Fairtrade Foundation, which was established in 1992 by a group of charities and NGOs and is part of the Fairtrade Labelling Organizations International, which coordinates 21 national initiatives across Europe, North America, Japan, and Australia/New Zealand. The Fairtrade mark is a certification label awarded to products sourced from the developing world that meet international standards of fair trade. Under Fairtrade labeling, there are two sets of standards, one that applies to small farmers, and one for workers in plantations and processing factories. These standards (www.fairtrade.org.uk, accessed March 8, 2012) stipulate that traders must:

- Pay a price to producers that covers the costs of sustainable production and living: the Fairtrade minimum price.
- Pay a "premium" that producers can invest in social, environmental, and business improvements: the Fairtrade Premium.
- Make partial advance payments when requested by producers.
- Sign contracts that allow for long-term planning and sustainable production practices.
- Implement traceability systems and record-keeping systems.
- Ensure subcontractors' compliance with the relevant requirements.

By April 2010, the foundation was working with over 440 producer organizations in 55 countries, with benefits extending to approximately 5 million farmers, workers, and family members. Sales of Fairtrade mark goods have expanded rapidly in the United Kingdom, up almost fifty-fold from £16.7 million in 1998 to £799 million in 2009. The mark can be found on 13 categories of foodstuffs, of which coffee, bananas, chocolate/cocoa, and tea are the most important, and four nonfood products including cotton and cut flowers. Altogether over 7,275 products are Fairtrade certified now, with 1,137 new Fairtrade products certified in 2009 alone. Leading Fairtrade brands in the United Kingdom include Cafédirect (coffee and tea) and Traidcraft (food, drinks, crafts and gifts), and some supermarkets such as Tesco are now developing their own Fairtrade brands. While

Fairtrade is based on a central code/standard, it also signifies to the consumer that a product meets that standard through its labeling system.

In an even more specific form, multi-stakeholder organizations can be established to regulate *labor standards and conditions* in global commodity chains. The U.K.'s Ethical Trading Initiative (ETI) is a multi-stakeholder organization established in 1997, and funded jointly by its membership and the U.K. government's Department for International Development (DFID) (the Fair Labor Association is a similar initiative in the U.S. context). By mid-2011, the ETI was constituted by 72 corporate members with a combined turnover of over £107 billion (including leading retailers and suppliers selling into the U.K. market, such as Asda, The Gap, Tesco), 15 NGOs (including many charities such as Christian Aid and Oxfam), and representatives from nine international trade unions representing almost 160 million workers worldwide. The ETI has established a Base Code for labor conditions in supply chains consisting of the following nine provisions (www.ethicaltrade.org, accessed on March 8, 2012):

- Employment is freely chosen
- Freedom of association and the right to collective bargaining are respected
- Working conditions are safe and hygienic
- Child labor should not be used
- Living wages are paid
- Working hours are not excessive
- No discrimination is practiced
- Regular employment is provided
- No harsh or inhumane treatment is allowed

While most of the retailers use large international, independent social auditors such as Bureau Veritas, some pass on the auditing role to key suppliers, and others seek the involvement of NGOs with an on-the-ground presence in key source areas (Hughes, 2005). Interestingly, the ETI explicitly chooses not to pursue a label, arguing that it is simply not possible for a retailer or a brand-name merchandiser to know about labor conditions at all points on every commodity chain, and also that several of the Base Code provisions rely partly on government action in source countries (e.g., the right to form a labor union).

Apart from these broad institutional initiatives, recent decades have also witnessed the rise of *private standards* imposed on different commodities, particularly agrifoods. These standards allow individual firms (buyers) to be more hands-off through a new form of collective governance of supplier chains. Through a market-based mode of self-regulation among suppliers, these buyers are able to discipline at a distance their suppliers elsewhere who are expected to conform to normal (good or best) farming practices. Global-GAP (good agricultural practices), for example, is a scheme launched as a business-to-business standard by a coalition of mainly British and Dutch retailers in 1997 to harmonize existing disparate and chain-specific supermarket protocols on food safety and quality (www.globalgap.org, accessed on March 8, 2012). Since then, the scheme has expanded significantly beyond an exclusively retailer-based model to a more open multi-stakeholder network that includes retailers, producers, input suppliers, and certification bodies.

In coffee production, there is a wide range of schemes currently in existence to promote sustainable production (Table 8.3). The seven listed schemes clearly illustrate the degrees of variation described earlier. For example, they may apply primarily to environmental or economic conditions (bird-friendly vs. Fairtrade, respectively); involve a label or be primarily a code of conduct (Fairtrade vs. Utz Kapeh); and be global or regional in their coverage (Global-GAP vs. bird-friendly). Moreover, schemes may be initiated by single private companies (such as Starbucks' CAFE Practices), groups of companies (e.g., Global-GAP), civil society organizations (e.g., Fairtrade) or multi-stakeholders organizations (such as 4C) encompassing a range of domains. A growing proportion of global coffee falls under these schemes that are rapidly becoming mainstream: while the overall level was estimated at just 1% of total production in 2003, by 2009 Utz Kapeh covered 5% of the total, Rainforest Alliance 1.5%, Bird Friendly 1.2%, and Fairtrade 1%. In the same year, Starbucks' CAFE practices applied to 81% of their coffee purchases – equivalent to 136,000 tons of coffee – with a target of 100% coverage by 2015.

At the more general and international level, ISO9000 (quality) and ISO14001 (environment) provide excellent examples of *generic management systems standards* that can be applied to any organization, large or small, whatever its product or service, in any sector of activity, and whether it is a business enterprise, a public administration, or a government department. The Geneva-based International Organization for Standardization (ISO) is perhaps the best-known global standards organization, and brings together 162 national standards institutes from 105 member countries. Its central objective is to facilitate international trade and investment by harmonizing national standards with international ones. While many ISO standards are technical, the ISO9000 standards (of which ISO9001:2008 is the current version) and ISO14001 (of which ISO14001:2004 is the current version) standards are procedural, outlining a comprehensive set of quality and environmental management practices. Their purpose is to provide external assurance to customers by demonstrating that a supplier is conforming to a formalized quality and environmental management system. The requirements of ISO9000, which are essentially a set of written rules, cover a wide range of activities such as a firm's customer focus, quality planning, product design, review of incoming orders, and monitoring of customer perceptions about the quality of the goods and services it provides. ISO14001 requires an organization of any size

Table 8.3 Certification schemes for sustainable coffee

Name	Year Established	Actors or Organizations Setting the Standards	Characteristics	Geographic Coverage
Fairtrade	1988 (in Netherlands)	Fairtrade Labelling Organizations International (FLO)	Minimum guaranteed price paid to registered small farmers' organizations that match standards on socioeconomic development	Global
Rainforest Alliance certified (shade-grown)	1996	Rainforest Alliance	Certifies farms on the basis of sustainability standards; covers environmental protection, shade, basic labor and living conditions, and community relations	Primarily Latin America
Utz Kapeh	1997	Utz Kapeh Foundation	Includes standards on environmental protection and management, and labor and living conditions	Global
Bird-friendly coffee (shade-grown)	2000	Smithsonian Migratory Bird Center (SMBC)	Minimum standards on vegetation cover and species diversity needed to obtain use of label; also covers soil management	Latin America
Global-GAP Code for Green Coffee	2003 (Eurep-GAP until 2005)	Global-GAP Organisation	Designed to offer assurances about how coffee is grown on the farm with respect to environmental impacts, the use of chemical inputs, and worker health and safety	Global
Starbucks CAFE (coffee and farmer equity) practices	2004 (guidelines established in 2001)	Starbucks Corporation	Guidelines designed to ensure high-quality coffee, protect the environment, and promote fair relationships between farmers, workers, and communities	Global
Common Code for the Coffee Community (4C)	2005	4C Association (Germany)	To achieve baseline levels of social, ecological, and economic sustainability for all types of coffee production systems	Global

Source: Adapted from Daviron and Ponte (2005), Table 5.11; and Neilson and Pritchard (2009), Table 6.1.

Table 8.4   Regional share of ISO9001:2000/2008 certificates (December 2009)

Region	Number of Countries	Number of Certifications	Share of World Total (%)
Africa/West Asia	69	77,408	7.3
Central and South America	32	36,551	3.4
North America	3	41,947	3.9
Europe	50	500,319	47.0
Far East	22	398,288	37.4
Australia and New Zealand	2	10,272	1.0
World	178	1,064,785	100.0

Source: www.iso.org (accessed on July 27, 2011).

or type to do the following: identify and control the environmental impact of its activities, products, or services; improve its environmental performance continually; and implement a systematic approach to setting environmental objectives and targets and monitoring progress toward achieving them (for more details, see www.iso.org, accessed on March 8, 2012).

The system can be certified in different ways: by a firm itself, by a customer firm, or by a third-party accreditor. Adoption of the standard is growing rapidly. ISO9001:2000, which was established in December 2000, expanded from 44,388 certifications in December 2001 to 1,064,785 in 178 countries and economies by December 2009. Interestingly, the global geography of the uptake is highly uneven, with Europe and East Asia together accounting for over 84% of total certifications (see Table 8.4). At the country level, China, Italy, Japan, Spain, Russia, and Germany are the top six adopters (in that order), with China alone accounting for 257,076 certifications in late 2009 (24.2% of the global total). Mapping ISO9000 certification in this way is an interesting indicator of rates of trade growth and international connections with the global economy. Most importantly here, however, ISO9000 is illustrative of the way in which a global standard can transmit expectations about appropriate forms of undertaking business along a global commodity chain.

We have now introduced several regulatory initiatives that seek to standardise and improve a wide range of business practices along commodity chains. However, while the world of standards continues to broaden and deepen in terms of range and reach, we should not make the mistake of assuming that such initiatives always ensure the desired outcomes. The reality is often a complicated mosaic of intended and unintended consequences that may reflect pre-existing power relations within the global economy (see Box 8.5).

FURTHER THINKING

Box 8.5   *The limits to standards?*

Section 8.4 has introduced the increasingly complex world of standards involved in the ongoing regulation of commodity chains. However, there is always the danger that the initiatives themselves can become incorporated into, or even commodified as part of, the capitalist system – that is, as another way for business firms to make money. Hence, it is important to think carefully, for instance, about the motives of firms for participating in standards schemes, with many being based on the *voluntary* participation of corporations. Another potential pitfall may lie with the simple issue of *who* pays for these standards and regulatory initiatives. The rise of standards-based schemes has fueled the emergence of a new category of commodity chain participant, the independent auditor – such as Scientific Certification Systems (SCS), for example, who audit the Starbucks scheme – many of whom themselves are profit-seeking firms. As a form of private agrifood standard, the case of Global-GAP (see Table 8.3) is a retailer-driven attempt to impose a market-based mode of self-regulation that pitches individual farmers' skills against each other. In effect, the global standard serves as a significant *barrier to entry* that increases production and certification costs to the extent that only the best organized and most well-funded farmers can make production for developed-country markets into a profitable exercise. Other critiques of such regulatory initiatives are more explicitly *geographical*. Many of the codes of conduct or general standards are by necessity quite crude abstractions and simplifications. A simple statement such as "living wages are paid" sounds promising on a first read, but in reality a living wage will vary dramatically from place to place, and even between different social groups within the same locality. Calculating meaningful wage rates would be hugely expensive and time-consuming. Equally, by focusing primarily on formal employment relations, these initiatives may make less of an impression in sectors and places where there is a significant informal (and often female) component to the workforce. This speaks to the wider issue that standards – and understandings of what is acceptable or unacceptable – are geographically specific. As a result, the implementation of standards may have ambiguous impacts in the areas they are meant to help or promote (see also Box 8.4). The result may be standards that are detached from sectoral and geographical contexts and hence serve to worsen the position of certain groups of people in particular places. Overall, while there are many positive aspects to different regulatory schemes of standards and codes of conduct, we must interrogate them with the same critical economic-geographical perspective that we would any aspect of the economic system.

8.5   Where Does a Commodity Chain End? From Waste to Commodities Again

Our discussion has so far assumed a linear structure of commodity chains such that inputs, materials, actors, and institutions in these chains can be clearly identified and regulated. It is important to question, however, the assumption of the linearity of commodity chains. The purchase and primary use of a commodity is often seen to be the end point of the production system or commodity chain. While this assumption provides useful limits to our analysis, in reality many commodity chains are interconnected – the end point of one commodity chain may well be the beginning of another commodity chain. This circularity and interconnectedness in commodity chains becomes obvious when we examine waste and/or the recycling of materials.

In general, processes of waste disposal and recycling have received little attention in academic studies and the popular media, and are assumed to be the final stage in a linear and sequential process. Exciting recent research by economic geographers, however, has shown how processes of disposal and recycling can serve at the starting point for new commodity chains as materials move through complex circuits of use and reuse. Two examples will be drawn upon here, both involving Bangladesh in South Asia where three-quarters of the world's merchant ships are dismantled and recycled as raw materials for industrial production. In 2009, an estimated value of US$1 billion worth of ships was imported to Bangladesh for such recycling activity, representing some 25% of all ships scrapped in the world. Sitakunda Beach, a 20-km stretch of shoreline just north of Chittagong, Bangladesh, is one of the world's foremost locations for ship-breaking (Gregson et al., 2010, 2012). Thousands of men working for some 37 ship-breaking yards literally take large ships apart piece by piece. There may be 20 to 25 ships being dismantled along this stretch of coast at any one time. The furnishings, fittings, and furniture are stripped out and then the ship itself is broken up, with the resulting plates and metal bars being taken to the nearby rerolling mills, which provide 90% of the steel used in Bangladesh.

This though, is not the end of the story, but rather the beginning of others. For example, the steel is used in the reinforcing rods for the concrete being poured on building sites across the country, and the motors, boilers and compressors are reconditioned and used in the export garment sector. The ship furniture is repaired, reconditioned, and renovated by over 70 furniture units in the town of Bhatiary on the Dhaka-Chittagong highway, in an industry employing an estimated 10,000 workers. This furniture then takes on a new life as a desirable commodity for the emerging middle classes of Chittagong and other Bangladeshi cities, accounting for up to 40% of all reconditioned household goods sold in the country. The breaking up of the ship effectively serves as the starting point for a plethora of new commodity chains emanating outward from the beach at

Sitakunda. It also compensates for the lack of indigenous production in these commodities in Bangladesh.

A similar story can be told about electronic waste, with an estimated 60,000 workers in Dhaka being involved in the capture and creation of value from unwanted electronics products (Lepawsky and Billah, 2011). Money can be generated by a range of different processes including resale, refurbishing, remanufacturing, repair, and dismantling. In terms of remanufacturing, for example, old computer monitors can be converted into low-cost television sets and video game monitors. Even simply dismantling an old computer monitor and selling the constituent parts can generate profits of 230%. Once again, the recycling of electronic waste marks the start of a new round of commodity stories and journeys. What both these examples serve to illustrate is that we perhaps need to reconsider the notion of the end point of a commodity chain not simply as "waste," but to think in more sophisticated ways about the ongoing transformations of materials in the different overlapping commodity chains that underpin the various products that we consume.

8.6 Summary

As an economic system, capitalism serves to conceal the intense connections between distant producers and consumers that underpin its operation. The result is that answering even such a deceptively simple question as "Where does your breakfast come from?" requires substantial "geographical detective work" (Hartwick, 2000, p. 1178). While labels may reveal the country of origin of a particular product, they tell us little about the working conditions under which it was produced. In this chapter, we have seen how the notion of the commodity chain allows us to chart the geographical journeys taken by commodities, as they are transformed from initial raw materials and ideas into finished products and services, thereby serving to connect producers and consumers. The precise nature of the journey taken will vary tremendously from commodity to commodity. Each and every commodity chain is delineated by a particular sequence of value-adding activities, a distinct geographical configuration, different combinations of modes of governance, and various institutional contexts. Commodity chains are thus important – and inherently geographical – *organizational features* of the contemporary global economy; they are the hidden social relations that enable capitalism to extend its global reach. Conceptually, the commodity chain is an extremely important integrative idea that allows us to reveal the interconnections between the many actors – states, environments, and workers – that we considered in depth in Part II of this book.

The careful mapping of commodity chains opens up the potential for intervening in the nature and operation of those chains. We have reviewed examples of initiatives through which consumers, governments, labor unions, and businesses, either working individually or together, can seek to alter the conditions under which certain commodities are produced. In particular, the rise of different forms of standards is highly significant. While some of these schemes have had noticeable impacts, it is also important to recognize their potential limitations. On the one hand, they are susceptible to being incorporated into the profit-seeking capitalist mainstream. On the other hand, unless attempts to re-regulate commodity chains are alert to their inherent geographic variation and tension, they in turn will have uneven, and perhaps unintended, outcomes. Finally, we explored the difficulty of delineating when and where exactly a commodity chain ends. When we consider the waste products from one particular chain, we can reveal how they provide the material inputs for new commodity systems. In the next chapter we move on to consider how technological developments are altering the geography and operation of contemporary production systems.

Notes on the references

- Gereffi (1994) and Gereffi et al. (2005) are two seminal statements on the global commodity/value chain perspective. See Bair (2009) and Gibbon et al. (2008) for two recent collections of work in this genre, and Dicken et al. (2001) and Coe et al. (2008) for geographical critiques.
- For more details on the coffee, catfish, clothing, and fresh vegetable commodity chains discussed in this chapter see Daviron and Ponte (2005), Duval-Diop and Grimes (2005), Tokatli (2008), and Dolan and Humphrey (2004), respectively.
- For other excellent commodity chain studies, see Cook (2004) on papaya, Neilson and Pritchard (2009) on coffee and tea, Franz and Hassler (2010) on organic peppers, Ivarsson and Alvstam (2011) on furniture, Oro and Pritchard (2011) on beef, and Patel-Campillo (2011) on cut flowers.
- Nadvi and Waltring (2004), Nadvi (2008), and Ouma (2010) are excellent studies of global standards in different industries.
- Gregson et al. (2010, 2012) and Lepawsky and Billah (2011) offer some outstanding materials on how "waste" is recycled in different commodity chains.

Sample essay questions

- How does a commodity chain approach enable us to reconnect distant producers and consumers in the global economy?
- How and why are commodity chains governed in different ways?
- In what ways might the institutional contexts of a commodity chain affect its structure and operation?

- What are the pros and cons of adopting standards-based approaches to commodity chain regulation?
- When and where does a commodity chain end?

Resources for further learning

- http://www.globalvaluechains.org: This site contains a wealth of conceptual and empirical material on global value chains.
- http://www.sourcemap.org: An open source site that literally allows users to map commodity chains for different products.
- http://www.unido.org: The United Nations Industrial Development Organization (UNIDO) website also offers a wide range of data and reports on different commodity chains and the potential they offer for economic development in different localities.
- http://www.ico.org/index.asp: The website of the International Coffee Organization provides a range of information on the coffee industry, and in particular, its evolving regulatory structures.
- http://www.fairtrade.org.uk: The U.K.'s Fairtrade Foundation is one of the most well-known attempts to improve the economic returns offered to commodity producers.
- http://www.fairtracing.org: The website of a research project seeking to support ethical trade.
- http://www.iso.org/iso/en/ISOOnline.frontpage: The International Standards Organization's website is rich in information on a variety of international standard programs.
- http://www.thewasteoftheworld.org: The website of The Waste of the World program, which is concerned with recycling and waste economies within global recycling networks. The project spans sites in South Asia (Bangladesh and India) as well as the United Kingdom and continental Europe.

References

Bair, J., ed. (2009). *Frontiers of Commodity Chain Research*. Stanford: Stanford University Press.

Baird, I., and Quastel, N. (2011). Dolphin-safe tuna from California to Thailand: localisms in environmental certification of global commodity networks. *Annals of the Association of American Geographers*, **101**, 337–55.

Coe, N., Dicken, P., and Hess, M. (2008). Global production networks: realizing the potential. *Journal of Economic Geography*, **8**, 271–295.

Cook, I. (2004). Follow the thing: papaya. *Antipode*, **36**, 642–64.

Daviron, B., and Ponte, S. (2005). *The Coffee Paradox: Global Markets, Commodity Trade and the Elusive Promise of Development*. London: Zed Books.

Dicken, P. (2011). *Global Shift*, Sixth Edition. London: Sage.

Dicken, P., Kelly, P. F., Olds, K., and Yeung, H. W. C. (2001). Chains and networks, territories and scales: towards an analytical framework for the global economy. *Global Networks*, 1, 89–112.

Dolan, C., and Humphrey, J. (2004). Changing governance patterns in the trade of fresh vegetables between Africa and the United Kingdom. *Environment and Planning A*, 36, 491–509.

Duval-Diop, D. M., and Grimes, J. R. (2005). Tales from two deltas: catfish fillets, high-value foods, and globalization. *Economic Geography*, 81, 177–200.

Franz, M., and Hassler, M. (2010). The value of commodity biographies: integrating tribal farmers in India into a global organic agro-food network. *Area*, 42, 25–34.

Gereffi, G. (1994). The organization of buyer-driven global commodity chains: how U.S. retailers shape overseas production networks. In G. Gereffi and M. Korzeniewicz, eds. *Commodity Chains and Global Development*, pp. 95–122. Westport, Conn.: Praeger.

Gereffi, G., Humphrey, J., and Sturgeon, T. (2005). The governance of global value chains. *Review of International Political Economy*, 12, 78–104.

Gibbon, P., Bair, J., and Ponte, S., eds. (2008). Special issue on governing global value chains. *Economy and Society*, 37, 315–459.

Gregson, N., Crang, M., Ahamed, F., Akhter, N., and Ferdous, R. (2010). Following things of rubbish value: end-of-life ships, 'chock-chocky' furniture and the Bangladeshi middle class consumer. *Geoforum*, 41, 846–854.

Gregson, N., Crang, M., Ahamed, F., Akter, N., Ferdous, R., Foisal, S., and Hudson, R. (2012). Territorial agglomeration and industrial symbiosis: Sitakunda-Bhatiary, Bangladesh, as a secondary processing complex. *Economic Geography*, 88, 37–58.

Hartwick, E. R. (1998). Geographies of consumption: a commodity chain approach. *Environment and Planning D*, 16, 423–437.

Hartwick, E. R. (2000). Towards a geographical politics of consumption. *Environment and Planning A*, 32, 1177–1192.

Hughes, A. (2005). Corporate strategy and the management of ethical trade: the case of the U.K. food and clothing retailers. *Environment and Planning A*, 37, 1145–1163.

Humphrey, J., and Schmitz, H. (2004). Chain governance and upgrading: taking stock. In H. Schmitz, ed., *Local Enterprises in the Global Economy*, pp. 349–377. Cheltenham: Edward Elgar.

Ivarsson, I., and Alvstam, C. G. (2011). Upgrading in global value-chains: a case study of technology-learning among IKEA-suppliers in China and Southeast Asia. *Journal of Economic Geography*, 11, 731–752.

Kessler, J., and Appelbaum, R. (1998). The growing power of retailers in producer-driven commodity chains: a "retail revolution" in the U.S. automobile industry? Unpublished manuscript, Department of Sociology, University of California at Santa Barbara.

Lepawsky, J., and Billah, M. (2011). Making chains that (un)make things: waste-value relations and the Bangladeshi rubbish electronics industry. *Geografiska Annaler B*, 93, 121–139.

Nadvi, K., and Waltring, F. (2004). Making sense of global standards. In H. Schmitz, ed., *Local Enterprises in the Global Economy: Issues of Governance and Upgrading*, pp. 53–94. Cheltenham: Edward Elgar.

Nadvi, K. (2008). Global standards, global governance and the organization of global value chains. *Journal of Economic Geography*, 8, 323–44.

Neilson, J., and Pritchard, B. (2009). *Value Chain Struggles: Institutions and Governance in the Plantation Districts of South India*. Oxford: Wiley-Blackwell.

Oro, K., and Pritchard, B. (2011). The evolution of global value chains: displacement of captive upstream investment in the Australia–Japan beef trade. *Journal of Economic Geography*, **11**, 709–29.

Ouma, S. (2010). Global standards, local realities: private agrifood governance and the restructuring of the Kenyan horticulture industry. *Economic Geography*, **86**, 197–222.

Oxfam (2003). *Mugged: poverty in your coffee cup*. Oxford: Oxfam International.

Patel-Campillo, A. (2011). Transforming global commodity chains: actor strategies, regulation and competitive relations in the Dutch cut flower sector. *Economic Geography*, **87**, 79–99.

Ponte, S. (2002). The "latte revolution"? Regulation, markets and consumption in the global coffee chain. *World Development*, **30**, 1099–1122.

Tokatli, N. (2008). Global sourcing: insights from the global clothing industry – the case of Zara, a fast fashion retailer. *Journal of Economic Geography*, **8**, 21–38.

Watts, M. (1999). Commodities. In P. Cloke, P. Crang, and M. Goodwin, eds., *Introducing Human Geographies*, pp. 305–315. London: Arnold.

CHAPTER 9

TECHNOLOGICAL CHANGE
Is the world getting smaller?

Goals of this chapter

- To demonstrate how certain technologies can be used to transcend time and space
- To understand how technologies can be used to generate both new products and new production processes
- To appreciate the variable spatial impacts of technology on economic systems
- To recognize how the geography of technology production is itself highly uneven

9.1   Introduction

In a cramped room in Changsha, China, over 100 employees work ten-hour shifts earning virtual gold, games items, and skill levels within the *World of Warcraft* computer game for clients in North America and Western Europe. They work for a firm called Wow7gold, which earns annual revenues of over US$1.5 million. The company also has sales, advertising, and research departments, where highly skilled female graduates can earn up to US$400 a month working as customer service operators. There are estimated to be at least 50,000 such firms in China with an average size of 2–10 computers. Regulations on working hours and conditions are lax. The young, largely unskilled "playborers" earn around US$130–200 a month, and although communal meals and accommodation are provided by their employer, given the long shifts, this equates to an hourly wage of just US$0.50. For busy clients in developed countries, however, employing a firm such as Wow7gold is a time- and cost-efficient way of circumventing the more mundane aspects of progressing through virtual worlds.

What is being described here is part of a growing phenomenon known as "gold farming." What started out as a cottage industry in the late 1990s has now mushroomed into a significant sector employing some 400,000 people across many developing countries and generating estimated revenues of over US$1 billion each year. Gold farmers sit at an internet-connected computer and play online games such as *Everquest*, *Final Fantasy*, *Runescape*, and *World of Warcraft* for the purposes of generating "real" money. Several such games operate virtual economies that use in-game currencies: workers play these games to gain currency that they can then sell over standard online payment systems such as PayPal to their clients – other players of the game, often on the other side of the world. The tasks undertaken may also include "power-leveling" – improving the skill/combat level of their customer's character – or the selling of valuable in-game items. China is the main home to this activity, with its dynamic coastal cities accounting for an estimated 80–85% of global employment, but gold farming for the world market also occurs in Mexico, Romania, Russia, and Indonesia, while domestic market-oriented services have been detected in India, Malaysia, South Korea, and the Philippines. With the annual subscriptions to games such as *World of Warcraft* growing by over 10% a year, gold farming seems likely to increase in scale and significance in the coming years (see Heeks, 2008, and *The Guardian*, 2009, for more on the phenomenon).

At first glance, the rapid expansion of gold farming appears to illustrate clearly the transformative potential of new information and communications technologies such as the internet. Over 50 million people now engage in online gaming, with 20 million of those paying subscriptions generating US$2.5 billion in annual revenues. Gold farming is an entirely new form of economic activity created within these online worlds, connecting workers and consumers together in real-time across the global economy. It will almost certainly prove to be a significant and lasting economic-geographical phenomenon, alongside previous waves of successful online businesses such as eBay, Google, Amazon, and Yahoo. To some observers, it offers yet another indication of the end of geography and the eradication of distance as a barrier to economic transactions in the contemporary era. And yet, we should be cautious about overemphasizing the ability of technology to reshape or erase the geography of the economy.

The continued importance of geography can be seen on two levels. First, even an apparently global phenomenon such as gold farming is nowhere near as global and omnipresent as one might think. As with all cyberspace activities, it is underpinned by a series of very real and uneven geographies. For example, online gaming is still predominantly a developed-country phenomenon, relying as it does upon both the availability and affordability (in local terms) of internet access, which is still very uneven at a global scale. Equally, the availability of workers both willing and able to undertake such work is highly geographically variable. As such, the gold farming phenomenon is connecting together particular groups of people with very different access to economic resources and opportunities. Moreover, the industry depends on a real-world infrastructure of computers, servers, wires, and cables that constitute the very architecture of the online economy. Second, the gold farming example only illustrates the potential impact of one kind of technology (the internet) on one particular segment of the economy (online gaming). Across the economy as a whole, the geographical impacts of a wide range of new technologies have been varied and uneven. Some technologies have allowed the reorganization of economic activities at a global scale. Others, however, have reinforced preexisting geographic inequities, or indeed, have *increased* the propensity for the clustering of economic activity in particular places. We have already seen in Chapter 8 how the geography of different activities within commodity chains may be concentrated or dispersed, or indeed a combination of both: here, then, we look at the role technology plays in shaping these geographies.

We unravel the complexity of geography-technology interrelations in four substantive sections. First, we critique simplistic notions concerning the ubiquity of new technologies and the extent to which they are eradicating distance, and therefore location, as economic concerns (Section 9.2). In this section we also introduce some key ideas that form the basis of a more nuanced view of technological change and its impacts. Second, we look at how a range of space-shrinking transportation and communication technologies have underpinned the globalization of economic activity in recent decades (Section 9.3). Not all technologies are space-transcending, however: in Section 9.4 we look at product and process technologies that are reshaping *how* things are made. Such technologies, in turn, have their own varied geographical implications. Finally, we close by revealing another uneven geography of technology – namely, that of the design and production of technologies themselves (Section 9.5).

9.2 The Universalization of Technology?

Ever since computers first became widely available in the 1980s, commentators and futurologists have heralded the transformative potential of new information and communication technologies. Such accounts are often explicit about the ability of new technologies to change profoundly the geographies of social and economic life, using appealing metaphors such as the "end of geography" and the "death of distance." For example:

> The death of distance as a determinant of the cost of communicating will probably be the single most important force shaping society in the first half of the next century. Technological change has the power to revolutionize the way people live, and this one will be no exception. It will alter, in ways that are only dimly imaginable, decisions about where people work and what kind of work they do, concepts of national borders and sovereignty, and patterns of international trade. (Cairncross, 1997, p. 1)

An extreme but nonetheless influential contribution to this popular discourse even went as far as to suggest that the world has become "flat":

> It is now possible for more people than ever to collaborate and compete in real time with more other people on more different kinds of work from more different corners of the planet and on a more equal footing than at any previous time in the history of the world – using computers, email, fibre-optic networks, teleconferencing and dynamic new software. (Friedman, 2007, p. 8)

Inherent in such accounts is a notion that new technologies, and their associated benefits, are *universally* available and that there will be some kind of "leveling-out" of economic activities at the global scale as a result.

While there are undoubtedly grains of truth in these depictions, real-world realities do not match up well to what is being described. The world economy is far from being a flat landscape of equal access and opportunity. Equally, the predicted global village of paperless offices, teleworkers, and the like has simply not materialized. Instead, new technologies have been overlain on, and interacted with, the preexisting patterns of uneven power relations and differentiated development in the global economy. The falling costs of transportation and communication have not been felt equally everywhere, large cities continue to dominate the global economy, and poor people and regions continue to lag behind as new technologies are adopted first in wealthy countries. Even when they get involved in the digital economy, these poorer people are often doing the lower-skill forms of technologically-mediated work that have been outsourced by developed economies. Nonetheless, it is true to say that technologies have played a part in fostering new patterns of intense interconnections between particular nodes in the global economy. Ultimately, however, it is difficult to talk in general terms about technologies and their geographic impacts.

In short, we need to develop more nuanced interpretations of technological change and its geographical impacts. There are four key elements to thinking about these issues more effectively. First, technology must be seen not as a technical process that can itself drive economic processes – a fallacy known as technological determinism – but rather as a *social process* through which individuals and organizations deploy technologies to achieve certain ends. In the case of firms, technologies are usually used as part of the drive to make profits in a competitive capitalist environment. Technologies themselves do not cause change: instead, we should think of how technologies facilitate and enable different kinds of change through their use by different types of economic actors.

Second, we need to distinguish between different kinds of technology. A key difference here is between what we might call "space-shrinking" transport and communications technologies that help overcome the frictions of space and time, and technological changes in production processes that may, or may not, as we shall see, have geographical ramifications. Both types of technology are integral parts of the "machinery" of capitalism outlined in Chapter 3. Space-shrinking technologies enable capitalists to speed up the spatial circulation of capital in their pursuit of profits. Equally, production process technologies are crucial for increasing productivity and facilitating innovation in that same search for maintaining, and increasing, profitability.

Third, we need to think about different *levels* of technological change (Freeman and Perez, 1988). While *incremental innovations* are ongoing small-scale changes to existing products and production processes – such as adding more features to a mobile phone or using computer software to replace paperwork in an office environment – *radical innovations* are discontinuous events that significantly change an existing product or process. Examples here could include the invention of a new kind of television technology (e.g., flat-screen television) or the implementation of robot technology to replace human workers on automobile production lines. Changes in *technology systems* are more extensive changes in technology that have an impact on several parts of the economy, and may have the potential to create new industries. These changes are linked to the emergence of generic technologies with wide applications, such as biotechnology or a new energy technology. Changes in the *techno-economic paradigm* are revolutionary upheavals that have economy-wide effects. Clearly the advent of information technology in the contemporary era represents a new techno-economic paradigm: it is hard to think of an economic activity that has not been affected in some way by computers and the internet. The power of this latest paradigm is derived from the way in which it enables users to both transmit and process information, activities that were previously conducted using separate technologies (e.g., telephones and free-standing computers).

Fourth, technological change needs to be placed in a long-term or *evolutionary* perspective. The information technology paradigm has come after a series of other earlier technological revolutions linked to periods of economic growth and prosperity, often known as long or Kondratiev waves, after the Russian economist who studied them in the 1920s (see Figure 9.1). Similar to the idea of spatial divisions of labor in Chapter 3, these different transitions have been overlaid on one another sequentially over time. The different spatial outcomes of each wave of new technology have interacted with patterns of activity left over from the preceding wave. This observation starts to hint at a potential challenge to the alleged end of geography. London, for example, was established as a site of global financial power through the activities of the British Empire long before the information technology revolution (see Chapter 7). It was therefore extremely well placed to benefit from the advantages conferred by new technologies. This is known as *path dependency*, with the fortunes of certain places under conditions of technological change being partly determined by preexisting conditions.

Already then, we can start to see the simplifications inherent in the arguments of commentators such as Cairncross and Friedman. There are different kinds of technologies, different levels of technological change, and even the same

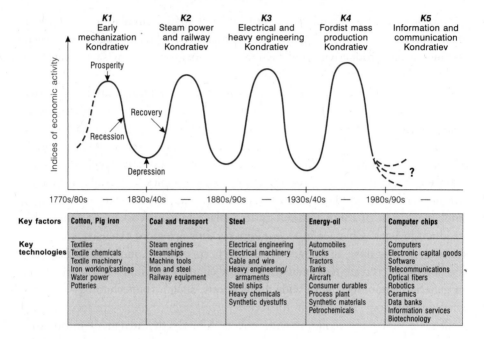

Figure 9.1   Kondratiev long waves and their characteristics
Source: Adapted from Dicken (2011), Figure 4.1.

technology may be used differently by different people in different places. We now move on to look at the two key categories of technology – space-shrinking, and product and process technologies (Dicken, 2011) – and their geographic impacts in more detail.

9.3   The Space-Shrinking Technologies

The functioning of the contemporary world economy depends on a wide range of space-shrinking technologies that connect the constituent elements of commodity chains – that is, firms, workers, governments, and consumers – together in different ways (see Chapter 8). What these technologies have in common, though, is the ability to help their users partially overcome the constraints of space and time. There are two important kinds of space-shrinking technologies that we need to introduce here: *transportation systems*, the means by which material goods (and people) are transferred between places, and *communications systems*, which enable the transmission of various kinds of information (e.g., text, numbers, images, video-clips, music files, computer programs) between places. Over the last century, dramatic transformations of both systems have occurred, with profound implications for the geography of the world economy. We will now briefly consider the most important of these innovations.

Transportation Technologies

In terms of *transportation technologies*, it barely needs to be stated that the world has "shrunk" in highly significant ways, particularly over the last two centuries. During the second Kondratiev wave in the mid-19th century (see Figure 9.1), the advent of steam power and the use of iron and steel in railways and shipbuilding were vital elements in supporting the emergence of truly global trading empires. Two key transportation developments in the 1950s gave the development of a global economy fresh impetus, however:

- *The advent of commercial jet aircraft*: While it seems obvious today that the extensive air travel network is an integral component of the global economy – 2009 saw 74.1 million aircraft flights in total – it is worth remembering that it was only in the late 1950s that jet travel took off as an affordable and widely available mode of transport. On the one hand, the air travel network is important for transporting different groups of economically significant people (e.g., tourists, managers in transnational corporations, highly skilled migrants, contract workers, students), both within and between countries. In 2009, the world's airports handled 4.8 billion passengers and the top three in terms of international flights were London Heathrow (61 million), Paris (53 million), and Hong Kong (45 million). On the other hand, air travel is also vital in circulating a wide range of high-value (e.g., documents, electronic components) and fresh products (e.g., fresh fruit, vegetables, cut flowers). Almost 80 million tons of air freight were shifted in 2009. Figure 9.2 shows the world's leading air cargo airports. While many of the world's largest cities feature on the map (e.g., Hong Kong, Shanghai, Paris, and Tokyo) other places are more

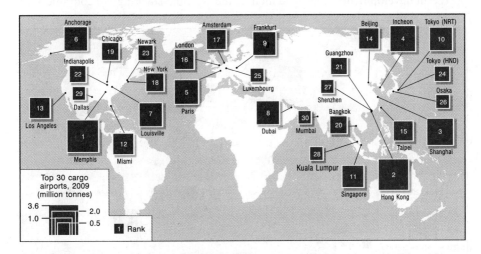

Figure 9.2   The top 30 cargo airports in 2009
Source: Data from www.aci.aero, accessed February 15, 2011.

surprising, but their importance reflects their role in the distribution networks of leading logistics firms. Memphis, Tennessee, for example, is FedEx's key global hub, while Louisville, Kentucky, fulfills the same function for UPS. Anchorage is used by both in facilitating their connections with China and the rest of East Asia (all data from www.aci.aero, accessed November 4, 2010).

• *The advent of containerization*: Another highly significant development since the mid-1950s has been the adoption of standardized 20- and 40-foot-long metal containers for land and sea freight, thereby vastly simplifying the transport and transhipment (i.e., transfer from ship to ship, or ship to train) of a huge range of commodities (see Box 9.1 for more on how this came about). Everyday, millions of containers are moving around at sea or on land, or standing in a warehouse somewhere waiting to be delivered. There are currently 9,000 container ships plying the world's oceans, with the largest capable of carrying 11,000 twenty-foot containers at a time. It is estimated that containers account for 90% of the world's traded goods by value, and in 2008 the total number moved was 153 million. The global movement of containers is dominated by five flows: intra-Asia (28.3 million units in 2008), transpacific (20.6 million), Europe-Asia (18.6 million), intra-Europe (9.5 million) and transatlantic (6.3 million), providing an interesting window on the nature of global trade patterns. Another perspective is provided by Figure 9.3, which maps the top container ports in 2009, a list headed by Singapore, Shanghai, Hong Kong, Shenzhen, and Busan (South Korea). Ports in East and Southeast Asia have grown rapidly in importance over the past decade – with 16 of the top 25 ports now being found in the region – while leading European and North American ports have slipped in relative terms (all data from www.shipsandboxes.com, accessed November 4, 2010). More recent developments in the field include the "double-stacking" of containers by

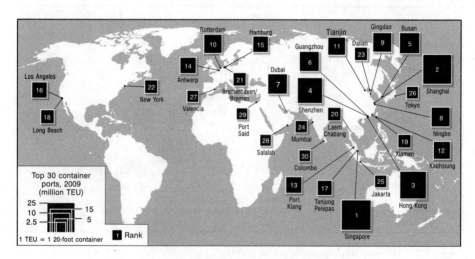

Figure 9.3   The top 30 container ports in 2009
Source: Data from www.container-mag.com, accessed February 15, 2011.

CASE STUDY

Box 9.1   The Box

Marc Levinson's much-lauded 2006 book, *The Box: How the Shipping Container Made the World Smaller and the World Economy Bigger*, is a highly engaging and readable account of the rise of the simple shipping container from one man's idea into a foundational part of the global trading system. His story starts on April 26, 1956, as a crane lifted 58 aluminium truck bodies onto an old tanker ship called the Ideal-X in Newark, New Jersey. Five days later, the boxes were unloaded in Houston, Texas, onto 58 trucks that took them to their destinations. In a stroke the cost of loading a ship fell from an estimated US$5.83 per ton for loose cargo to just US$0.16 per ton for containers. The idea was the brainchild of truck company owner Malcolm Maclean, forged in response to competitive conditions and growing congestion on America's east coast. Levinson's account tells how from these humble beginnings the container industry grew steadily in the 1960s and 1970s to become the global standard for freight transhipment, overcoming a range of obstacles along the way. That there should have been resistance to container usage is not surprising given that its growth lead to the demise of many ports and shipping lines and wholesale changes in seafarer and dock-working occupations. On the other hand, the expansion in container shipping created opportunities for new specialist ports such as Felixstowe (England), Tanjung Pelepas (Malaysia), and Busan (South Korea), among many others, and helped drive the growth of industrial complexes in Los Angeles and Hong Kong. That this story is so interesting stands in stark contrast to the central object of the account. As Levinson describes in typically eloquent fashion: "What is it about the container that is so important? Surely not the thing itself. A soulless aluminium and steel box held together by welds and rivets, with a wooden floor and two enormous doors at one end: the standard container has all the romance of a tin can. The value of this utilitarian object lies not in what it is, but in how it is used. The container is at the core of a highly automated system for moving goods from anywhere, to anywhere, with a minimum of cost and complication on the way" (pp. 1–2). More open to debate, perhaps, is the extent to which the container drove increases in global trade and the integration of the global economy from the 1970s onward, or was merely one contributing factor among many.

rail and the development of ever larger container ships in the pursuit of economies of scale. In February 2011, for example, shipping giant Maersk placed an order for ten ships capable of carrying 18,000 twenty-foot containers from Korean manufacturers Daewoo in a deal worth US$2 billion.

While innovations such as these have undoubtedly helped to "shrink" the world in relative terms, it is important to remember that the shrinkage has been extremely uneven. Transportation developments have served to pull the world's leading cities and economies closer together, while other less industrialized and rural areas continue to be far less well connected, if not being pulled further apart. This uneven shrinking of the world further exacerbates global uneven development. The major global cities, for example, are far better interconnected than the individual cities are with rural parts of their home countries.

Communications Technologies

Meanwhile, huge advances in *communications systems* parallel these developments in transportation technologies. Key developments include:

- *Satellite and optical fiber technology*: Since the mid-1960s, when the first geostationary communications satellite was launched, global communications have been transformed. While networks of satellites can now transmit voices, images, and other forms of data across the globe almost instantly, optical fiber systems have increasingly challenged satellite technology since the 1970s due to their ability to transmit huge quantities of information at very high speeds. An extensive network of ocean-bed optical fibers now criss-crosses the globe. While the connections are strongest between North America, Western Europe, and East Asia, in 2009 the Seacom cable system was completed, connecting South and East Africa to the global grid for the first time. Taken together, satellites and fiber optics offer affordable and high-volume data transfer capability, and the capacity of fiber optics in particular continues to increase exponentially (see Table 9.1).
- *The internet*: Clearly the internet has revolutionized communications through the development of an interactive, mass-user, computer network system. The

Table 9.1   Expansion of submarine cable capacity, 1979–2005

Year	Number of Submarine Cables	Number of Countries Linked	Total Bandwidth at Full Capacity (Mbps)
1979	55	35	321.4
1989	90	51	46,614.8
1999	245	92	6,852,133.8
2005	291	120	101,594,123.4

Source: Adapted from Malecki and Wei (2009), Table 1.

internet facilitates cheap and reliable interpersonal and interorganizational email and information sharing on an unprecedented scale. The expansion of internet use over the last 15 years has been astounding, rising from 16 million in 1995 to 361 million in 2000, 939 million in 2005, and 2,267 million, or 33% of the global population, by late 2011. Despite these growth rates, access to the internet is still somewhat uneven globally. In late 2011, Asia, Europe, and North America accounted for 45%, 22%, and 12% of global users, respectively, while Africa only accounted for 6%. The highest and lowest penetration rates are to be found in North America and Africa, at 79% and 13.5% of the population. The geography of internet *hosts* – a good indicator of where the information on the internet is being produced – is highly skewed, with the top three countries, the United States (316 million hosts), Japan (40 million), and Germany (23 million), accounting respectively for 57%, 7%, and 4% of the global total of 554 million hosts in July 2008. Flows of data traffic on the internet are also highly uneven as Figure 9.4 shows, with links between North America, Europe, and Asia dominant, in part reflecting the underlying fiber optic infrastructure described above (data from www.internetworldstats.com, accessed April 17, 2012).

- *Mobile telecommunications*: The emergence of the internet has been paralleled by significant developments in mobile telephony. Most important, over the last two decades, the mobile phone has gone from being an expensive novelty to a seemingly ubiquitous presence in everyday life: mobile phone subscriptions globally leapt from 1 billion in 2002 to 5.3 billion by the end of 2010, the latter representing 77% of the world's population. While in 2003 61% of the world's population were covered by a mobile signal, by 2009 the figure had risen to 90%. In general, however, mobile telephones are more widely used in wealthy countries. In 2010, for example, the average number of mobile subscriptions per 100 people was 120 in Europe compared to 41 in Africa and 68 in Asia, while the average across all developing countries was 68. Growth rates in mobile phone usage have been very strong in the developing world for several years, with the level of mobile telephone use now exceeding that of fixed lines in many countries (in particular in the Asia Pacific, Central Asia, and Sub-Saharan Africa), with the latter technology effectively being bypassed. Their potential for fostering economic development is also increasingly widely recognized, for example through programs in Africa allowing people who may not have a bank account to transfer funds through text messages (*The Guardian*, 2008). Newer mobile phones allow wireless connection to the internet, enabling users to access information and be in communication while on the move and in a greater variety of locations; penetration of such broadband devices was 51% in the developed countries at the end of 2010, but only 5.4% in developing countries (data from www.itu.int, accessed January 4, 2011).

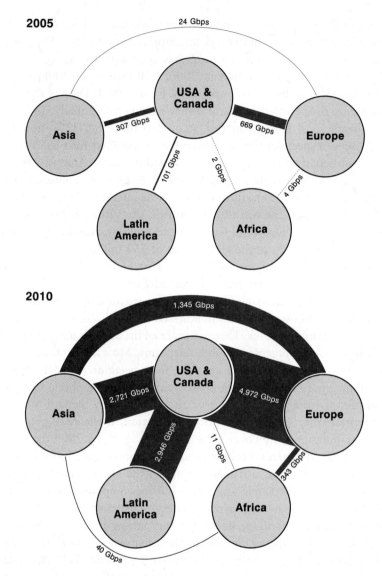

Figure 9.4   Interregional internet bandwidth, 2005 and 2010
Source: www.telegeography.com, reproduced with permission, accessed 22 June 2009.

- *The electronic mass media*: Often overlooked in these debates, the electronic mass media play a crucial role in helping products and services to transcend space and time through their role in constructing markets and expectations in countries around the world. Radio and television are both extremely powerful media in this regard, carrying a wide range of advertising for branded products, for example. Indeed, leading television stations with global networks such as

CNN, MTV, and BBC World have become significant brands in their own right. Since the 1980s, a combination of deregulation and new outlets such as cable and satellite television has lead to the proliferation of television channels available in many countries. Again, however, it is vital to bear in mind inequality in access to the media across the globe. In 2010, 98% of households in developed countries had televisions – with 100% coverage in many – while the figure for developing countries was 72%. This latter figure masks great unevenness, however, with several African countries such as Chad, Ethiopia, Niger, and Rwanda still having less than 10% of households with televisions (data from www.worldbank.org and www.itu.int, accessed November 4, 2010). Increasingly, we also need to include the internet as a mass medium, particularly given the growing importance of advertising on leading websites such as Google and Facebook. In the United Kingdom in 2009, for example, internet advertising revenues of £3.5 billion exceeded those from television advertising for the first time.

Taken together, the space-shrinking attributes of transport and communications technologies have had significant impacts on the geography of the world economy. On a simple level, they allow us to buy goods from around the world in our local supermarkets. Not only do they facilitate the spatial separation of producers and consumers of goods and services, however, but they also allow the *disintegration* of different elements of the production process (or commodity chain) both within and between countries.

It is not just manufacturing processes that are being affected in this way, however. Space-shrinking technologies are also allowing the spatial disintegration of service activities. The massive growth in *call centers* in recent years is a clear indication of these trends. Call centers represent a form of firm–customer interface whereby a wide range of support services (e.g., sales and marketing, technical support, claims enquiries, market research, reservations, information provision) are provided via telephone from dedicated centers to an often widely dispersed customer base. While the establishment of call-center operations in low-cost locations has long been a popular business strategy, the process has intensified in recent years, particularly in terms of the international relocation of these activities. This process has become known as the *offshoring* of services. In addition to call centers, two other important kinds of offshore services are back offices (covering activities such as claims and accounts processing, payroll processing, data processing, invoice administration) and information technology services (e.g., software development, application testing, content development). Leading sectors involved in offshoring include financial services, telecommunications, transport services, health services, and the public sector. India has been one of the key beneficiaries of these trends (see Box 9.2).

CASE STUDY

Box 9.2   Offshore Services in India

The Indian offshore services industry started to take off in the early 1990s, when companies such as American Express, British Airways, and Swissair set up back-office operations in India. By 2008, Indian software and service exports had reached a level of US$47 billion – up dramatically from US$12 billion in 2004 and just US$0.6 billion in 1994 – and the sector employed over 2 million people. The main markets for these services were the Americas (60%) and Europe (31%). Investors are attracted by the availability of a large, cheap, well-qualified English-speaking workforce who can easily be connected to home markets through international telecommunications. Although being eroded over time, salary differentials are still of paramount importance: one 2007 study found that the average annual call center agent salary in India was US$2,500 compared to US$35,000 in the United States and US$27,300 in the United Kingdom. It is important to delineate the Indian industry along two dimensions. On the one hand, it is constituted by three distinct segments: IT services (US$27 billion of exports in 2008), business process outsourcing (BPO) – a broad category of activities encompassing call centers and other back-office functions (US$12.8 billion) – and engineering services, R&D, and software products (US$7.3 billion). India attracts the major share of such offshored activity at the global scale: in the BPO segment, for example, it attracted 63% of the global total in 2008, although the level is declining in the face of competition from other locations such as China, the Philippines, and Central and Eastern Europe. On other hand, the Indian industry encompasses at least three different kinds of operations. First, there are directly owned subsidiaries of foreign corporations; for example, Dell has four customer service centers in India employing several thousand people in total. Second, there are the directly owned subsidiaries of foreign firms that are themselves involved in providing offshore services for clients; for example, IBM employed an estimated 130,000 staff in India in 2010 – making it the second largest private-sector employer in the country – and Accenture's network of delivery centers employed over 50,000 staff in the same year. Third, services can be *outsourced* to Indian firms – for example, Delta Airlines' contract with Wipro to handle some of its call center reservations. The general trend has been toward the outsourcing model, and leading Indian providers such as HCL (Hindustan Computers Ltd.), TCS (Tata Consultancy Services), Wipro Technologies, and Infosys Technologies are huge transnational corporations in their own right, with extensive international operations. For example, by 2010 HCL had 71,000 staff, a turnover of US$5 billion, operations in 29 countries including the United States, United Kingdom, China, and Brazil, and 500 service centers across 170 cities in India. On the ground, India's offshore service sector is concentrated in five city-regions: the National Capital Region (Delhi, Gurgaon, and Noida), Bangalore, Mumbai, Chennai, and Hyderabad (data from www.nasscom.org and www.hcl.in, accessed November 4, 2010).

Bringing Things Together: The Logistics Revolution

Using powerful combinations of transportation and communications technologies to connect together globally dispersed elements of commodity chains has become strategically significant. *Logistics* refers to the process of planning, implementing, and managing the movement and storage of raw materials, components, and finished goods from the point of origin to the point of consumption. It encompasses a wide range of tasks, including shipping, planning, warehousing, inventory management, export documentation, and customs clearance. Crucially it is purely about the movement of goods, but also involves the collection and processing of knowledge about the distribution system. The growing level of integration and management of these services across the whole commodity chain is described by some commentators as heralding a *logistics revolution* (Bonacich and Wilson, 2008).

This revolution has involved bringing together the new transport and communications technologies described above with new process technologies that allow for enhanced speed, flexibility, and reliability in logistics provision. A broad shift has occurred from systems based on large stockpiles of components and products and infrequent deliveries to *lean distribution* systems based on the flexible delivery of smaller volumes of goods. In addition to the containerization described earlier, these lean distribution systems are underpinned by technological advances in four interrelated areas:

- *Transportation terminals*: Significant technological changes have seen new terminals constructed that can operate at very high levels of turnover. Better handling facilities have lead to significant improvements in throughput, and port facilities are also increasingly supported by inland terminals (e.g., the Prem Nagar Dry Port in Lahore, Pakistan) connected by high-capacity corridors.
- *Distribution centers*: Contemporary distribution centers (DCs) do not hold inventory for long and are characterized by very rapid turnover of goods. They are designed for throughput and have specialized loading and unloading bays and sorting equipment. The aggressive use of cross-docking, for example, allows goods to be brought in, selected, repacked, and dispatched to stores with

limited, if any, time in inventory. Some of the constituent technologies of DCs are high-speed conveyors with advanced routing and switching capabilities, reliable/accurate laser scanning, and powerful software/IT systems.

- *Electronic Data Interchange (EDI)*: Integrated IT systems and common software platforms across the supply network allow the instantaneous and secure transmission of large quantities of data concerning sales, product specifications, orders, invoices, shipment tracking, and the like. Increasingly, however, the internet has become the standardized platform through which such information is exchanged. Tight control is maintained through the use of bar codes and RFID (radio frequency identification devices) tags that allow materials to be tracked and traced throughout the commodity chain.

- *E-commerce*: The advent of the internet has initiated changes to logistics as part of a wider shift toward e-commerce. Both business-to-business and business-to-consumer relationships have been reworked with the rise of new kinds of electronic marketplaces and intermediaries such as Amazon and eBay. Physical logistics infrastructures have not, however, become less important as a result; instead, a wide range of different approaches to fulfilling e-commerce orders have developed. Particularly challenging is the timely fulfilment of low-volume, high-frequency deliveries to customers' homes. To give one example, while some grocery retailers fulfill online orders direct from their distribution centers and warehouses, others prefer to collect products from their store shelves before delivering them.

Overall, and looking across these different technologies, what is apparent is that effective logistics provision necessitates a mastery of both *real* and *virtual* space.

Two sectoral examples are illustrative of the importance of highly efficient and integrated logistics systems. First, fast and efficient air cargo services are absolutely critical in moving components, part-assembled products, and finished goods through the complex electronics production networks of Southeast and East Asia (Leinbach and Bowen, 2004). These services include not only the airport-to-airport carriage of goods, but also a variety of land-based operations including the transport of products to and from airports and the management of warehouses adjacent to airports to maintain timely and efficient air transportation. Second, temperature-controlled logistics operations are crucial to international trade in a variety of horticultural sectors such as fresh fruit, vegetables, and cut flowers. These integrated *cool chains* utilize refrigerated warehouses and trucks to allow transport of perishable goods across thousands of miles in line with international quality and safety standards. The Swiss group logistics group Kuehne + Nagel, for example, uses refrigerated trucks, RFID technology, and a cold storage facility at Nairobi airport to ensure that cut flowers are shipped from Kenya to Amsterdam and other destinations in perfect condition within 24 hours of being picked.

Given the complexity of contemporary logistical requirements – and in particular the necessity of reliable, small, fast, and frequent shipments using both air and ground capacity – most large firms outsource such activities to dedicated providers (Aoyama et al., 2006). Moreover, they often ask for a seamless and integrated service across many countries. As a result, today's leading logistics firms have come a long way from their roots as basic road haulage and air/sea freight businesses, developing into complex, full-service transnational corporations. Large global operators have emerged through ongoing processes of merger and acquisition within the industry as firms have sought to achieve the necessary scale and scope to serve global clients. Three of the world's leading logistics firms are profiled in Table 9.2. The scale of the companies is immediately obvious: with several hundred thousand employees, bases in over 200 countries, thousands of operating facilities and turnovers in the tens of billions of dollars, by many measures they are among the world's largest companies. Also notable are their extensive transport fleets: in 2008 FedEx, for example, was operating 650 planes and over 80,000 land vehicles in moving 7.5 million packages daily. All three firms have different divisions reflecting the key business areas of packages, freight forwarding, and supply chain management services. As outsourcing trends continue, some logistics providers are taking over manufacturing-related tasks such as packaging, labeling, and inventory control; some have even started to carry out

Table 9.2 Leading logistical providers – key facts and figures in 2008

	DHL*	FedEx	UPS
Headquarters	Bonn, Germany	Memphis, Tennessee	Atlanta, Georgia
Revenue (2008, US$ billion)	57.3	35.5	51.5
Employees	300,000	280,000	415,000
Aircraft	350	654	570
Vehicles	72,000	80,000	107,000
Average daily shipments	Na	7.5 million	15.5 million
Sorting/handling facilities	23 million sq. ft.	9 million sq. ft.	35 million sq. ft.
Countries	220+	220+	200+
Facilities (offices, depots, etc.)	4500+	4000+	2700+
Key divisions	DHL Express DHL Global Forwarding, Freight DHL Supply Chain	FedEx Express FedEx Ground FedEx Freight FedEx Services	UPS (Packages) UPS Freight UPS Supply Chain Solutions

*Does not include data for Deutsche Post, the German postal service: total revenues for Deutsche Post DHL Group in 2008 were US$75 billion.
Source: Corporate websites and annual reports.

light assembly work for large customers, for example in the consumer electronics industry.

9.4   Product and Process Technologies

The competitive capitalist economy dictates that not only must firms connect the different parts of the production process together efficiently – ranging from accessing resources and inputs right through to a variety of after-sales support services – but they must also strive to improve constantly the rate at which they develop new products and services and the efficiency of their production processes. It is to these issues, and their geographical dimensions, that we now turn.

Product Innovation

Developing a stream of new products and services is central to maintaining profitability in the contemporary capitalist economy. While expanding the geographic market for existing offerings is another option, there are finite limits to such markets and hence product innovation focuses on trying to open up new kinds of market. The idea that demand for any given product will decline over time is encapsulated in the notion of the *product life cycle*. This simplified model suggests that products follow a typical path from initial innovation through phases of early development, growth, maturity, decline, and obsolescence and that demand waxes and then wanes through this transition. Importantly, the technological requirements of production are seen to vary through the cycle. While the start may be characterized by short production runs and trial-and-error experimentation with new techniques, by the mature phase long production runs and stable technologies dominate. In terms of labor requirements, while scientific, engineering, financial, and management skills are required to get the cycle up and running, during the mature phase the product can be mass-produced by semi- and unskilled workers. In seeking to overcome the constraints imposed by the product life cycle, firms can either introduce a new product once the existing one starts to decline, extend the life of the existing product by making minor modifications or finding new uses for it, or find newly efficient ways of producing the existing product. In geographical terms, the model is often taken to suggest that production can in time be decentralized down the urban hierarchy, or to peripheral areas, once the product matures and production cost considerations become paramount.

   Although useful to a point, there are some clear limitations to this model. First, it is clearly a gross generalization to suggest that all products follow a similar trajectory over time. In reality the speed and nature of the transition vary greatly: while the cycle is very fast in the consumer electronics industry (e.g., mobile phones, televisions), it may barely apply in sectors where the emphasis is on customization and quality (e.g., hand-made furniture, craft products). Second, and relatedly, the model has obvious limitations when it comes to intangible service products, such as insurance policies, where the technological and labor implications of new product development are very different. Third, the model perhaps underestimates the degree to which even standardized and mature products can constantly be rejuvenated through innovation later in the life cycle. The classic example here is the automobile, a long-standing product that has being modified and improved over time.

Moving beyond this model, however, there are perhaps five interrelated signature characteristics of contemporary product innovation (drawing on Malecki and Moriset, 2008):

- In general terms, product life cycles are becoming shorter and rates of innovation are increasing. This is particularly obvious in consumer goods segments such as electronics and clothing where the lifespans of the latest iPhone or fashion item, for example, are measured in weeks and months rather than years. New offerings overlap with existing products in complex ways as firms seek to offer a suite of related products for different price brackets and tastes.

- This speeding up has been facilitated by a wide range of new forms of software at all stages of the life cycle, including computer-aided design (CAD), enterprise resource planning (ERP), and customer relationship management (CRM) technologies. Such techniques not only facilitate making adjustments to existing products, but also minimize the investment that is necessary in real productive capacity early in the process; the Boeing 777, for example, was famously designed and prototyped in an entirely digital environment.

- Relatedly, it is perhaps no longer appropriate to think of product innovation in terms of a linear process in which distinct research and development teams pass a finished design on to manufacturing departments for production. Instead, product innovation should be thought of in more cyclical and iterative terms, with IT and more fluid organizational structures allowing innovation to be an ongoing rather than discrete process.

- An increasingly crucial element in this more fluid process in many industries is client/customer feedback (Grabher et al., 2008). "Data from customer enquiries must result in short notice in production improvements or even in design alterations. IT-enabled *lifecycle engineering* results in permanent, virtually real-time feedback processes and adjustments in the whole production cycle" (Malecki and Moriset, 2008, p. 66, emphasis in original). In certain contexts – for example, open-source software and online user forums – customers can play a full and active role in product innovation.

- Finally, it is important to reflect upon how many product firms are increasingly selling combinations or bundles of products and services that serve to blur traditional distinctions. Examples would include automobile companies seeking to sell after-sales service, insurance, and credit, many tourism activities, and mobile telephones that come with subscription packages. Such developments necessitate redefining the notion of product innovation to incorporate novel and rapidly-shifting product–service combinations.

Production Processes and Their Geographies

We now move on to look at production processes – that is, how products are made. All firms need to select an appropriate technique of production, and in particular the precise combination of labor (workers) and capital (buildings, machinery, computers, and telephones, for example) that will be used. As noted earlier, we are currently in the midst of an evolving techno-economic paradigm being driven by new information technologies. In the same way as these technologies are revolutionizing communication patterns, they are also driving profound transformations in production processes. There is much debate, however, about the nature of these production processes, which can be thought of as *post-Fordist* in relation to the mass production system known as *Fordism*. Fordism signifies a range of industrial practices first associated with the workplace innovations of the American automobile manufacturer Henry Ford in Detroit, Michigan, during the second decade of the 20th century.

The Fordist production system has four important elements. First, it is characterized by a distinctive division of labor – that is, the separation of different work tasks between different groups of workers – in which unskilled workers execute simple, repetitive tasks and skilled technical and managerial workers undertake functions related to research, design, quality control, finance, coordination, and marketing. Second, it is a system in which the manufacture of parts and components is highly standardized. Third, it is organized not around groups of similar machinery, but with machines arranged in the correct sequence to manufacture a product. Finally, the various parts of the production process are linked together by a moving conveyor belt – the assembly line – to facilitate the quick and efficient fulfillment of tasks. Together, these four attributes can provide *economies of scale* whereby the large volume of production reduces the cost of producing a single product. Prices can then be reduced, leading to increased sales and the potential development of a *mass market* for the commodity in question (see Box 9.3; see also Figure 9.1).

Over the last three decades, Fordism has been augmented by new modes of production, the chief characteristic of which is production *flexibility*. At the heart of this enhanced flexibility is the use of information technologies in machines and their operation to allow more sophisticated control over the production process. Most importantly, computerization enables the rapid and efficient switching from one part of a manufacturing process to another, which in turn allows firms to tailor products to the requirements of individual customers. Rather than engaging

FURTHER THINKING

Box 9.3   Fordism – More than a Production System?

The term Fordism has come to connote much more than just the production-line technologies of Henry Ford. For many commentators, it signifies a broader, society-wide model of economic growth that was prevalent in North America and Western Europe – and therefore is sometimes known as Atlantic Fordism – in the period stretching from the end of World War II until the early 1970s. This was a period of strong economic growth in those regions associated not only with mass production and the emergence of large, vertically integrated corporations, but also particular sets of national institutional and political conditions. This view recognizes that Fordism was not simply a form of production, but also depended on a particular political and social system that gave it stability and tried to manage its inherent limitations. In its most developed form, this comprised both the macroeconomic management associated with Keynesian policies and the social support mechanisms of the welfare state. It was also underpinned by a social contract, brokered by the state, in which firms and labor unions sought to link annual wage increases to the productivity gains achieved through mass production techniques, with the result that workers had sufficient income to sustain mass consumption. These national agreements were supported by new postwar supranational organizations such as the International Monetary Fund (IMF) and World Bank that helped create a relatively stable global economy. The limits to the system became apparent in the early 1970s, when falling levels of profits, increased global competition, and rising oil prices initiated a period of intense economic restructuring, broadly labeled as the crisis of Fordism, during which the previously stable relations between firms, governments, and labor unions unravelled. This heralded the arrival of the *post-Fordist* era. While the Fordist production system still exists in many sectors and places, the associated sociopolitical system has been progressively dismantled, particularly in the United States and United Kingdom, since the late 1970s.

in mass production, firms are now able to make a wide variety of products for different market niches without compromising the cost savings traditionally associated with large-volume production.

That being said, it is not possible to identify one kind of post-Fordist production regime. Instead, it is more effective to identify three different kinds of industrial system that coexist in the contemporary economy (Dicken, 2011; see Table 9.3). First, Fordism is alive and well in certain sectors where scale economies remain crucial, for example agriculture and food processing, routinized services

Table 9.3 Comparing contemporary production systems

Characteristic	Fordist Mass Production	Flexible Specialization	Japanese Flexible Production
Technology	Complex, rigid single-purpose machinery; standardized components; difficult to switch products	Simple, flexible tools/machinery; nonstandardized components	Highly flexible (modular) methods of production; relatively easy to switch products
Labor force	Narrowly skilled professionals conceptualize the product; semi/unskilled workers execute production in simple, repetitive, highly controlled sequences	Mostly highly skilled workers	Multi-skilled, flexible workers, with some responsibilities, operate in teams and switch between tasks
Supplier relationships	Arms-length supplier relationships; stocks held at assembly plant as buffer against disruption of supply	Very close contact between customer and supplier	Very close supplier relationships in a tiered system; just-in-time delivery of stocks requires close supplier network
Production volume and variety	Very high volume of standardized products with minor "tweaks"	Low volume and wide (customized) variety	Very high volume; total partially attained through the production of range of differentiated products
Geographical tendencies	Dispersal of branch plant operations to peripheral regions/developing countries in search for cost savings	Tight clustering of customers, suppliers, and their partners	Clustering of production plants, key suppliers, and logistical providers

Source: Adapted from Dicken (2011), Figure 4.13.

activities, and the electronic component industries. Second, in other instances, flexible production technologies have led to the resurgence of craft-based production driven by innovative, small independent firms. At the heart of this so-called *flexibly specialized* production system are skilled craft workers using flexible machinery to produce small volumes of customized goods. These developments can be observed in craft industries producing specialized shoes, jewelry, clothing, furniture, ceramics, and the like. Third, there are forms of *flexible production* emanating originally from Japan that are distinctive from both the Fordist and flexibly specialized systems. Key here is the combination of information technologies with flexible ways of organizing workers. Another important attribute of this system is the application of information technologies to supply chains, allowing the replacement of the expensive and inefficient "just-in-case" systems inherent to Fordism with "just-in-time" systems (see Table 9.4). The overall result is perhaps best thought of as *mass customization*, with firms able to produce substantial volumes of a large range of products by combining mass-produced components in different ways. While developed in the automobile industry, it has subsequently spread into other consumer goods industries (such as personal computers and mobile phones) and sectors such as chemicals and steel. Dell Computer provides an interesting case study of these technological and organizational dynamics (Box 9.4).

Table 9.4   The characteristics of "just-in-case" and "just-in-time" systems

"Just-in-Case" System	"Just-in-Time" System
Components delivered in large, but infrequent, batches	Components delivered in small, very frequent, batches
Very large and costly buffer stocks held to protect against disruption in supply or discovery of faulty batches	Minimal stocks held, only sufficient to meet the immediate need
Quality control based on sample check after supplies received	Quality control built in at all stages of the supply chain
Large warehousing spaces and staff required to hold and administer the stocks	Minimal staff and warehousing required
Use of a large number of suppliers selected primarily on the basis of price	Use of a small number of preferred suppliers within a tiered supply system
Remote relationships between customers and suppliers	Very close relationships between customers, suppliers, and logistics operators
No incentive for suppliers to locate close to customers	Strong incentive for suppliers to locate close to customers

Source: Adapted from Dicken (2011), Table 5.2.

CASE STUDY

Box 9.4   Production Process Innovation: The Case of Dell Computer

Dell, the American computer company, was formed in 1984. In 1994, it was the world's tenth largest computer firm, with a global market share of 2.4%. By 2004, it was the world's largest personal computer manufacturer, with a market share of 18%, a position that had slipped slightly by 2008 when its 15% share put it in second place globally. This remarkable growth has been achieved not through product innovation – the personal computer was not invented by Dell and it has long been a largely standardized product – but by production and organizational innovations concerning the ways in which it buys components, assembles computers, and sells them. There are three distinctive elements to Dell's business model. First, the company only sells direct to its customers, bypassing almost entirely distributors, resellers, and retailers. Second, Dell customizes all its products to meet the specific needs of its customers (i.e., mass customization). Third, it has used the internet to establish not only its system of direct sales, but also the procurement and assembly operations underlying the mass customization. In short, by combining direct sales and procurement with an internet infrastructure, Dell has been able to create an innovative and profitable business model that squeezes out unnecessary intermediaries and product inventory. The impacts of the internet-based

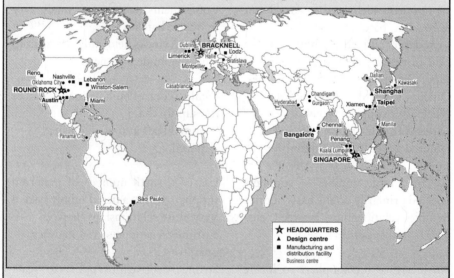

Figure 9.5   Dell's global operations, mid-2009
Source: Data from www.dell.com.

just-in-time procurement system have been particularly profound, reducing inventory from an average of 32 days in 1994 to just three days by 2002. On the ground, Dell's operations are organized around a network of ten assembly plants, six design centers, and a range of customer contract centers (see Figure 9.5). The Dell model nicely illustrates the variable geographical impacts of technology being described in this chapter. On the one hand, Dell is able to maintain an extensive business operation, meeting the needs of a global customer base through the intensive use of internet technology. On the other hand, the assembly operations are at the heart of regional production clusters, combining local and global suppliers and logistics centers, and exhibiting the need for proximity inherent in just-in-time production systems. For more, see Fields (2006).

But what are the geographical configurations of these three competing and overlapping production systems? In its heyday, Fordist production was associated with a series of great industrial regions in North America and Western Europe, most notably the manufacturing belt in the northeastern United States and a zone of development in Europe stretching from the Midlands in the United Kingdom through northern France, Belgium, and Holland to the Ruhr in what was then West Germany. By the 1960s and 1970s, the developing crisis in Fordism (Box 9.3) and the increased standardization of products and manufacturing processes (related to the product life cycle introduced above) led to a decentralization of production, initially to peripheral regions in the core countries – for example, southern and southwestern states in the United States, or Scotland, Wales, and the North in the U.K. context – and then increasingly, from the late 1970s onward, to overseas locations as part of wider globalization dynamics. In this way, Fordist production drove the formation of *branch plants* for undertaking the production process itself – as opposed to research and development, or sales and marketing, for instance – which could be relocated either nationally or internationally to benefit from variations in labor wages, skills, and levels of unionization. Fordist production is now to be found predominantly in developing countries – for example, the call centers of India, the garment factories of Bangladesh and India, and the consumer goods industries of coastal China.

Intriguingly for economic geographers, the flexible specialization and Japanese flexible production systems both seem to exhibit a newfound tendency for firms to agglomerate, or group together, in particular places or *clusters*, often deliberately beyond the traditional Fordist regions. Flexible specialization, for example, has long since been linked to the development of specialized *industrial districts*, a form of cluster based on the manufacture of one particular product type, and exhibiting high levels of subcontracting relationships between dense networks of small and medium-sized firms with a high reliance on familial connections and artisanal skills. The industrial districts of central and northeastern Italy are perhaps the

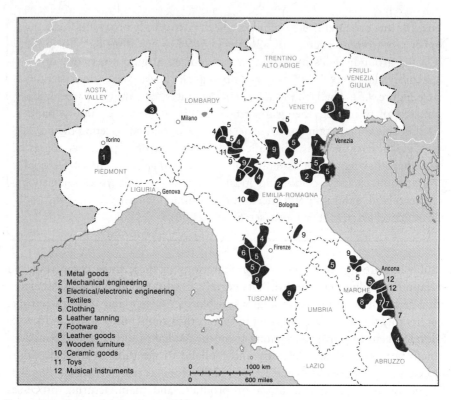

Figure 9.6   Industrial districts in Italy

Source: Adapted from Amin, A. Industrial districts, in Sheppard, E. and Barnes, T. (eds.):
A companion to economic geography, Blackwell.

preeminent examples. These districts, scattered across the provinces of Tuscany, Emilia-Romagna, and Veneto (see Figure 9.6) have collectively become known as the "Third Italy" in contrast to the traditionally industrial north and agrarian south of the country. Each has an established international reputation for the production of a particular kind of customized, design-intensive, and high-quality product. Santa Croce, a leather tanning district in Tuscany 40 km east of Pisa, is one such case (Amin, 2000). From the 1970s onward, in response to burgeoning international demand for quality Italian leather products, this small area has developed into a production district of several thousand employees and over five hundred artisanal firms and subcontractors. These firms are engaged in a complex and fine-grained division of labor undertaking the 15–20 separate stages of leather tanning. The district is also home to the warehouses, export/shipping agents, buyers, chemical companies, and haulage firms needed to sustain the industry as well as a range of supportive non-firm institutions.

     Flexible mass production also has inherent tendencies to the concentration of economic activity in clusters, in large part due to the imperatives of just-in-time systems. As Table 9.4 illustrates, the scale and frequency of component deliveries inherent in the system requires close relationships between customers, suppliers,

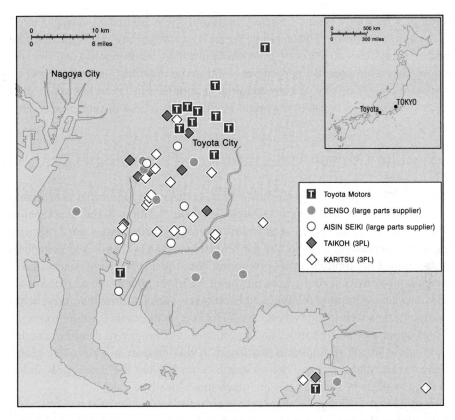

Figure 9.7 Location of Toyota Motors, large-parts suppliers, and third-party logistics providers around Toyota City in 2006

Source: Adapted from Kaneko, J. and Nojiri, W. (2008) The logistics of Just-in-Time between parts suppliers and car assemblers in Japan, *Journal of Transport Geography*, 16: 155–173, Figure 6.

and logistics operators and therefore creates a strong incentive for co-location. Figure 9.7, for example, illustrates the close proximity of key suppliers and logistics firms to Toyota's plants in Toyota City, Japan. The three different kinds of plant are connected by tightly choreographed "milk runs" of delivery vehicles that constantly circulate between them in the most efficient manner. While Toyota City is a powerful example of these dynamics, detailed research into Japan's auto industry – arguably the home industry and country of these production methods – suggests that such dynamics should not be overstated (Kaneko and Nojiri, 2008). First, the industry is far from inventory free, with independent logistics providers keeping buffers of certain components in their distribution centers. Second, Japan's auto industry is perhaps best characterized by a form of medium- or long-distance just-in-time production. Suppliers and assemblers both within, and across, different regions are connected together by highly efficient long-distance haulage operations in combination with advanced warehouses near assembly plants that sort, compile, and dispatch components by the appropriate deadline. The whole system is tightly managed by integrated IT systems across all the actors involved. This example serves as an important reminder that new production systems do not erase prior configurations, but instead are overlain on, and interact with, preexisting patterns of industrial organization. In this case, the long-standing geography of auto component suppliers in Japan has adapted to, rather than being completely reworked by, the needs of just-in-time production.

9.5   The Uneven Geography of Technology Creation

So far in this chapter we have profiled how, contrary to arguments proclaiming the "death of distance," new technologies have facilitated the development of new uneven geographies of economic activity. Technologies are both used unevenly – depending on who has the resources to access them – and contribute to unevenness by simultaneously promoting both the dispersal and concentration of economic activities. And yet, to this point we have not considered perhaps the most uneven geographies relating to technology – namely, those that underlie the creation of new technologies themselves. Here we shall briefly consider examples of a transport, communication, and production technology to demonstrate that while many of the technologies discussed in this chapter are relatively global in their reach and coverage, the geographies underlying the innovation and/or production of those technologies are much more localized.

First, we saw earlier in the chapter how air travel underpins important elements of the global economy. Of the global fleet of approximately 20,000 jet airliners, some 70% are made by just two companies: Boeing and Airbus. In terms of larger jets carrying over 100 people, their combined market share is 100%. Although both have global sales and service operations and suppliers in many countries, their research and development and final assembly operations are extremely geographically concentrated. Boeing's commercial airlines division is based near Seattle, Washington (U.S.), where the company assembles the 737, 747, 767, 777, and 787 aircrafts. Boeing employed some 74,000 people in and around Seattle in early 2011, and is at the heart of an aerospace industry in Washington State that generates annual revenues of over US$30 billion and incorporates a cluster of some 650 companies. Airbus started life in the 1970s as a consortium of U.K., French, German, and Spanish aerospace firms and is now owned by the European aerospace group EADS. Like Boeing, Airbus makes a wide range of jet types including the A320 and the huge double-deck wide-body A380. It has a workforce of 52,500, the vast majority of whom are located in Airbus' locations in France, Germany, and the United Kingdom. The most important sites are Toulouse (approximately 16,000 workers) and Hamburg (10,000 workers) where final assembly takes place, although the company opened a joint venture assembly line for the A320 in Tianjin, China, in 2008. Overall, the result of several decades of merger and acquisition activity is that the production of a key component of the global economy – the jet passenger aircraft – lies in the hands of just two companies. Their two prime locations and headquarters, in Seattle and Toulouse, are unique concentrations of research and manufacturing expertise. The small jet industry has undergone similar processes of consolidation, leaving Bombardier (Montreal, Canada) and Embraer (Sao Paulo, Brazil) as the two key players. In short, whenever you travel on a jet aircraft, it will almost certainly have been assembled in one of just six cities globally (see Bowen, 2010, for more).

Second, we can consider the information and communication technologies that underpin the internet. Silicon Valley is a collection of towns spread out along Highway 101 as it runs south from San Francisco to San Jose, California. The region has been a globally-significant area of technological innovation since the 1980s when companies associated with both IT hardware and software were at the heart of the personal computer revolution. Silicon Valley has retained its leadership in these areas, and continues to be the pre-eminent global location for IT research and development in the internet era. The area is home to some 1.3 million workers, of whom 273,000 work in information products and services (including 110,000 in software and internet services) and a further 141,000 in innovation and specialized services. The Valley's dominance of the internet industry encompasses microchip firms (such as Intel, headquartered in Santa Clara, and AMD, Sunnyvale), personal computer manufacturers (including Apple, Cupertino; HP, Palo Alto), internet router manufacturers (e.g., Cisco, San Jose; Juniper Networks, Sunnyvale), software firms (e.g., Oracle, Redwood Shores; Adobe, San Jose), internet browser firms (e.g., Mozilla, Mountain View; Safari, Cupertino), internet search engines (e.g. Google, Mountain View; Yahoo!, Sunnyvale) and key website firms (e.g., Facebook, Menlo Park; eBay, San Jose). Most other leading IT companies, even if they do not originate from the area, will have a base there. The important point here is that while the internet is increasingly taken to be a ubiquitous and "placeless" phenomenon, many of the key players and technological innovators in the industry have their headquarters and research operations within this relatively small urban belt. We shall return to the example of Silicon Valley in Chapter 12 as we seek to explain this concentration of activity.

Finally, in terms of production technologies, the industrial robot industry provides an interesting example. By the end of 2010, there were some 1.1 million industrial robots installed worldwide, mainly in the automobile component and assembly industries (43% of the total), but also in other advanced manufacturing sectors, especially the electrical machinery industry. The distribution of these robots through the global economy is highly uneven, in part reflecting the structure of the industries where they are placed: in 2010, 175,000 were to be found in North America, 574,000 in Asia, and 344,000 in Europe. Japan's total of 316,000 alone accounted for almost 30% of the global total, although the proportion is falling slowly due to strong growth in usage in China and South Korea in particular. While robot usage is therefore extremely uneven, the figures on robot *production* are even more staggering. Despite robots being pioneered and first commercialized in the United States in the 1960s, today fully 70% of global industrial robots are designed and built in Japan, in an industry worth over US$5 billion per year to the Japanese economy and US$3.5 billion in export revenues. The Japanese Robotics Association has over 130 member companies, and three of the industry's main global players – Fanuc, Yaskawa, and Kawasaki – are Japanese (data from www.worldrobotics.org, accessed November 4, 2010). Japan's dominance of this fast-growing and high-value industry reflects several factors. On the demand side, Japan's domestic market for robots has been strong, with much faster adoption of robots than in other countries, and strong uptake among small and medium-sized firms as well as large corporations. On the supply side, Japan's robot firms are much larger than the early U.S. firms they have subsequently overtaken, and are also part of the large *keiretsu* groupings that provide both an initial market for the robots and a ready source of research and development funding. Moreover, the industry has been heavily supported by government programs targeting the adoption of, and research into, industrial robotics.

In sum, the internet, aircraft, and robotics industries offer powerful examples of how the development and commercialization of particular technologies can become associated with particular places that in turn can derive huge economic benefit from their dominance. We will return in Chapter 12 to looking at the internal dynamics within such localities that underlie and drive innovation and growth.

9.6   Summary

This chapter has explored the interactions between new technologies and economic geography. New transportation and communication technologies have transformed the workings of the global economy by allowing the fragmentation and separation of production processes over space. However, they have done so in ways that are hugely uneven in social and geographical terms. Even a seemingly placeless technology such as the internet is in reality full of spatial inequalities and contradictions of different kinds. In some cases, new space-shrinking technologies have served to reinforce existing patterns of inequality, rather than allowing more people and places to interact with the global capitalist system as some accounts of technological change might suggest. New information technologies have also transformed how products and services are made, engendering greater production flexibility in different forms of post-Fordist production. These changes have created a propensity for related firms to agglomerate in clusters in order to reduce the costs of transacting with one another. And, as we have just seen, the development and commercialization of the underlying technologies can also be geographically uneven.

So, can technology eradicate distance? In some instances, the answer is that technology can help to overcome at least geographic separation. The ability to move people, money, products, and technologies quickly, efficiently, and cheaply around the world has been highly significant in enabling the globalization of economic activity. Yet, in other instances, technologically-mediated communication and interaction is not enough. In industries where certain "sticky" forms of knowledge are important, people still need to locate together to immerse themselves in the buzz of day-to-day interaction, sociability, and innovation. That is a topic we will return to in more detail in Chapter 12.

Notes on references

- Dicken (2011, Ch.4) provides a clear account of different kinds of technology and their geographical impacts.
- See Zook (2005) and Malecki and Moriset (2008) for more on the emerging geographies of the internet world.
- Leinbach and Bowen (2004) offer a detailed account of the centrality of air cargo services to high-tech production networks, while Rodrigue et al. (2009) offer an excellent overview of transport geography more generally. For more on the global airline industry, see Bowen (2010).
- Fields (2006) provides a detailed case study of Dell's use of technology and the variable impacts on the geography of its global operations.
- Kaneko and Nojiri (2008) give a fascinating insight into the role of just-in-time production techniques within Japan's automobile industry.

Sample essay questions

- Why is it important to take an evolutionary perspective to technological change?
- Why do the space-shrinking technologies have varied and uneven geographical impacts?
- Does the internet have geographies?
- What forces are driving the changing nature of product innovation?
- What are the geographical implications of the supposed shift from Fordism to post-Fordism?

Resources for further learning

- http://www.worldbank.org and http://www.itu.int: The World Bank and International Telecommunications Union offer a wide range of data on the worldwide usage rates of information technologies such as fixed and mobile telephones.

- http://www.unctad.org/en/docs/wir2004_en.pdf: UNCTAD's 2004 *World Investment Report*, "The shift towards services," provides a wealth of information on the offshoring of service activity and can be downloaded for free.
- http://www.dell.com: Personal computer manufacturer Dell's homepage contains a range of information on the company, its strategies, and its history (click on "About Dell"). Likewise, the homepages of leading logistics and trading companies such as FedEx, DHL, and Li & Fung provide a wealth of information on their global activities.
- http://news.bbc.co.uk/1/hi/in_depth/business/2008/the_box/default.stm: This site provides a wealth of fascinating information and background on the British broadcaster's attempt to monitor the progress of one container for a year as it crisscrossed the globe during 2008–09.

References

Amin, A. (2000). Industrial districts. In E. Sheppard and T. Barnes, eds., *A Companion to Economic Geography*. Oxford: Blackwell, pp. 149–168.

Aoyama, Y., Ratick, S. J., and Schwarz, G. (2006). Organizational dynamics of the U.S. logistics industry: an economic geography perspective. *The Professional Geographer*, 58, 327–40.

Bonacich, E., and Wilson, J. B. (2008). *Getting the Goods: Ports, Labor and the Logistics Revolution*. Ithaca, N.Y.: Cornell University Press.

Bowen, J. (2010). *The Economic Geography of Air Transportation*. London: Routledge.

Cairncross, F. (1997). *The Death of Distance: How the Communications Revolution Will Change Our Lives*. Cambridge, MA.: Harvard Business School Press.

Dicken, P. (2011). *Global Shift: Mapping the Changing Contours of the World Economy*, Sixth Edition. London: Sage.

Fields, G. (2006). Innovation, time and territory: space and the business organization of Dell Computer, *Economic Geography*, 82, 1191–46.

Freeman, C., and Perez, C. (1988). Structural crises of adjustment, business cycles and investment behaviour. In G. Dosi, C. Freeman, R. Nelson, G. Silverberg, and L. Soete, eds., *Technical Change and Economic Theory*. London: Pinter, pp. 38–66.

Friedman, T. L. (2007). *The World Is Flat: The Globalized World in the Twenty-First Century*, Second edition. London: Penguin.

Grabher, G., Ibert, O., and Flohr, S. (2008). The neglected king: the customer in the new knowledge ecology of innovation. *Economic Geography*, 84, 253–80.

Guardian, The. (2008, June 17). Cash in hand: why Africans are banking on the mobile phone, p. 25.

Guardian, The. (2009, March 5). Welcome to the new gold mines, Technology Supplement, p. 1.

Heeks, R. (2008). Current analysis and future research agenda on "gold farming." *Development Informatics Working Paper 32*, IDPM, University of Manchester.

Kaneko, J., and Nojiri, W. (2008). The logistics of Just-in-Time between parts suppliers and car assemblers in Japan. *Journal of Transport Geography*, 16, 155–73.

Leinbach, T. R., and Bowen, J. (2004). Air cargo services and the electronics industry in Southeast Asia. *Journal of Economic Geography*, **4**, 299–321.

Levinson, M. (2006). *The Box: How the Shipping Container Made the World Smaller and the World Economy Bigger*. Princeton, NJ: Princeton University Press.

Malecki, E. J., and Moriset, B. (2008). *The Digital Economy: Business Organization, Production Processes and Regional Developments*. London: Routledge.

Malecki, E. J., and Wei, H. (2009). A wired world: the evolving geography of submarine cables and the shift to Asia. *Annals of the Association of American Geographers*, **99**, 360–82.

Rodrigue, J-P., Comtois, C., and Slack, B. (2009). *The Geography of Transport Systems*, Second edition. London: Routledge.

UNCTAD. (2004). *World Investment Report 2004: The Shift toward Services*. New York: UNCTAD.

Zook, M. (2005). *The Geography of the Internet Industry*. Oxford: Blackwell.

CHAPTER 10

THE TRANSNATIONAL CORPORATION

How does the global firm keep it all together?

Goals of this chapter

- To question the claim that transnational corporations are really global
- To understand how firms organize complex global activities
- To explore the variety of organizational forms used by transnational corporations
- To appreciate the various limits to the global reach of large firms

10.1 Introduction

A transnational corporation (TNC) is simply a firm that controls and coordinates other firms' activity in more than one country. This control and coordination is possible because a TNC owns its subsidiaries abroad or because it is the single largest buyer of outputs from these foreign firms. Nike provides an excellent example of a contemporary TNC. Headquartered in Beaverton, Oregon, the company focuses on the design, R&D, and marketing of sports products. It is the world's largest and leading athletic brand, with revenues of $20.9 billion in 2010–11. Apart from manufacturing its Air-Sole cushioning materials and components in two wholly-owned facilities in Beaverton, Oregon, and St. Charles, Missouri, Nike outsources all the production of its thousands of products to contract factories worldwide: as of July 2011, Nike had contracts with over 700 factories in 46 countries that employed almost 800,000 workers. Through this seemingly straightforward outsourcing arrangement, Nike was able to achieve profit margins of 44–46% between 2006 and 2011 (www.nikebiz.com, accessed on July 29, 2011). These impressive margins, however, mask the enormous complexity and potential risks inherent in its global outsourcing arrangement.

Nike's success and unique outsourcing model was coming under increasing scrutiny by the beginning of the 2000s mostly as a consequence of mounting pressure from the media and civil society activists concerned about sweatshop and child labor. NGOs and other activists targeted Nike in particular for three reasons: complaints and exposés had highlighted poor working conditions for its factory workers; its high profitability and thus the apparent affordability of implementing improvements; and its "demonstration effect" as an industry leader. As a result, in 2005 Nike became the first TNC in its industry to disclose its list of contract factories in the United States and worldwide. Asian factories have been a particular focus of attention. In 2010, for example, Vietnam accounted for 37% of Nike's global footwear production, with 40 active contract factories employing some 84,000 workers. Nike ran into trouble with working conditions in its contract factories almost as soon as it moved into Vietnam in 1995. In 1998, several of its contract factories in Chien Bien Hoa experienced serious labor abuses, with young female workers being punished with running laps around factories. While things seemingly improved after Nike stepped in with factory audits, nearly a decade later Nike's labor troubles returned. Some 13,000 workers of Tae Kwang Vina, a South Korean–owned contract factory for Nike, went on strike for five days in November 2007 to protest against low wages, significantly disrupting footwear supply to Nike in advance of the very busy Christmas season (www.ft.com, accessed on July 29, 2011). In April 2008, more than 20,000 workers at Ching Luh, a Taiwanese-owned factory in Vietnam, went on strike to demand a 20% pay rise to cope with inflation, and better canteen lunches (news.bbc.co.uk, accessed on July 29, 2011).

The labor challenges faced by Nike in Vietnam vividly illustrate the enormous organizational complexity and operating risks faced by today's transnational corporations (TNCs) when they expand internationally. The lack of adequate management supervision and control systems over its contract factories has underpinned Nike's labor woes in Vietnam and elsewhere in Asia. The influence of its head office culture of freedom and innovation clearly diminishes with the number of time zones traversed. What we have here, then, is a case of a global firm finding it increasingly difficult to control and coordinate adequately its constituent elements (most not directly owned) in different continents and regions, with potentially disastrous consequences (see also Rothenberg-Aalami, 2004). Other more recent high-profile cases are Toyota's safety troubles with sticky brake pedals in its U.S.-manufactured automobiles in late 2009 and Apple's bad press due to the suicide attempts by 18 workers in 2010 in its Taiwanese-owned manufacturing service provider in southern China. On the whole, though, many firms do successfully coordinate immensely complex global operations, some with hundreds of thousands of employees and turnovers the size of the economies of small countries. How do they do it? How does a TNC keep it all together?

This chapter adopts an economic-geographical perspective to describe and explain the strategic organization of these TNC activities in five main sections.

By questioning the false claim that global corporations are truly global in their reach in Section 10.2, we argue that TNCs are not placeless and that such a view clearly both exaggerates and simplifies the corporate power of TNCs. Rather, they are deeply local entities in the sense that they are constantly looking for new ways of organizing production in different territories and locations in order to stay ahead of their competitors. Our focus on the TNC is a deliberate one because of its enormous corporate power, capacity for switching activity from one place to another, and extensive global reach. In Section 10.3, we revisit the idea of value chains and propose the concept of production networks for understanding TNC activities. The section serves to reinforce our understanding of the structure and organization of commodity chains introduced in Chapter 8. The next two sections (10.4 and 10.5) seek to explain the internal and external divisions of labor in which TNCs are involved. In Section 10.4, we introduce the idea that the diverse activities coordinated and controlled by TNCs can be understood as *intra-firm* networks: managing different TNC functions across borders forms the core issue in this section. In Section 10.5 we examine the different modes of *inter-firm* relationships that are increasingly important in sustaining the competitive advantage of TNCs. Section 10.6 considers the different cultural obstacles faced by TNCs in globalizing their economic activities, arguing that TNCs are far from immune to different sociocultural contexts.

10.2   The Myth of Being Everywhere, Effortlessly

On its global website, HSBC (Hongkong and Shanghai Banking Corporation) claims categorically that it is the "world's local bank" (www.hsbc.com, accessed on July 29, 2011). Headquartered in London, HSBC is one of the world's top five largest banking and financial services groups and the most globalized among them, operating a global business network of around 7,500 offices in 87 countries and territories. Its shares are listed and traded on stock exchanges in London, New York, Hong Kong, Paris, and Bermuda. Some 221,000 individuals or institutions from over 127 countries and territories are its shareholders. Despite its humble origins as a British bank founded in Hong Kong and Shanghai in 1865 to finance the growing trade between China and Europe, HSBC has now developed into a global corporation that operates in almost every corner of today's global economy. HSBC's global reach is not just a matter of its geographical locations. More importantly, it gives substance to the myth of the global corporation. In this myth, a TNC is often viewed simply as an organizational "black box" that converts inputs into outputs and creates profits for its shareholders. There is no compelling reason why a TNC should be different in its organizational makeup or corporate behavior in different places and geographical contexts. After all, TNCs are supposed to respond merely to competitive signals and cost changes in global markets. They are popularly conceived as having the managerial and organizational power to easily control and command their worldwide operations from one central corporate headquarters.

To many protagonists of this myth, TNCs are the same everywhere. For over two decades, well-known business gurus and consultants have denounced any possible influence of particular places on the formation and behavior of TNCs. In his bestseller *The Borderless World*, for example, Kenichi Ohmae (1990, p. 116) proclaims that "country of origin does not matter. Location of headquarters does not matter. The products for which you are responsible and the company you serve have become denationalized." Ohmae's caricature is rather misleading, despite coming from a business guru who is very familiar with the distinctive Japanese culture of business organization. In addition, the media often celebrates a global convergence in corporate practices as a natural phenomenon. The central tenet of this popular myth is fairly simple: global competition drives global firms to become more and more alike in their business practices and corporate organization. For TNCs from different regions and countries, there is supposedly no holding back the tide of globalization; any irregular or idiosyncratic business and organizational practices are erased. In short, only a standardized and efficient form of business organization will survive global competition. From this perspective, HSBC is expected to be no different from Citibank, Nike will be similar to Adidas, and Toyota should resemble Ford.

This interpretation of TNCs is not inconsequential as it leads to all kinds of misleading conceptions about what TNCs do and how they organize their business operations on a global basis. The myth of the global corporation is underpinned by its apparently enormous financial capital, massive productive capacity, huge levels of employment, and privileged access to technology and knowledge. In reality, most TNCs are indeed significant corporations that have the necessary capital to fund their international expansion. Their huge financial muscle often allows them to bargain hard with host country governments and global suppliers and to outcompete most, if not all, local competitors in host markets. In developing countries that are desperately looking for ways to industrialize their economies, the dependence on TNCs for job creation means that TNCs have substantial corporate power to negotiate and bargain with these developing countries, particularly those characterized by weak or failed states (see Chapter 4). As TNCs gain experience through globalization, they develop relatively more sophisticated marketing and distribution know-how and technological activities in different host countries.

Upon closer inspection, however, the notion of the effortless global reach of the placeless TNC does not withstand scrutiny. Global presence does not necessarily mean that organizing on a global basis is easy or the same everywhere. On the contrary, as TNCs become more globalized, they face greater difficulties in holding everything together within a single corporate entity, as the Nike case aptly shows. TNCs not only need to configure carefully their functions and roles in order to adapt to different local economies and places, an idea known as "placing firms" (Dicken, 2000) and exemplified in HSBC's branding as the "world's local bank"; their entry into diverse host markets also necessitates different organizational forms and processes, as we will see shortly. To make sense of the diversity of TNCs and their operations, we must seek answers to the following three highly important and yet interrelated questions:

- How do TNCs organize their diverse activities in different parts of the world?
- How do TNCs engage and interact with other firms when they operate internationally?
- How do TNCs keep everything together in the face of intense market competition, civil society opposition, and political barriers?

10.3   Value Activity and Production Networks: The Basic Building Blocks of TNCs

In this section we will build on the idea of the commodity chain introduced in Chapter 8 as a necessary first step toward developing an understanding of contemporary TNCs. Here, we note that while the commodity chain is a useful concept for understanding value-adding activities that underpin the global travels of particular commodities such as coffee or T-shirts, the concept is more limited when it comes to analyzing and explaining the organizational dynamics of a global entity such as the TNC. We need a more sophisticated framework that probes into the network forms of value-adding activity organized and coordinated by the TNC in conjunction with its many global partners, suppliers, service providers, and customers. In Economic Geography, the *global production networks* (GPN) perspective (see Box 10.1) explains how organizing the global production of

---

KEY CONCEPT

Box 10.1   From commodity chains to Global Production Networks *(GPNs)?*

Chapter 8 talked almost exclusively in terms of commodity *chains*, understood as a sequential progression from sourcing and production through to distribution and consumption. But is this really the best metaphor for representing the complexity of contemporary production systems typically orchestrated by transnational corporations? What might be the advantages of a more open-ended "network" approach to economic systems? A central problem with representing these systems as chains is the inherent assumption of linearity – that is, the step-wise progression from a grower/producer to an end consumer through a series of intermediaries. This conceptualization of production and consumption processes has important limitations. In reality, the basic linear structures of a commodity chain are enmeshed within complex and diffuse networks of relationships involving a wide range of corporate (e.g., TNCs, financiers, management consultants, logistics providers) and institutional (e.g., standard setting bodies, NGOs, labor unions) actors. These networks of relationships clearly influence the nature of value-adding processes and, yet, may be underplayed in "chain" analyses. Moreover, a focus on the linear addition of value to commodities may obscure important connections between nonsequential stages of the system. In particular, it may downplay the complex exchanges of information within a production system that may play a vital role in its nature and operation. Information about quality standards may be passed directly from a retailer to a manufacturer, for example, or through an intermediary such as a logistics firm or importer/exporter. These insights suggest that it might be productive to move from talking about commodity chains to production networks.

The global commodity chain concept discussed in Chapter 8 has provided a crucial starting point for the development of the global production networks (GPN) framework in Economic Geography. The GPN framework is explicitly concerned with how economic value is created, enhanced, and captured within the different spatial configurations fundamentally underpinning a TNC's global operations. Moreover, by drawing actors such as firms and non-firm institutions into a common analytical framework, the GPN analysis seeks to provide a dynamic conceptual apparatus that is sensitive to multiple scales and power relations. What though are GPNs? How are they organized, and how do they emerge in different industries? A brief summary is useful here. Since the 1960s, firms from industrialized economies have increasingly been taking production activity across borders. Through this process of internationalization, they have become transnational corporations (TNCs). These TNCs are not autonomous and vertically integrated organizations; rather, they are constituted by a mesh of intra-firm and inter-firm networks comprising a large assortment of other actors and organizations. As TNCs become much more global in their scale and scope of operations, their networks are also more global in nature, leading to the emergence of GPNs. A GPN is thus defined as a network system that is coordinated and controlled by a globally significant TNC and involves a vast network of overseas affiliates, strategic partners, key customers, and non-firm institutions. These lead firms are key shapers of the global economy; they are market leaders in terms of their brand names, technology, products/services, and marketing capabilities. For more on the GPN approach, see Coe et al. (2008).

a consumer product such as a pair of sports shoes or a mobile phone is far from a simple process. Two key issues are at stake here for each TNC – namely, the configuration and coordination of the production network. Configuring the production network is a matter of deciding where each value-added activity along the value chain should be located. This configuration is important in the formation of the various spatial structures of commodity chains. Nike, for example, has to decide if it wants the manufacturing of all sports shoes to take place in one country from which it can ship products to the rest of the world. This would be a highly geographically concentrated production network configuration. On the other hand, it may decide to locate production in its major markets in different regions, thereby following a more dispersed production network configuration. Geographical outcomes can therefore vary.

Coordinating the production network, however, is mostly an organizational matter that shapes the network's governance. Depending on the nature of the value activity (producer- or buyer-driven), a TNC has to tackle the issue of how to coordinate a wide range of activities spanning different countries and locations. Nike, for example, may choose to centralize most of its corporate decisions at the headquarters located in Oregon, with its regional offices located in various macro-regions implementing these decisions within the respective markets. Equally, hypothetically at least, it might decide to operate as a much looser confederation of autonomous national subsidiaries, which meet the needs of the various markets in which they are located with only a light guiding hand and financial oversight being provided by central headquarters. Different TNCs may thus coordinate their production networks via different mechanisms that range from being highly centralized to highly decentralized. Taken together, each TNC will develop its own particular way of configuring and coordinating its global production network(s).

The above discussion has assumed that all value-added activities are conducted by the same TNC and its worldwide subsidiaries. What if the TNC, such as Nike, is only engaged in some parts of the production network and develops contractual relationships with other firms as part of a wider system? A distinction should therefore be made between two types of transactions in the production network:

- *Internalized transactions*: These are value-added activities that occur within the legal and organizational boundaries of a particular TNC. We use the term "hierarchy" to describe these intra-TNC transactions as they are subject to the internal governance system of the TNC. For example, when Samsung's mobile phone division "buys" microprocessor or memory chips from Samsung's own semiconductor division, this is an internalized transaction within Samsung and it is not necessarily subject to the open market mechanism of price competition. This internalized transaction is also more likely to be subject to decisions made by the corporate headquarters.
- *Externalized transactions*: These are business relationships that exist between independent firms, some of which may be TNCs. We describe these inter-firm

relationships as "markets" because they are subject to pressures of price competition and other factors beyond the governance system of a particular firm. For example, when Nokia buys lithium batteries for its mobile handsets from its Japanese supplier Sanyo, this transaction is market-based, and both Nokia and Sanyo are considered to have engaged in an inter-firm transaction.

At this point, it is useful to provide a fuller example of a TNC – Nokia – and its global production networks. How might we make sense of Nokia and the geography of its mobile phone business? Following the above line of thinking, we can describe and break down the mobile phone's entire production, starting with Nokia, the lead firm that coordinates the phone's global production network. Arguably the best-known Finnish TNC today, Nokia first started as a forest industry enterprise in southwest Finland in 1865, only seriously moving into mobile communications at the beginning of the 1980s and launching its first mobile phone device in 1986 (www.nokia.com, accessed on August 1, 2011). By the turn of the new millennium, Nokia had emerged undoubtedly as the global market leader in mobile communications. Between 2003 and 2009, Nokia had a worldwide market share of 30–39% in the mobile devices business, at least 15% ahead of its closest competitors such as Samsung, LG, Motorola, and Sony Ericsson. By the end of 2010, Nokia's market share had fallen to 28.9%, compared to 17.6% for its main competitor Samsung. Both market leaders faced immense challenges at both ends of the market from, for example, unbranded Chinese devices in the low-cost segment and Apple's iPhone and other smartphones in the high-end market (www.gartner.com, accessed on August 1, 2011). To stay as the global market leader, Nokia has spent hundreds of millions of dollars in research and development (R&D) for mobile communications, invested considerable money and effort on marketing its products, and no longer manufactures in its home country but has instead established worldwide production facilities. For every Nokia mobile phone that reaches the final consumer, numerous actors are involved: the *designer* responsible for the phone's look and feel; the *engineer* for the technological specifications; the *financier* for funding the project; the *worker* for assembling it; the *driver* for handling logistics; the *salesperson* for facilitating the act of buying; the *government official* for ensuring safety and product standards, and so on.

In short, there are many ways of organizing the production network of a particular product or service. Each TNC may be involved in many such products and/or services that can be analyzed through a global production networks framework. To cite another leading TNC, Hewlett Packard (HP) has to coordinate on a global basis the production networks of hundreds of its products ranging from personal computers and mobile devices to imaging and printing equipment and enterprise business products and solutions. These production networks often overlap in terms of corporate functions (e.g., common technologies between personal computer and electronic medical devices) and geographical locations (e.g., China and Singapore) to form complex sets of internal (intra-firm) and external (extra-firm) relationships that cannot be captured by the simple notion of a commodity chain. In the next two sections, we look in turn at these different kinds of network relationships internal and external to the TNC.

10.4   Organizing Transnational Economic Activities 1: Intra-firm Relationships

Building on the notion of the production network as a basic building block of the global economy, we will now explore how TNCs organize their economic functions and divisions of labor internally (drawing extensively on the ideas in Dicken, 2011). We argue that the internal organizational structure of TNCs varies according to the different corporate strategies being pursued and their distinctive corporate cultures (see Box 10.2). To support our argument, we draw on examples from both manufacturing and service sectors, and pay special attention to the role of geography in shaping diverse organizational formations. We leave the role of people and social relations in shaping TNC activities to Part 4 of this book. In this chapter, our primary intention is to illustrate the *spatiality* of the organization of TNC activities.

KEY CONCEPT

Box 10.2   Corporate cultures

One of the principal differences between TNCs is the distinctive, firm-specific practices and behavioral norms that can be broadly conceptualized as *corporate cultures*. To go beyond the "black box" conception of the TNC often found in mainstream economics (see Section 10.2), we need to understand corporate cultures in two ways. First, we can consider how TNCs differ from each other because of their corporate cultures. For example, even among what most consumers consider to be similar retail giants, such as Wal-Mart and Costco in the United States, there are significant differences in the ways they treat their employees and customers (Herbst, 2005). Second, we can open up the TNC by viewing it as an organization that is internally heterogeneous and contested by different intra-firm interest groups. These groups and actors enjoy varying degrees of power and access to resources. They represent different fragments and thus different subcultures of a TNC. It is worth noting that TNCs are becoming increasingly interested in understanding themselves in terms of cultural and subcultural practices. Indeed, as self-knowledge becomes ever more critical in managing across borders, we can expect greater importance to be

---

attached to the understanding of how corporate cultures and subcultures influence learning and competitive outcomes in today's TNCs.

What, then, exactly *is* corporate culture? How do we go about identifying the distinctive cultural traits of an organization? We can see corporate culture operating in four ways (here we follow Schoenberger, 1997):

- *Ways of thinking*: This refers to the ways in which thinking at a particular company is constrained, directed, or focused. Inevitably, over time, organizations develop elements of "group think" in which certain ideas and practices are accepted uncritically. It is often easier for an outsider or newcomer to spot these elements than it is for someone who has been immersed in a particular environment for a long time. In the financial sector, for instance, some banks may see themselves as "safe" institutions making conservative investment decisions and loans, while others may take on higher risks for greater returns.

- *Material practices*: When ways of thinking are converted into *action*, corporate culture is manifested in the everyday business practices. In short, this relates to what a firm should be doing, and how it should be doing it. In particular firms and workplaces this encompasses a wide range of activities, including conversations, meetings, work tasks, divisions of labor, production processes, service delivery, and so on. Over time, practices become routinized and unthinkingly accepted as a "natural" part of how things are done.

- *Social relationships*: Implicit in ways of thinking and material practices are the relationships that exist within a firm between its various employees. How do people work together, get along, and respond to orders from superiors or suggestions from subordinates? These relationships form the social glue that holds a company together and enable a complex division of labor to work successfully. There are, however, different models that might be adopted, ranging from informality and friendship, team spirit and collegiality, through to competitiveness, rigid hierarchy, and mutual animosity. These different cultures of relationships within the firm affect the way in which work is distributed, obligations are met, orders are followed through, and strategy is executed.

- *Power relations*: The final dimension of corporate culture emerges from internal social relationships. It concerns the ways in which power is distributed and wielded within a company. Obviously any firm, even the smallest enterprise with just a few employees, has a hierarchy comprising those who make decisions and those who mostly just follow orders or prescribed work patterns. In a large firm, these power relations are complex and may comprise long chains of command. But the way in which power is allocated and exercised will vary across different firms. In some settings, the hierarchy will be clear and a strong vertical chain of command will exist. In others, a more flexible and flattened reporting system is implemented.

The exact internal divisions of labor within TNC networks are often the combined results of organizational/technological forces and location-specific factors. Different parts of the TNC may have different locational needs, and their geographical outcomes have to be considered on a case-by-case basis. Generally, three key kinds of TNC organizational units can be discerned:

- *Corporate and regional headquarters*: These are the nerve centers of TNCs where important strategies are formulated and decisions are made (e.g., Nike's HQ in Beaverton, Oregon, or HSBC's HQ in London). These decisions apply to a wide range of corporate functions such as corporate finance and investment decisions, market research and development, product choice and market specialization, human resource development, and so on.
- *Research and development (R&D) facilities*: These activities are found in both manufacturing and service TNCs. R&D facilities encompass activities such as product development, new process technologies, operational research, and the like. They provide important knowledge and expertise to keep the TNC competitive in the global marketplace.
- *Transnational operating units*: These units cover a wide range of activities from manufacturing plants and facilities to sales and marketing offices, fulfillment centers, and after-sale service centers.

We can use the example of BMW to illustrate the complex spatial organization of intra-TNC networks (see Figure 10.1). First, headquarters play a highly important role as the apex of management and financial control, and the processor and transmitter of market and production information within the TNC. Such corporate functions thus require a strategic location in major cities that provide access to high-quality external services (e.g., management consulting or advertising), and the presence of skilled labor and excellent infrastructural and communications support. As shown in Figures 10.2 and 10.3, BMW is headquartered in Munich, the third largest city and a major business and financial center in Germany. Leading global cities such as this act as important control and command centers in the global economy by virtue of their hosting of the headquarters of major TNCs. Most of the Fortune Global 500 TNCs are headquartered in a small group of global cities including Chicago, Frankfurt, London, New York, Paris, Shanghai, and Tokyo (see Box 7.3).

Second, the location of R&D facilities can be highly critical to the success of TNCs because of the importance of continuously developing new products and/or services. And yet despite decades of partial internationalization of R&D activity, the most significant R&D activities remain located in TNC home countries.

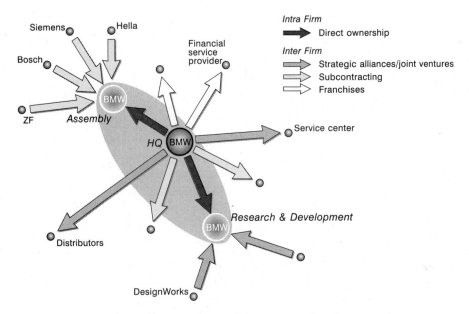

Figure 10.1 Different forms of organizing transnational operations
Source: The authors.

Figure 10.2 The BMW headquarter office in Munich, Germany

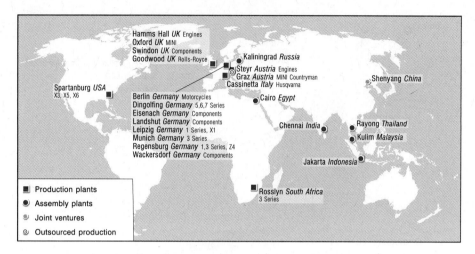

Figure 10.3   BMW's global production networks
Source: Reproduced with permission by BMW Group.

Different types of R&D activities have different locational requirements, however. In general, there are three main types of R&D activities (after Dicken, 2011):

- *Internationally integrated R&D lab*: This center normally supplies the core technologies and knowledge for the entire TNC's operations worldwide. It is the brain behind a TNC's innovative products and/or services that contribute to its competitive advantage.
- *Locally integrated R&D lab*: These laboratories apply the fundamental technologies and knowledge developed in the internationally integrated R&D lab in order to develop innovative new products for the host market in which the lab is located. Such labs are oriented toward local market and regulatory requirements that are not necessarily found in other markets.
- *Support lab*: This is the lowest level of R&D facility and is primarily concerned with adapting parent company technologies for the local market and providing technical backup.

In the case of BMW, the most important R&D centers for the group's innovation, technology, and car IT research continue to be located in, or near to, its Munich headquarters (Figure 10.2). It has also established other R&D labs such as DesignWorks in California and Singapore to engage in design work and emission tests, for example. A Tokyo lab supports the group's R&D efforts by tapping into the technological and innovative environment of Japanese carmakers. A newly opened design center in Singapore reflects the group's anticipation of Asia's emerging influence on design. What is clear in the case of BMW is that

a) Globally concentrated production

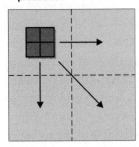

All production occurs at a single location. Products are exported to world markets.

b) Host-market production

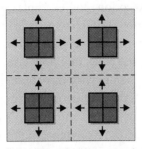

Each production unit produces a range of products and serves the national market in which it is located. No sales across national boundaries. Individual plant size limited by the size of the national market.

c) Product-specialization for a global or regional market

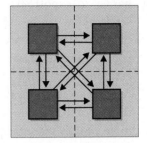

Each production unit produces only one product for sale throughout a regional market of several countries. Individual plant size very large because of scale economies offered by the large regional market.

d) Transnational vertical integration

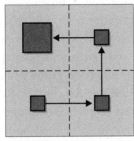

Each production unit performs a separate part of a production sequence. Units are linked across national boundaries in a 'chain-like' sequence – the output of one plant is the input of the next plant.

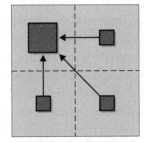

Each production unit performs a separate operation in a production process and ships its output to a final assembly plant in another country.

Figure 10.4 Geographies of transnational production units
Source: Dicken, P. (2011), Global Shift: Mapping the Changing Contours of the World Economy, Sixth Edition, London: Sage, Figure 5.14.

there are different R&D patterns for different market orientations. The firm's Munich-based R&D activities cater to the global market, whereas its California, Singapore, and Tokyo centers have more specific local/regional market and regulatory orientations.

Third, the most spatially mobile part of a TNC's global production network is often its production operations (with respect to either tangible products or services). Drawing upon Dicken (2011), we highlight in Figure 10.4 four main ways of organizing transnational production units among TNCs. Each cell in the figure represents a geographical unit, such as a country.

- *Globally concentrated production (Figure 10.4a)*: This mode of transnational production applies to some resource extraction and manufacturing industries.

It does not generally work well for service industries, as services are much harder to export. An exception, for example, may be advanced producer services that can be exported through experts being sent from corporate head offices to serve clients worldwide. In this mode, a TNC can exercise tight control of its subsidiaries and tends to follow a spatial pattern of geographical concentration. Most TNCs start with this strategy in their early years, when they are likely to locate all production in their home countries and export to the rest of the world. Today, we still find Toyota's Lexus series of automobiles to be entirely manufactured in Japan and exported to the rest of the world.

- *Host-market production structure (Figure 10.4b)*: This mode of organization is preferred where there are considerable barriers to trade in host countries, meaning that exports may not be the most efficient channel to reach host markets. In service industries, the demand for a local presence and/or regulatory requirements also explain why service TNCs, ranging from professional services (e.g., law and accounting) to consumer services (e.g., retailing and hotels), often set up operations in each host market. In this case, centralized corporate control of local subsidiaries and operations is likely to be rather difficult since considerable autonomy needs to be granted to each national unit or even regional units (such as HSBC in Asia vs. HSBC in America). This is because each operation in the host country is predominantly local in its business orientation and very sensitive to local demand. No sales occur across national boundaries as products and services are highly customized to the tastes and preferences of local markets.

- *Product specialization for a global or regional market (Figure 10.4c)*: This mode of transnational production tends to be applicable mainly to manufacturing industries – for example, the electronics, automobile, and petrochemical industries. As product mandates are given to macro-regional production units, a great deal of autonomy rests with these operations supplying global or macro-regional markets. Hewlett Packard's regional headquarters in Singapore, for example, coordinates the global mandate for all HP printers. Its design and engineering teams based in Singapore have a long history of excellence in designing and developing industry-leading imaging and printing products. In November 2010, the first-ever global design center for HP's Imaging and Printing Group was opened in Singapore. Its high-end computer servers, however, remain coordinated by its global headquarters in Palo Alto, California.

- *Transnational vertical integration (Figure 10.4d)*: This mode is the most developed and coordinated organizational structure, although it is very demanding in terms of management expertise and control in order to ensure that production responsibilities are well specified and coordinated across the TNC's global operations. Sony and Sharp, for example, have developed extensive production networks for their LED and LCD televisions that span Japan (core components), China (assembly and testing), and many Southeast Asian countries (modular components and subassembly).

One of the main advantages of this transnational production mode is to exploit spatial variations in production costs. A TNC can establish offshore production for some of the more labor-intensive manufacturing operations. For example, many American TNCs in the electrical, electronics, automobile, textiles, and garment industries have set up assembly operations in the *maquiladoras* in Mexico (see Box 10.3) to take advantage of the latter's lower production costs

CASE STUDY

Box 10.3 Transnational production in the maquiladoras of northern Mexico

The term *maquiladora* refers to an assembly plant in northern Mexico that manufactures export goods for the U.S. market. They were instituted by the Mexican government in 1965 when it started the Border Industrialization Program to try and tackle the severe economic and social problems of the towns along the Mexico–U.S. border. Under the auspices of the North American Free Trade Agreement (NAFTA), which was driven primarily by American interests and came into effect in 1994, U.S. firms manufacturing in the *maquiladoras* across the Mexican border can benefit from much cheaper labor costs in Mexico, virtually nonexistent taxes and custom fees, and less stringent environmental regulations. The *maquiladoras* have thus become offshore factory spaces for American firms and TNCs from other countries (such as Japan and countries of Western Europe) selling into the U.S. market. Leading employers include Song Corp (Japan), Lear Corp (U.S.), Yazaki (Japan), Johnson Controls (U.S.), and Jabil Circuit (U.S.). The deal arguably benefits American consumers and firms far more than low-wage workers in Mexico. *Maquiladoras* are prevalent in Mexican cities such as Tijuana, Ciudad Juarez, and Matamoros that lie directly across the border from the U.S. cities of San Diego (California), El Paso (Texas), and Brownsville (Texas). Factories in the *maquiladoras* primarily produce electronic equipment, automobile parts, clothing, plastics, furniture, and appliances. In April 2011, Mexico had about 1.8 million jobs in 5,087 *maquiladora* companies established throughout Mexico and registered under the Maquiladora Manufacturing Industry and Export Services (IMMEX) Program.

Labor-intensive assembly work remains the cornerstone of employment in *maquiladoras*. In September 2010, for example, some 1.5 million of the 1.8 million jobs in *maquiladoras* were direct labor and technicians, with the remaining in administrative positions (www.maquilaportal.com; accessed on August 1, 2011). Many *maquiladoras* might be considered sweatshops, employing young women for as little as 50 cents an hour, for up to ten hours a day, six days a week. The cost of living in border towns is often 30% higher than in southern Mexico, and many of the *maquiladora* women are forced to live in shantytowns that lack electricity and water surrounding the factory cities (see also Box 13.2 on devaluing third-world women). While some of the companies that own the *maquiladoras* have sought to increase their workers' living and working conditions, much of the workforce is nonunionized. Some *maquiladoras* are responsible for significant industrial pollution and environmental damage to the northern Mexico region. Competition from China has weakened the attraction of *maquiladoras* for inward investors in recent years. The 2008 global financial crisis has also taken a toll on the *maquiladoras*. Between October 2007 and July 2009, *maquiladora* and export manufacturing companies lost 378,436 jobs in Mexico, although 218,194 had been recovered by July 2010. However, the future of the *maquiladoras* remains unclear, as they are heavily dependent on the sluggish U.S. market and NAFTA: in 2011, Mexico was the third largest supplier for the U.S. market in non-petroleum goods after China and Canada.

and tax exemptions under the North American Free Trade Agreement (NAFTA). As assembly work represents the most labor intensive and the least technologically sophisticated part of the production chain, it can be easily transferred to low-cost locations. However, assembly operations in these overseas subsidiaries often have little autonomy to develop new products or market strategies. Instead, they tend to have to follow strictly the production plans set by the corporate headquarters.

In reality, we should note that these four organizational structures are ideal-typical modes of transnational production. A TNC may use one or more ways of configuring its worldwide operations. BMW, for example, has moved a long way from vertically integrated production based in Munich during its early days. Figure 10.5 illustrates the technological and organizational complexity and large number of components (>20,000) of its modern 5 series sedan that necessitates the globalization of its production networks. BMW has therefore developed sophisticated production networks spanning different regions in the European Union and Southeast Asia (see Figure 10.3). While its plants in Bavaria, Germany, continue to manufacture the full range of BMW cars for the entire European Union market (Figure 10.4c), BMW has a full production plant in the United Kingdom to serve the right-hand-drive British market (Figure 10.4b). It has also established full-scale, local production plants in the United States

Figure 10.5　Making a BMW: Munich and beyond
Source: Reproduced with permission by BMW Group.

(Spartanburg) and South Africa (Rosslyn) to serve both continents. BMW's parts and components manufacturing takes place in its supplier and innovation park in Wackersdorf in Eastern Bavaria, Germany. The establishment of an international logistics center in Wackersdorf is critical because it distributes BMW's international parts and components from its engine plant in Steyr (Austria) and over 800 suppliers throughout Europe to its foreign production plants in the United States and China and assembly plants in Egypt, India, Indonesia, Malaysia, Thailand, and Russia (see Figure 10.3). Managed by a third-party logistics provider, the Wackersdorf center sends out daily shipments of over two million parts (www.bmw-plant-wackersdorf.com, accessed on August 1, 2011).

More specifically, BMW's assembly operations in East and Southeast Asia resemble the transnational vertical integration mode in Figure 10.4d. In this mode of transnational production, its core parts and components are produced in Germany and shipped to its plants in Asia for assembling with some local parts and components provided by either local firms, or foreign suppliers that have followed BMW to these overseas assembly locations. The local content of a BMW car assembled in Thailand, for example, exceeded 40% of value added in July 2010, with the exception of its latest 7 series (www.nationmultimedia.com, accessed on August 1, 2011). However, this included labor costs and sourcing within Southeast Asia, not just Thailand. In China, BMW has operated a production plant since May 2004 in Shenyang to serve the rapidly growing Chinese market. From June 2012, it will build the new 3 series in China with 174 locally produced components and local content of over 40% (www.bmwgroup.com, accessed on August 1, 2011). To sum up, the example of BMW represents well the global production strategies, locational impacts, and organizational characteristics of all major automobile manufacturers.

Even though the automobile industry depends on a large number of independent parts and component suppliers that may be large TNCs in their own right (e.g., Denso, Bosch, Delphi, Lear, and ZF), the major car makers continue to organize their transnational production networks with a great deal of *vertical integration* – a pattern of industrial organization in which the parent TNC directly owns and controls most of its operations along the production chain. Compared to other modern industries such as electronics and consumer products (e.g., garments and shoes), tightly coordinated and controlled intra-firm networks are common within the automobile industry. This occurs despite the extensive use of independent suppliers by major car makers through inter-firm relationships, the theme of the next section.

10.5   Organizing Transnational Economic Activities 2: Inter-firm Relationships

As products and services become increasingly sophisticated, few TNCs today can produce everything by themselves. The BMW sedan in Figure 10.5 is made up of at least 20,000 parts, a HP personal computer can be broken down to a few thousand parts, a Nokia mobile phone or an Apple iPhone has a few hundred components (see Figure 10.6; Dedrick et al., 2010), and even an ordinary Nike sports shoe has at least 10–20 parts. In the service sector, an advertising project requires creative talents, financial consultants, technologists, public relations experts, and clerical workers among many others. Running a hotel business may bring together property owners, international hotel management companies (e.g., Marriott), independent food and beverage operators, service suppliers, and so on. In short, most TNCs are simultaneously controlling their own intra-firm networks of foreign subsidiaries and affiliates at the same time as managing a dense web of inter-firm networks of independent contractors, suppliers, business partners, and strategic allies. When these intra-firm networks overlap with inter-firm networks in different geographical contexts, we can imagine each TNC as being part of a complex set of organizational relationships (see Figure 10.1 again). In this section, we focus on three main modes of inter-firm relationships that lie somewhere between market (buying goods and services from independent firms) and hierarchy (intra-firm relations):

- Subcontracting and outsourcing
- Strategic alliances and joint ventures
- Franchising and cooperative agreements

Our focus on these three forms reflects both their significance in organizing TNC activities and in contributing to international trade and investment patterns. UNCTAD's (2011) *World Investment Report* noted that these investment flows tend to be based on non-equity modes (NEM) such as those explained in this section. First, the rapid rise of subcontracting and outsourcing during the

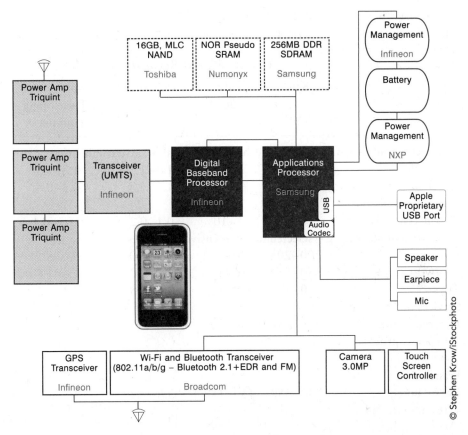

Figure 10.6 Apple iPhone 3G: its components and key suppliers
Source: Based on http://sigalonmobile.soup.io/tag/iSuppli and
http://www.iphonekiller.com/iphone-3g-s-teardown

1990s and the 2000s has helped many TNCs to ease competitive pressures and to focus on their core businesses (e.g., Nike and its 700 contract factories). Second, the role of strategic alliances and joint ventures is critical in industries where investment costs are high and business horizons are uncertain (such as pharmaceuticals). Third, franchising and cooperative agreements (e.g., licensing) have emerged to be very successful modes of organizing international expansion in the service industries (retailing, hotels, and fast-food chains, for example). Table 10.1 shows that cross-border NEM activity coordinated by TNCs was very significant in 2010 in a wide range of industries, from high-tech to labor intensive and from manufacturing to services. Here, we are interested in the various strategies and governance mechanisms through which TNCs engage in these inter-firm relationships. In the context of TNCs, the discussion focuses primarily on the international dimensions of these relationships. As in the preceding section, we draw upon different empirical examples to illustrate our discussion.

Table 10.1 Non-Equity Modes (NEMs) of cross-border activity by TNCs in selected industries, 2010 (in billions of dollars and millions of employees)

	Estimated NEM-Related Worldwide			
	Sales	Value Added	Employment	Employment in Developing Economies
Contract manufacturing – selected technology/capital intensive industries				
Electronics	230–240	20–25	1.4–1.7	1.3–1.5
Automotive components	200–220	60–70	1.1–1.4	0.3–0.4
Pharmaceuticals	20–30	5–10	0.1–0.2	0.05–0.1
Contract manufacturing – selected labor intensive industries				
Garments	200–205	40–45	6.5–7.0	6.0–6.5
Footwear	50–55	10–15	1.7–2.0	1.6–1.8
Toys	10–15	2–3	0.4–0.5	0.4–0.5
Services outsourcing				
IT services and business process outsourcing	90–100	50–60	3.0–3.5	2.0–2.5
Franchising				
Retail, hotel, restaurant, and catering, business and other services	330–350	130–150	3.8–4.2	2.3–2.5
Management contracts – selected industry				
Hotels	15–20	5–10	0.3–0.4	0.1–0.15

	Estimated NEM-Related worldwide		
	Fees	Associated Sales	Associated Value Added
Licensing			
Cross-industry	17–18	340–360	90–110

Source: UNCTAD (2011), Table IV.4, p. 133.

International Subcontracting

Subcontracting (also known as outsourcing) involves engaging independent firms to produce goods specifically for the principal firm. There are two types of international subcontracting: commercial and industrial. Commercial subcontracting occurs when the principal firm (the buyer) subcontracts most, and perhaps all, production to another firm (the supplier) in another country. At the beginning, the buyer may supply product specifications in what is called original equipment manufacturing (OEM) so that the finished products are exactly the same as if the buyer had produced them. Over time, the supplier may learn and develop new technology and expertise so that it can improve its value-added activity by engaging in some original design manufacturing (ODM). In this latter mode of subcontracting, the supplier will discuss product specifications with the buyer and design and make the product in question to meet those technological and marketing requirements. For buyers that subcontract their entire production, they will specialize in brand management and the marketing of products bearing their brand name.

This mode of international subcontracting is most common in today's electronics industry, particularly the manufacturing of personal computers, consumer electronics, and household appliances (see Table 10.1). Most standard personal computers, televisions, refrigerators, digital devices, and the like are no longer manufactured by the TNCs whose brand names appear on the product. Instead, a large group of independent contract manufacturers or electronics manufacturing service (EMS) providers are designing, producing, and sometimes distributing for these brand-name TNCs. These independent subcontractors are mostly based in newly industrialized economies and emerging developing countries, whereas the brand name TNCs are still mostly from advanced industrialized economies.

Take the example of notebook or laptop personal computers. Table 10.2 reveals that in 2010, the top ten global brand-name TNCs were subcontracting the production of large proportions of their notebook products through original design manufacturing (ODM) agreements with leading Taiwanese contract manufacturers such as Quanta, Compal, and Wistron. In 2010, these three relatively unknown Taiwanese contract manufacturers alone accounted for some 68% of the world's total notebook shipments of 187 million units (www.researchinchina.com, accessed on August 1, 2011). The top six Taiwanese

Table 10.2 Subcontracting of the world's top notebook brand-name companies to Taiwanese firms, 2010

Company	2010 Shipment (thousands)	Largest Client	Second Largest Client	Third Largest Client
Quanta	52,100	HP	Lenovo	Apple
Compal	48,100	Acer	Dell	Toshiba
Wistron	27,500	Dell	Acer	Lenovo
Inventec	16,150	Toshiba	HP	Lenovo
Pegatron	15,450	Asus	Apple	Dell
Hon Hai	10,000	Asus	Dell	HP

Source: www.researchinchina.com, accessed August 1, 2011.

suppliers provided some 173 million or 92% of these shipments. These figures included notebooks manufactured in China by Taiwanese-owned production facilities. As Taiwan is rapidly emerging as the world's leading center of electronics manufacturing, the same phenomenon of subcontracting to Taiwanese companies is occurring with electronics products such as tablet computers, computer monitors, flat panel and plasma televisions, mobile telephone handsets, and personal digital assistants (PDAs).

To illustrate how this international subcontracting works, we can look at the example of a customized Apple computer notebook computer (see Dedrick et al., 2010 for a case study of the HP nc6230). A customer in the United States or the United Kingdom may order an Apple MacBook or MacBook Pro via Apple's website or through a retail channel (e.g., local distributor). This order will be received by Apple Computer Inc., the brand-name buyer, which in turn will subcontract the notebook production to a system manufacturer or EMS provider in Taiwan (e.g., Quanta, Pegatron, or Hon Hai Precision). Upon receiving its daily order from Apple and other brand-name customers, Taiwan's Quanta will regularly place orders of parts and components with its key suppliers, such as Intel (chipsets), NVidia (graphics card), Seagate or Hitachi (hard disks), Samsung or Infineon (memory chips), Delta (power supplies), and Chunghwa Picture Tubes (computer screens). As Quanta has already moved its entire notebook manufacturing facility to Kunshan and Shanghai, China, it will ask its key suppliers to ship parts and components directly to its warehouse and assembly plants in Kunshan and Shanghai (see Yang, 2009). Once the notebook is assembled in China, it will be sent via express service to the customer in the United States, the United Kingdom, or elsewhere. Typically, this process – from the initial order by the consumer to receipt of the finished product – takes no more than seven days within the Asia-Pacific region and 14 days worldwide. This manufacturing system is a form of the integrated production shown in Figure 10.4d: in short, the Apple MacBook embodies a complex geography of international production and trade flows spanning many different locations in the global economy.

Unlike commercial subcontracting, industrial subcontracting takes place when the suppliers carry out only OEM production on behalf of its key customers – that is, manufacturing only. Industrial subcontractors do not engage in ODM and final distribution on behalf of their customers. In the footwear industry, such a non-equity mode of activity by TNCs is very common (see Table 10.1). Nike's international subcontracting practices provide an excellent example. The geographical implications of international subcontracting are highly variegated. First, subcontracting may lead to the development of highly localized manufacturing clusters in developed countries (see Chapter 12). Second, international subcontracting has led to the development of enclaves of export-oriented production in developing countries – a geographical phenomenon described in Box 10.4 as the new international division of labor. These export enclaves or satellite

KEY CONCEPT

Box 10.4   *Transnational corporations and the new international division of labor*

During the 19th century, the world economy was dominated by the industrialized economies of Western Europe and the United States that specialized in the manufacturing of goods on the basis of raw materials extracted from developing countries. Supported by imperialism and colonial relations, this "traditional" form of international division of labor prevailed until at least the 1960s. From the 1960s onward, a new international division of labor (NIDL) supposedly started to emerge in which certain developing countries took on new roles within the global economy through the investment strategies of TNCs. The NIDL described the establishment by European, North American, and Japanese TNCs of a global manufacturing system based on establishing labor-intensive export platforms in so-called newly industrializing economies (NIEs). The NIDL theory, proposed most explicitly by Fröbel et al. (1980) in their study of garment manufacturing, argued that falling rates of profitability in manufacturing production in core countries were combining with market saturation and under-consumption as market growth lagged behind productivity increases. In response, TNCs used their global reach to relocate production from the industrial core to low-cost production sites in the NIEs; the manufactured goods were then in turn exported back to core markets from the offshore branch plants. Crucially, the system depended on new technologies that allowed production fragmentation, thereby creating tasks that could use often young and female semiskilled or unskilled workers in the periphery. While the more capital-intensive parts of production processes continued to be located in advanced industrialized economies, NIEs across South America, Eastern Europe, Africa, and Asia became bases for labor-intensive phases of production. The appearance of designated export processing zones (EPZs), often offering a range of tax incentives, from the 1960s onward were the most obvious physical manifestation of the emerging NIDL.

The NIDL thesis, however, has been extensively critiqued. In particular, it overlooks the following important aspects of the contemporary global economy:

- The continued importance of the "traditional" form of international division of labor through significantly more investment in raw material extraction and agribusiness.
- Foreign direct investment flows remains dominated by investments between developed countries.

- A rapid growth of outward investment and TNCs from the newly industrialized economies, complicating the simple NIDL model.
- A substantial proportion of world's foreign direct investment now occurs in the service sector, primarily aiming to serve host markets rather than being for re-export.
- The significance of cheap labor in attracting manufacturing functions is clearly overstated. Many other institutional factors such as conducive government policies and appropriate business climates are equally if not more important.

The reality of the global economy today thus reflects overlapping divisions of labor, some of which are newer than others. The geographies of these divisions of labor are necessarily more complicated than the simple core-periphery model underlying the NIDL thesis. For more, see Coe (2011).

production clusters host labour-intensive manufacturing activity that forms an integral part of the global production networks of most TNCs. Collectively, these enclaves constitute the manufacturing base of the global economy. Prime examples are the *maquiladoras* in Mexico (automotive components, electrical assembly, and clothing), Dongguan in China (one of the world's largest shoe production centers), Penang in Malaysia (electronics), Rayong in Thailand (automobiles and electronics) (see Figure 10.7), and Slovakia's Presov (clothing). In the service industries, extensive outsourcing in the software industry has led to the rise of Bangalore in India and Dublin, Ireland, as centers of software production and data processing. The outsourcing of back-office operations has also contributed to the global coordination of call center activities (see also Section 9.3).

Third, extensive macro-regional integration can take place if the production networks of TNCs span different countries within the same region and yet strong inter-firm relationships serve as the thread to stitch together different places within the region. Box 10.5 explains how this geographical process of macro-regional economic integration works in relation to the presence of TNC production networks in Southeast Asia.

Strategic Alliances and Joint Ventures

Apart from subcontracting, inter-firm networks can also be formed when TNCs engage in strategic alliances and joint ventures with other firms. These two organizational forms are different from mergers and acquisitions (M&As). Firms participating in strategic alliances or joint ventures do not experience ownership change and thus remain independent of the alliance or joint venture partner.

Martin Hess, with permission

Figure 10.7 The automotive cluster, Rayong Province, southern Thailand

CASE STUDY

Box 10.5 Global production networks and macro-regional integration in Southeast Asia

Most commentators on the global economy agree that the globalization of TNC activities is producing a disintegrated global production system and an integrated global trading system. However, on a closer look, it becomes apparent that many TNCs are increasingly organizing their production activities within specific macro-regions of the global economy. This growing *macro-regionalization* of the global economy can be explained by three interrelated factors. First, the scope and complexity of many TNCs' operations means that they need to coordinate and integrate their activities at a macro-regional, rather than truly global, scale. Second, economic resources such as labor, capital, and technology are increasingly mobile within specific regional contexts, for example the European Union. Third, macro-regional free trade arrangements having been growing in number and significance at the same time as globalization processes have been accelerating (see also Section 4.5).

We can illustrate these dynamics by looking at the example of macro-regional integration in Southeast Asia, an area covered by the Association of Southeast Asian Nations (ASEAN) grouping. There, integration processes are primarily being driven by TNCs from Japan and, more recently, South Korea, and have been given further impetus by the implementation of the ASEAN Free Trade Agreement (AFTA) since the mid-1990s. Evidence of macro-regionalisation can be seen in two sectors in particular (UNCTAD, 2004):

- *The automobile and automotive component industries*: Japanese and other automakers are consolidating their production in Southeast Asia and are adopting macro-regional production network strategies and plant specialization to serve the AFTA market. Toyota and Nissan, for example, have networks of operations linking different functions (e.g., regional HQ, finance, training, parts suppliers, and assembly) in different ASEAN countries. Ford also adopts a regional strategy of specialization to serve the ASEAN market: while the plant in Thailand makes pickup trucks, the plant in the Philippines produces passenger cars.
- *The electronics industry*: Similar dynamics can be observed in this sector. Samsung (South Korea), for example, has developed extensive macro-regional production networks in ASEAN. Samsung Corning (Malaysia) provides tube glass as a major input to Samsung Display's factory for color picture tubes in Malaysia. It also sells intermediate products and components to Samsung Electronics (Thailand) and affiliates in Indonesia (color TVs), Malaysia (computer monitors), and Vietnam (color TVs). Sony from Japan has also developed similar regional production networks for producing electronics goods.

To sum up, these production networks serve to strengthen macro-regional integration through production and supply linkages and intra-firm sourcing of key parts and components from affiliates located in the same macro-region.

In contrast, firms in M&As necessarily undergo ownership change. Some well-known merged TNCs are Royal Dutch Shell (oil), Fuji-Xerox (electronics), GlaxoSmithKline (pharmaceuticals), Unilever (consumer products), AOL-Time Warner (media and entertainment), PricewaterhouseCoopers (PwC) and KPMG (accounting), Delta Airlines and United Airlines (air transport).

In some fiercely competitive industries, firms tend to engage in a very specific type of inter-firm collaboration: strategic alliances. Only some business activities are involved in these function-specific strategic alliances, and no new equity capital will be involved. This cooperative approach to competition can be thought of as "co-opetition," a mix of cooperation and competition. In the semiconductor and pharmaceutical industries, for example, strategic alliances are common because of intense competition, the high costs of R&D and new product development, and the rapid rate of technological change. These pressures raise the investment stakes in these two industries beyond the financial means of any individual TNC. Cooperation through strategic alliances becomes the most effective means of competing on a global basis. Many of the TNCs in these strategic alliances are mutual competitors in some product segments and allies in other.

In service industries, function-specific strategic alliances are also common. Most major airlines, for example, tend to participate in one of the following strategic alliances: Star Alliance, One World Alliance, and SkyTeam. Through these alliances, participating airlines can, for example, enjoy code-sharing in their computer reservation systems that allows for cross-loading of passengers and thus reduces the excess capacity of any individual airline. Founded in 1997, the Star Alliance network is the world's leading airline alliance, with the highest number of member airlines, daily flights, destinations, and countries flown to (www.staralliance.com, accessed on August 1, 2011). The basic idea of Star Alliance is complementarity in operations and global reach. By sharing codes and customer loads, the 27 members of Star Alliance can offer their 603.8 million passengers the convenience and possibility of a smooth travel experience and access to over 1,160 airports in some 181 countries. The combined revenue of member airlines reached $150.7 billion in October 2010. Another benefit of this strategic alliance is that it affords anti-trust immunity to its member airlines in markets that are subject to such anti-monopoly regulation (e.g. the U.S.). Apart from convenience in booking, studies have shown that cooperation among member airlines in Star Alliance could lead to up to 27% reduction in inter-airline ticketing fares.

In the retail industry, Wal-Mart, already the world's largest retailer and with its own buying office in China, decided on January 28, 2010 to enter into a strategic alliance with Hong Kong-based Li & Fung in order to tap into the latter's global sourcing expertise to cut costs, improve quality, and spread its own global exposure. As part of its global sourcing consolidation, Wal-Mart bought about $2 billion worth of goods in the first year of a new sourcing unit known as WSG, established by Li & Fung and dedicated to servicing Wal-Mart's global sourcing needs. The sourcing business is expected to grow to $10 billion per year by 2015 (online.wsj.com, accessed on August 1, 2011). While Li & Fung had been supplying Wal-Mart as a wholesaler, this would be its first direct-sourcing deal for the global giant. In this strategic alliance between a global giant and one of the world's largest supply chain managers, both parties benefit from economies of scale in global sourcing and firm-specific expertise: Wal-Mart in retailing (see also Chapter 11) and Li & Fung in sourcing (Chapter 8).

When two or more firms decide to establish a separate corporate entity for a specific purpose, alliance joint ventures are often the organizational outcome. Here, partners need to invest new equity capital in the joint venture. Sometimes, a partner may use other assets (e.g., land or goodwill) to substitute for capital investment and this is known as a cooperative joint venture. As joint ventures are formed for partners to share financial risks, to benefit from inter-firm synergy, and to develop new products or markets, they are a popular form of inter-firm relationship that can be found in most industries. In many developing countries, government regulations requiring a minimum shareholding to be held by host-country citizens or firms further promotes the use of equity or cooperative joint ventures as the preferred entry mode for foreign TNCs. Clearly many automobile TNCs engage in joint ventures with each other. For example, BMW's manufacturing operation in Shenyang, China, is a joint venture between the BMW Group and Brilliance China Automotive Holdings Ltd., and produces vehicles solely for the local market and contributes to developing and penetrating the Chinese market. In South America, Africa, Eastern Europe, and Asia, equity joint ventures are very common among foreign automobile manufacturers seeking to develop new markets.

Just like international subcontracting, the geographical implications of international strategic alliances and joint ventures are manifold. We are witnessing the rise of project-based teamwork in many industries in which strategic alliances occur. As face-to-face interaction within these project teams tends to take place between cities and science hubs located in different countries, certain kinds of places become more strongly interconnected in the global economy to the exclusion of others. For example, in the semiconductor industry there are intense transnational flows and interactions between Taiwan's Hsinchu and Silicon Valley. As most of the activities in these alliances and ventures concern high-value-added projects, places hosting these activities become more prosperous and competitive (e.g., R&D clusters in Europe and North America). Spatial uneven development is often exacerbated.

Franchising and Cooperative Agreements

Franchising refers to an organizational form in which the TNC owner of a registered trademark or intellectual property rights agrees to let a franchisee (often outside the home country) use that trademark or rights provided that the franchisee follows the guidelines and requirements laid down by the TNC. There is thus an inter-firm relationship established between the franchisor (a TNC) and the franchisee (a local firm). Familiar examples are McDonald's, KFC, and Burger King fast-food restaurants (see Figure 10.8), Starbuck's coffee outlets, and 7-Eleven convenience stores. We are all aware of the global presence of these registered trademarks because of the extent to which the TNCs have used franchising as a preferred method of internationalization. Their outlets in many countries are *not* necessarily owned by these TNCs, but rather are often operated by local franchisees.

Figure 10.8 Fast-food chains in the Caribbean

As evident in Table 10.1, this non-equity mode of inter-firm network is particularly popular in the service sector for two key reasons. First, most service TNCs may not have sufficient capital or may not want to incur the substantial costs of expanding into many markets at the same time. This often occurs in the retail and fast-food industries. Second, some service TNCs may not want to be exposed to risks arising from unfamiliarity with local cultures, social relations, and the practices of local customers. Franchising provides a convenient and low-cost alternative for service TNCs to have a local presence through their franchisees, who are likely to be local entrepreneurs familiar with host-country markets.

On the other hand, cooperative agreements encompass a wide range of inter-firm relationships that range from licensing agreements to non-equity forms of cooperation. These agreements can be found in both manufacturing and service industries. In the manufacturing sector, a TNC may decide to license out its patented technology in return for royalty payments. For example, through owning the DVD format patents, Philips (the Netherlands), Sony (Japan), Pioneer (Japan), and LG (South Korea) will receive royalties for every DVD player manufactured by their licensees (https://www.ip.philips.com, accessed on August 1, 2011). In service industries, for example the hotel sector, two firms may come together to form a cooperative agreement in training, combining their training teams to offer human resource development programs to their combined employees. Each can reap the benefits of the combined human resource practices of both hotels.

Through both franchising and cooperative agreements, TNCs can rapidly internationalize their market presence and promote the consumption of their products and/or services in these host markets (see Chapter 15 for more on consumption). In the retail and restaurant business, the global presence of certain restaurant outlets (e.g., McDonald's and KFC) and consumer products (e.g., Coca-Cola) has often been taken as a direct evidence for the globalization of economic activity, known variously as "McDonaldization" or "Coca-Colonization." The pervasive presence of iconic franchised activities throughout the world leads us to obvious questions: Is there any barrier to the global reach of TNCs? Is the entire world economy really open to these giant corporations? This brings us to the penultimate section on the possible limits to global reach.

10.6   Are There Cultural Limits to Global Reach?

The above discussion of global production networks has focused on the corporate logics of TNCs and the different organizational structures and strategies that they may pursue as they expand internationally. Table 10.3 summarizes the costs and benefits associated with the different organizational forms we have discussed. We want to argue further that, contrary to the common perception of TNCs as all-powerful economic institutions, there are significant cultural limits to their global reach in today's highly competitive world economy. These limits extend far beyond other well-known limits to the global dynamics of TNC activity, such as the rise of anti-globalization movements (see Box 10.6).

As we saw earlier, TNCs from around the world may vary in their corporate cultures and conventions. Sometimes, corporate culture can lead TNCs into irrational or misguided decisions that are ultimately harmful to their best interests – a failure that has been described as "the cultural crisis of the firm" (Schoenberger, 1997). The stakes involved in understanding corporate cultures and conventions are therefore very high. Taken together, the four dimensions described in Box 10.2 provide a way of characterizing corporate culture and highlight the ways in which everyday business practices become stabilized within a set of conventions that may vary from TNC to TNC. While this variation between different TNCs is commonly understood, corporate culture can also be internally contested by different segments within the same TNC. Indeed, the TNC itself is not a coherent and homogeneous "black box" that behaves according to some predetermined logic. Rather, its competitive market behavior can represent outcomes negotiated by different groups within the TNC. Often, what is at stake is the definition of corporate culture – for example, will the view of research scientists prevail or will

Table 10.3 Different forms of organizing transnational operations – costs and benefits

Organizational Form	Costs to the TNC	Benefits to the TNC	Geographical Impacts
1. Fully owned subsidiaries	• Heavy capital investment • Potential conflicts with host governments and local firms • More cross-border managerial and organizational issues	• Full managerial control • Protection of trade secrets and proprietary technology • Consistency in production and market services	• Potential benefits to local economies through spin-offs and technology transfer • Spatial transfer of organizational cultures
2. International subcontracting	• Disruption in supply chains • Leakage of proprietary technology and intellectual property rights (IPRs) • Lack of direct managerial control of production	• Cost competitiveness • Flexibility in inventory management • Reduction in investment risks	• Development of export-processing zones in developing countries • The rise of the new international division of labor • Potential upgrading in some host regions
3. Strategic alliances and joint ventures	• Lack of control of technology and IPRs • Problems in managing partners	• Sharing of risks • Access to technology and knowledge of partners	• Deepening of linkages in local economies • Development of extra-local network ties
4. Franchising and cooperative agreements	• Potential infringement of IPRs • Costs of monitoring franchisees and managing ties	• Low or no capital investment required • Rapid expansion of market penetration • Financial returns to trademarks, brands, and IPRs	• Rapid spatial development of franchising outlets • Diffusion of cultural norms and practices to franchisees • Homogenization of consumption

CASE STUDY

Box 10.6   Transnational corporations and anti-globalization movements

The global reach of transnational corporations has increasingly faced serious and organized resistance in recent years. After protests at the Seattle ministerial meeting of the World Trade Organization (WTO) in 1999 the anti-globalization movement quickly gathered momentum with subsequent organized demonstrations against global capitalism and global corporations in Prague (2000), Quebec City (2001), and Genoa (2001) among many others. Participants in this movement stand in opposition to trade liberalization and the deregulation of financial markets, which together are seen as incentivizing TNCs to maximize profits at the expense of work conditions, environmental sustainability, and national independence and sovereignty. In June 2010, G-20 heads of government met in Toronto to discuss the global financial system. In the days before the meeting, NGOs and civil society groups staged a wide range of protests on issues such as poverty, fuel consumption, gay rights, and indigenous rights. During the summit, the protesters numbered over 10,000 people, and major riots broke out on June 26. A Nike clothing store was the first property reportedly damaged.

Apart from these protests at major meetings, new organizations and sectoral watchdogs have emerged. The rise of the World Social Forum since 2001 as a critical response to the World Economic Forum, a pro-TNC organization of business and political leaders, has further politicized the global reach of TNCs. These organized movements have exerted counterpressures on the unlimited expansion of TNC activity to an extent never seen before. TNC behavior has thus become much more cautious in order to avoid being targeted by these social movements. One of the best examples is Nike. As a consequence of intense criticism from social and labor movements on the working conditions in its subcontractors' factories during the 1990s, Nike has attempted to develop extensive subcontractor monitoring and assessment systems. Through its partnership with the Global Alliance for Workers and Communities, it aims to give greater voice to its contract factory workers through regular field studies and focused interviews. Severe doubts remain, however, about the ability of such schemes to improve dramatically the conditions of factory workers. While its own employment code imposes stringent labour standards to ensure that its contractors do not use any form of forced labor (prison, indentured, bonded, or otherwise) or child labor (workers below 16 years of age) and do not engage in discriminatory employment practices, Nike delegates much of this responsibility for labor standards and compliance to the contractors themselves. This delegation leaves the door wide open for labor abuses and poor working/employment conditions in production facilities. Clearly, even a global corporation like Nike finds it hard to monitor all its contractors and their labor standards across numerous localities and national regulatory contexts.

the engineering division set the agenda? Will the finance people or the marketers win out? Or will one regional division assert its priorities over another?

At the intra-firm scale, corporate slogans such as "fierce global competition," "lack of competitiveness," and "cost pressures" are commonly produced as mantras to guide thinking and action. The board of directors of a TNC may call for organizational reengineering based on their perception of the worsening corporate bottom line or the need for a new business model. However, the issue of competitiveness is sometimes far from life-threatening to the survival, let alone profitability, of the TNC. The argument for cost-cutting measures put forward by one group may not be accepted by others. A labor union may pose a counter-argument of exploitation that demystifies the competitiveness discourse. Even within the management structure of a TNC, alternatives may be articulated by different senior executives, each vying for the power to define the company's purpose and self-image. Thus, far from being the homogeneous and unitary entity, the TNC may in fact be a fundamentally incoherent and unstable amalgamation of competing interests and viewpoints.

This discussion of corporate culture has significant implications for the notion of global convergence of business practices described earlier (Section 10.2). It demonstrates that even if globalization and global standards are supposedly clearly laid out in the corporate vision of a particular TNC, they can be undone by the sheer lack of internal coherence. Ensuring the cooperation of different actors in a TNC is typically difficult because different groups tend to have diverse aims and objectives, some of which may be in conflict with a well-defined corporate vision. As TNCs grow larger and spread themselves further geographically, this internal diversity and contestation and its potential clashes with national cultures and business systems (see Box 4.3) are likely to increase. Just as Chapter 6 provided examples of how workers in different parts of a TNC can unite and contest organizational change, managers too may be pulling in different directions.

Clearly, the notion of the global economy masks huge variation in cultural norms and patterns of social behavior. The continued vitality of local cultures, alternative consumption practices, and kinship and social relationships in different localities can also pose serious problems for the globalizing efforts of TNCs. Figure 10.9 offers a vivid example of Nike's failure to appreciate local culture in its December 2004 worldwide advertising campaign. Titled "Chamber of Fear,"

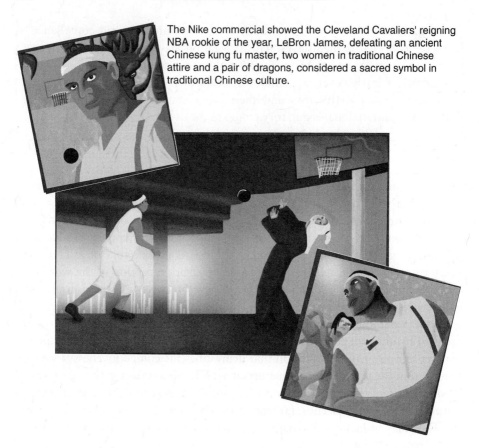

The Nike commercial showed the Cleveland Cavaliers' reigning NBA rookie of the year, LeBron James, defeating an ancient Chinese kung fu master, two women in traditional Chinese attire and a pair of dragons, considered a sacred symbol in traditional Chinese culture.

Figure 10.9 "Chamber of Fear" and the cultural limits to global reach: China's ban on a Nike television commercial in December 2004
Source: Redrawn from original TV screen shots available on
www.chinadaily.com.cn/english/doc/2004-12/07/content_397920.htm, accessed on August 2, 2011.

what was originally meant to be a sleek and dynamic advertisement showing the meeting of a modern basketball player and an ancient Chinese kung fu master backfired and was seen as insulting to the Chinese. The insult arises from the advertisement's insensitive treatment of the Chinese subjects: namely, the kung fu master, Chinese dragons, and women fighters, who were all defeated by LeBron James dressed in Nike products! Even though James was a fan of Bruce Lee's martial arts and the commercial was based on Lee, it is hard to expect the Chinese consumer to empathize with the subject of the advertisement: "Buy Nike." This lack of cultural sensitivity was a costly lesson for Nike as the advertisement was banned in China, one of its major markets for sports products (see also the report on April 25, 2005 in *Business Week*, www.businessweek.com, accessed on

August 1, 2011). This incident is symptomatic of many other corporate failures in managing joint venture partners, suppliers, marketing, and so on that can arise out of insufficient appreciation of social and cultural differences. TNCs that appreciate and can effectively adapt to these local differences tend to reap the benefits.

10.7 Summary

This chapter has systematically examined the transnational corporation, traditionally something of a "black box" in mainstream economics and a form of economic actor that shapes our everyday life in important ways. We can usefully understand the TNC as a system of both internal (intra-firm) and external (inter-firm) production networks. Interpreting TNCs in this way has many advantages. First, it allows us to appreciate the great variety and diversity in the global organization of production networks orchestrated by leading TNCs. Our analysis has demonstrated various ways in which these intra- and inter-firm networks can be organized. These range from wholly-owned subsidiaries and affiliates in intra-firm networks to subcontracting, joint ventures, and strategic alliances through inter-firm networks. Second, it shows how different localities and regions are connected into globally extensive production networks that operate across different national and macro-regional formations. A global production network (GPN) approach thus allows us to be sensitive to geographical variations in the global economy that are both a cause and an outcome of TNC globalization strategies. It also reveals how economic value is produced, captured, and retained in (or extracted from) different places within global production networks. Third, conceptually, the global production network approach brings together a greater range of actors and stakeholders than the conventional economic focus on firm networks alone.

With this great diversity in organizational forms, geographies of transnational production, and active stakeholders, can the TNC really keep everything together? This chapter has argued that effective globalization is not easy or even natural for the TNC, as the opening case of Nike illustrates so well. To set up extensive global operations requires the TNC to take seriously the geographical differences that exist in political, economic, and social-cultural realms. As TNCs globalize into different regions of the global economy, these geographical differences are further accentuated to produce greater organizational challenges and problems of maintaining legitimacy, alternatively known as the "liability of foreignness." This in turn increases the vulnerability of TNCs to a wide range of potential risks such as industrial action, sabotage, consumer boycotts, punitive regulation, financial exposure, and logistical bottlenecks. Many of these risks not only are increasing in tandem with global reach, but are also beyond the control of any individual TNC. In summary, there are inherent limits to the global reach of TNCs, particularly in the cultural and political realms.

Notes on the references

- Dicken (2011) offers the most authoritative account of TNCs as movers and shapers in the global economy from an economic-geographical perspective. See Beugelsdijk et al. (2010) for a striking contrast of how researchers in economics and management view TNCs from a rather different economic-geographical perspective. Though dated, Harrison (1994) makes one of the most compelling arguments about why TNCs often win out over SMEs in the age of global flexibility.
- For a full account of the global production networks framework, see Henderson et al. (2002), Hess and Yeung (2006), and Coe et al. (2008). See Yeung (2009) for an analysis of the organizational relationships between TNCs and global production networks.
- Grabher (2006) and Yeung (2008) provide extensive reviews of the geographical literature on networks and organizations.
- There are many TNC- and industry-specific case studies. For example, see Rothenberg-Aalami (2004) for a full account of Nike's global production activities. For some recent geographical studies of the wood industry, see Murphy and Schindler (2011); for the automobile industry, see Dicken (2011, Chapter 11), Hassler (2009), and Isaksen and Kalsaas (2009). For service sector examples, see Wrigley et al. (2005), Coe and Lee (2006), Durand and Wrigley (2009), and Tacconelli and Wrigley (2009) on retail TNCs, Beaverstock et al. (2010) on professional recruitment firms, and Hall (2009) and Jones (2010) on education and business services, respectively.
- For an excellent historical-geographical study of the early globalization initiatives by U.S. TNCs between 1890 and 1927, in particular Singer Manufacturing Company, see Domosh (2010).
- Schoenberger (1997) provides a classic study of corporate culture in Economic Geography.

Sample essay questions

- Why do TNCs organize transnational activities differently across different industries and host regions?
- What is the distinctive role of geography in the global configuration of TNC production networks?
- What factors may lie behind a TNC's decision to engage in inter-firm relationships?
- What are global production networks (GPNs), and what role do TNCs play in coordinating such GPNs?
- What are the cultural constraints that may limit the global reach of TNCs?

Resources for further learning

- http://www.unctad.org/en/Pages/Publications/WorldInvestmentReports (1991–2009).aspx: The World Investment Report website contains the most accurate data and comprehensive analysis on the global reach of transnational corporations.

- www.hsbc.com: The HSBC global corporate website gives a good indication of what it means to be "the world's local bank."
- www.bmwgroup.com: The BMW corporate website contains valuable information about its global production networks.
- pcic.merage.uci.edu: The Personal Computing Industry Center at the University of California, Irvine, conducts some of the best research into the globalized industry, including detail case studies of many lead firms in that industry and their transnational activities.
- www.csr-asia.com: This website provides excellent insights into TNCs and their corporate social responsibility issues in Asia.
- www.globalexchange.org: This website offers a bottom-up view of advancing a people's globalization. It has a very interesting section on "sweatfree communities."

References

Beaverstock, J. V., Faulconbridge, J. R., and Hall, S. J. E. (2010). Professionalization, legitimization and the creation of executive search markets in Europe. *Journal of Economic Geography*, **10**, 825–43.

Beugelsdijk, S., McCann, P., and Mudambi, R., eds. (2010). Special issue on place, space and organization – economic geography and the multinational enterprise. *Journal of Economic Geography*, **10**, 485–618.

Coe, N. M. (2011). Unpacking globalization: changing geographies of the global economy. In A. Leyshon, R. Lee, L. McDowell, and P. Sunley, eds., *The Sage Handbook of Economic Geography*. London: Sage, pp. 89–101.

Coe, N. M., and Lee, Y. S. (2006). The strategic localization of transnational retailers: the case of Samsung-Tesco in South Korea. *Economic Geography*, **82**, 61–88.

Coe, N., Dicken, P., and Hess, M. (2008). Global production networks: realizing the potential. *Journal of Economic Geography*, **8**, 271–95.

Dedrick, J., Kraemer, K. L., and Linden, G. (2010). Who profits from innovation in global value chains? A study of the iPod and notebook PCs. *Industrial and Corporate Change*, **19**, 81–116.

Dicken, P. (2000). Places and flows: situating international investment. In G. L. Clark, M. A. Feldman, and M. S. Gertler, eds., *The Oxford Handbook of Economic Geography*, pp. 275–91. Oxford: Oxford University Press.

Dicken, P. (2011). *Global Shift*, Sixth Edition. London: Sage.

Domosh, M. (2010). The world was never flat: early global encounters and the messiness of empire. *Progress in Human Geography*, **34**, 419–35.

Durand, C. and Wrigley, N. (2009). Institutional and economic determinants of transnational retailer expansion and performance: a comparative analysis of Wal-Mart and Carrefour. *Environment and Planning A*, **41**, 1534–55.

Fröbel, F., Heinrichs, J., and Kreye, O. (1980). *The New International Division of Labour*. Cambridge: Cambridge University Press.

Grabher, G. (2006). Trading routes, bypasses, and risky intersections: Mapping the travels of "networks" between economic sociology and economic geography. *Progress in Human Geography*, **30**, 163–89.

Hall, S. (2009). Ecologies of business education and the geographies of knowledge. *Progress in Human Geography*, **33**, 599–618.

Harrison, B. (1994). *Lean and Mean: The Changing Landscape of Corporate Power in the Age of Flexibility.* New York: Basic Books.

Hassler, M. (2009). Variations of value creation: automobile manufacturing in Thailand. *Environment and Planning A,* **41,** 2232–47.

Henderson, J., Dicken, P., Hess, M., Coe, N. M., and Yeung, H. W-C. (2002). Global production networks and the analysis of economic development. *Review of International Political Economy,* **9,** 436–64.

Herbst, M. (2005). The Costco challenge: an alternative to Wal-Martization? www.workinglife.org, accessed on December 15, 2010.

Hess, M., and Yeung, H. W-C. (2006). Whither global production networks in economic geography? Past, present and future. *Environment and Planning A,* **38,** 1193–1204.

Isaksen, A., and Kalsaas, B. T. (2009). Suppliers and strategies for upgrading in global production networks: the case of a supplier to the global automotive industry in a high-cost location. *European Planning Studies,* **17,** 569–85.

Jones, A. (2010). Theorizing global business spaces. *Geografiska Annaler,* **91B,** 203–18.

Murphy, J. T., and Schindler, S. (2011). Globalizing development in Bolivia? Alternative networks and value-capture challenges in the wood products industry. *Journal of Economic Geography,* **11,** 61–85.

Rothenberg-Aalami, J. (2004). Coming full circle? Forging missing links along Nike's integrated production networks. *Global Networks,* **4,** 335–54.

Schoenberger, E. (1997). *The cultural crisis of the firm.* Oxford: Blackwell.

Tacconelli, W., and Wrigley, N. (2009). Organizational challenges and strategic responses of retail TNCs in post-WTO-entry China. *Economic Geography,* **85,** 49–73.

UNCTAD. (2011). *World Investment Report 2011: Non-Equity Modes of International Production and Development.* Geneva: UNCTAD.

Wrigley, N., Coe, N. M., and Currah, A. (2005). Globalizing retail: conceptualizing the distribution-based transnational corporation (TNC). *Progress in Human Geography,* **29,** 437–57.

Yang, C. (2009). Strategic coupling of regional development in global production networks: redistribution of Taiwan PC Investment from Pearl River Delta to Yangtze River Delta, China. *Regional Studies,* **43,** 385–408.

Yeung, H. W-C. (2008). Perspectives on inter-organizational relations in economic geography. In S. Cropper, M. Ebers, C. Huxham, and P. Smith Ring, eds., *The Oxford Handbook of Inter-Organizational Relations,* pp. 473–501. Oxford: Oxford University Press.

Yeung, H. W-C. (2009). Transnational corporations, global production networks, and urban and regional development. *Growth and Change,* **40,** 197–226.

CHAPTER 11

SPACES OF SALE
How and where do we shop?

Goals of this chapter

- To recognize the importance of retailing within the capitalist system
- To appreciate the changing geographies of retailing
- To understand how retail spaces of different kinds are actively designed and managed
- To reflect on the ways in which the advertising industry serves to create and shape demand for goods and services

11.1   Introduction

In October 2010, Edwina Dunn and Clive Humby announced that they were stepping down from the day-to-day management of the little-known company bearing their name. Established in 1989, by 2011 dunnhumby employed 1,300 staff across 30 offices in Europe, Asia, and the Americas, and had a turnover of £300 million. The company has become a market leader in collecting and analyzing information about 350 million consumers and their shopping patterns, in turn using this information to advise over 150 clients on new ways of interacting with customers and enhancing brand value. Dunnhumby's majority shareholder and most important client is the U.K. food retailer Tesco. Since 1995, the companies have worked together to introduce and develop Tesco's *Clubcard*, the card-based loyalty scheme that monitors members' shopping patterns in return for rebates on money spent in Tesco stores. The scheme, used regularly by some 15 million shoppers in the United Kingdom, is widely credited with fueling the retailer's explosive growth over the last two decades – quite an accolade when one considers Tesco's current size and status as the world's third largest retailer (after Wal-Mart and Carrefour). One in every eight pounds spent in a British retailer goes to Tesco, and almost one in three in the grocery market. The company made £3.8 billion profits on turnover of £67.6 billion in 2010, employing 294,000 people across 2,715 stores in the United Kingdom and another 198,000 in 2,500 stores in 12 overseas markets. *Clubcard* has accompanied Tesco as it has expanded overseas; in South Korea it is known as *Familycard*, in China it is *Membercard*, and in Malaysia it is *Biz Clubcard*.

Dunnhumby's core databases are designed to profile every customer in the United Kingdom, whether they shop at Tesco or not. Pulling together a variety of data sources, they classify individual consumers according to variables such as their wealth, travel patterns, credit levels, amount of free time, lifestyle, environmental ethic, charity contributions, and susceptibility to promotions and new products. Combining this detailed customer information with *Clubcard* data enables Tesco to develop individually-targeted promotional and incentive schemes, saving an estimated £350 million a year on expensive blanket marketing schemes. *Clubcard* sends out £400 million worth of vouchers each year in nine million different targeted combinations, and in 2009 – during a deep economic recession – dunnhumby's databases were used to stem the exodus of shoppers to discount retailers by doubling the level of *Clubcard* points awarded (*The Guardian*, 2010). In addition to Tesco, dunnhumby's services are used by other leading retailers such as Kroger and Home Depot (both U.S.), Casino (France), and Metro (Germany), along with a range of manufacturers including Coca-Cola, Kraft, Mars, P&G, Shell, and Unilever. The information is heavily analyzed geographically, of course, bringing two important benefits to clients. First, it allows retailers to gauge the most appropriate locations for new stores by comparing local social and economic conditions with the profile of their wider customer base. Second, it enables them to adjust their product mix both spatially and temporally. For example, in Brixton, London, an area settled by immigrants from the Caribbean, Tesco sells plantains, a type of savory banana. Tesco stores in central London do not sell them, but instead are based around selling sandwiches to office workers in the middle of the day, and prepared "ready" meals in the evening. In short, technologies such as *Clubcard* help large retailers adapt their offering to meet the precise requirements of local markets.

As introduced in our discussion of commodity chains in Chapter 8, retailing refers to the distribution and sale of commodities to final consumers. The example of dunnhumby and Tesco serves to introduce us to several important aspects of contemporary retail dynamics. The sheer number and coverage of Tesco stores alerts us to the fact that retailing is a necessarily geographically more extensive activity than production. It illustrates the growing influence of large transnational retailers on the range of products we buy and where we are able to buy them. It reveals how this power is in part based upon access to highly detailed information about how and where we shop. It speaks to processes of retailer globalization and the development of similar retail experiences in many countries around the world.

And yet, as the comparison of Tesco in Brixton and central London shows, the retail experience remains highly geographically variable, even between the stores of the same corporation in the same city. Retail spaces such as supermarkets are shaped by the economic, social, and cultural spaces of which they are part, and retailers must respond to these geographical variations in order to be successful. This chapter seeks to explore these geographies of retailing and also to consider the ways in which specific retail spaces are structured and organized in order to induce us to purchase commodities.

The argument proceeds in four stages. First, we look at the traditional way of exploring retail location in Economic Geography – namely, through using the tools of central place theory. Although this approach contains some insights, we consider some of its limitations in explaining recent developments in retail (Section 11.2). Second, we look at the changing geographies of retailing in the contemporary era, encompassing the ongoing globalization of retail activity, shifting patterns of urban and regional retail restructuring, and the growing importance of online retailing (Section 11.3). Third, we zoom in to look at different retail spaces – the street, the mall, the store, and markets/fairs – and how they are designed, configured, and used (Section 11.4). Finally, we start to move from thinking about retailing as a largely economic phenomenon to one that is embedded in wider sociocultural processes of consumption – a journey that is completed in Chapter 15 – by looking at the role of the advertising industry in creating and shaping consumer demand for goods and services (Section 11.5).

11.2   Explaining Retail Geographies: Central Place Theory and Beyond

The exchange of money for commodities, performed in retail spaces of various kinds, is an essential element of the capitalist system. Of course, many commodities, known as *intermediate* goods and services, are sold between businesses. Retailing, however, is generally taken to refer to the distribution and marketing of commodities to the general public for final consumption. The nature of retailing means that it exhibits a necessarily more extensive economic geography compared to production operations. Put simply, while not every urban center will have a car or plastics factory or insurance company offices, all except the very smallest will have retail outlets of some kind. Explaining the spatial distribution of retail activity has been a long-standing project in the field of Economic Geography, most notably through an approach called *central place theory*. A central place is simply a place that offers goods and services for sale. Here we first introduce the basic tenets of this approach before thinking about some of its limitations for explaining contemporary retail dynamics.

Perhaps the best-known model of central places was that developed – initially in the 1930s, but translated into English in 1966 – by the German economic geographer Walter Christaller. He was interested in explaining the spatial distribution of towns and cities of various sizes. Like other important theorists of the time such as Johann-Heinrich von Thünen (see Chapter 1) and Alfred Weber (see Chapter 12) his approach was based on the twin assumptions of the isotropic plane – a flat, featureless world onto which economic life was projected – and individual rational consumers with access to "perfect" information about the world around them. Christaller's model was derived from two key variables: the distance a consumer will travel to purchase a particular good (the *range* of a good), and the minimum level of business required for a retail operation to be viable (the *threshold* of a good). As rational actors, retailers would locate as close as possible to their customers, while customers would visit the nearest available center, thereby both minimizing transport costs and maximizing the expenditure on goods and services.

Using an imaginary model of an isotropic plane with uniformly distributed population, Christaller was able to demonstrate that this requirement of centrality would lead to an orderly spaced hexagonal network of central places, reflecting the fact that hexagons are the most efficient geometrical figures for serving territory without overlap. The central places were organized into a hierarchy reflecting the number of goods and services with similar range and threshold values, of which Christaller identified seven orders (see Figure 11.1, which depicts a three-level system). A higher-order place would offer all the goods and services

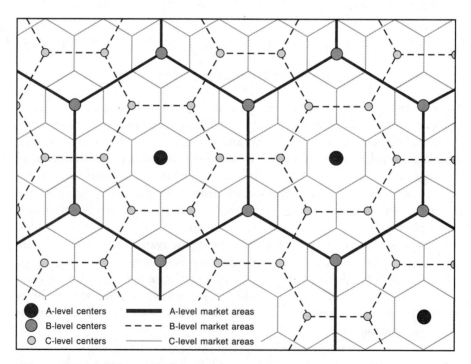

Figure 11.1   Christaller's hexagonal central place theory pattern

of the central places below it in the hierarchy, in addition to higher-order goods. Each central place was located within a hinterland, with those at the lowest level of the hierarchy having the lowest number of retailers and serving the smallest hinterlands, and thus being the most widely dispersed (the C-level centers in Figure 11.1). Christaller could model different urban hierarchies by mobilizing three different principles. The *marketing* principle minimized the number of centers, the *transport* principle minimized the road length required to join the central places together, while the *administrative* principle determined that hinterlands could only be nested hierarchically within each other. A more flexible variation of Christaller's model was subsequently offered by August Lösch (1939, in English 1954). Rather than grouping all functions into seven orders, Lösch allowed all goods and services to have a separate range, threshold, and hexagonal hinterland, and relaxed the assumption that higher-order centers would always contain all lower-order functions. The result was a much wider range of potential market landscapes that seemed to map somewhat more closely onto real-world structures.

The models of Christaller and Lösch were extremely influential in driving a body of quantitative modeling work that was at the forefront of Economic Geography research from around the mid-1950s until the mid-1970s in the Anglo-American context (e.g., Berry, 1967; and Barnes, 2012). In subsequent decades, however, a series of ongoing dynamics have led many to question the contemporary applicability of these theories. Widespread car ownership and the availability of cheap and efficient transportation have reduced the influence of transportation costs. Rising levels of disposable income have been distributed highly unevenly in spatial and social terms. The consumer population has become increasingly differentiated along age, gender, ethnicity, and other lines. Retailers, as active capitalist firms, have sought to respond to complex patterns of population decentralization and recentralization by developing multiple retail formats across a wide range of different locations. These strategies have, in turn, been heavily shaped by the state regulatory frameworks that are in place in different territories. Put simply, the complexity of contemporary retail geographies has moved considerably beyond the isotropic plains and neatly ordered hierarchies of Christaller, Lösch, and their more recent adherents. These changes have stimulated a more recent wave of Economic Geography research – which we draw on in this chapter – that has as its central tenet the increasingly important role played by retailers within the wider economy.

Two core geographical dimensions run through our analysis in this chapter. First, the geographies of retailing are multi-scalar, encompassing the globalization of retail activity, the shifting urban and regional distribution of retailers, and the micro-management of individual retail spaces. Second, the sheer range of retail spaces has proliferated in recent times. The central place shops of Christaller and Lösch's models now sit alongside, and interact with, a wide range of other formal, informal and electronic retail spaces, including out-of-town superstores, suburban malls, temporary markets, yard/tailgate or car boot sales, catalogue shopping, and a huge variety of online retail offerings. A central theme in what follows is that influential corporate interests lie behind these restless retail geographies and that there is a growing concentration of power in the hands of large retail corporations. Retailers have not simply responded to changing demand conditions; rather, they have sought to develop and expand their markets as part of their profit-making strategies.

11.3   The Shifting Geographies of Retailing

Retailing is a huge economic sector in its own right. In the United Kingdom in recent years, for example, it employed 3 million people – or one in nine of the total workforce – across 200,000 businesses with 320,000 separate outlets. In the United States, 14.4 million people were employed in over 1.1 million retail outlets. The changing geographies of retailing are therefore an important component of the contemporary economic landscape. In this section we consider these changing geographies in three stages. First, we will profile the ongoing globalization of retail activity and its wider impacts. Second, we explore the shifting intra-national geographies of retailing, and in particular, dynamics at the urban scale. Third, we move beyond physical retail forms to consider the growing significance of online retailing.

The Globalization of Retailing

Perhaps the most important geographical outcome of the rising power of retailers has been the extensive globalization of retailing over the last 20 years, as retailers have sought new opportunities in which to invest the profits secured from their home markets (Coe, 2004). The period since the mid-1990s has seen the emergence of a select group of transnational retailers that has used aggressive merger and acquisition activities, backed up by subsequent rapid organic growth, to assume dominant market positions across a range of countries. The leading transnational retailers, ranked in terms of their foreign sales, are detailed in Table 11.1. A number of important observations can be made from these data. First, they give a sense of the *scale* of international retail operations, with the retailers deriving between US$10 and 100 billion in sales from foreign markets in 2010. Second, they give a sense of the *scope* of international retailing, with many of the leading players having store operations in 20–30 countries, a level of international expansion comparable with many manufacturing sectors. Third, they reveal that the leading retail transnationals – with some exceptions such as the world's largest retailer, Wal-Mart – are Western European. While many of the world's very largest retailers hail from the United States (e.g., Home Depot, Kroger, Target, Sears, and Safeway), they can achieve that size without straying far beyond the borders of their home country. Fourth, they show that the leading

Table 11.1 Leading transnational retailers, ranked by sales outside home market, 2010

Rank	Name of Company	Country of Origin	Type of Retailer	International Sales (US$m)	International Sales (% of total)	No. of Countries of Operation
1	Wal-Mart	U.S.	Food and general merchandise	109,539	26	16
2	Carrefour	France	Food and general merchandise	79,248	59	23
3	Metro	Germany	Food and general merchandise	55,112	61	34
4	Lidl & Schwarz	Germany	Food and general merchandise	40,132	52	26
5	Auchan	France	Food and general merchandise	38,064	59	12
6	Aldi	Germany	Food and general merchandise	33,576	53	18
7	Tesco	U.K.	Food and general merchandise	31,085	32	14
8	IKEA	Sweden	Furniture	29,097	94	38
9	Ahold	Netherlands	Food and general merchandise	25,605	65	10
10	Rewe	Germany	Food and general merchandise	24,303	34	13
11	Delhaize	Belgium	Food and general merchandise	21,721	78	6
12	Costco	U.S.	Food and general merchandise	19,581	25	9
13	Seven & I	Japan	Convenience stores	17,803	31	18
14	H&M	Sweden	Clothing	16,430	94	36
15	Amazon	U.S.	General merchandise	15,497	45	7
16	PPR	France	Luxury goods	13,112	67	84
17	Best Buy	U.S.	Electrical	13,086	26	15
18	Inditex	Spain	Clothing	12,060	72	74
19	Tengelmann	Germany	Food and general merchandise	10,347	70	11
20	Kingfisher	U.K.	Home improvement	9,516	59	8

Source: Derived from IGD's *Retail Analysis* (igd.com), Deloitte/Stores 2011*Global Powers of Retailing* and Annual Reports. Revenue figures for the financial year that corresponds most closely to calendar year 2010. Exchange rates used are for December 31, 2010 and are taken from www.xe.com. Key rates: $1 = £0.64 = €0.75.

transnational retailers tend to be food retailers or general merchandisers, rather than specialty providers (e.g., selling toys or computer goods).

The globalization of retailing is not a new process, dating back as far as the late 1800s. Foreign expansion really took off in the 1960s, however, and until the 1990s it was largely dominated by investments between the leading economies of North America, Western Europe, and Japan. Since the mid-1990s the expansion has taken on an entirely new geographical configuration. Figure 11.2 shows Tesco and Wal-Mart's global store distributions in early 2011. Tesco has stores in six markets in Eastern Europe and five in East Asia in addition to operations in the United Kingdom, Ireland, and the United States. Wal-Mart has operations in China, India, Japan, and the United Kingdom in addition to extensive operations throughout the Americas. In general, international retail investment by the leading

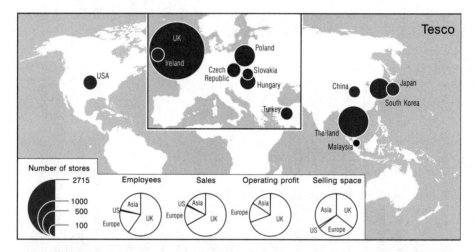

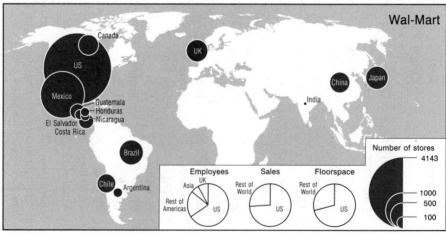

Figure 11.2   The global distribution of Tesco and Wal-Mart stores in 2011
Source: Data from company annual reports.

retail transnationals is now targeted toward countries in so-called *emerging* regions, namely South and Central America, East Asia, and Eastern Europe. Tesco's 2007 entry into the U.S. market with the "fresh & easy" brand was interesting in that it reversed this trend; progress has been very slow, however, compared with the company's European and Asian investments (see Lowe and Wrigley, 2010, for more). Africa will perhaps be the next region to see significant inward investment, with Wal-Mart's 2011 purchase of the leading South African retailer Massmart seen as an important development.

Three further characteristics distinguish this latest phase of retail globalization. First, the sheer *speed* of the international expansion is notable. For example, in 1990 Carrefour was to be found in five countries outside of France, and Wal-Mart and Tesco were only present in their home market. By 2010 Carrefour was present in 23 markets, Wal-Mart in 16, and Tesco in 14. Second, the *scale* of investment currently being undertaken is unprecedented. Static "snapshots" such as Table 11.1 and Figure 11.2 may not give a real sense of the size of the phenomenon. For example, over the period 1994–2010, Tesco went from having no foreign floorspace to having 65% of its global total in its overseas territories. These huge expansion processes are creating retail giants with foreign operations of unprecedented size. Third, the *impacts* of this expansion on the retail structures of the host countries have been significant. Table 11.2 profiles Poland's retail structure in 2010, revealing that seven of the leading grocery retailers (including the top four) are foreign, or more specifically Western European, owned. Those transnational retailers account for 44% of the total grocery retail market and

Table 11.2   Top grocery retailers in Poland, 2010

Company	Ownership	Grocery Market Share (%)	Number of Stores	2010 Sales ($ millions)
Biedronka	Portugal	12.5	1,649	6,392
Lidl & Schwarz	Germany	7.2	532	3,722
Tesco	U.K.	6.9	385	3,548
Carrefour	France	6.0	336	3,100
Eldorado	Poland	5.7	1,496	2,916
Auchan	France	5.5	48	2,837
Spolem	Poland	5.6	4,140	2,667
Metro	Germany	4.2	54	2,000
Lewiatan	Poland	3.4	2,146	1,715
Eurocash	Poland	2.8	575	1,429
Leclerc	France	1.8	58	921

Source: www.igd.com, accessed June 13, 2011.

Table 11.3   Top grocery retailers in Thailand, 2010

Company	Ownership	Grocery Market Share (%)	Number of Stores	2010 Sales ($ millions)
Tesco Lotus	U.K.	11.4	773	4,539
CP All (7-Eleven)	Thailand	7.7	5,720	3,003
Big C Group (Casino)	France	5.9	79	2,346
Carrefour Thailand	France	2.4	39	941
Central Retail Corporation	Thailand	1.7	129	683
Siam FamilyMart	Thailand	0.4	562	173

Source: www.igd.com, accessed June 13, 2011.

own over 3000 stores in Poland. A similar story can be told for many leading economies across Eastern Europe, Latin America, and East Asia. Table 11.3, for example, shows the leading grocery retailers in Thailand, where Tesco has assumed the dominant market position since its 1998 entry. The implications for host economies stretch way beyond the competitive impacts on the domestic retail sector (Coe and Wrigley, 2007). Transnational retailers also drive important changes in terms of consumption patterns (e.g., consumers switching from local markets and small stores to supermarkets and hypermarkets) and supply chain dynamics (e.g., buying in bulk, removing intermediaries, demanding higher quality standards, and implementing new logistics technologies). They have also triggered a range of responses from state regulators who have tried to manage the expansion of retail transnationals in different ways, often through planning laws.

Leading transnational retailers, then, are coordinating increasingly extensive networks of stores at the global scale. However, we need to be cautious about inferring from these trends that the nature of retail spaces is becoming identical around the world. Clearly, there are tendencies in that direction, with elements of store formats and signage, for example, being common across all markets. A British shopper entering a Tesco shop in Prague (Czech Republic) or Bangkok (Thailand) would quickly see aspects of the store design that are very similar to stores in the United Kingdom. Equally, however, the same shopper would also notice significant variations in terms of the product range on offer, how goods are displayed to consumers, and the physical structure of the store. In Prague, for example, Tesco has a city-center department store stretching over several storeys. In Thailand, Tesco hypermarkets often include a range of other food, leisure, and community services, while in rural areas smaller Tesco stores offer external market space to local traders (see Figure 11.3). All these stores are quite different to the

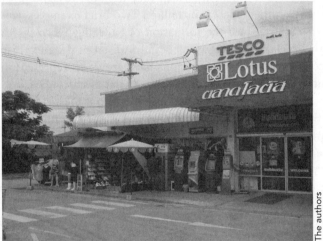

Figure 11.3   Tesco Lotus in Thailand

free-standing single storey supermarkets that dominate in the United Kingdom. Importantly, transnational retailers vary in the extent to which they try to offer a standardized retail experience across all the countries in which they operate. While some companies, such as Wal-Mart and IKEA, operate in a highly centralized and uniform manner, others such as Ahold pursue a more federal strategy, operating as a collection of relatively autonomous national subsidiaries. Meeting the needs of different national consumer markets is a challenging business, and the overall trend toward globalization masks a range of successes and failures across different retailers and national markets (see Box 11.1).

CASE STUDY

Box 11.1   Success and failure in global retailing

The recent history of retail globalization is littered with examples of both high-profile successes and failures across the leading retail transnationals. Wal-Mart, for example, one of the world's most well-resourced companies, has failed to establish itself and subsequently exited the markets of Germany (1998–2006), Hong Kong (1994–1996), Indonesia (1996–1998), and South Korea (1998–2006). Carrefour, one of the first retailers to expand into Southeast Asia in the early 1990s, announced in late 2010 that it was withdrawing from the region and that its hypermarkets in Thailand, Malaysia, and Singapore were up for sale. Transnational retailers work hard to achieve legitimacy within host markets through the *strategic localization* of their services. The extent to which they are able to achieve this has a strong bearing on their success or failure. Two brief examples can serve to illustrate this point. Since the formation of Samsung-Tesco in May 1999 through a partnership between Tesco and Samsung's fledgling distribution arm, Tesco's South Korean operations have grown to represent its largest market outside the United Kingdom. From an initial starting base of just two stores, by 2011 Samsung-Tesco had expanded to a multi-format network of over 400 stores and 22,000 staff, making it the second largest retailer in the market. Effective strategic localization is at the heart of this success story (Coe and Lee, 2006). By using a local partner and local managers, the company has been able to adapt to the needs of Korean consumers and forge strong local supply network relationships. In response to both competitive and regulatory pressures, Samsung-Tesco has progressively developed its operations over time through, for example, aggressive expansion of its smaller Express format as the hypermarket segment has become saturated, and through offering a wide range of other retailing services in order to build customer loyalty. By contrast, accounts of Wal-Mart's failure in Germany have focused on the mismatch between a "lean retailing" business model developed in the United States and the needs of the German market. In particular, Wal-Mart was unable to adapt its supplier and labor relations to conform to German norms and struggled financially as a result (for more, see Christopherson, 2007). An inability to adapt sufficiently its business model to the needs of a host market also underpinned Wal-Mart's exit from South Korea. It is too simplistic, however, to conclude that some retailers are good at international expansion while others are not; the reality is a complex picture of success and failure as retailers from different home contexts strive to adapt to widely varying host-economy conditions.

From Center to Suburbs, and Back Again?

Having established the contemporary importance of processes of retail globaliza-tion, we can now move on to consider the changing geography of retailing at the urban scale. In particular, we want to profile the shifting of retail investment from the inner city and downtown to the suburban periphery and back again, and in so doing, reveal how retail capital plays a central role in the constant remaking of the urban-built environment (a process introduced in Chapter 3).

Up until the 1950s, retailing was essentially a *central* urban activity. The post–World War II suburbanization of retail capital and related decline of city-center retailing were arguably pioneered – and indeed most evident – in the United States. The scale and significance of these dynamics can be illustrated through briefly looking at developments in and around Chicago, Illinois, over the period from 1950 to the mid-1970s (drawing on Wrigley and Lowe, 2002). By the end of the 1950s, four large open-air shopping centers with ample parking provision had appeared on the periphery of Chicago as department stores began to realize the potential of shifting their focus to the rapidly expanding middle class suburbs (see Figure 11.4). The 1960s saw the building of a series of enclosed shopping centers or *malls* in a ring around Chicago, fueled in part by the expanding urban expressway network. By the end of the decade, a total of 11 suburban shopping centers had combined retail sales to rival central Chicago. By 1974 the total was up to 15 shopping centers, and a second ring around Chicago was starting to emerge. Some of the key tenants of these new suburban centers were the very department stores, such as Sears Roebuck and J.C. Penney, that had once dominated downtown shopping districts. At the same time, central and inner city retail outlets were closing in their hundreds. Chicago's leading inner city shopping district, known as 63rd and Halsted, was severely hit, finding itself in the middle of a decaying area with falling income levels by the early 1970s. In just two-and-a-half decades, the geography of Chicago's retail provision had undergone a profound transformation.

These dynamics, often driven by powerful alliances of property developers, retailers, and financiers, have been repeated in hundreds of cities across the United States and Canada. A similar suburbanization of retail provision has occurred in many other countries, and particularly those of Western Europe. Such trends have been fueled by a series of overlapping dynamics: the increased mobility of consumers and high levels of car ownership; the emergence of portions of the population with large amounts of disposable income; population decentralization from urban to suburban areas; higher female participation in the workforce driving demand for quicker and more efficient means of shopping (e.g., the "one-stop-shop"); and the growth strategies of retailers as they seek enhanced economies of scale through larger and more accessible stores. The *extent* to which the North American trends described above have been replicated elsewhere has been dictated by the strength of planning regulations in different countries;

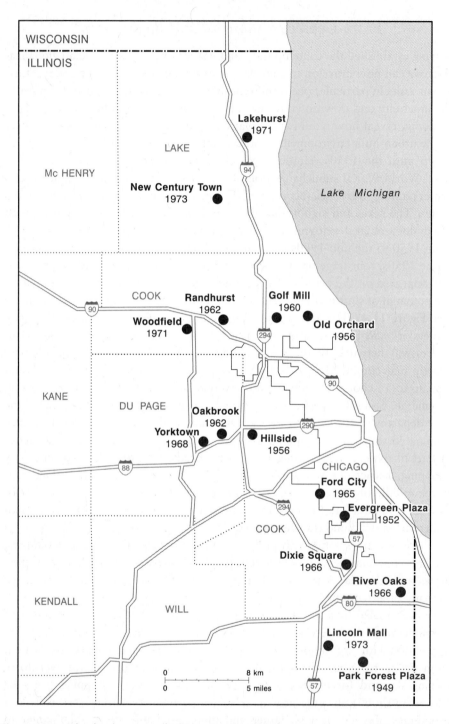

Figure 11.4 The development of Chicago's suburban shopping centers, 1949–1974
Source: Adapted from Berry, B. J. L., Cutler, I., Draine, E. H., Kiang, Y-C., Tocalis, T. R. and de Vise, P. (1976) *Chicago: transformations of an urban system*, Ballinger, Cambridge, Massachusetts, Figure 22.

most European countries, for example, have been more resistant to suburban greenfield developments than in the U.S. case. Moreover, in many fast-growing Asian cities – for example, Bangkok, Hong Kong, Seoul, and Shanghai – strong centralization of retail remains the dominant feature of the urban retail landscape. Regulatory conditions also critically affect the timing of any decentralization processes. In the United Kingdom, for example, there was a boom in the development of out-of-town food superstores in the 1980s and early 1990s when planning guidelines on building in suburban areas were favorable. By the mid-1990s, these trends were promoting concerns about the development of *food deserts* – low-income areas of British cities characterized by the lack, or poor quality, of food retail provision (Wrigley et al., 2003).

In most countries, however, suburban retail parks are now a well-established part of the retail landscape. In the United Kingdom, the initial wave of food superstores has been joined by three subsequent phases of out-of-town development (Fernie, 1995). First, there has been a wave of retail parks containing, for example, home maintenance, carpet, furniture, and electrical stores whose bulky products suit such accessible locations. Second, in contrast to the large-scale decentralization of city-center retailing seen in the United States, planning constraints have restricted such relocations (beyond the food superstores and retail parks just mentioned) to a small group of massive regional shopping centers that aim to serve more than their immediate urban hinterland. As shown in Figure 11.5, these are quite evenly spread across the country and, pioneers such as Brent Cross and the MetroCentre apart, were largely built in the 1990s. The 1998 Trafford Centre, for example, on the western outskirts of Manchester, has approximately 1.9 million square feet of retail, catering, and leisure space in addition to 11,500 car parking spaces. The centre has 8,000 employees working in 200 shops, including four large "anchor" tenants, 60 restaurants, a 1,600-seat food court, a 20-screen cinema, and several other leisure attractions. It attracts over 30 million visitors a year, and an estimated 5.3 million people live within a 45-minute drive. Figure 11.5 also shows that the leading shopping centers are owned and managed by a small group of property companies, revealing another form of corporation that plays an important role in shaping retail geographies. The third, and latest, wave of suburban retail in the United Kingdom takes the form of outlet shopping malls that offer clothing manufacturers' excess stock at reduced prices. Outlet malls originated in the United States; one of the most famous is Woodbury Common, a 800,000 square foot facility just one hour's drive from New York City, offering 220 different stores. The largest outlet mall in the United Kingdom – Cheshire Oaks in northwest England – has more than 140 stores selling a wide range of designer brands (see Figure 11.6, which also shows the tendency of some suburban malls to mimic the look and feel of central urban "high streets"). Again, however, planning restrictions and property market conditions have meant that the expansion of such outlets has been much slower than predicted. Overall, then, it is important to recognize that the suburbanization

Figure 11.5 Britain's largest shopping centers, 2011

Source: Adapted from *The Guardian* (2011) UK shopping mall supremo scents victory in battle over Trafford Centre, 12th January, p. 26.

Figure 11.6   Cheshire Oaks outlet mall

of retailing now constitutes many different kinds and formats of retailing, and that its extent varies significantly across different national regulatory contexts.

While suburbanization has had a massive and lasting impact on retail geographies, there are also notable recent trends working in the opposite direction. While the nature and extent of the trend will again vary from context to context, three brief illustrations can be offered here to show the kinds of dynamics that are occurring. First, in the U.S. context, since the mid-1970s when the level of retail decentralization was at its peak, considerable efforts have been made to regenerate the downtown districts of American cities. Central to such initiatives have been so-called festival marketplaces, stimulated by the success of the Faneuil

Hall Marketplace, which opened in Boston in 1976. The central idea here is to use combinations of architecture, cultural exhibits, concerts, and ethnic festivals to attract people to speciality markets, shops, and restaurants. A different, but closely related, form of development has seen downtown retail revitalization as part of multiple-use complexes comprising offices, hotels, leisure facilities, and convention centers (for example, the Marina Bay Sands integrated resort in Singapore). Similar schemes have subsequently been tried in many cities around the world, including those where retail decentralization has been less pronounced. Second, in many of the post-industrial cities of developed countries (e.g., New York, Manchester) inner-city retail has been boosted by gentrification processes, which have seen the return of young middle class professionals to live in new or renovated apartments in inner-city areas. This has created high-income areas that are attractive to shops, restaurants, nightclubs, and so on. These processes are also becoming apparent in certain developing country cities such as Shanghai. Third, large retailers have started to reinvest in the inner city, prompted in part by the gentrification trends just mentioned. In the United Kingdom, both Tesco and Sainsburys have established new small supermarket and convenience store formats to tap into these growing urban markets. These processes have partly been driven by the competitive responses of retailers to planning restrictions on large-format stores, a process also seen in foreign markets (e.g., the rapid recent growth of Tesco's small Express format in South Korea and Thailand).

The Online Spaces of Retailing

Many forms of retailing still rely on the co-presence of sellers and buyers within a dedicated physical space. Since the late 1990s, however, and using the technological developments outlined in Chapter 9, various forms of business-to-consumer e-commerce have emerged, presenting challenges to traditional retailers. The rapid early growth of "e-tailers" such as Amazon and online auction houses such as eBay led to predictions that online retailing was a disruptive technology that would transform the way we all shop. Growth in online sales has indeed been strong over the past 10–15 years in relatively wealthy markets with high internet penetration rates. In South Korea, for example, online sales expanded five-fold from US$4.9 billion in 2001 to US$25 billion in 2009. By the latter year, online sales were the second most important retail channel after hypermarkets, representing 11% of total retail sales (the comparable figure for the U.S. in 2011 was 4.6% of total retail sales). At the same time, online retailing has not expanded to the degree expected by early proponents, with the reality being an uneven geographical and sectoral uptake of online channels.

This uneven growth trajectory has revealed important things about the nature of retailing. It has become clear that while online retailers can save money by avoiding the need for store premises, there are still considerable costs associated with selecting and delivering goods ordered over the internet. As noted in

Chapter 9, the logistical challenges associated with high-frequency, low-value local deliveries are severe, meaning that purely online retailers still face significant startup costs. Tellingly, while Amazon, perhaps the world's best known online retailer, was established in 1994, it did not return any profit until 2001. The infrastructural requirements of running its operations are immense. Its main U.K. fulfillment center near Milton Keynes covers the area of several football pitches and employs over 1,000 people with many more hired at busy times such as December. Other constraints relate to the elements of tactility and sociality inherent in many forms of shopping – consumers like to see/feel the goods they are purchasing and to discuss them with family, friends, and retail staff – and concerns over the security of online financial transactions. As a result, the penetration of online retailing has been uneven across different kinds of products, ranging from just 3% of food and grocery purchases to 30% of music and videos in the United Kingdom in the late 2000s, for example. In some segments the advent of online retailing has been transformative. Online sales of books and CDs, for example – easily transported, nonperishable standardized products – have expanded rapidly, bankrupting many store-based retailers. The challenge for music retailers has been compounded by the advent of electronic music formats available through online stores such as iTunes. In the United States in 2010, digital music purchases accounted for 35% of sales, and iTunes alone accounted for 27% of the total. Online retailing is also effective for a range of niche and specialist goods (e.g., relating to hobbies) due to the internet's ability to match buyers and sellers over long distances, thereby increasing the effective size of the market. In terms of services, the internet has proved to be a very efficient channel in several areas including travel and insurance. Box 11.2 offers a profile of one highly successful online-only retailer, Asos.

The biggest challenge to online only retailers has, however, come from store-based retailers who have moved parts of their business online, remodeling themselves as multi-channel "bricks-and-clicks" operators. Major store-based retailers have considerable competitive advantages relating to their pre-existing warehouse and supply chain infrastructures, customer support centers, and product-return networks (Wrigley, 2009). They have been able to leverage these resources and economies of scale – for example, in terms of sourcing – by grafting e-commerce operations onto their existing businesses. Many of the leading online retailers now fall into this category, operating both stores and online operations. There are two basic models of fulfilling online orders, either direct from a dedicated distribution facility or by picking products from store shelves (Murphy, 2007). In the grocery market the latter has proved more successful, as distribution centers are not set up to handle the small volumes associated with individual customer orders. Some major retailers combine both models. Tesco, for example, fulfills grocery orders from stores but combines this with an extensive catalogue and warehouse delivery system for larger domestic goods. Many contemporary retailers have thus become hybrids, seamlessly operating across both physical and

CASE STUDY

Box 11.2   Asos – online clothing retailer

The U.K.-based fashion retailer ASOS (As Seen on Screen) offers a powerful example of the growth potential that exists for online-only operators that are able to find a distinctive market niche and business model. The company – the largest online independent fashion retailer in the United Kingdom – was created in 2000 and was able to raise capital through listing on the AIM (Alternative Investment Market) of the London Stock Exchange in 2001. It made its first profits in 2004. In the company's own words, ASOS is "a global online fashion and beauty retailer and offers over 50,000 own-label and branded product lines across womenswear, menswear, footwear, accessories, jewellery and beauty with approximately 1,500 new product lines being introduced each week." The target market is 16- to 34-year-olds, which it aims to tap into by promoting "celebrity-inspired" fashion, and the emphasis is placed on continuous innovation and updating of the product range. By 2011 the website had 5.3 million registered users and 3.0 million active customers (who had ordered in the past 12 months). ASOS's expansion has been startling. In 2006 the company made £1.3 million profit on sales of £20 million. In 2011, it made profits of £28.6 million on sales of £324 million, based on 8 million separate orders across 160 countries. Key elements of growth have been its own-label ranges (launched in 2004 for women, 2007 for men) and rapid international expansion. By 2011, international revenues were up to £140 million including sales of £18.6 million in the United States, and in that year the company launched dedicated websites for the American, French, and German markets. The company also offers free worldwide delivery of its goods. ASOS fulfils its orders from a purpose-built 500,000 square foot distribution center in Barnsley, South Yorkshire. For more on ASOS, visit: http://www.asosplc.com/

electronic space. Similarly, many consumers now use both forms of retail space in their everyday lives, sometimes even for the same product, combining in-store assessment of goods with the power of online price comparisons. In sum, online retail spaces have by no means replaced physical spaces, but rather are overlain on, and interact with, the changing geographies of store-based retailing. Moving forward, as the coverage, speed, and security of the internet continues to develop, it is likely that online retailing will steadily increase its share of the overall market but that multi-channel retailers will be at the heart of the growth.

To summarize the arguments of this section: retail geographies are restless and continually changing. These changes reflect both the growing size and power of retailers, and wider societal dynamics such as patterns of urban decay/renewal and the movement of high-income segments of the population. Due to the acute sensitivity of retailing to planning and regulatory conditions, the timing, nature, and extent of change varies substantially from place to place. Moreover, the tangible geographies of stores, and the logistics and distribution systems upon which all forms of retailing depend, increasingly coexist and intersect with a range of online retail spaces.

11.4   The Configuration of Retail Spaces

In the previous section we outlined some of the key changes to the geographies of retailing in contemporary societies. We now move to looking at specific retail spaces in more detail. This section serves three purposes. First, it demonstrates the range of different retail spaces that are present. Second, we look within different kinds of retail spaces to examine the way in which firms, designers, and planners try to manipulate the layout and design of those spaces in order to stimulate purchases. Geography is important both in terms of the management of space within these different retail arenas and in the way in which different places are invoked as part of the design strategy. Third, we highlight the tension between public and private space – in other words, the extent to which different retail spaces are truly open-access, freely usable spaces, or whether they are closely monitored and policed by corporate interests. In order to do this we will look in turn at examples of four kinds of retail space: the street, the mall, the store, and informal spaces such as markets/fairs.

The Street

For centuries, streets have played a crucial role in the public and economic life of cities. As Jane Jacobs (1961, p. 29) once famously observed, "Streets and their sidewalks, the most public places of a city, are its most vital organs. Think of a city and what comes to mind? Its streets. If a city's streets look interesting, the city looks interesting; if they look dull, the city looks dull." What is of particular interest to us here is the relationship between the street and retailing. It is no coincidence that the main street (or streets) of most, if not all, cities is devoted to the retail function: Oxford and Regent Streets in London, Madison and Fifth Avenues in New York, and Orchard Road in Singapore are a few iconic examples. This remains the case despite the processes of retail decentralization we have described above, and even though property prices and rental rates in these districts are some of the highest anywhere on the globe. This association between key streets and leading retailers has developed over time, reflecting the essentially urban nature of retailing for most of human history. It is also reinforced through planning processes that seek to maintain important retail districts in order to attract both national and international visitors. The importance of these "main streets," however, goes beyond the collection of stores they contain, as they become nationally and internationally recognizable symbols of shopping activity. In turn, retailers seek to feed off this reputation and benefit from the large numbers of visitors by maintaining a presence in these relatively expensive locations.

Beyond the prime importance of particular central urban streets for retailing in general, there is also a tendency for specific kinds of shops and restaurants to agglomerate in certain streets and districts of towns and cities. In part, they cluster together to benefit from agglomeration economies (see Chapter 12). However, as with the main streets just described, they also gather in particular streets for less tangible reasons to do with the image, reputation, and mix of attractions in certain localities (Crewe and Lowe, 1995). In this way, certain streets and districts become associated with specific forms of retailing, such as the famous cluster of electronics stores in the Akihabara area of central Tokyo (Figure 11.7). Central Manchester offers a useful window onto these dynamics (see Figure 11.8). While Market Street offers a broad range of national chain stores – including Boots, Esprit, H&M, HMV, JJB Sports, Marks & Spencer, Primark, and TK Maxx – the Northern Quarter is a network of streets associated with a range of small-scale independent retailers involved in fashion, music, arts, and handicrafts along with a broad range of bars and restaurants. These streets have a completely different look and feel and appeal to consumers of different ages and tastes; Market Street is a somewhat bland pedestrianized space while the Northern Quarter offers a grittier and more intimate shopping experience. While some of these associations between retail type and streets are long-standing, others are more dynamic. Manchester's best known street for luxury shopping, King Street in

Figure 11.7   Akihabara – Tokyo's electronics shopping district

(a)

(b)

(c)

Figure 11.8 Different retail streetscapes in Manchester. (a) The Avenue, Spinningfields (b) Market Street (c) The Northern Quarter
Source: The authors.

the old banking district, has recently been overtaken by a new cluster of luxury shops along The Avenue in the thriving Spinningfields commercial district of the city, which offers Calvin Klein, DKNY, Emporio Armani, Flannels, Kurt Geiger, Mulberry, and Ted Baker outlets, among others. It is not just the shops themselves that are retail spaces, therefore, but the street as a whole, leading some to suggest that the street itself is effectively a brand. The combination of different ensembles of shops and different types of consumers on particular streets creates different senses of place within urban areas. The street is largely shaped by two groups: urban planners, who determine what different zones of cities may or may not be used for, and property and retail corporations that strategically invest in the urban-built environment in order to generate profits.

The Mall

In some instances, the notion of the street can blur into our next category: the mall. On the one hand, to take a walk down Singapore's Orchard Road is to experience a retail streetscape dominated by the giant shopping malls that line the road. On the other, entering Bugis Junction in the same city leads the shopper into an interior, air-conditioned street as part of the mall, and many other malls evoke notions of the street in their design. Importantly, however, malls are enclosed, privately owned and more tightly controlled spaces than streets. As such, to understand them, we need to look beyond the surface "magic of the mall" (drawing on Goss, 1993, 1999). Put another way, by exploring the form, function, and meaning of shopping malls it is possible to better understand how developers, retailers, and designers actively encourage the purchase of goods and services. First, there is a wide range of seemingly mundane attributes of mall design that seek to promote consumer spending, including:

- The use of attractive central features to draw shoppers in particular directions and then direct them toward further options
- The configuration of escalators, and the strategic positioning of toilet facilities and exits, in order to make consumers walk past as many shop fronts as possible
- Limiting long straight stretches so consumers are unaware how far they are walking
- The use of signs and displays to encourage shoppers to keep walking
- Supplying rest points and seats for the weary consumer
- The use of plants and vendor carts to encourage circulation within the walkways
- The use of soft lighting and the absence of natural light to help suspend a sense of time
- The use of mirrors and reflective glass to create an illusion of space
- The manipulation of the ambient temperature

- The use of soothing background music to induce spending
- The delineation of relatively segregated zones of shops designed to appeal to different types/classes of consumer
- The presence of highly visible security staff to reinforce the safety of the environment
- Ongoing, regular cleaning to reinforce the cleanliness of the environment

In sum, what at first might seem to be a rather bland shopping mall is in reality a highly designed and strategic space, reflective of the power of its owners, designers, and tenants.

Second, and linking back to arguments made in Chapter 8, strenuous efforts are made to disconnect commodities from their real-world production systems and reposition them in a world of pleasure, fantasy, and magic promoted through architecture, interior design, and theming that evokes other times and places. The huge Mall of America in Minneapolis, Minnesota, provides a powerful example. The mall, which opened in 1992 and attracts 40 million visitors annually, covers 4 million square feet and contains over 520 specialty stores, four department stores, over 50 dining options, a 14-screen cinema, and many other leisure attractions, including the 7-acre indoor theme park, Nickelodeon Universe. The Mall of America endeavors to create a sense of public, market, and festival spaces through its design, and repeatedly evokes notions of travel, nature, primitiveness (through Native American references), childhood, and heritage in its retail and leisure attractions. The Rainforest Café, for example, offers a "safari" meal experience complete with passport, while there are many other "destination" restaurants such as the Cadillac Ranch All American Bar and Grill, Kokomo's Island Café, Little Tokyo, Cantina #1, and the Napa Valley Grill. The staging of events (such as fashion shows, cultural displays, and children's entertainment) and the provision of a range of services to facilitate expenditure (e.g., food courts, daycare facilities for small children) are also an integral part of the general strategy of encouraging "guiltless" spending. Overall, as Goss (1993, p. 40) suggests: "the shopping center appears to be everything it is not. It contrives to be a public, civic place even though it is private and run for profit; it offers a place to commune and recreate, while it seeks retail dollars; and it borrows signs of other places and times to obscure its rootedness in contemporary capitalism."

The Store

As we have already seen in this chapter, the store is the key retail space in contemporary capitalist society. Retailers actively use geography in stores to try and promote purchases in two ways. First, the interior layout and use of space within stores is designed to enhance sales levels. Many of the techniques used in a standard supermarket are now well known: placing attractive fresh produce near the store entrance, using the smells of fresh bread or roasting chicken to entice customers in, putting important products at eye level, stacking promotional goods at the end of aisles, the use of signage to attract attention to particular products, grouping functionally related products (e.g., stir-fry vegetables and sauces), placing "impulse" products (e.g., confectionery, magazines) near the checkouts, and so on. Similar techniques are used in all kinds of shops; window displays, for example, are critical in defining the target consumer for a clothing store. What these practices reveal is a conscious micro-management of the store environment on the part of the retailer to maximize sales. Furthermore, as we saw earlier, the precise range of products in a store may have already been shaped in line with the social and demographic profile of the area in which it is located. More broadly, retailers use the exterior and interior design of their stores as a strategic resource designed to distinguish themselves from their competitors and develop a brand image aimed at certain portions of the population.

Second, particular places are evoked within the store environment. We can illustrate this idea by looking at the coffee chain Starbucks (here we draw on Smith, 1996; see also Box 8.1). Starbucks outlets conform to a broadly similar design that seeks to evoke a particular sense of place. The design combines associations with European coffee houses and bars, and a North American Pacific Northwest lifestyle of relaxed living and outdoor activities (the latter reflecting the chain's roots in Seattle, Washington). These associations are manifest in a simple décor of green and brown color schemes with metal trims, the visibility of the machinery and process of coffee making, the prominence of educational information about coffee and its origins, and the use of large glass windows and bar-style seats that allow drinkers to observe and "consume" the streetscape in front of them. Creating a particular notion of place – albeit a hybrid one in this instance – is central to Starbuck's strategy of bringing coffee to a mass market. Even more intriguingly, BritishIndia is a Southeast Asian retail chain that started in Malaysia in 1994, selling a range of clothes and furnishings evoking the British rule of India in the 1800s. The chain has more than 30 stores across Malaysia, Singapore, Thailand, and the Philippines. With an emphasis on comfortable, elegant, high-quality products, the stores are explicitly aiming at middle-class and expatriate portions of the population. The stores are decorated with photos and other memorabilia of British India, creating a colonial feel. Here, then, is a fascinating example of the contemporary mobilization of notions of place (and indeed time), with a retailer in postcolonial Southeast Asia using images of a colonial past in another part of Asia to sell products and develop a distinct brand identity.

Informal Retail Spaces

Thus far we have tended to focus on formal retail spaces as perhaps the most prevalent and important spaces of sale in capitalist society. It is vital to recognize, though, that beyond the formal spaces of retail lie a wide variety of informal trading and exchange practices and spaces. These spaces are different from formal retail spaces in several important ways. First, while some are relatively permanent features of the built environment (e.g., charity shops), many are transient and temporary in nature (e.g., mobile street vendors and street markets). Second, while some will be entirely legal operations, others may operate at the very limits of the law, or even illegally. Third, prices are often agreed through negotiation and bargaining, rather than being fixed. In this sense, there may be more consumer involvement than in the standard shopping experience structured by retailers. Fourth, and relatedly, shopping in informal retail spaces can be a less predictable experience, as participants do not know what goods will be available. Instead, the shopping experience unfolds as an exciting and highly tactile (hands-on) search for bargains.

Informal retail activities are present in all countries. This is demonstrated in developed markets, for example, by the prevalence and popularity of tailgate or car boot /yard sales, and the number of charity and secondhand shops in central urban areas (Gregson and Crewe, 2003). Occasional markets also continue to be a persistent feature of the retail landscape, as seen, for example, in the weekly fresh produce markets of the towns and villages of the Provence region of France (see Figure 11.9). However, while informal retail spaces can thrive in developed market contexts, they tend to operate around the margins of the mainstream, formal retail sector. Informal retail spaces take on far greater significance in the context of developing world cities, where such exchanges may be more important than formal sites of retailing. Examples are found in the extensive street foodscapes of many Asian cities such as Bangkok, with small temporary stalls selling a huge variety of freshly prepared snacks and meals (Yasmeen, 2007). In such contexts, informal retailing is central to the livelihood strategies of many

The authors

Figure 11.9   Vegetable market in L'Isle sur la Sorgue, Provence, France

families and households. Estimates suggest that in the developing world, 37% of employment is provided by informal economic activity (ranging from 33% in Latin America to 54% in Africa) compared to 3% in high-income countries. While some of this work is constituted by small-scale manufacturing activity, a considerable proportion is derived from informal retailing activities.

South Africa – where nonstore retailing accounts for half of the country's informal economy and provides employment for an estimated 900,000 to 1,200,000 people – provides a compelling case study of informal retailing (Rogerson, 2008). There are two main kinds of informal retail in South Africa: street traders who sell goods such as foodstuffs, handicrafts, and traditional herbs either in temporary markets or at fixed sites in the street; and home-based enterprises that sell a wide range of goods – including cooked foods, cakes, sweets and snacks, clothes, fruit and vegetables, jewelry, furniture, live chickens, crafts, paraffin, and cosmetics – and are known as *spazas* or tuck shops in the South African context. The relative importance of these two kinds of informal retailers varies between different places. In the inner city of Johannesburg and Durban, for example, street traders dominate, while in township areas *spazas* are the most common form of operation. Different forms of regulation have been implemented across South Africa in response to the continued prevalence of informal retailing. In Johannesburg, the estimated 8,000–10,000 street traders sit uneasily with city leaders and their stated goal of having an established world city by 2030. Between 1999 and 2005 six markets were constructed to try to concentrate traders in particular areas, but the policy has been largely unsuccessful due to the out-of-the-way locations chosen for the markets. In Cape Town, an arguably more progressive approach has been adopted in which *spazas* have been targeted for business support in an effort to make them more efficient and sustainable small businesses. The regulatory approach also changes through time. For example, since 2000 Durban has had an informal economy policy that seeks to liaise with, and provide support to, street traders, but in recent years there have been severe police clampdowns on unlicensed traders in what can be characterized as "an ongoing struggle for access to the streets" (Skinner, 2008, p. 239). It is also important to note that formal and informal retailing are not insulated from one another but rather coexist and compete with each in complex and spatially variable ways. The ongoing formalization of South African retailing and the movement of chain stores into township areas, for example, have put increasing pressure on poorly resourced *spazas* in certain districts (Rogerson, 2008).

In sum, this section has explored some of the micro-geographies of retailing. A growing range of places now perform retail functions: theme parks, train stations, airports, even hospitals and schools in some instances, are all increasingly home to retail outlets of different kinds. Such spaces do not evolve by accident; rather, they are carefully designed and manipulated by planners, property developers, and above all retailers to induce us to consume goods and services. Although we can access them as members of the public, in most cases these are privately owned and controlled spaces that are knowingly shaped and reshaped by these interest groups. A range of informal retail activities and spaces also exist alongside these formal retail operations. The importance of such activities varies geographically: in some contexts they will be relatively insignificant, filling particular market niches in a largely formal retail sector, while in others they form the mainstay of the urban economy, representing the only way for millions of poor urban inhabitants to make a living.

11.5   Constructing Needs and Desires: The Advertising Industry

The sale of commodities to consumers does not simply depend on the availability of goods for purchase within different kinds of retail spaces. It also relies upon flows of information from producers of goods and services to potential customers. This information can come from many different sources: the media, peers, friends, and family members. In particular, however, *advertising* plays a critical role in constructing and disseminating knowledge about commodities and services (we have already seen in Chapter 8 how advertising may be used to efface the real origins of commodities). Manufacturers and retailers engage specialist advertising firms to design and manage these flows of information, with the result that advertising itself is a huge global business worth an estimated US$400–500 billion a year. The leading global spenders on advertising are consumer goods and automobile firms. In 2009, for example, spending was topped by Proctor & Gamble (US$8.7 billion) and Unilever (US$6.3 billion). These huge investments are designed to create and sustain demand for these firms' products, by seeking to make connections between the particular qualities of the commodities concerned and our senses of self and the lifestyles to which we aspire as consumers. Advertisers can thus be thought of as an important form of cultural intermediary that seeks to influence the many decisions we make as consumers about what to buy – from the everyday purchase of food and drink, to regular demands for clothing, books, and music, to less frequent large purchases such as cars, TVs, and furniture. One way they do this is through developing brands – sets of positive associations that consumers link to producers of particular commodities (see Box 11.3).

In the Fordist era (see Chapter 9) advertising was concerned with stimulating the mass consumption required to support mass production systems through educating consumers about the attributes and availability of various products. The global advertising industry in this period was dominated by U.S. agencies that developed worldwide campaigns for U.S. export products such as Coca-Cola. The advent of post-Fordism in the latter part of the 20th century, however, with its flexible production technologies and fragmented consumer markets, has posed significant challenges to this standardized export model of global advertising.

FURTHER THINKING

Box 11.3   *Geographies of branding*

Brands and branding are an integral part of these processes of creating demand. Naomi Klein, for instance, in her influential book *No Logo* describes brand consultancies as "factories, hammering out what is of true value: the idea, the lifestyle, the attitude. Brand builders are the new primary producers in our so-called knowledge economy" (2000, p. 196). Branding seeks to add value to goods and services by linking positive associations to the brand name – for example, quality, style, reliability, sophistication, or design – and thereby cultivating customer loyalty to that brand. Branding is a nonmaterial, creative element of the broader production process that seeks to differentiate products and services from those of competitors and make them more meaningful to consumers, in turn allowing firms to charge a premium. Branding often works through the use of symbols such as logos – think of Apple's iconic logo, Coca-Cola's swirling script, or Nike's swoosh – and/or people – for example, David Beckham (soccer) or Roger Federer (tennis). Recent research, however, has demonstrated how branding processes are also always geographical in nature.

Most important, branding has strong connections to the geographical origins of the goods and services in question. This is most often seen in terms of links to the country of origin, which may be either explicit – for example, British Airways or Nippon Steel – or implicit, such as the German quality and efficiency inherent in brands such as BMW and Mercedes-Benz. Equally, however, the geographical component of branding may operate on other spatial scales such as the regional (Scottish whisky, Californian wine) or the local (Newcastle Brown Ale, Parmigiano Reggiano, Beefeater Gin). Emphasizing the geographical origins of the product and associated qualities becomes an integral part of the brand-building process. The attachments created can be with a range of different elements of the "home" environment: economic (e.g., Japanese design and innovation – Sony), social (e.g., Scandinavian design – IKEA), political (e.g., Swiss neutrality and discretion – Credit Suisse), cultural (e.g., Italian style and design – Prada) or ecological (e.g., small-town values and environmental commitment – Ben & Jerry's ice cream). Branding processes are also highly geographical in the sense that they are unevenly implemented in social and spatial terms. Brands require a degree of adaptation to different consumer cultures, and sometimes branding will be about trying to weaken the geographical associations the brand conveys (e.g., McDonalds' links to unhealthy American fast-food culture). It can also be argued that by spatially targeting brands at differentiated groups of consumers in different places, for example by promoting luxury brands to rich enclaves within global cities, branding processes actually serve to reinforce patterns of uneven economic development. For more on the multiple geographies of branding, see Pike (2009, 2011).

Leading agencies now have to create demand for products and services across multiple geographically dispersed and heterogeneous markets (Faulconbridge et al., 2011). The contemporary global industry is still dominated by a handful of large holding groups, but they now originate from Europe and Japan in addition to New York, the industry's traditional "home" (see Table 11.4).

These holding groups – which generate billions of dollars in revenue, employ tens of thousands of employees, and may bring together dozens of different companies – have developed through ongoing processes of merger and acquisition

Table 11.4   The leading global communications groups and their advertising agencies

Group (HQ)	Revenues US$ billion (2009)	Global Employees (2008)	Main Global Agencies
WPP (Dublin)	13.6	90,182	J. Walter Thomson; Ogilvy and Mather Worldwide; Young & Rubicam; Grey
Omnicom (New York)	11.7	70,000	BBDO Worldwide; DDB Worldwide Communications; TBWA Worldwide
Publicis (Paris)	6.3	43,808	Leo Burnett Worldwide; Publicis Worldwide; Saatchi & Saatchi
Interpublic (New York)	6.0	43,000	Draft Foote Cone and Belding Worldwide; Lowe Worldwide; McCann Erickson Worldwide
Dentsu (Tokyo)	3.1	n/a	Dentsu Incorporated
Havas (Paris)	2.0	14,747	Euro RSCG Worldwide; Arnold Worldwide

Source: Adapted from Faulconbridge et al. (2011), Table 2.1.

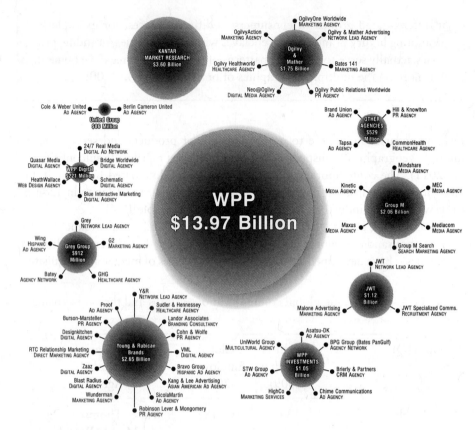

Figure 11.10   WPP group's organizational structure
Source: Adapted from www.adage.com/images/random/datacenter/2010/agency
familytrees2010.pdf, accessed August 29, 2011.

and are shaped by geographical variation in consumer preferences and cultures. Their loose corporate structure allows advertisers with expertise in particular national markets and different industries to develop tailor-made, culturally-appropriate campaigns. As Figure 11.10 demonstrates, for example, WPP Group is constituted by a wide range of companies encompassing specialisms in different sectors (e.g., CommonHealth or Ogilvy Healthworld), different kinds of media activities (e.g., public relations firm Cohn & Wolfe or market research firm Kantar), different advertising media (e.g., the digital agencies Schematic or Bridge Worldwide) and different territories (e.g., Bates 141 in Hong Kong or Blue Interactive Marketing in Singapore). As Hudson (2005, p. 72) describes, increasingly such groups "are devolving responsibility to local branches or agencies for creating adverts that are customised to local conditions – variations on a global theme, but tailor-made to fit local circumstances."

Two key points should be taken from this brief analysis of the advertising sector. First, the economic significance of advertising as an activity is an important reminder that selling goods and services is about more than simply offering them up for sale in shops or online; demand needs to be created for those commodities through communicating information about their attributes to final consumers. Second, and as the highly variegated internal structures of the advertising holding groups suggest, such efforts to propagate demand need to be seen as embedded in geographically specific consumption cultures. These varied cultures shape both how products are best advertised and sold, and also how consumers of different kinds respond to such initiatives in very different ways. This idea, that retailing and the business of selling are part of much broader and cultural process of *consumption*, provides the theme for Chapter 15.

11.6   Summary

After a brief review of "traditional" approaches to studying retail location, this chapter has investigated the various geographies inherent in contemporary retailing in three stages. First, we have explored the changing patterns of where retailing takes place. At the global scale, store and sourcing networks are increasingly coordinated by a cadre of large retail transnationals. At the national scale, the suburb has risen to challenge the city center as the preeminent site of retailing in contemporary society, but this general trend conceals complex patterns of decentralization and recentralization that vary across different regulatory contexts. The physical spaces of retailing also increasingly coexist with online retail spaces. Second, we have looked more closely at a variety of different retail spaces – specifically the street, the mall, the store, and informal retail outlets – to explore not only their significance but also how they can be designed and configured in a bid to induce the purchase of goods and services. These practices involve not only the shaping of physical space, but also the use of a wide range of geographical references and images within the built environment.

Third, we have introduced the role of advertisers as cultural intermediaries that are also involved in shaping our wants and desires for the goods and services we purchase. Overall, our focus in this chapter has been on the powerful corporate interests – notably retailers, property developers, and advertisers – that shape the geographies of retail and structure the ranges of goods that are available to end consumers in particular places. Consumers, though, are not mere pawns in this game, but rather actively engage with and use retail spaces to suit their own purposes. We will explore how retailing is embedded within broader, and geographically variable, consumption processes in Chapter 15.

Notes on the references

- Wrigley and Lowe (1996; 2002) and Wrigley (2009) provide student-friendly reviews of the best geographical work on retailing.
- Coe (2004) and Coe and Wrigley (2007, 2009) offer overviews of the globalization of retailing from an economic-geographical perspective. For a wide ranging review of Wal-Mart's activities, see Brunn (2006).
- See Murphy (2007) and Malecki and Moriset (2008, ch. 5) for more on the possibilities and perils of online retailing.
- Rogerson (2008), Skinner (2008), and Turner and Schoenberger (2011) offer more conceptual and empirical material on informal retail spaces.
- For two challenging but rewarding readings on the modern shopping mall, see Goss (1993, 1999).

Sample essay questions

- What are the limitations of traditional approaches to studying retail location?
- What forces are driving the globalization of retail activity?
- Why are there growing tendencies for the recentralization of retailing at the urban scale?
- How are different retail spaces manipulated in order to try and induce consumer purchases?
- In what ways are the geographies of retailing shaped by national and local regulation?

Resources for further learning

- Deloitte's annual *Global Powers of Retailing* report provides a detailed overview of the world's largest 250 retailers (http://www.deloitte.com/view/en_US/us/Industries/Retail-Distribution/index.htm).
- The websites of leading global retailers such as Tesco (http://www.tesco.com), Wal-Mart (http://www.walmartstores.com), and Starbucks (http://www.starbucks.com) provide a range of information on their global store operations.
- Similarly, the sites for regional or mega-malls such as Manchester's Trafford Centre (http://www.traffordcentre.co.uk/) and Minneapolis' Mall of America (http://www.mallofamerica.com/) give some sense of what is on offer at such attractions.
- Leading shopping streets, such as London's Oxford Street (http://www.oxfordstreet.co.uk/home.html) and New York's Madison Avenue (http://www.

madisonavenuenyc.com/) also seek to promote their own brand of shopping experience.

- The *Age of Persuasion* radio series provides a fascinating account of how advertising works, as told by an industry insider: http://www.cbc.ca/ageof persuasion/episode/.

References

Barnes, T. (2012). Reopke lecture in economic geography: notes from the underground: why the history of economic geography matters: the case of central place theory. *Economic Geography*, 88(1), 1–26.

Berry, B. J. L. (1967). *Geography of Market Centers and Retail Distribution*. Englewood Cliffs, NJ: Prentice-Hall.

Berry, B. J. L., Cutler, I., Draine, E. H., Kiang, Y-C., Tocalis, T. R., and de Vise, P. (1976). *Chicago: Transformations of an Urban System*. Cambridge, MA.: Ballinger.

Brunn, S. D., ed. (2006). *Wal-Mart World: The World's Biggest Corporation in the Global Economy*. New York: Routledge.

Christaller, W. (1966 [1933]). *Central Places in Southern Germany*. Englewood Cliffs, NJ: Prentice-Hall.

Christopherson, S. (2007). Barriers to "U.S. style" lean retailing: the case of Wal-Mart's failure in Germany. *Journal of Economic Geography*, 7, 451–69.

Coe, N. M. (2004). The internationalisation/globalisation of retailing: towards an economic-geographical research agenda. *Environment & Planning A*, 36, 1571–94.

Coe, N. M. and Lee, Y. S. (2006). The strategic localization of transnational retailers: the case of Samsung-Tesco in South Korea. *Economic Geography*, 82, 61–88.

Coe, N. M., and Wrigley, N. (2007). Host economy impacts of retail TNCs: the research agenda. *Journal of Economic Geography*, 7, 341–71.

Coe, N. M., and Wrigley, N., eds. (2009). *The Globalization of Retailing* (2 vols.). Cheltenham: Edward Elgar.

Crewe, L., and Lowe, M. (1995). Gap on the map? Towards a geography of consumption and identity. *Environment and Planning A*, 27, 1877–98.

Deloitte. (2011). *2011 Global Powers of Retailing* (downloaded from www.deloitte.com, 14 June 2011).

Faulconbridge, J. R., Beaverstock, J. V., Nativel, C., and Taylor, P. J. (2011). *The Globalization of Advertising: Agencies, Cities and Spaces of Creativity*. London: Routledge.

Fernie, J. (1995). The coming of the fourth wave: new forms of retail out-of-town development. *International Journal of Retail and Distribution Management*, 23, 4–11.

Goss, J. (1993). The "magic of the mall": an analysis of the form, function, and meaning in the contemporary retail built environments. *Annals of the Association of American Geographers*, 83, 18–47.

Goss, J. (1999). Once-upon-a-time in the commodity world: an unofficial guide to the Mall of America. *Annals of the Association of American Geographers*, 89, 45–75.

Gregson, N., and Crewe, L. (2003). *Second-hand Cultures*. Oxford: Berg.

Guardian, The. (2010, October 30). Clubcard couple head for checkout, p. 49.

Guardian, The. (2011, January 12). U.K. shopping mall supremo scents victory in battle over Trafford Centre, p. 26.

Hudson, R. (2005). *Economic Geographies*. London: Sage.

Jacobs, J. (1961). *The Death and Life of the Great American Cities*. New York: Vintage Books.

Klein, N. (2000). *No Logo*. London: Flamingo.

Lösch, A. (1954 [1939]). *The Economics of Location*. New Haven, CT.: Yale University Press.

Lowe, M. (2000). Britain's regional shopping centres: new urban forms? *Urban Studies*, 37, 261–74.

Lowe, M., and Wrigley, N. (2010). The "continuously morphing" retail TNC during market entry: interpreting Tesco's expansion into the United States. *Economic Geography*, 86, 381–408.

Malecki, E. J., and Moriset, B. (2008). *The Digital Economy: Business Organization, Production Processes and Regional Developments*. London: Routledge.

Murphy, A. J. (2007). Grounding the virtual: the material effects of electronic grocery shopping. *Geoforum*, 38, 941–53.

Pike, A. J. (2009). Geographies of brands and branding. *Progress in Human Geography*, 33, 619–45.

Pike, A. J., ed. (2011). *Brands and Branding Geographies*. Cheltenham: Edward Elgar.

Rogerson, C. M. (2008). Policy and planning for changing informal retailing in South Africa. *Africa Insight*, 37(4), 230–47.

Skinner, C. (2008). The struggle for the streets: processes of exclusion and inclusion of street traders in Durban, South Africa. *Development Southern Africa*, 25, 227–42.

Smith, M. D. (1996). The empire filters back: consumption, production, and the politics of Starbucks coffee. *Urban Geography*, 17, 502–24.

Turner, S., and Schoenberger, L. (2011). Street vendor livelihoods and everyday politics in Hanoi, Vietnam: the seeds of a diverse economy? *Urban Studies*, 49, 1027–1044.

Wrigley, N. (2009). Retail geographies. In R. Kitchin and N. Thrift, eds., *The International Encyclopaedia of Human Geography*, 9, pp. 398–405. Oxford: Elsevier.

Wrigley, N., and Lowe, M., eds. (1996). *Retailing, Consumption and Capital: Towards the New Retail Geography*. Harlow: Longman.

Wrigley, N., and Lowe, M. (2002). *Reading Retail: A Geographical Perspective on Retailing and Consumption Spaces*. London: Arnold.

Wrigley, N., Warm, D., and Margetts, B. (2003). Deprivation, diet, and food-retail access: findings from the Leeds "food deserts" study. *Environment and Planning A*, 35, 151–88.

Yasmeen, G. (2007). *Bangkok Foodscapes: Public Eating, Gender Relations and Urban Change*. Bangkok: White Lotus.

PART IV

PEOPLE, IDENTITIES, AND ECONOMIC LIFE

CHAPTER 12

CLUSTERS
Why do proximity and place matter?

Goals of this chapter

- To understand why proximity still matters for many different kinds of economic activity
- To explore the different forces, both economic and sociocultural, that bind economic actors together in particular places
- To appreciate the range of different clusters found in the global economy
- To reflect on the limitations of the cluster concept for understanding contemporary economic geography

12.1 Introduction

You may well not have heard of Silicon Valley's Sand Hill Road, a visually unremarkable street in Menlo Park, California. Since the early 1970s, however, when pioneering firms such as Kleiner Perkins Caufield & Byers and Sequoia Capital were established, Sand Hill Road has been the center of the global venture capital industry. There are now dozens of venture capital funds clustered in the area (see Figure 12.1). Venture capital firms are private fund managers who invest in early-stage companies – usually in high-technology sectors – by acquiring equity shares in new corporate ventures through successive rounds of investment. It is a high risk and reward form of corporate financing that funds ideas that cannot usually be supported through traditional bank loans. The United States accounted for 79% of global venture capital investment in 2009. California, in turn, dominates the U.S. industry with a 50% share of venture capital investments, while Silicon Valley constituted 80% of the California

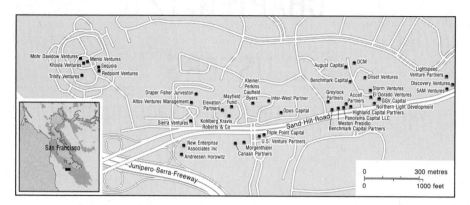

Figure 12.1   Venture capitalists on Silicon Valley's Sand Hill Road
Source: Data from www.portfolio.com and en.wikipedia.org/wiki/Sand_Hill_Road, accessed November 15, 2010.

total. Put another way, the cluster of firms centered on Sand Hill Road invested one-third of the global total of venture capital in 2009, or approximately US$7 billion. In aggregate terms the firms were managing around US$80 billion in investments (data from NVCA, 2010). When rental rates peaked in the area in mid-2000, they were higher than Manhattan or London's West End. Although in the decade since then several leading firms have set up new offices in places such as China, India, and Israel to try and tap new market opportunities, Sand Hill Road's position as the home of the global venture capital industry remains essentially unchallenged.

How can we understand this long-standing cluster of venture capitalists along Sand Hill Road? First, in an era of advanced and instantaneous telecommunications, it illustrates the continued importance of "being there," of having a physical presence among the networks of lawyers, investors, inventors, and entrepreneurs that constitute the social fabric of Silicon Valley. The close physical proximity of talent and capital in the towns of the area is central to its competitive success. As one commentator observes, "You've got to be face to face. Here people bump into each other in the market, they see each other at restaurants, they grab a quick coffee at Starbucks or somewhere. That's how deals happen. It's an ecology" (*BBC News*, 2010). Second, it demonstrates how over time, certain places become associated with certain forms of economic activity; they become known as the "place to be." These reputational effects reinforce the strength of the cluster, and start to explain the importance of having an address on Sand Hill Road in particular, rather than just being located in the wider Silicon Valley region. In short, having the right address gives the firms located there a certain credibility that in turn enables them to raise the capital on which their business depends.

Third, the attributes of venture capital as an economic activity are important to understand. In previous chapters (8 and 9) we have seen how certain forms of production can be dispersed at the global scale and managed over great distances through the use of technology. Venture capital investments, however, are based upon localized networks of know-how and know-who that allow the rapid and effective circulation of particular forms of knowledge. Selecting new industries for investment and finding promising firms requires detailed observation of market and technological trends over time, and working with firms after the initial investment also necessitates frequent ongoing interactions based on trust. In the risky and volatile world of venture capital and high-technology industries, the place-specific iterative relations between technologists and financiers are absolutely crucial (Zook, 2004). Fourth, Sand Hill Road offers a window onto the way in which locally-specific ways of doing business may develop over time. Silicon Valley is in part unique because of the particular business practices or cultures that have developed there over a matter of decades, distinguishing it from other high-tech clusters around the globe.

This chapter, then, is about understanding the tendency of certain kinds of economic activities to cluster together in particular places. Clustering is not specific to Silicon Valley or the venture capital industry; rather, it is a generic characteristic of the capitalist economy and is seen in all sectors and places. The argument proceeds in five stages. In Section 12.2, we explore the need to move beyond the intra-firm factors considered in traditional location theory models and incorporate relationships external to the firm in explaining location decisions. In Section 12.3 we look at the various contractual connections between firms that may bind them together in a cluster. Third, we broaden our canvass to look at a range of sociocultural forces that may also provide the glue holding a cluster together (Section 12.4). As part of this analysis, we will look at how regional cultures may build up over time as a result of long-standing patterns of intense interaction. Fourth, we detail the wide range of clusters that coexist within the global economy and that are driven by different combinations of types of interaction (Section 12.5) – the venture capitalists of Sand Hill Road constitute just one cluster type among many. Finally, we start to question the notion of clusters by thinking about how much economic activity is actually undertaken through complex combinations of face-to-face contact and technologically-mediated interaction (Section 12.6). In other words, achieving proximity does not necessarily mean being in the same physical location all the time.

12.2   Industrial Location Theory

During the Industrial Revolution, Sheffield, England, emerged as the world's foremost steel-making city, a position it held into the early 20th century. The industry prospered from Sheffield's combination of local iron ore, coal, and water power as well as key innovations in metalworking that built upon the city's long-standing reputation as a center for making cutlery. In textiles, Manchester similarly emerged as a global center for the cotton industry, with 2,000 mills employing some 360,000 workers having sprung up across northwest England by 1860. Further afield in the northeast United States and northwest Europe, large-scale production of textiles and iron and steel also developed rapidly in coalfield regions, with burgeoning canal and rail networks facilitating the transportation of both raw materials and finished products. Lowell, Massachusetts (U.S.), Lille (France), and Ghent (Belgium), for example, are other places synonymous with early industrialized textile production.

It was against this backdrop that the first attempts to understand why firms locate in particular places emerged. Here we will focus on the work of Alfred Weber, a German economist whose 1909 classic work *The Theory of the Location of Industries* was published in English in 1929. He developed his theory in the context of strong industrial growth in the coalfield areas of the Ruhr region of Germany in the late 19th century. His aim was to develop a general theory of location that could explain the spatial distribution of production activities, and the approach he developed was a *least cost model* of industrial location. This seemed to be entirely appropriate in a period in which the dominant industries were locating close to energy and raw material sources and to transportation networks.

Weber started from the observation that manufacturing plants tended to locate where total transportation costs were minimized. He proposed that transportation costs were shaped by two factors: the *weight* of the materials to be shipped along with that of the final products to be taken to market, and the *distances* over which both the materials and the end products had to be transported. From these two elements Weber derived a basic unit of cost, the ton-mile, with the locational decision then becoming a search for the point when the total ton-mileage of a given production and distribution process was at a minimum. His locational triangle was a key representation of the industrial location problem (see Figure 12.2a). The triangle models a situation in which a single manufacturing plant needs to access two sources of raw material inputs and one market location. Weber then used both mechanical and mathematical models to calculate the least cost location within the triangle. This process highlighted the significance of whether the manufacturing process was *weight-losing* – in which case the optimum location was drawn toward the raw materials sources – or *weight-gaining* – in which case it was drawn toward the market site. So while metal smelting, for example, is likely to take place near where the relevant ores are extracted, the production of soft drinks or cars is likely to be located close to market.

In addition to modeling the effects of raw materials, markets, and transportation costs, Weber also introduced variable labor costs into his analysis. Through an index of labor cost, he measured the average cost of labor required to produce a given unit weight of a product; industries with a higher index were presumed to be more sensitive to spatial variations in labor costs. His model predicted that firms would move toward a source of cheap labor if the savings per unit of production were greater than the increase in transportation costs incurred by

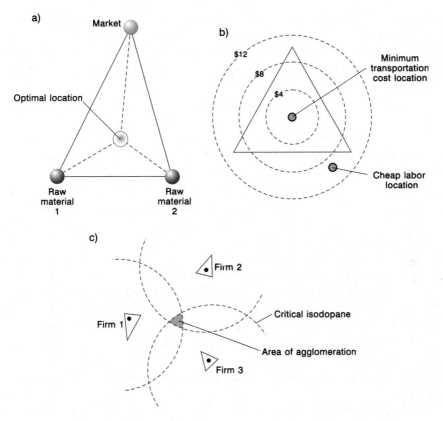

Figure 12.2   Weber's industrial location theory

moving. This was calculated through using *isodapanes*, or lines of equal cost; the *critical* isodapane represented the places where the savings from a given location factor were just enough to outweigh the additional transport costs of being in that position. If the cheap labor location was inside the critical isodapane, then the factory should move to that location (Figure 12.2b).

Weber's approach has been extensively critiqued by economic geographers, particularly since the 1970s as such abstract modeling techniques have slipped out of fashion. In many ways it is easy to see why. His model unfolds on a featureless surface that bears little resemblance to real places. It assumes that industrial decision makers have access to perfect information and act in an entirely rational and economistic manner. It represents raw materials, markets, and cheap labor as being located at single points in space, which is clearly not the case. It prioritizes transport costs that are now a far less significant component of a firm's total costs, particularly in the service sectors that dominate the structure of many advanced economies. That being said, the basic insight that industrial location is shaped by the trade-off between different production costs still has tremendous contemporary resonance, particularly in the notion of the new international division of labor (see Box 10.3). The rise of China, India, Brazil, and others as global manufacturing hubs reflects the way in which low production costs can offset transportation costs. Moreover, state economic policies in many localities across the world are clearly driven by a desire to attract inward investment through offering lower production costs.

Importantly, however, Weber was also one of the first to consider the nature of the *agglomeration economies* that form the focus of much of the rest of this chapter. While industries do indeed cluster, to differing degrees, at the sources of raw materials, at transportation and transhipment terminals, and at cheap labor locations, as predicted by industrial location theory, clustering offers further economies for the firms that take part in it. In contrast to the *internal* economies that can be accrued within a firm by producing at larger volumes (known as "economies of scale"), for example, these are economies *external* to these firms that can be derived from connections to other firms and organizations (described as inter-firm transactions in Chapter 10). Put another way, the connections of a firm to other participants in the wider cluster can lead to additional savings unavailable to those managing their own resources. For Weber, these agglomeration economies exerted a similar "deviational" or pulling force on the least cost location generated by proximity to cheap labor. Whether firms moved in response to them depended upon whether the savings outweighed the additional transport costs accrued from moving. As Figure 12.2c demonstrates, agglomeration only occurred when the critical isodapanes of the relevant firms intersected.

The analysis of agglomeration was arguably one of the weaker elements of Weber's work. His work reveals little about the nature of agglomeration economies, and his view of industrial relocation is extremely simplistic and mechanistic. Moreover, agglomeration economies are different from other variables such as resource or labor costs in that they depend on the alignment, or not, of locational decisions taken by multiple firms. Nonetheless, in developing the idea of agglomeration economies, Weber's work signaled what has become a very significant strand of recent economic geographical research. As we will see in the sections that follow, however, the way in which agglomeration economies are conceptualized and understood has developed significantly in recent decades.

12.3 Binding Clusters Together: Agglomeration Economies

At first glance, the continued agglomeration of economic activity in an era of advanced transportation and communications technologies seems rather counterintuitive. To put it another way, why do "sticky places" hosting these clusters continue to exist in the "slippery spaces" created by new technologies (Markusen, 1996)? Traditional explanations of agglomeration – reprising the famous economist Alfred Marshall's (1890) ideas about the industrial districts of Britain in the late 19th and early 20th centuries – focus on four main sets of factors:

- The development of *intermediate industries* providing specialized inputs. The co-location of manufacturers with suppliers of component goods and services serves to reduce transport costs.
- The growth of a pool of *skilled labor* as workers acquire the skills required by local industry, thereby reducing the costs for firms and workers to find employees and jobs, respectively.
- The emergence of *dedicated infrastructure* and various other collective resources that reduce the costs for individual firms. This encompasses elements of the built environment – property, transport, communications, power supply, and so on – as well as a wide range of services, such as those pertaining to education, training, and health.
- The facilitation of both formal and informal *face-to-face contact* through the concentration of economic activity, in turn leading to the transfer of information about all aspects of economic activity, thereby enhancing the diffusion of ideas and innovations.

Taken together, these four forces generate agglomeration economies (i.e., cost savings) for individual firms locating within a particular cluster. These agglomeration economies can either derive from the concentration of firms in the same or related industry (such as a pool of suppliers or specialized labor) or from the general clustering of different industries in large urban areas (e.g., access to common educational or transport services).

We will now look at two kinds of agglomeration economies that help to bind firms together in specialized clusters. The first can be termed *traded* interdependencies (Storper, 1997) and can be accrued through firms co-locating in a cluster alongside suppliers, partners, and customers with which they have formal trading relationships. Proximity to these firms reduces the transaction costs of transportation, communication, information exchange, and searching and scanning for potential customers. The greater the spatial dispersion of firms, the more onerous these costs will be. The consequence is that groups of firms with high interaction costs will choose to co-locate, thus creating a tendency to agglomeration. Hence, we can see that tendencies to agglomerate are inherent in both of the post-Fordist production variants we presented in Chapter 9. On the one hand, the small firm complexes of the flexible specialization model require very close contact with nearby suppliers to facilitate the fine-grained division of labor between firms. On the other hand, under the Japanese flexible production system, suppliers often choose to locate close to customers in order to meet the needs of just-in-time supply systems. It is thus common to find clusters of suppliers surrounding assemblers and manufacturers in sectors where this production system is found (e.g., automobiles in Toyota City, personal computers in China's lower Yangtze River delta). In both cases, a key process underlying the agglomeration of firms is *vertical disintegration*, whereby firms shed many activities and purchase them instead from their suppliers in order to focus on *core competencies* (Scott, 1988). This disintegrative process creates more inter-firm transactional relationships within a particular sector, and hence a strong propensity for spatial agglomeration to minimize the costs associated with sustaining those relationships.

We can illustrate these arguments with a brief exploration of the film and television production agglomeration centered in, and around, Hollywood, Los Angeles (Scott, 2002; see Figure 12.3). Hollywood has long been the center of the global film industry that sells some 5.7 billion tickets and is worth $30 billion each year. The industry employs approximately 200,000 people in Southern California (over 60% of the national total for the sector) and is constituted by more than 5,500 firms. In 2008 California produced 480 films and television shows, generating wages of US$16.7 billion (for more, see http://www.mpaa.org/policy/industry). Figure 12.4 characterizes the nature of this agglomeration in schematic terms. At the core of the system are the immensely powerful Hollywood studios (known

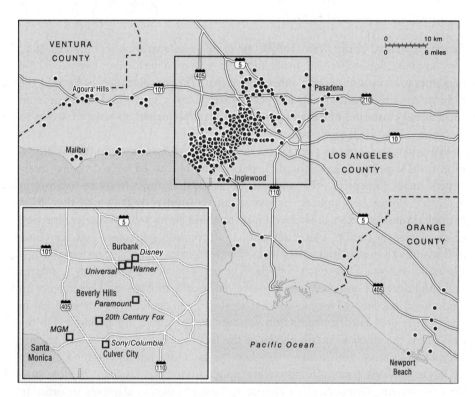

Figure 12.3   The Hollywood film production cluster: location of production companies and major studios
Source: Adapted from Scott A. J. (2002) A new map of Hollywood: the production and distribution of American motion pictures, Regional Studies 36: 957–975, Figure 4.

as the "majors") – Warner Brothers, Paramount, Sony/Columbia, 20th Century Fox, Buena Vista (Disney), and Universal – all of which are now part of global media conglomerates. While in the 1950s the studios were vertically integrated and used to undertake the whole production process in-house, regulatory and technological change meant that by the 1980s, a "new" Hollywood had emerged and still prevails today. The new production system is characterized by vertically disintegrated production networks providing powerful traded interdependencies among three groups of firms: the studios, a large base of independent production firms, and an even larger range of specialized service firms (e.g., script writing, lighting, costumes, catering, and cameras). The studios, however, retain control over the entire production networks through their domination of finance and distribution activities.

As Figure 12.4 shows, there are three other layers to the agglomeration, which suggest that we may also need to look beyond the notion of traded interdependencies. First, there is a local labor market containing a large number of skilled individuals. This labor market is constantly being renewed by the in-migration of talents from all over the world. Second, there is a rich institutional environment comprised of many organizations and associations representing firms

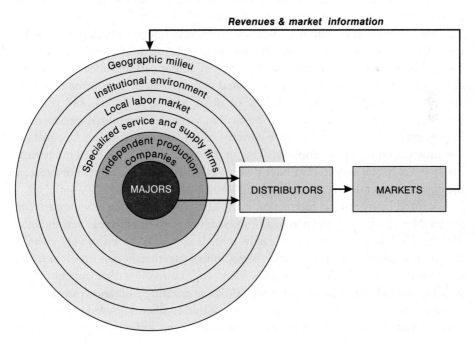

Figure 12.4   Schematic representation of the Hollywood film production agglomeration in Los Angeles
Source: Adapted from Scott A. J. (2002) A new map of Hollywood: the production and distribution of American motion pictures, Regional Studies 36: 957–975, Figure 3.

(e.g., the Alliance of Motion Picture and Television Producers), workers (e.g., the Screen Actors Guild), and government departments (e.g., the Entertainment Industry Development Corporation). These institutions help oil the wheels of the industry and represent its common interests. Third, the whole agglomeration is embedded in a particular place and landscape that, through path-dependent processes, has developed a reputation and image as *the* key place to make films (see also Box 3.5 on California's historical development). Taken together, these various dimensions have created a cluster of dense extra- and inter-firm relationships that offers strong agglomeration economies to its participants, thereby sustaining Hollywood's leadership in the industry and its strong exports to countries around the globe.

12.4   Untraded Interdependencies and Regional Cultures of Production

Hollywood is constituted by more than just tangible transactional relationships between directly related firms, however. It is also a place of gossip, information exchange, deal-making, trust-building, and reputation forging between key decision makers and financiers within the global industry. As such, revealing the economic relationships within an agglomeration may not be enough to explain its formation, scale, and significance. We also need to bring into view the social and cultural bases of economic clusters, or what we might call *untraded interdependencies* (Storper, 1997). These are informal connections that tie firms together, and are constituted by intangible sets of skills, attitudes, habits, and conventions that become associated with particular forms of specialized production. In particular, clusters can facilitate patterns of intense and ongoing face-to-face communication between people working in the same or closely related industries. These interactions may in turn have a very real impact on the success of firms in the cluster by facilitating innovation and knowledge exchange.

Localized interpersonal interaction is thought to be particularly important for the transfer of what is known as *tacit knowledge*. Tacit knowledge is best thought of as know-how that cannot be easily told. It can only be effectively created and shared through people actually doing things together in real life. It is often contrasted with *codified knowledge* – ideas and know-how that can be made tangible through, for example, writing it down or creating a diagram (an example being the book you are reading right now). While codified knowledge (such as a movie production) is able to travel across space, tacit knowledge (e.g., making blockbuster movies) is much stickier geographically, and may only be accessible by being immersed in the untraded social and cultural interactions of a locality. This is turn may attract more skilled workers, further driving the dynamic and creative social environment of particular places (see Box 12.1).

KEY CONCEPT

Box 12.1   The creative class and spatial clustering

In his influential book, *The Rise of the Creative Class*, Richard Florida (2002) charts the rapid emergence of a so-called creative class as the key contemporary employment dynamic affecting the U.S. economy. Florida argues that the creative class tends to cluster in particular places, which he terms *creative centers*. These centers are thriving economically as they offer an attractive quality of place to creative class workers combining elements of what's there (e.g., the built and natural environment), who's there (e.g., social diversity and high levels of interaction), and what's going on there (e.g., vibrancy of street and outdoor life). He argues that it is the "people climate" of places that attracts firms and investment as much as the other way around. Florida's overall prognosis is that for places to succeed economically, they need to offer all three T's of Technology, Talent, and Tolerance. Constructing a *creativity index* from various measures of these dimensions, he found that San Francisco, Austin, San Diego, Boston, and Seattle were the leading creative centers, emerging from a new geographic sorting of classes that is bypassing many places in the United States. Intriguingly, in contrast to accounts of regional economic development that look purely at industrial structure and innovation levels to explain growth rates, Florida argues that a vibrant, diverse, and bohemian social scene is a necessary condition for creativity-led growth.

According to Florida's analysis, the creative class accounts for roughly 30% of the entire U.S. workforce. There are two main components. First, there is a super-creative core of scientists, engineers, professors, entertainers, actors, writers, and the like (12% of the total workforce). Second, there is a wide range of creative professionals working in high-tech sectors, financial services, the legal and healthcare professions, and business management (18% of the total workforce). The class is defined in economic terms as workers who add economic value through their creativity, and covers many of the most highly paid professions in the economy. What the disparate activities supposedly have in common are creative class values based on notions of individuality, meritocracy, and diversity/openness to difference.

It is worth noting, however, that Florida's vision is a vigorously contested one. Concerns have been raised, for example, as to whether his dimensions of creativity are the cause of economic growth, or simply side-effects, and about what happens to the remaining 70% of the workforce if the creative class are given priority in urban policy initiatives. For more see Peck (2005) and Tokatli (2011).

Untraded Interdependencies at Work: The Case of Motorsport Valley

But what specific form do these untraded interdependencies take in reality? The case of Britain's Motorsport Valley provides an interesting window into these dynamics. The term refers to a dense agglomeration of motorsport activity in southern England, and more specifically, an area with a radius of 50 miles around the counties of Oxfordshire and Northamptonshire (northwest of London) (see Figure 12.5). The majority of the British motorsport industry – which has some 40,000 employees, 25,000 of which are highly skilled engineers, and 4,500 firms – is located in this area. The industry has an annual turnover of £6 billion (or US$9.6 billion), of which 60% is accounted for by exports, and spends an incredible 30% of turnover on research and development activities. Over 75% of the single-seat racing cars used around the world are U.K.-built, and eight of the twelve Formula 1 teams that competed in the 2010 season were U.K.-based. Motorsport Valley acts as the global hub for the industry, with drivers, expertise, engineers, components, money, and sponsorship coming from all over the world (for more, see www.the-mia.com/The-Industry). What is of interest to us here is the ways in which specialist knowledge circulates among the hundreds of small firms that constitute the Motorsport Valley *outside* of formal business transactions – that is, as untraded interdependencies. It is possible to identify six interacting ways that knowledge is disseminated (drawing on Pinch and Henry, 1999):

- *Staff turnover*: The ongoing circulation of key personnel – engineers, drivers, designers, and mechanics – means that crucial information about how things are done is transferred between firms and teams. Analysis of career trajectories has revealed that an average designer or engineer will move firm every 3.7 years, and eight times during their career.
- *Shared suppliers*: Another mechanism for knowledge transfer is through the links that motorsport teams have with numerous component and services suppliers. While confidentiality agreements are in place to limit the transfer of successful ideas from one team to another through a common supplier, in reality there is a tendency for technical know-how to leak, as suppliers want to provide their clients with the best and most up-to-date advice.
- *Firm births and deaths*: Motorsport is a very expensive and risky economic enterprise. As a result, the industry is characterized by high levels of firm births and deaths. Each of these individual events represents an opportunity for staff (re)mixing and hence the (re)diffusion of knowledge.
- *Informal collaboration*: Racing car companies are on the one hand extremely competitive and secretive, yet on the other hand are involved in a tightly regulated collective endeavor that necessitates interaction and involvement in working groups. Collective discussion of the regulations, and how to respond

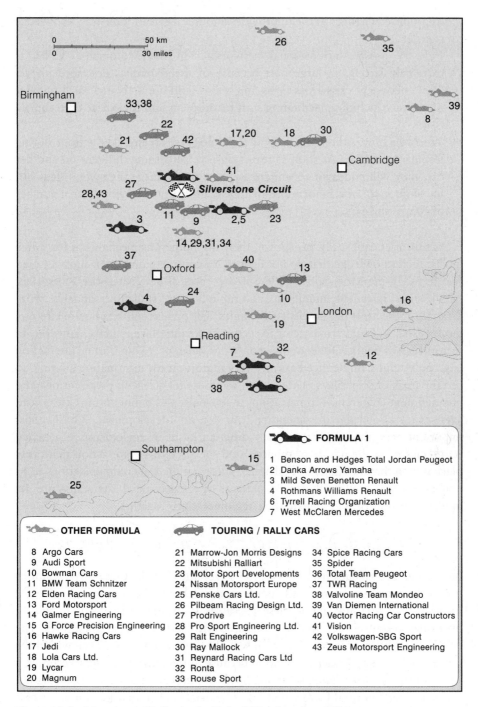

Figure 12.5 Motorsport Valley in the United Kingdom, late 1990s

Source: Adapted from Pinch, S. and Henry, N. (1999) Paul Krugman's geographical economics, industrial clustering and the British motor sport industry, Regional Studies, 33: 815–827, Figure 1.

to proposed changes in them, provides another mode of inter-firm knowledge transfer.

- *Industry gossip*: Interpersonal networking in the Motorsport Valley is extremely strong, in large part because of the dynamics described above. These networks criss-cross firms and teams, and are activated for a number of reasons, including facilitating staff recruitment and for advice on technical issues.
- *Trackside observation*: Racing car technology is revealed on the track during testing and racing. As soon as teams observe something different on another car, they will endeavor to imitate and test the changes themselves. It is not uncommon for new innovations to appear almost simultaneously on a number of competing cars.

As this brief case study has shown, there are numerous mechanisms for transferring tacit knowledge that do not depend on contractual relationships between two firms. In the Motorsport Valley, strong rates of new enterprise generation, high levels of inter-firm interaction, and rapid flows of skilled personnel between firms are at the heart of knowledge dynamics. The precise combination of effective mechanisms will vary from sector to sector, and from place to place, however. In sectors such as finance, for example, non-workplace spaces of informal interaction (e.g., pubs, clubs, and coffee bars) are very important for information-sharing. In certain areas of scientific research (such as computer science), presentations and interaction at conferences, publications in journals, and online forums are crucial avenues for knowledge dissemination, transfer, and development. What is most important here, though, is that many different kinds of agglomeration generate economies from the nature of the localized social-cultural interaction that takes place within them. In an era of advanced information and communications technologies, it may well be that these patterns of intense social interaction are the most significant form of glue holding agglomerations together.

Untraded Interdependencies at the Regional Scale

But what if we step back from a particular sector and think about the preconditions for the development of untraded interdependencies at the urban or regional level? Another way to think about these processes is through the extent to which, over time, certain places become associated with certain industrial cultures. The existence of different local or regional sets of business practices – even within the same country – has attracted attention because of the purported link to economic success through, for example, varying levels of innovation and entrepreneurship. The social and cultural factors that lie at the heart of regional economic success can be encapsulated in the notion of *institutional thickness* (Amin and Thrift, 1994). The term "institution" is used here to refer to a wide variety of non-firm organizations that are an integral part of regional economies, ranging from educational and healthcare institutions (e.g., schools and hospitals) to political and economic institutions (e.g., governments and central banks). Some regions appear to be endowed with a beneficially "thick" range of institutions that in turn engenders pro-growth economic cultures and a strong regional capacity for innovation (Storper, 1997). This thickness is constituted by four overlapping elements:

- The range of institutions seeking to promote growth in a particular region.
- The level and effectiveness of the cooperation among those institutions.
- The emergence of a collective voice for the region when bargaining for resources with authorities at the national and supra-national scales.
- The development of a sense of common enterprise among the firms and institutions within the regional economy.

In the most favorable circumstances, the outcome of institutional thickness will be a regional economy characterized by dynamic, flexible institutions, ongoing innovation, high levels of trust, and effective knowledge circulation. Some geographers have used the term *learning region* to describe these successful territorial ensembles of production and innovation. Important regional institutions may emerge in the form of research centers and universities, other educational and training institutions, local associations of producers, chambers of commerce, business associations, and technology-promotion agencies. The ongoing interaction of these institutions with local businesses and workers gradually creates regionally specific industrial practices or industrial cultures.

These arguments are well illustrated by a comparative study of Silicon Valley in California and Route 128 near Boston (drawing upon Saxenian, 1994; see also Figure 12.6). The key question is why, during the 1980s and the early 1990s, certain industrial sectors (software and electronics, in particular) flourished in California's Silicon Valley, while along Route 128 in Massachusetts they declined. Despite similar histories and technologies, Silicon Valley developed a decentralized but cooperative industrial system while Route 128 came to be dominated by independent, self-sufficient corporations. The two regions in the same country apparently had different *regional cultures*, characterized by more cooperative *inter*-firm relationships in Silicon Valley, and greater reliance on *intra*-firm vertical integration along Route 128.

As early as the 1940s, the Boston area in Massachusetts hosted a sizable group of electronics manufacturers. At around the same time, the Santa Clara Valley (today's Silicon Valley) in California was mostly an agricultural region famous for its apricot and walnut orchards. Much was to change in the ensuing four decades. But how did Massachusetts, home to world-class educational institutions such as Harvard and MIT, lose out to Silicon Valley by the 1980s as the world's leading region in the electronics industry? In short, engineers and other professionals in Silicon Valley were much more entrepreneurial in their business culture. Moving

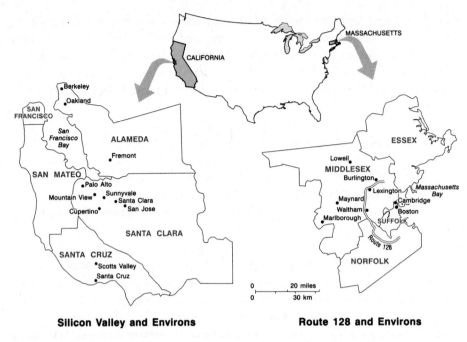

Silicon Valley and Environs   **Route 128 and Environs**

Figure 12.6   Silicon Valley and Route 128 corridor, United States

frequently from job to job, or quitting a job to start a new firm, were both seen as perfectly acceptable in Silicon Valley's cultural environment. In Massachusetts, however, the industrial culture was very much big firm–centered in the sense that engineers and professionals were expected to remain working for the large firms that hired and nurtured them. The kind of labor market flexibility and mobility that prevailed in Silicon Valley was generally seen as disloyal and ungrateful in this context.

As a consequence of this difference in regional industrial cultures, there was much more cooperation and collaboration among Silicon Valley firms than among their counterparts in Massachusetts. Silicon Valley firms collaborated with one another in formal and informal ways: developing alliances, contracting for components and services, and simply sharing information. Employees of different firms mixed frequently at local business and social gatherings. In contrast, Massachusetts firms were highly secretive and self-contained, and employees had fewer inter-firm contacts. The industrial culture in Silicon Valley was much more relaxed about sharing of information and skills, whereas firms in Massachusetts anxiously sought to protect their intellectual property and prohibited inter-firm sharing of information and knowledge. As we saw earlier, venture capitalists in Silicon Valley have also played a critical role in transferring skills and knowledge among firms, while in Boston, where more traditional sources of finance were used, the banks supplying such capital simply did not have technical expertise to advise and promote start-ups.

In sum, these contrasts between Silicon Valley and Route 128 illustrate the differing regional industrial cultures that can emerge, and the implications of these differences for entrepreneurship and innovation. The contrasts emerged gradually over time but eventually became very clear distinctions in the networks of knowledge circulation and support for small enterprises that existed in each place. Once such practices are entrenched, they tend to endure, and success stories such as Silicon Valley become examples that other localities try to emulate. Indeed, fostering a local culture of innovation and learning has become a celebrated cause for many local and national governments (see Box 12.2).

CASE STUDY

Box 12.2   High-tech clusters in Asia

Over the past two decades, a wide range of local, regional, and national governments across Asia have sought to develop high-technology clusters based in particular around information and biotechnologies. We will use two brief examples here to illustrate the differing size and scope of these initiatives. Singapore's *one-north* is a 200-hectare development a few kilometers to the west of central Singapore and close to the Science Park, National University of Singapore, National University Hospital, and Singapore Polytechnic (see Figure 12.7). Conceived in the mid-1990s and started in 2001, one-north is a 20-year, US$7 billion multiphase project that seeks to offer world-class research facilities and business parks to support growth in the biomedical, information technology, media, and engineering sectors. The project has several elements: Biopolis, a set of facilities for the

Figure 12.7   Fusionopolis and one-north, Singapore

biomedical sciences, which opened in two phases in 2003 and 2006 and was home to 2,000 workers by 2010; Fusionopolis, designed to house information technology, media, and engineering activities, the first phase of which was opened in 2008 and housed 2,500 workers by 2010; Vista Xchange, a forthcoming office, retail, and leisure complex; Mediapolis, a proposed media hub; and Wessex Estate, envisaged as a living and working area for creative workers. The scheme has been managed by the Jurong Town Corporation (JTC), a quasi-governmental property developer, although it also overseen by several other government departments including the Economic Development Board and Urban Redevelopment Authority of the Singapore government. Overall, the plan is to integrate business, research, residential, retail, and leisure activities: "the one-north masterplan consciously integrates educational institutes, residences and recreational amenities with research facilities and business parks, to create an ideal work-live-play-learn environment where creative minds congregate and groundbreaking ideas are born" (http://www.jtc .gov.sg/product/one-north, accessed November 26, 2010).

One-north is dwarfed, however, by the development of China's *Zhongguancun Science Park* in Beijing, or Z-Park, a sprawling complex of ten industrial parks covering over 100 square kilometers primarily to the northwest of the city center. The scale of Z-Park is astonishing: by late 2008 the district was home to an estimated 20,000 high-tech companies, including over 1,500 foreign firms, employing some 1 million people and with a combined turnover of US$150 billion of which US$20 billion was accounted for by exports. Z-Park is the biggest and oldest of China's 50+ national high-tech zones. It dates to 1980, when a Chinese Academy of Sciences researcher returned from a trip to Silicon Valley and established a private technology consulting firm, and was officially established as a national science park in 1988 (http://www.zgc.gov.cn/english, accessed November 26, 2010). The role of the state in relation to the Z-Park has changed over time: in the 1980s there was little state involvement, in the 1990s the Park was "discovered" and supported heavily by Beijing's municipal government, while by the 2000s Zhongguancun had achieved national prominence and was promoted by the central government as a cluster of global significance. Haidian Park is the most important of the ten parks, focusing on R&D and the incubation of firms in industries such as electronics, information technology, biopharmaceuticals, and a variety of new energy and materials sciences. Lenovo, Digital China, Kingsoft, and Tencent are major tenants. Haidian District, where the park is located, is home to 39 universities – including Peking University and Tsinghua University – and over 200 research institutes, 41 national engineering centers, and ten national technical centers. For more on the two cases of one-north and Z-Park, see Wong and Bunnell (2006) and Zhou (2008), respectively.

12.5 Toward a Typology of Clusters?

There is a wide variety of different kinds of agglomerations in the contemporary economy, held together by varying combinations of traded and untraded interdependencies. A large city such as New York, for example, offers general agglomeration economies at the metropolitan scale to a wide variety of different economic activities. Look within the city, however, and you will find many more focused clusters of activity offering additional specialized agglomeration economies of both the traded and untraded varieties to firms in the financial sector (Wall Street district), advertising (Madison Avenue), and the fashion industry (midtown Manhattan), to name but a few. Recognizing this variability in terms of geographical scale, sectoral composition, and causal dynamics is vital if economic geographers are to understand effectively the contemporary space economy. In this context we can usefully think of the following eight-fold typology of significant cluster types:

- *Labor-intensive craft production clusters*: These are often found in industries such as clothing, where work is characterized by sweatshop conditions and often very high levels of immigrant labor. Firms are involved in tight subcontracting networks, and may also use home-workers. Examples include the garment production districts of Los Angeles, New York, and Paris.
- *Design-intensive craft production clusters*: These refer to dense agglomerations of small and medium-sized firms specializing in the high-quality production of a particular good or service. They are characterized by a highly disintegrated production system in which individual firms perform specialized and narrowly defined roles. Examples include the renowned towns and districts of the Third Italy – for example, Arezzo (gold jewelry), Carpi (knitwear), Sassuolo (ceramics), and Ancona (shoes) – and Jutland, Denmark (furniture) and Baden Württemberg, Germany (machine tools).
- *High-technology innovative clusters*: These are characteristic of "new" after-Fordist sectors such as computers and biotechnology. These clusters tend to have a large base of innovative small and medium-sized firms and flexible, highly-skilled labor markets. They have often grown up in areas with little history of industrialization and unionization. Examples include Silicon Valley and Route 128, Boston, districts in the United States; Grenoble, France; and Cambridge, United Kingdom.
- *Flexible production hub-and-spoke clusters*: In these clusters, a single large firm, or small group of large firms, buys components from an extensive range of local suppliers to make products for markets external to the cluster. These clusters represent the spatial logic of just-in-time production systems. Examples include Boeing in Seattle, United States, and Toyota in Toyota City, Japan.

- *Production satellite clusters*: These clusters represent congregations of externally owned production facilities. These range from relatively low-tech assembly activity, through to more advanced plants with research capacity, but all are relatively stand-alone. Firms have generally co-located to access the same labor market conditions or financial incentives. Examples are to be found across the export processing zones (EPZs) of the developing world. The electronics industry on the Malaysian island of Penang is a specific instance of this kind of cluster.

- *Business service clusters*: Business services activities such as financial services, advertising, law, accountancy, and so on are often concentrated in the central districts of leading or "global" cities (e.g., New York, London, and Tokyo), and in some cases, their hinterland regions (e.g., software and computer services in the Western Arc of counties outside London).

- *State-anchored clusters*: A disparate category of clusters has developed due to the location decisions of government facilities, such as universities, defense industry research establishments, prisons, and government offices. Examples include agglomerations that have developed due to government research investment (Colorado Springs, U.S.; Taejon, South Korea; M4 Corridor, U.K.; Singapore's Fusionopolis in Figure 12.7) and universities (Madison, Wisconsin, U.S., and Oxford/Cambridge, U.K.).

- *Consumption clusters*: There are also strong propensities to cluster – often in central urban areas – in a wide variety of consumer service activities including retailers, bars and restaurants, and cultural and leisure facilities. The theater districts of the London's West End and New York's Broadway offer specific examples, as do the many retail and entertainment districts of Tokyo (e.g., Shibuya) and Shanghai (e.g., Xin Tian Di). Tourist activities more generally also tend to congregate around different kinds of amenities (for more see Chapters 11 and 15).

In reality, of course, applying any such typology is not an easy task. Some places are constituted by clusters that are *hybrid* forms merging the characteristics of two or more of these types. For example in Figure 12.8, Silicon Valley is effectively a cluster of clusters, variously exhibiting the traits of a high-technology innovative cluster, a flexible production hub-and-spoke cluster (due to the presence of large manufacturers such as Intel), and a state-anchored district (due to the significance of huge levels of government defense spending). In addition to this complexity, Silicon Valley is not just defined by local interactions. Its firms are also very well plugged into a variety of important national and global connections with partners, suppliers, and customers. In other words, clusters are not necessarily shaped by local forces only. We now move on to consider how we might conceptualize clusters as being defined by patterns of *both* local and nonlocal interaction.

Figure 12.8   A multifaceted cluster? High-tech business in Silicon Valley, California

12.6   Rethinking Proximity

As this chapter has made clear, economic geographers continue to be fascinated by the importance of spatial proximity – that is, people being in the same place at the same time – in enabling socioeconomic interaction. Yet how do we reconcile the apparent importance of spatial proximity with what we know about globalization and the power of various information and communication technologies to transcend space and time? In order to make sense of this apparent paradox, we need to refine our analysis in three ways.

First, we need to recognize that the concentration of certain economic activities in particular places is part of the very same processes that are seeing other kinds of activities dispersed across the global economy. As we saw in chapters 8 and 9, within the production system for a particular product or service, there are some operations such as corporate decision making that need to be located in agglomerations, while other activities such as production, sales and marketing, and customer support can be more spatially disparate. Put another way, the spatial decentralization of economic activity facilitated by space-shrinking technologies *requires* the very kinds of control and coordination that can only be achieved through locating certain important corporate functions in key global cities. More specifically, this dispersion creates different challenges of integration and coordination for firms (Amin and Thrift, 1992). Information has to be acquired about what is going on across the production system as a whole. Social interaction is required to allow important interpersonal relations such as trust to develop. Finally, firms somehow have to keep up with innovations in both products and processes within their industry. Clusters offer a solution to these problems by acting as centers of tacit knowledge transfer and control within wider global networks, for example through the interaction of corporate headquarters, government departments, and industry-specific media. Clusters act as centers of sociability, places where interpersonal relationships are created and maintained. And they act to provide a critical mass to help generate innovations through the intense ongoing interaction of numerous users and suppliers of technologies. Such processes can be seen to underpin the continued dominance of leading financial centers such as London and New York, for example, which if anything has increased over time despite processes of globalization and innovation in information technologies (see Chapter 7; also Thrift, 1994). Clusters, then, should not be considered on their own, and purely in terms of the local relationships they contain, but as *nodes in global networks*, where certain kinds of local relationships are used to facilitate and support other kinds of nonlocal (and possibly global) relationships (see Box 12.3 for an extended critique of the cluster concept).

FURTHER THINKING

Box 12.3   Clusters – the mesmerizing mantra?

Although the idea of clusters has become extremely influential in both academic and policy circles, its ascendance has not been without critique. Taylor (2010) goes as far as to argue that the notion of clusters has become a "mesmerizing mantra" that obscures meaningful understanding of the growth dynamics of local agglomerations. There are several elements to these critiques:

- *Slippery scales*: The geographical scale at which clustering is seen to occur varies widely from small districts within cities (e.g., Soho in London and Tribeca in New York) through city-regions as a whole (e.g., Hong Kong and Shanghai) to small national economies (e.g., Singapore and Ireland). Precision is required as to the precise spatial scale at which clustering dynamics take place or the concept risks becoming meaningless.
- *Capitalist imperatives*: Many clustering accounts focus on the positive aspects of community, collaboration, shared goals, and win-win interactions. The capitalist imperatives of competition, prices, profits, and unequal power relationships – equally if not more important to the success of a cluster – can seemingly be neglected.
- *The time dimension*: While studies may describe the characteristics of clusters at particular points in time, they are generally far less effective at explaining (a) how the cluster was initiated in the first place and

(b) how the cluster has evolved over time. For example, some clusters may become trapped in outmoded forms of production while others are nimble enough to adapt to market and technological changes.

- *Overprioritizing "the local"*: There is a real danger that cluster accounts overemphasize the importance of local economic interconnections at the expense of connections over wider geographies – for example, at the national or international scales. Often the significance of local connections is presumed rather than demonstrated empirically, while there is counter-evidence that forms of "sticky" tacit knowledge can be effectively transferred across clusters – that is, within TNCs.

- *Underplaying entrepreneurship*: Studies of clusters often underplay the role of entrepreneurship and new firm formation processes in sustaining clusters. Localized processes of firm fragmentation and spin-offs can be a key contributor to clustering processes. At the same time, the real-world problems of commercializing new ideas are often absent in cluster approaches.

- *Cluster interactions*: Views also vary on the nature and importance of links between individual industry clusters within wider agglomerations. The notion of *related variety* has been developed, for example, to capture the potential benefits of loosely connected clusters in cognate industries driving wider growth. Such variety can offset the inherent danger of lock-in and stagnation in overly specialized clusters.

- *Chaotic conceptions*: Some of the concepts underpinning cluster theory, for example notions of "institutional thickness" and "social capital,"are in some ways quite fuzzy – that is, difficult to operationalize and measure. Critics of institutional thickness, for example, point out that the links between thickness and economic success are far from clear: many struggling regions have very rich institutional infrastructures whereas strong growth regions may have quite the opposite.

See Cumbers and MacKinnon (2006) for more on the various pros and cons of the cluster concept.

Second, we can further open up our economic-geographical interpretation of clusters by thinking of spatial proximity as just one of several different kinds of proximity (or a sense of nearness) that enable the global economy to function. Additionally, we can think of the following:

- *Institutional proximity*: Nearness created through operating within the same legal and institutional frameworks as others (e.g., within the German economy and the EU).

- *Cultural proximity*: Nearness derived from a shared cultural heritage and linguistic background (e.g., the Irish, Jewish, or Chinese diasporas; for more see Chapter 14).
- *Organizational proximity*: Nearness engendered through both written rules and codes, and unwritten ways of doing things within a particular firm or institution (e.g., the corporate culture of a large transnational firm – see Chapter 10).
- *Relational proximity*: Closeness derived from informal interpersonal relations (e.g., long-standing friends in distant places).

In some situations, these other forms of proximity may be as important as, if not more important than, physical proximity. Relational proximity, for example, may facilitate the same transfer of tacit knowledge that is argued by many to depend on physical proximity. While relational proximity may depend to a certain extent upon face-to-face interaction, it can also be achieved through modern communications technology, and through the mobility of knowledgeable individuals. In this way, nearness is achieved through combinations of interpersonal and electronic communication, and travel. Similarly, organizational proximity allows TNCs to bridge across clusters in different national contexts through both the use of technologies – for example, knowledge management systems designed to act as repositories of knowledge and share best practices – and the circulation of key staff such as managers and trainers between different sites. Firms also develop partnerships with firms in other clusters to achieve similar benefits. In reality, the "buzz" of local clusters is also combined with various global pipelines (see Figure 12.9) that connect and transfer knowledge between clusters.

Third, we can think about the different temporalities or timeframes of clustering processes. Put another way, do firms have to be permanently located in clusters in order to benefit from the associated agglomeration economies and knowledge spillovers? Recent research has suggested that there are forms of *temporary* cluster as well as the more permanent forms usually recognized in the literature (see Figure 12.10). On the one hand, *project-based* forms of economic organization are becoming increasingly prevalent across a range of industries, with specialized teams of workers being brought together, from both within and across different firms, for a given time period to work on a particular task (see Box 12.4 for more). On the other hand, it can be argued that professional gatherings such as conferences, conventions, and trade fairs can be viewed as temporary clusters in that they exhibit some of the knowledge transfer mechanism found in clusters, but in a more short-lived and intensive form (Bathelt and Glückler, 2011). Firms can use such events to monitor the latest technological and market developments and to build relationships to help access distant markets and knowledge pools – in this way, trade fairs help to forge new global pipelines. Some of the interaction is preplanned and strategic, but much will be accidental, with chance meetings occurring in the corridors, cafés, bars, and restaurants surrounding the formal

FURTHER THINKING

Box 12.4 Project working

While projects – the bringing together of a set of people to fulfill a complex task – are a long-established way of working in industries organized around one-off schemes such as architecture, advertising, construction, ship-building, and film-making, there is evidence that they are spreading into new sectors (e.g., automobiles, chemicals), while many new industries (e.g., software, new media, and business consulting) are dominated by project working. Projects can be thought of as temporary social systems in which people of varied professional and organizational backgrounds work together in an environment that is insulated from the other ongoing activities of the firm or organization. Projects can vary greatly in their constitution, duration, and location. In terms of their constitution, they may be made up of a range of different skilled professionals from within a single corporation or from a range of different firms. In the latter case, the project team can be thought of as a virtual firm. In terms of their duration, a project may range from a few weeks (e.g., the filming of a TV movie-of-the-week on location) to many months and even years (e.g., a large-scale software project for a major multinational client).

In terms of the location, a project team may be based within a firm's office network, on a client's site, at a neutral site, or even in cyberspace (as when a team of far-flung specialists works simultaneously or in sequence on a particular piece of software or an architectural blueprint). Many will combine both face-to-face and electronically mediated forms of communication. While project members may be drawn from within a single office building or a national subsidiary of a company, firms are increasingly assembling *transnational* project teams to meet the needs of their clients. Whatever the precise configuration of different project teams may be, two things are certain. First, projects represent an increasingly pervasive form of working that is extremely flexible in both time and space, and one that facilitates the transfer of various forms of tacit know-how within and across organizations. Second, projects pose an important challenge to economic geographers by blurring the boundaries of key analytical categories such as "firms" and "clusters." Indeed, even just positioning where economic activity is taking place can be problematic in the world of project working. See Grabher (2002) and Sunley et al. (2011) for more on project working in the advertising industry and Tokatli (2011) for the fashion design industry.

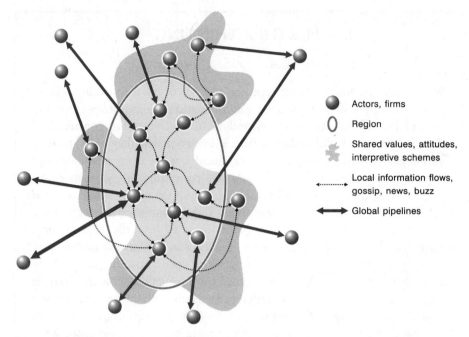

Figure 12.9   Local "buzz" and global pipelines
Source: Adapted from Bathelt, H., Malmberg, A. and Maskell, P. (2004) Clusters and knowledge: local buzz, global pipelines and the process of knowledge creation, Progress in Human Geography, 28: 31–56, Figure 1.

		Time dimension	
		Semi-permanent	Temporary
Focus of knowledge exchange	Strongly focused	Stable inter-firm networks	Inter-firm projects
	Broadly focused	Clusters	Trade fairs, conventions, professional gatherings

Figure 12.10   Temporary clusters?
Source: Adapted from Maskell, P., Bathelt, H. and Malmberg, A. (2006) Building global knowledge pipelines: the role of temporary clusters, European Planning Studies, 14: 997–1013, Table 1.
Copyright © 2004, Sage Publications.

meetings. The leading global events are truly massive: for example the CON-EXPO construction industry trade fair in Las Vegas in 2008 attracted 2,182 exhibitors and 144,600 attendees from over 130 countries across five days, while the Light + Building architecture and building technology event in Frankfurt in 2010 drew in 2,177 exhibitors and 180,000 visitors, again over a five-day period. Below these big headline global gatherings, however, are thousands of regional, national, and macro-regional meetings all performing a similar function in terms of the circulation of knowledge and market intelligence within different business groupings. Trade fairs in turn are big business for popular destination cities such as Las Vegas, which may play host to several hundred such meetings a year.

The Vancouver film and television industry offers an interesting window on the dynamics we are describing here (Barnes and Coe, 2011). The west coast Canadian city is widely considered to be the third largest production center in North America after Los Angeles and New York, generating annuals revenues of around CDN$1 billion and 23,900 full-time equivalent jobs across some 2,000 independent small and medium-sized firms. These firms and workers come together in different configurations to produce the 250 or so film and television shows made in the city each year. As its nickname "Hollywood North" indicates, however, Vancouver's fortunes are closely connected to those of the Hollywood film and television complex, with inward investment accounting for between 60% and 80% of total spending each year. The genesis and emergence of the Vancouver film industry has largely depended on the spending decisions of the large U.S. studios and media conglomerates introduced earlier in the chapter. These firms often have complete financial and creative authority over the projects filmed in Vancouver, with control exercised indirectly through subcontracting networks rather than through direct ownership.

In some instances, key production personnel are sent to Vancouver to oversee or manage particular projects through a temporary subsidiary. In others, the complete production process is subcontracted to a Vancouver production company. The small firm networks of the Vancouver film industry thus rely ultimately on funding from U.S. studios and TV networks. The key group that engages with financial decision makers in Los Angeles are Vancouver-based producers. Over the past thirty years, many have built strong personal relationships with Hollywood studio executives. In this sense, they, along with studio managers and a few government officials, are crucially embedded in extra-local personal relationships that facilitate the continued flow of capital from Los Angeles to Vancouver. In turn, the initiation and nurturing of these relationships takes place at film festivals and industry meetings across North America, including, for example, the Vancouver and Los Angeles film festivals. On face value then, Vancouver's film and television industry looks like a relatively typical and stable cluster of interconnected firms. In reality, however, it is heavily project-based, is constituted by both local and nonlocal relationships, relies on the circulation of key individuals between the city and Los Angeles, and requires ongoing participation in other temporary forms of cluster. Connecting back to our discussion of Hollywood earlier in the chapter (Section 12.3), we also need to reframe our thinking about that cluster; rather than seeing it is as self-contained, it should be thought of as the preeminent control and finance node in the global networks of film and television production.

12.7   Summary

It is well over a century since the economist Alfred Marshall (1890, p. 225) famously argued the following:

> When an industry has thus chosen a locality for itself, it is likely to stay there long: so great are the advantages which people following the same skilled trade get from near neighbourhood to one another. The mysteries of the trade become no mysteries; but are as it were in the air ... Good work is rightly appreciated, inventions and improvements in machinery, in processes and the general organization of the business have their merits promptly discussed: if one man starts a new idea, it is taken up by others and combined with suggestions of their own; and thus it becomes the source of further new ideas.

These words are still remarkably relevant today. The forces driving the clustering of economic activities are as powerful as ever, and the economic landscape continues to be characterized by overlapping and interconnected clusters of specialized activity. Large urban areas offer firms powerful agglomeration economies deriving from the access to shared resources and markets, while within urban centers, specialized clusters allow firms to reap further economies within their particular area of activity. Such economies can be both traded and untraded in nature. Post-Fordist production techniques have created a renewed propensity for related firms to agglomerate in order to reduce the costs of transacting with one another. In many industries, however, firms also come together to benefit from the local "buzz" and ongoing processes of social interaction that allow crucial forms of knowledge to circulate. As Marshall evocatively describes, such benefits seem to be in the air of successful clusters. Over time, such dynamics can stabilize into regional cultures of production, such that certain places are associated with certain ways of doing business. The precise combination of agglomeration economies will differ from sector to sector and from place to place, with the result that there is a wide variety of cluster forms in the contemporary global economy. For example, a central business district of a global city and a coastal export processing zone both represent clusters of economic activity; they are very different, however, in their constitution and in terms of the forces driving and sustaining agglomeration.

In turn, it is important to recognize the potential limitations of the clustering concept. Clusters need to be seen as open nodes in global networks, with globalization processes meaning that clusters are more intensely interconnected than ever before. Economic activity clearly does not only take place through local economic and social interactions. Rather it is constituted by complex combinations of local and nonlocal relationships, of face-to-face contact and electronically mediated communication, of people who largely remain in one place and those who circulate between places, and of tangible production processes and intangible exchanges of knowledge. Unpacking this complexity is one of the key challenges of Economic Geography. It is clear, however, that purely economistic interpretations underestimate the extent to which the economy is a social system at the same time as it is an economic one.

Notes on the references

- Dicken and Lloyd (1990) offer a comprehensive yet accessible guide to the classical theories of industrial location.
- Storper (1997) explores the various economic, social, and cultural factors behind regional agglomerations.
- The following studies provide more detail on Hollywood (Scott, 2002, 2005), Motor Sport Valley (Pinch and Henry, 1999), Vancouver (Barnes and Coe, 2011), London (Sunley et al. 2011), and New York/Paris (Tokatli, 2011). For an excellent overview of clusters in the Latin American context, see Pietrobelli and Rabellotti (2007).
- Two edited collections – Cumbers and MacKinnon (2006) and Asheim et al. (2006) – offer wide-ranging discussions and examples of the cluster concept from an Economic Geography perspective.
- Bathelt and Glückler (2011) develop the notion of trade fairs as temporary clusters in more detail.

Sample essay questions

- What are the different factors that may cause agglomerations to develop?
- In what ways, and to what extent, is spatial proximity important to economic processes?
- How can we explain the continued dominance of global cities within the wider space economy?
- What are the limitations to the cluster concept for explaining contemporary economic development processes?
- What role can temporary clusters play in processes of knowledge generation and transfer?

Resources for further learning

- http://www.isc.hbs.edu/econ-clusters.htm: Explore Michael Porter's website on clusters and cluster development at Harvard Business School.
- http://www.creativeclass.com: See this site for a wide range of materials on Richard Florida's controversial ideas about the creative class.
- http://www.lboro.ac.uk/gawc: See the *Globalization and World Cities* (GaWC) website hosted by Loughborough University for a huge range of resources on world cities and the networks that connect them.

- http://www.jointventure.org: The Joint Venture Silicon Valley Network provides a range of perspectives on this world-renowned high-tech cluster.
- http://www.cityoflondon.gov.uk/Corporation: A good website for finding out more about one of the world's preeminent finance clusters. http://www.thecityuk.com is another rich source of data on the City of London.
- http://www.unido.org/index.php?id=o4297: The website of UNIDO's clusters and networks development program.
- http://www.clusterobservatory.eu: Explore clusters on the European scale at the website of the European cluster observatory.

References

Amin, A., and Thrift, N. (1992). Neo-Marshallian nodes in global networks. *International Journal of Urban and Regional Research*, **16**, 571–87.

Asheim, B., Cooke, P., and Martin, R., eds. (2006). *Clusters and Regional Development: Critical Reflections and Explorations*. New York: Routledge.

Barnes, T., and Coe, N. M. (2011). Vancouver as media cluster: the cases of video games and film/TV. In C. Karlsson and R. G. Picard, eds., *Media Clusters: Spatial Agglomeration and Content Capabilities*, pp. 251–77. Cheltenham: Edward Elgar.

Bathelt, H., and Glückler, J. (2011). *The Relational Economy: Geographies of Knowing and Learning*. Oxford: Oxford University Press.

Bathelt, H., Malmberg, A., and Maskell, P. (2004). Clusters and knowledge: local buzz, global pipelines and the process of knowledge creation. *Progress in Human Geography*, **28**, 31–56.

BBC News. (2010, August 27). 165 University Avenue: Silicon Valley's "lucky building."

Cumbers, A., and MacKinnon, D., eds. (2006). *Clusters in Urban and Regional Development*. London: Routledge.

Dicken, P., and Lloyd, P. E. (1990). *Location in Space: Theoretical Perspectives in Economic Geography*, Third Edition. New York: Harper & Row.

Florida, R. (2002). *The Rise of the Creative Class*. New York: Basic Books.

Grabher, G. (2002). The project ecology of advertising: tasks, talents and teams. *Regional Studies*, **36**, 245–62.

Markusen, A. (1996). Sticky places in slippery spaces: a typology of industrial districts. *Economic Geography*, **72**, 293–313.

Marshall, A. (1890). *Principles of Economics*. London: Macmillan.

Maskell, P., Bathelt, H., and Malmberg, A. (2006). Building global knowledge pipelines: the role of temporary clusters. *European Planning Studies*, **14**, 997–1013.

NVCA. (2010). *National Venture Capital Association 2010 Yearbook*. Arlington, VA: NVCA.

Peck, J. (2005). Struggling with the creative class. *International Journal of Urban and Regional Research*, **29**, 740–70.

Pietrobelli, C., and Rabellotti, R., eds. (2007). *Upgrading to Compete: Global Value Chains, Clusters and SMEs in Latin America*. Cambridge, MA: Harvard University Press.

Pinch, S., and Henry, N. (1999). Paul Krugman's geographical economics, industrial clustering and the British motor sport industry. *Regional Studies*, **33**, 815–27.

Scott, A. J. (1988). Flexible production systems and regional development: the rise of new industrial spaces in North America and Western Europe. *International Journal of Urban and Regional Research*, **12**, 171–86.

Scott A. J. (2002). A new map of Hollywood: the production and distribution of American motion pictures. *Regional Studies* **36**, 957–75.

Scott A. J. (2005). *On Hollywood: The Place, the Industry.* Princeton, NJ: Princeton University Press.

Storper, M. (1997). *The Regional World.* New York: Guilford Press.

Sunley, P., Pinch, S., and Reimer, S. (2011). Design capital: practice and situated learning in London design agencies. *Transactions of the Institute of British Geographers*, **36**, 377–92.

Taylor, M. (2010). Clusters: a mesmerising mantra. *Tijdschrift voor Economische en Sociale Geografie*, **101**, 276–86.

Thrift, N. (1994). On the social and cultural determinants of international financial centres: the case of the City of London. In S. Corbridge, R. Martin, and N. Thrift, eds., *Money, Power and Space*, pp. 327–55. Oxford: Blackwell.

Tokatli, N. (2011). Creative individuals, creative places: Marc Jacobs, New York and Paris. *International Journal of Urban and Regional Research*, **35**, 1256–71.

Wong, K. W., and Bunnell, T. (2006). "New economy" discourse and spaces in Singapore: a case study of one-north. *Environment and Planning A*, **38**, 69–83.

Zhou, Y. (2008). *The Inside Story of China's High-Tech Industry: Making Silicon Valley in Beijing.* Lanham, MD: Rowman and Littlefield.

Zook, M. A. (2004). The knowledge brokers: venture capitalists, tacit knowledge and regional development. *International Journal of Urban and Regional Research*, **28**, 621–41.

CHAPTER 13

GENDERED ECONOMIES
Does gender shape economic lives?

Goals of this chapter

- To show how certain kinds of work are excluded in conventional economic analyses
- To demonstrate the gendered nature of the workforce and of workplaces
- To reveal the geographical connections between workplaces and homes
- To examine how gender shapes livelihoods and entrepreneurship in diverse contexts
- To discuss the possibility of a feminist economic geography

13.1 Introduction

In 2011, just one major high-tech company in the United States was headed by a woman. The company, Xerox, is well known as the maker of copying and printing machines, but has also expanded into other forms of information processing such as claims reimbursement, electronic toll transactions, and call centers. Leading Xerox and its 136,000 employees in 160 countries was Ursula Burns, the company's CEO and chairwoman.

Burns' story is one of remarkable upward mobility. Born in 1958, she was raised by her single mother in a low-income New York City housing project. Having studied engineering at Columbia University, she joined Xerox in 1980 as a summer intern. Over the next 30 years she worked her way through various departments in the company, becoming CEO in 2009 and chairwoman of the board in 2010. In 2011, Burns was one of just 14 women to head a Fortune 500 company (the largest companies listed on the U.S. stock market). Burns' ascent to the top of the corporate ladder is inspiring in many ways, especially given the challenges faced by many others with a similar background. But her success does raise a number of important questions.

First, we might pause to ask what we actually mean, not just by "success" but also by "work" in general. When we look at an individual such as Ursula Burns, it is easy to say that her success is reflected in the job title that she holds. But paid employment is not the only kind of work that we do. In media interviews, Burns herself has frequently emphasized the central role played by her mother in laying the foundation for her success. No one would doubt that raising a daughter as a single parent in a low-income neighborhood was hard work, but it is a type of work that is not always acknowledged or valued. Furthermore, it is a type of work that often falls disproportionately to women. The first set of questions to ask, then, is what do we mean by work and what falls outside of conventional definitions? A particularly important geographical question concerns *where* such unpaid work is performed, and how it connects across space with forms of paid work. In particular, we will ask how unpaid work in the home, often done by women far more than men, is linked to opportunities to engage in paid employment outside the home.

A second question we might ask is why women are so underrepresented in the high-tech sector. Ursula Burns' position as the only head of a major high-tech company in the United States reflected a much wider pattern across the workforce. In 2010, only 7.2% of electrical and electronics engineers and 10.3% of computer hardware engineers in the United States were women. At the same time, women accounted for 91.1% of registered nurses and 96.1% of secretaries and administrative assistants. Similar patterns, with some variation, are repeated all over the world. Clearly, then, the labor market is sorted according to gender in important ways. But what gives different jobs a certain gender "coding"? How are the places in which work is done given masculine or feminine meanings? Why might the fast-paced and competitive world of high-tech firms be less inviting for women? And how does the gendering of work vary across different places around the world?

A third set of questions prompted by Ursula Burns' experience is why the vast majority of top corporate jobs in different sectors seem to be dominated by men. Despite accounting for almost half of the workforce in each context, women comprise just 14.4% of senior corporate officers and executives in the United States, 17.7% in Canada, and 8.0% in Australia (www.catalyst.org, accessed October 15, 2011). If we look inside academic institutions a similar pattern exists. In Canadian universities in 2009, for example, just 19% of university presidents and 22% of full professors were women, but women represented 47% of assistant professors and 54% of temporary lecturers. Does a glass ceiling still exist that hinders women's upward mobility in the workforce? Why is this so?

Finally, the most singular feature of Ursula Burns' story is that she is the first and only African-American to lead a U.S. business giant. We will address the economic aspects of ethno-racial identity in the next chapter, but there is an important point to note here. A person's identity is multifaceted. Everyone comes with a gender, sexuality, ethnic origin, race, immigration status, language group, regional origin, and so on. All of these facets combine together to shape their experiences in the labor force. It is impossible to say whether an African-American woman from a working class family in New York is any one of these identities first and foremost. All of them shape the economic life of an individual such as Ursula Burns, so the intersection of gender with other dimensions of identity will be an important point to remember. A related point is that the effects of gender on economic experiences will be uneven across space, as gender itself has a varied and uneven geography. Femininity and masculinity, as the norms of behavior conventionally associated with women and men, are culturally constructed in different ways in different places.

This chapter addresses these questions in five main sections. The first section asks how gender is usually understood in relation to economic life and how this perspective might be rethought using a geographical approach (Section 13.2). We then start our geographical analysis by examining what kinds of *unpaid* work are done in domestic spaces, who does such work, and how this varies across the world (Section 13.3). The following Section (13.4) then explores how gender is played out in the *waged* labor market, determining the type of work available to women, and how it is performed in the workplace. A recurring pattern is the devaluation of the work done by women, and the association of women's work with domestic work and "soft skills." Building on these ideas, we then consider how women in the workforce tend to find their professional lives in tension with their domestic responsibilities, and we explain the fundamentally geographical nature of this home–work linkage (Section 13.5). The next section moves away from the issue of gender and paid work to consider how gender shapes entrepreneurship and livelihoods in diverse contexts (Section 13.6). Most of this chapter involves expanding our idea of the spaces in which economic activity is conducted (e.g., bringing the "home" and the "body" scales into view; see Figure 1.4), and exploring how gender differences are fundamental to understanding how the productive economy is linked to the supposedly noneconomic sphere. In concluding, however, we consider how a feminist economic geography might go further than simply taking notice of a gender-differentiated economic world, and might actually see that world differently (Section 13.7).

13.2   Seeing Gender in the Economy

The role of gender in economic life is often brushed aside as reflecting fundamental differences between men and women. When asked to explain why men and women tend to hold different kinds of jobs, for example, the usual explanation will start with physical differences between the sexes. The argument would be that men are more likely to work in construction, mining, and warehousing, for example, because of superior physical strength. To a limited extent there is some truth in this idea, but it hardly explains why men dominate among electricians and drivers, while nurses and elementary school teachers are overwhelmingly female. The conventional explanation might then move on to suggest that there are certain aptitudes associated with gender – women are supposedly good at detailed and painstaking work, while men have a better grasp of how to use advanced technological apparatus. But here too the argument collapses when specific examples are considered. The intricate work of watch repair and tailoring a suit, for example, are conventionally male domains, and yet they utilize what might be seen as female traits in other sectors or contexts. And although elementary school teachers or nurses are overwhelmingly female, there are enough men doing those jobs to demonstrate that their sex is perfectly capable of performing such work.

Explanations of why women are less represented in the upper echelons of the workforce, and are often paid less than men, can also be unpacked. A common rationale would be that women tend to curtail their ambitions, or pull out of the workforce entirely, when they decide to have children. This in turn affects their ability to work their way up the career ladder. But this argument is also rather unsatisfactory. If workplaces and employment relations were more commonly structured in such a way that family commitments could be combined with career development, then there would be no need to make such a choice. The idea that women make the choice to curtail their own careers is therefore ignoring the circumstances in which those choices are made. These circumstances can indeed be changed, for example through generous parental leave allowances, workplace daycares, and so on.

We need, then, to look elsewhere to find explanations of why gender might play such a profound role in economic life. In this chapter, we emphasize the geographical processes that underpin a more nuanced explanation. Some key points will form a basis for the arguments presented:

- The space of the home, where the work of cooking, cleaning, and child-care takes place, is almost universally construed as a feminine domain. But this is clearly a cultural understanding rather than something "natural" or fundamental.
- The greater responsibility shouldered by women within domestic spaces means that the relationship between home and work for women is often necessarily closer than for men, and this relationship in turn shapes various aspects of engagement with the waged labor market.
- The association of women with certain kinds of domestically-related work means that certain jobs and places of work become associated with femininity, while others become masculine domains.
- Finally, since gender, femininity, and masculinity are concepts that we define culturally, there are inevitably differences in the way gender affects economic life in different cultures around the world; and even within the same culture, their meanings may vary with age, generation, race, or region.

13.3   Gendered Patterns of Unpaid Work

We started this chapter with an example of a woman doing a high-profile, well-paid, and powerful job. But we also noted that hard work happens in other settings too, for example in raising a child, maintaining a household, or volunteering at a charity. And yet, we seldom count such work as economically important. In this section we ask why unpaid work tends to get neglected and what the consequences might be.

As we saw in Chapter 2, economic activity is generally measured using a system of national and international accounts. One of the most important measures is Gross Domestic Product (GDP), which represents the total monetary value of goods and services produced in a country. It is much more than a technical accounting tool; it also gives the general public a sense of economic trends and their collective material well-being. We also learned in Chapter 2 that GDP leaves a lot out of the picture. In short, those aspects of our lives that cannot be counted, do not count. Work that involves cleaning our homes, or renovating them, maintaining a car, raising children, caring for the elderly, preparing meals, and so on simply does not count unless money changes hands.

The first question we might ask is how much work is carried out on everyday tasks that do not get counted as economic output? Recent studies, based on large-scale surveys, suggest that a significant portion of our lives is spent on unpaid labor. On average, across the world's largest and wealthiest countries, around 43% of working minutes everyday are spent on unpaid tasks (OECD, 2011). Interestingly, there are significant differences around the world. For example, people in Japan work some of the longest hours in total, but carry out unpaid work for one of the shortest proportions of those hours, at around 30% of their total working time. In Germany, on the other hand, total working hours are shorter, and at almost 48%, a much higher proportion of them is spent on unpaid tasks. In both Turkey and Australia, just over 50% of work time is spent on unpaid activities. This tells us something about varied cultural patterns of life and work in different settings. The general pattern, however, is that a great deal of work is being done that does not get counted as productive economic activity.

What kinds of tasks and functions are being carried out by people doing unpaid work? Figure 13.1 shows the breakdown of unpaid labor tasks in various countries. Here again we see some intriguing differences, which again reflect how unpaid work might vary with different cultural contexts. A great deal of time is spent caring for household members in Ireland, for example, while a higher than average amount of volunteer work is done in New Zealand and Turkey.

If, in some cases, around half of all work time is being spent on unpaid activities, what is all of this work potentially worth? One way of calculating such a figure is through a "replacement cost" approach, using a nation's average hourly wage for unregistered informal activities such as a house cleaner or babysitter.

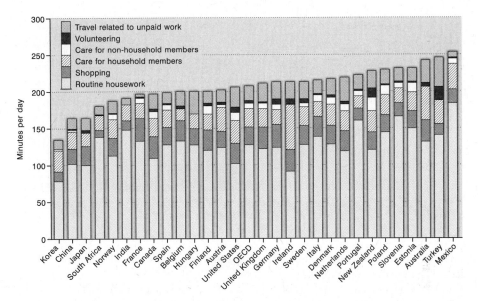

Figure 13.1 Main categories of unpaid work in various countries, ranked according to time spent on unpaid work (in minutes per day)
Source: OECD (2011), Society at a Glance 2011: OECD Social Indicators, OECD Publishing. http://dx.doi.org/10.1787/soc_glance-2011-en.

We then assume that all of the unpaid labor being done could be contracted out to people at this wage level. The results are quite startling. In most countries, the value of unpaid work measured in this way is between 20% and 40% of the national economy. In the United States, unpaid work is worth almost a quarter or 24% of national output, while in Britain, the figure is closer to one-third or 32%. It is also worth adding that these are quite conservative estimates. If, instead of using the average wage of informal work in a given country, we used just the average national hourly wage, then unpaid work would account for 35–75% of GDP in the countries shown (OECD, 2011).

Given the potentially huge component of national economic output represented by unpaid work, it is important to note that if this work *were* to be transferred into the domain of waged labor, it could have a significant effect on conventional measures of a national economy. If an entire country decided to use professional home cleaners rather than scrubbing and polishing themselves, or went to a laundromat rather than washing and ironing clothes at home, the effect could be remarkable, regardless of the fact that there would have been no net gain in actual production or well-being. Where more and more unwaged work is being shifted into the waged economy, as has happened a great deal in recent decades such as in childcare, then a rather false impression is created of greater overall output. One estimate, based on Canadian data from 1961–1992, suggested that GDP growth had been overestimated by 0.6–0.8% per year on account of shifts from unpaid to paid work. This is a significant error, when one considers that

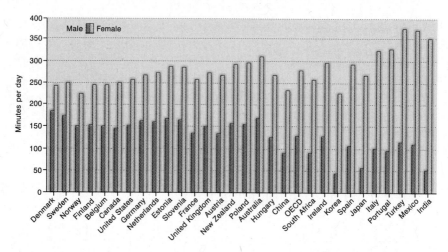

Figure 13.2 Minutes spent on unpaid work per day in various countries, ranked according to male–female disparity
Source: OECD (2011) Society at a Glance 2011: OECD Social Indicators, OECD Publishing. http://dx.doi.org/10.1787/soc_glance-2011-en.

many industrialized countries grow by only 2–4% per year (Statistics Canada, 1995).

The most important issue for this chapter, however, is the question of *who* is doing all of this work? The answer, overwhelmingly, is that it is being done by women. Figure 13.2 shows that in most countries women do twice or even three times as much unpaid labor as men. There is, however, a great deal of unevenness in gender roles around the world. In Scandinavian countries, men do less unpaid work than women but the gender gap is much smaller. In countries with lower rates of female participation in the waged labor force and different cultures of gender relations, we see a very wide gap. In India, Korea, and Japan, for example, we see men doing just a small fraction of the unpaid labor done by women. Overall, however, we see a general global pattern in which unpaid work tends to fall primarily to women.

While there is an uneven global geography of how paid and unpaid work gets done, what we are really seeing here is the result of individual negotiations and power relations within households. We saw in Figure 13.1, for example, that "routine housework" and "caring for household members" accounted for most of the time spent on unpaid work. There is then, a distinct geography to the inclusions and exclusions of national income accounting. The spaces of the private home are neglected, while the spaces of the public and nondomestic sphere are carefully counted. This in itself would not be a particularly gendered exclusion were it not for the fact that, as we have seen, the prevailing gender regime in most societies has the domestic sphere as a primarily feminine space, and the public sphere as a masculine domain. Thus, the uncounted portion is largely the work done by women in the home.

By looking inside the household, we are adding a new dimension to our analysis of economic processes. The key process shaping our working lives can no longer just be seen as the relationship between an employer and an employee. Instead, we now see a new set of power relations coming into view – between men and women. This set of power relations has been conceptualized as *patriarchy* (see Box 13.1).

Whatever view we take about the relationship between patriarchy and capitalism, we have established an important starting point by expanding our view of the spaces in which economic activity is seen to exist – beyond factories, farms,

---

KEY CONCEPT

Box 13.1   Patriarchy

Patriarchy refers to a system in which men are assumed to be superior to women and therefore to have authority over them. There has been substantial debate over whether patriarchy is an entirely separate system of domination and control from capitalism, or if the two are actually enabled by each other. The subordination of women is very useful to capitalist enterprises seeking controllable and compliant employees: around three-quarters of production-line labor in the world's export processing zones is young women, for example. It is also debatable whether patriarchy is universal, or has culturally distinctive manifestations. The feminist sociologist Sylvia Walby (cited in McDowell, 1999), identifies six structures of patriarchy:

- *Household production*: Where a division of labor is created in which women do large amounts of unpaid work in support of their family.
- *Waged work*: In which women are segregated into certain jobs, for example as secretaries, nurses, and elementary school teachers.
- *The state*: Where men dominate government institutions and produce gender-biased legislation and policies.
- *Violence against women*: Both physical and emotional.
- *Sexuality*: Where male control is exerted over women's bodies, for example in abortion legislation in many jurisdictions.
- *Cultural institutions*: Where various other forms of patriarchy are discursively reproduced, for example in the media and how it depicts society.

Patriarchy may thus be evident either in the public sphere of institutions and legal frameworks that differentiate in favor of men or in the private sphere, in the ways in which relationships between men and women are conducted. Clearly, however, the two are connected.

and shopping malls, to include gardens, kitchens, and playrooms. This is critical because, as we will see later, the association of femininity with domestic spaces and domestic work shapes a great deal of what women experience when working *outside* the home. Primary responsibility for unpaid domestic labor also affects the types of jobs that women are assumed to be suited for in the waged workforce, and the ability of women to balance paid work with unpaid work. The patterns of who does unpaid work in the home are therefore closely connected with opportunities in the labor force. More broadly, we can start to see the possibility of a feminist interpretation of the economy – one that sees work, *skilled* work, happening in spaces that are not usually assumed to have productive activities going on. We will come back to this point in the final section of the chapter. For now, though, we will turn to the formal waged economy to see how women have been incorporated into the labor force.

13.4   Gendering Jobs and Workplaces

Women and the Paid Workforce

The primary role that women play in the unpaid work of the domestic sphere has implications far beyond the home. The first, and most obvious, influence is the limitation that expectations of motherhood and marriage place on women. While they vary across cultures, these expectations often keep them from entering the workforce at all. Even in societies that consider themselves to be the most liberal and emancipated, it is undoubtedly true that the role of the primary parent and domestic worker is still a gendered one.

Participation in the labor force is defined as the number of people in the waged workforce, either employed, self-employed or looking for work, as a percentage of the total population of working age. Figure 13.3 shows that the trend over the last few decades has been toward greater participation for women. Only in China has the participation of women in the formal labor force declined slightly. But participation there was already exceptionally high. The pattern is still, however, very uneven. Canada, the United States, Brazil, and Australia now have some of the highest levels of female participation, whereas India, Italy, and Saudi Arabia are much lower.

In the developed industrialized countries, high levels of women's participation in the labor force represent the continuation of a trend starting in the mid-20th century. In some newly industrializing countries, on the other hand, the trend is more recent. In Malaysia, for example, the female participation rate grew from just over 30% when the country achieved independence from British colonial rule in 1957, to around 44% by 2009. Moreover, this increased participation is not due to larger numbers of women entering traditionally female occupations, such as running market stalls or working as elementary school teachers. Instead,

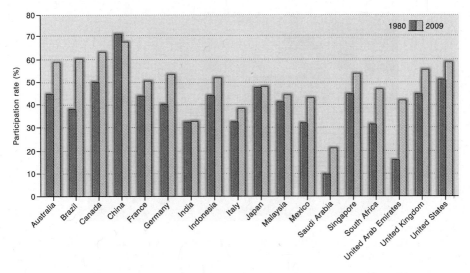

Figure 13.3 Female labor force participation in selected countries, 1980 and 2009
Source: World Bank (2011).

much of this participation is due to the rapid development of export-oriented garments and electronics sectors in the 1970s and the 1980s, which drew young rural women into assembly factories.

We thus see higher levels of women's participation in both the developed and developing worlds. These parallel changes are different in nature but ultimately part of the same process. The key is the restructuring of the global economy in recent decades (see Box 10.4 on the new international division of labor). One component of this has been the deindustrialization of the developing world and a structural shift toward service-sector employment. With this change, the secure and well-paid industrial work of the Fordist era, which supported male-breadwinner families and stay-at-home mothers, has largely disappeared. As levels of pay and job security have declined, dual income families are increasingly the norm in developed countries and larger numbers of people are employed in part-time work or temporary jobs. This change also came about at the same time as women's roles in broader society were changing from the 1960s onward, with greater participation in the public sphere and higher levels of education increasingly available. Partly because of this and partly causing it too, there has been a trend toward later marriage and childbirth, and the greater social acceptability of doing neither.

At the same time, certain kinds of manufacturing jobs from the developed world have migrated to Mexico, China, India, Indonesia, and other developing countries. In these places, it has been women who have been the rank and file of the industrial workforce. In many industrial estates around the world, where clothing, toys, electrical appliances, and computers are assembled, at least

Figure 13.4   Women wait for their shift to start at an Indonesian electronics factory

three-quarters of the workforce is female (see Figure 13.4). Asking why this is so would elicit different answers depending upon who was asked. Factory managers usually note a series of personality traits that are assumed to be distinctively feminine, and that are useful to them given the specific labor process involved: a high tolerance for boring, repetitive work; manual dexterity and precision (the infamous "nimble fingers"); and a submissiveness toward authority, usually in the form of older supervisors or male managerial staff.

Critics of labor conditions in the developing world would frame the situation somewhat differently, arguing that patriarchal power structures are recreated as a disciplining mechanism in the workplace so that the workforce is kept under tight control. They would also point out that workers are usually paid a very low wage, often so little that it barely covers their cost of living. Employers are able to do this essentially because their workforce is often young and single, without a family to support. Indeed in many cases, support is provided by rural families to assist industrial workers, rather than the other way around. These contrasting perspectives have led to much debate about whether employment in the factories of the developing world is actually liberating for young women, or whether it just represents further capitalist exploitation and oppression (see Table 13.1).

A final point that should be made about the increasing participation of women in the workforce is that their household work is not usually taken up by men, but by other women. Many countries that have seen growing female participation have also accommodated migration flows to supply low-cost domestic workers who will perform domestic cleaning and childcare functions for working mothers. In some countries, domestic workers arrive from poorer rural areas, while in cities such as Singapore and Dubai they arrive as temporary foreign workers. Singapore now has around 200,000 such workers, with many coming from the Philippines, Indonesia, and Sri Lanka.

Table 13.1 Contrasting views on the emancipatory potential of industrial employment for women

Female Industrial Work as Emancipation	Female Industrial Work as Oppression
• Working hours and conditions are better than in agriculture or domestic work. • Wages are good in terms of local purchasing power. • Some employers are looking for longer-term commitment and skills development for employees. • Employment provides participation and respect in the public sphere and a chance to associate with other women. • Income gives a greater social role and respect within the household, and greater control in terms of delaying marriage and childbirth.	• Low wages and benefits, sometimes below subsistence levels. • Often debilitating or dangerous working conditions and harsh factory discipline. • Insecure employment – jobs may disappear with changing international circumstances; women may be laid off if they marry or have children. • Few opportunities for upward mobility or skills development. • Women are confined to particular job categories, especially those involving repetitive, boring, manual work. • Patriarchal dominance and discipline is often reproduced inside the factory. • Waged work may not mean a reduction in household responsibilities; and wages may become the property of male household heads.

Overall, then, we see a process of increasing women's participation in the workforce worldwide. But women in the developing world find themselves incorporated in particularly subordinate roles. There is, therefore, an important intersection between issues of gender, ethnicity, and nationality in the valuation of working people, as Box 13.2 illustrates. Thus, while women's access to the labor market was once considered the litmus test for gender equality, it is now recognized that femininity is actually being used in the workplace as a form of devaluation, often in conjunction with race and nationality.

The Gendering of Jobs

We saw in Section 13.3 that women tend to take primary responsibility for unpaid household work, but the assignment of particular kinds of work to men or women also happens outside the household. In this section we will examine how certain jobs become labeled as masculine or feminine.

FURTHER THINKING

Box 13.2   *Devaluing the third world woman*

In the contemporary global economy, women in the developing world appear to be especially favored as employees by international investors in the manufacturing, tourism, and hospitality businesses. They are also favored in the industrialized world as the nannies and maids who carry out the domestic chores of those women who have entered the waged workforce. But being favored does not necessarily mean they are valued. Their work is generally lowly paid and they are treated as unskilled. In fact in some cases they are viewed as being *incapable* of skilled work, and the intersection of ethnicity, gender, and nationality creates strong stereotypes. These include the ability to do tedious and repetitive work without complaint, to work with close attention to detail, and to acquiesce to managerial control without the resistant pride of masculinity.

This devaluation has been especially and tragically evident in the factories of Mexico's northern border, known as *maquiladoras*. In the 1990s, the border town of Cuidad Juarez gained notoriety when almost 200 female corpses were discovered in the surrounding desert over a 5-year period. It has been argued that these murders were indicative of the worthlessness accorded to women's bodies, lives, *and* work, in the factories of the region. Within the factories, they are seen as unskilled, untrainable, and docile, and yet women constitute about 70% of labor-intensive assembly line workers. Managers and supervisors see themselves as the brains behind the operation, with women simply acting as robotic manual workers. As a result, it is seen as pointless to try to train such workers, and in any case they are viewed as far too inclined to resign or switch jobs to make it worthwhile. In essence, female Mexican workers are seen as entirely disposable and replaceable – a piece of the factory that eventually will become worn out (with stiff fingers, tired eyes, aching backs) and discarded. Outside the factory, newly independent women are often seen as sexually promiscuous, dangerously liberated, and on the same level as the prostitutes for which the city has long been known. In this way, it is argued that it was the devaluation of Mexican womanhood in general that was happening in Cuidad Juarez, and not just of factory employees in particular. The low pay, poor conditions, and lack of upward mobility for women factory workers were part of a more general devaluation of Mexican womanhood. Where industrial production called for a "disposable" kind of worker, northern Mexico became an ideal location. For more, see Wright (2006).

The United States, as we saw in Figure 13.3, has a comparatively high level of female participation in the workforce, with close to 60% of women now in the labor force. And yet, if we start to look at the kinds of jobs they hold, we see that gender makes a significant difference to the opportunities available to women and men. Table 13.2 shows the top ten jobs held by women in the United States. There are a number of points that we can take from this information.

One point is that many of the jobs listed have a high incidence of part-time work, especially among retail sales persons and supervisors. This not only reflects the concentration of many women in some of the more precarious segments of the labor market, but also highlights the fact that many women may be juggling paid work in the labor force with unpaid work at home. Working part time is sometimes a necessary outcome of many women being responsible for childcare and other work at home.

As we look down the list of job titles in the table it also becomes very clear that women are heavily concentrated in certain kinds of jobs that tend to reflect the traditional domestic roles of women. Just as unpaid work by women often involves childcare, food preparation, shopping, and house cleaning, we see these roles being reproduced in the paid work most commonly found by women. It is also important to note that women comprise a very large proportion of those employed in these jobs – over 90% in the case of secretaries and registered nurses. Clearly, these are jobs that have been "coded" as feminine.

Not all occupations are so clearly gendered as secretarial and nursing work, but across the labor market there is still a very clear sorting or *segmentation* of the workforce into those jobs coded as male jobs, and those as female (see Box 13.3 for more on segmentation). Some, as we have noted, are gendered because they are extensions of what are seen as "naturally" feminine domestic work. But others are gendered because they demand the performance of a particular form of femininity or masculinity. This is especially so in jobs that require some form of social interaction with a customer or a client. These are increasingly widespread in service-oriented economies and require not just technical skills, but also an embodiment of certain characteristics. At the top of the list is gender and sexuality, but linked to this are bodily features such as height, shape, weight, skin tone, grooming, dress, comportment, accent, elocution, emotional presence, and so on; in other words, the full range of ways in which the body is presented in a social interaction (McDowell, 2009).

The key point is that gendered identities are *performed* in the workplace. These performances create certain roles for men and for women, which are difficult to transcend. Moreover, in most jobs it is not just masculinity or femininity that is seen as embodying the requirements of the work, but particular *kinds* of masculinity or femininity. These are then enacted in selective recruitment, training strategies, evaluation, and promotion criteria, as well as more subtle techniques for ensuring adherence to organizational norms. The demeanor, including the gender and sexuality of the provider of a service, is thus crucially important, and indeed is a fundamental part of the "product." For example, it is

Table 13.2   Top 10 occupations of women employed in the United States, ages 16 and over, 2010

Occupation	Number of Women Employed	% Working Full-Time	% of All Employed in This Occupation Who Are Women	Women's Median Weekly Earnings for Full-Time Employees (US$)	Men's Median Weekly Earnings for Full-Time Employees (US$)
Secretaries and administrative assistants	2,962,000	78	96.1	657	725
Registered nurses	2,590,000	76	91.1	1,039	1,201
Elementary and middle school teachers	2,301,000	85	81.8	931	1,024
Cashiers	2,291,000	41	73.7	366	400
Retail salespersons	1,705,000	43	51.9	421	651
Nursing, psychiatric, and home health aides	1,700,000	68	88.2	427	488
Waiters and waitresses	1,470,000	37	71.1	381	450
First-line supervisors/ managers of retail sales workers	1,375,000	71	43.9	578	782
Customer service representatives	1,263,000	80	66.6	586	614
Maids and housekeeping cleaners	1,252,000	50	89.0	376	455
All Occupations	**65,638,000**	**68**	**47.2**	**669**	**824**

Source: Calculated from U.S. Department of Labor, Bureau of Labor Statistics, 2011 Employment & Earnings Online: Household Survey Data. http://bls.gov/opub/ee/2011/cps/annual.htm#empstat, accessed October 4, 2011.

KEY CONCEPT

Box 13.3   Theories of labor market segmentation

The process by which a labor market is divided up into different parts is known as segmentation. These segments might be geographically separated, but more commonly the term is used to describe different types of jobs, and usually to contrast better, higher paid, and more stable work from less desirable occupations. Once matched to different segments of the labor market, it generally becomes quite difficult for individuals to move outside of them. If people were randomly distributed across different segments, we would expect that the characteristics of people in particular occupations would reflect the characteristics of the population as a whole. Clearly, however, this is not the case. Gender, ethnicity, and immigration status represent three primary characteristics by which the labor force is differentiated, with women, ethnic minorities, and immigrants disproportionately occupying subordinate positions in the labor market.

The purpose of labor market segmentation theory is to explain such patterns of over- or underrepresentation. It rejects the view that job allocation is the outcome of the skills, qualifications, and experience (i.e., the human capital) of applicants. Instead it argues that each segment has its own distinctive set of rules that govern access to it and mobility within it. The historical, institutional, and geographical contexts for a labor market are hence more important than the human capital endowments of individuals. Early versions of segmentation theory identified primary and secondary labor markets, with the former characterized by stable, well-paid jobs generally occupied by white males, and the latter comprising unstable, low-paid jobs with few prospects for internal promotion. In recent years, segmentation theory has become more subtle, going beyond this dual labor market structure to consider multiple segments. In addition, the identities of employees themselves, as well as the types of jobs they occupy, have received attention. Thus the context provided by families, schools, neighborhoods – all of the circumstances in which a workforce is socially reproduced – are important in shaping labor market outcomes. As a result, labor markets vary greatly from place to place, not just because social relationships tend to be localized, but also because particular intersections of state control at the city, regional, national, and even international levels regulate the labor market in place-specific ways. For more, see Peck and Theodore (2010).

PRNewsFoto/Singapore Airlines/AP/Wide World Photos

Figure 13.5   Flight attendants: feminine work that becomes part of the product

common for airlines to emphasize the attentive and caring demeanor of its flight attendants (see Figure 13.5). Similar requirements are found in a wide range of occupations, including investment banking, real estate, bartending, healthcare, and academia – that is, any job that requires an interactive component.

Valuing Gendered Work

Returning to Table 13.2, we can compare the last two columns to reveal one of the most constant features of labor markets all over the world – the gap between women's and men's wages. What is especially notable here is that these columns compare just full-time workers who are doing exactly the same job. Yet, in every one of these ten jobs, we find men earning significantly more than women.

The issue is, however, a broader one than women getting paid less for the same work. More generally, just as women's work in the home tends to be invisible and ignored, classically feminine jobs in the workforce tend to be poorly paid and accorded a low status. In Table 13.2, only two of the occupations listed had female weekly earnings that exceeded the overall male median wage of $824.

How do we explain this pattern? First of all, as we have already observed, many of the jobs in Table 13.2 are extensions of conventionally feminine roles in domestic spaces – for example, teaching, nursing, and secretarial support. In the domestic sphere they are provided without any financial reward, and this devaluation of such work carries over into public spaces.

A second cause of devaluation of women's work revolves around how the skill involved is assessed and therefore how much it should be rewarded. The notion of "skill" is highly problematic as it is actually quite difficult to identify the tasks involved in most jobs that would enable us to rank them in terms of skill. In general, however, tasks associated with the home, which "everyone" does, are perceived to be low in skill requirements. The use of machine technology in particular appears to be a key basis for differentiation between skilled and unskilled work, but the relationship is far from consistent. The *place* of work is also significant.

A study of the garment industry in the Dominican Republic illustrates this point (Werner, 2012). Between 1986 and 2005 a series of changes in U.S. trade policy allowed the import of clothing such as jeans and shirts made wholly in the Caribbean, with just the design transmitted by the American buyer. The result was that some garment firms in the Dominican Republic started to "upgrade" their production processes, taking care of a wide range of functions such as cutting, laundry, finishing, and product development. This meant that the firms needed to have a "sample room" in which prototype clothing is created with great attention to detail by workers who know a range of different sewing operations. Interestingly, the workers in the sample room tend to be almost exclusively men. The work is defined as male in part because craft production in tailoring workshops has always been a male activity, while home sewing was widely done by women. Thus, while regular sewing was understood to be a "naturalized" skill that everyone could acquire, the multiple skills of tailoring were seen as a craft learned in a workshop. Here, then, we see a skill set associated with feminized domestic work being devalued relative to skills acquired by men in the workplace, even though the physical labor being carried out is essentially the same.

There are, then, cases where women may do the same work as men, but it will be paid less. This brings us to a third reason for the devaluation of women's work: the assumption that women are working in order to earn "extra" money for the household, that their earnings are secondary, and that they have less need than men to derive satisfaction, professional self-esteem, or identity from their work. In some societies the assumption that the main breadwinner will be male is quite explicit. In Singapore, for example, men are officially regarded as the head of a household and the principal earner (Oswin, 2010). Thus, in the country's civil service, health benefits for family members are only channeled through male employees, on the assumption that female employees have husbands whose jobs bring such benefits. In other societies, the secondary and supplemental nature of women's earnings is implicit. The effect, however, is to justify lower wages for women even if the work might objectively be seen as quite skilful. Thus, we have moved away from a rationale that says women are doing less skilful work, to one that sees work done by women as less important *because* it is being done by women.

Creating Gendered Workplaces

It is not just domestic spaces, and the jobs associated with them, that become gendered. Workplaces can also become very masculinized or feminized environments, resulting in the exclusion of women (or men) from contexts in which they are not seen to fit.

One example is provided by the high-technology sector. A study of firms in the Cambridge Science Park in the United Kingdom found that high-tech workplaces were male environments in a variety of ways (Gray and James, 2007). Women at such firms are more likely to be part-time employees because of pressure to combine work and family responsibilities. They are therefore less likely to socialize with colleagues during working hours as they focus on getting their work done. As a result, the social atmosphere of the firm becomes more male dominated, with lengthy discussions revolving around football, cars, and other classically male concerns. In turn, the maleness of the social atmosphere further discourages women from attending social events such as coffee mornings, parties, or after-work drinks. This might seem like a minor issue, but it has two major consequences. The first is that female employees are participating less in the informal exchange of knowledge in the firm and so their ability to contribute is reduced. This may reduce the overall productivity of the firm by creating exclusionary spaces of knowledge exchange and innovation.

The second consequence is that women are participating less in the networking and knowledge gathering that are essential to career advancement both within their current firm and in the broader labor market. The maleness of work-based social networks may place limits on women's upward mobility, and this further augments the general maleness of the atmosphere in the high-tech sector.

Similar examples are also found in the financial sector. Research in Singapore, for example, shows that women in the investment banking sector must conform to a particular set of bodily norms in the workplace. In an interview, one executive noted that:

> For women, you have to be on it. You have to look really good, unless you are really that great in your work, and there are very few of those. Female counterparts dress perfect. They dress to impress on the trade floor, especially those with good bodies . . . they can be Chinese Singaporeans or expats or whatever. Hot is hot! (Ye and Kelly, 2011, p. 702)

Interestingly, the bodily types apparently required of women in this instance are exemplified with reference to "Chinese Singaporeans" and "expats" (by which the interviewee meant "white"). The dress and self-presentation of others, such as Muslim Malay women, was implicitly excluded. This represents a reminder that femininity intersects with other dimensions of identity, but it also highlights the fact that workplaces become associated not just with masculinity or femininity, but also with specific kinds of gender performance. The quotation strongly implies that in the case of the banking industry there are heterosexual gender identities that must be exhibited to gain acceptance.

This section has largely focused on the patriarchal structures of male dominance that exclude women or devalue their work. It is, however, important to note as well that economic life is no longer quite as rosy for men in the labor market. While secure employment might once have been the expectation of most young men, economic restructuring in the developed world is increasingly removing the kinds of jobs that they would once have occupied. Box 13.4 examines this phenomenon in the United Kingdom in particular.

FURTHER THINKING

Box 13.4   Redundant masculinities

In the developed world, urban labor markets have become increasingly polarized. At one end there are high-status, well-paid, and technology-intensive jobs in what is often termed the "knowledge economy." They usually require high levels of formal education, and many are in the business services sector such as law, accountancy, marketing, and various kinds of consultancies, as well as creative employment in design, entertainment, advertising, and journalism. Senior bureaucratic jobs in the public sector would also enjoy some of the same privileges. Although there are barriers to women in some of these fields, as we have seen in this chapter, they are increasingly open, at least at the entry level, to men and women equally. The other pole comprises low-paid, low-status, and insecure jobs, largely in the service sector. These jobs exist, for example, in the retail, clerical work, food services, cleaning, and hospitality sectors. In these jobs, contact with the public is often a requirement, and many are socially constructed as feminine jobs. The classic attributes of masculinity effectively disqualify young men from such jobs.

Meanwhile, traditionally masculine jobs in the manufacturing and resource sectors are declining in number as global restructuring takes such employment abroad to locations with lower labor costs. Secure employment at a factory, in a mine, or at a pulp and paper mill is now difficult to find in Western economies. The result, in British society for example, is that the traditional sources of employment for young working class men with low levels of formal education are now very scarce. And with those jobs the traditional sources of employment that affirmed masculine identities have also disappeared. Not only are young men finding themselves redundant – that is, out of work – but the "laddish" masculinity from which they develop their self-esteem and identity is also increasingly redundant and out of date. For some this will mean a "portfolio career" in which they move from one short-term job to another. For many, masculine identities will no longer be derived from the work that they do, but rather from the consumption decisions that they make – for example, the clothes they wear, the music they listen to, and the drinks they consume. In short, while economic restructuring has brought women into the workforce in subordinate roles, in some contexts it has also led to the exclusion of young men and the removal of traditional bases for masculine identity. For more, see McDowell (2003).

13.5   Home, Work, and Space in the Labor Market

We have seen how the association of domestic roles with femininity makes certain jobs seem to be "women's work," and how certain workplaces become exclusionary masculine places. We also know that women do indeed shoulder most of the burden for household work. This means that there is a key link between home and work that must be examined in order to understand how women participate in the labor force. It is a link that is created in the practical everyday lives of women as they try to do all the work, paid and unpaid, that is expected of them.

Geographical considerations play a significant part in the daily integration of home and waged work. Several studies of specific cities in the United States and elsewhere have used statistical data to show how women's labor market entry is different from men's in several ways:

- Women are often constrained in the distance they can commute because of the need to be at home when children arrive from school, to collect them from school, or to incorporate shopping trips or other activities into their days (Rapino and Cook, 2011). As a result, studies have shown that women tend to have shorter commuting times than men. There are, however, exceptions. A study in New York found that black women and immigrants in particular commuted much further than either white women or men, largely because ethnic minorities resided in urban centers far from the major centers of employment, which had moved to the suburbs (Preston et al., 1998). The quality of public transportation can also be a factor. For example, a study of women in Sydney, Australia, found that suburban women were entrapped partly because they did not have access to a car to reach employment from areas poorly served by public transport (Dowling et al., 1999). The geographical structure of a city can also be a factor. A study in Montreal found that although women commute shorter distances in general, they tend to commute *further* than men to reach jobs in the Central Business District. Different centers of employment in a city may therefore have different gender-based patterns of commuting (Shearmur, 2006).
- In searching for jobs, there is some evidence that women tend to use personal networks, while men rely more on formal job search processes such as advertisements or employment agencies. Also, women's job search information networks tend to consist overwhelmingly of other women, and such social networks are often neighborhood-based (Hanson and Blake, 2009). This means that the job opportunities they lead to are often highly localized and limited, thereby reproducing the unintended concentration of women in certain jobs, and in workplaces that are close to home.
- Employers may locate in particular neighborhoods in order to access specific pools of labor, and may deliberately recruit through existing employees in order to ensure a locally resident workforce. This might in some cases be a

tactic to ensure a stable, undemanding, and nondisruptive workforce. These concentrations of female employment in the suburbs of large cities have been referred to as *pink collar ghettoes*.

- Women may find their jobs after a household has decided where to live. This decision is often made on the basis of the husband's employment. Thus, women tend to be looking for work locally without actually having selected their residential location with a view to the possibilities for employment.

- As communications technologies advance, it is possible that for some types of employment such technologies may provide solutions that allow home–work integration and participation in the labor market without a commute. For example, research in the Irish high-tech industry has shown that practices such as teleworking and video conferencing can alleviate the gendered barriers to participation in the sector (James, 2011).

The key point to note here is that both gender and space shape peoples' experiences in the labor market. The home is often treated as a place of feminine responsibility; networks used for job searching are gendered; and residential location may not match employment location. All of these factors demonstrate how the geographies of labor markets and home–work linkages can lead to uneven employment outcomes for men and women. There is, however, unevenness in the manifestation of these issues. The precise effects will depend on urban residential structure, transport facilities, the role of ethno-racial segregation, cultures of masculinity and femininity, and the possibilities for technological solutions. Thus, there is no universal pattern regarding the interplay between gender and space, and it will have its own uneven geographical manifestation in different cities around the world. Without a doubt, though, the different experiences of men and women in the labor market are closely tied to the gendered engagements with urban space. Perhaps the most important point, however, is that the integration of separate spaces of home and work has to be central to any understanding of how gender and geography affect the labor market. The economic processes of the labor market do not just happen at work, but happen in the context of much wider social and cultural processes in specific homes, neighborhoods, and cities.

13.6   Entrepreneurship and Livelihood Strategies

Two broad assumptions have underpinned much of the discussion in this chapter so far. The first is that women, who are increasingly participating in the labor force, are doing so when hired by an employer. The second assumption is that they are working outside the home. Both of these assumptions may be increasingly inaccurate.

Women are not just entering the formal economy through waged work. There are numerous examples from around the world of women creating their own entrepreneurial ventures. It is estimated that around one-third of the world's entrepreneurs are women, although this applies only to those with formally registered businesses. The ratio would be far higher if the full range of informal activities were also included. The reasons for engaging in self-employment are diverse, but the kinds of labor market barriers and inequalities discussed earlier are certainly part of the explanation. In other words, many women engage in their own entrepreneurial ventures because other opportunities for them are blocked.

Where women enter formal entrepreneurial ventures, there are a few important points to note. First, starting up a new enterprise requires the use of networks to raise capital, acquire know-how, develop markets, and, in some cases, recruit personnel. Just like entry into the waged labor market discussed earlier, these networks tend to be gender-specific and place-specific, with women tapping into different kinds of contacts, with different geographical patterns, than men (Hanson and Blake, 2009). Where women are excluded from networks, then entrepreneurship may be limited. But when networks exist to support female entrepreneurship, it may flourish. One of the best-known examples of a place-based entrepreneurship network is the Self-Employed Women's Association (SEWA) of India, which grew out of a union of textile workers in the 1970s. It provides not just skills training and micro-finance for women involved in various forms of self-employment, but also a sense of confidence and empowerment that enables women to overcome gender-based discrimination and envision themselves as entrepreneurs (Hanson, 2009). The uneven existence of such networks and support systems means that entrepreneurship among women is often dependent on the characteristics of place, developing most easily where such support is present.

A second point is that geographically uneven opportunities for entrepreneurship may be dependent on how gender is understood in different places. In the U.S. context, for example, assumptions about what types of business might be legitimate and appropriate for women can shape a bank's willingness to lend money to get the business started. A study of Worcester, Massachusetts, found that bank officers made gender-based decisions concerning the viability of proposed enterprises. A female-owned auto mechanic shop would be unlikely to get funded, for example, based on the bank's assumption that clients would be unlikely to patronize the business (Blake, 2006).

A third point to make is that engaging in entrepreneurship may blur the boundaries between spaces of home and work. This is not only, of course, a characteristic of female entrepreneurship. Small businesses established by men may also begin in basements, garages, and spare rooms. But given what we know about the burden of unpaid labor in the home, the necessity of combining paid and unpaid work in the home is likely to be far greater for women. In the United Kingdom, this has lead to a phenomenon termed "mumpreneurship" that highlights the integration of household work and entrepreneurial work within the same space and time. Recent research has shown that "mumpreneurs" are operating in a number of sectors including business consultancy, training, public relations, online retailing of children's and other products, therapy services, website development, and tourism (Ekinsmyth, 2011).

Businesses operating within a system of monetary exchange are perhaps the most common form of female entrepreneurship in the developed world, but in rural areas of developing countries the picture might be quite different. There, women may be supporting their family's livelihood through activities that do not involve buying or selling, but rather gathering, harvesting, and foraging. For many poorer families in the developing world, food, fuel, construction material, and medicines might be gathered from forests, rivers, or coastal waters. Women and children, in particular, are key participants in these livelihood activities, which go unrecorded and unrecognized as economic production. It is these kinds of activities that are particularly vulnerable to environmental change and the privatization of common property resources (see Chapter 5). One example, in the Philippines, concerns the conversion of mangrove forests along tropical shorelines into fishponds cultivating shrimp and other fish for export markets. Gathering shellfish, crabs, snakes, and other resources from mangrove forests is largely the domain of women and children from the poorest of local families (see Figure 13.6). When the forests are cut away to develop aquaculture ponds, they become male-dominated enterprises.

Although the home enterprises of India, the "mumpreneurs" of Britain, and the mangrove forest foragers of the Philippines might seem worlds apart, one characteristic unites them all. In each case, women engaged in these activities are redefining the priorities of profit maximization and growth that are usually taken to be fundamental in a capitalist enterprise. Instead, other values are entering the equation: balancing work and home, fostering self-esteem in a patriarchal

Figure 13.6 Women and children gathering resources in mangrove forest areas of Aklan Province, the Philippines

culture, seeking quality of life, and producing enough for the family rather than the maximum for the business. Together, then, they suggest the possibility of a different rationality in economic affairs or perhaps even a feminist Economic Geography. We will examine this possibility in the final section of this chapter.

13.7   Toward a Feminist Economic Geography?

Throughout this chapter we have examined how men and women experience economic processes differently in terms of paid employment, entrepreneurship, and livelihood, as well as unpaid work for the benefit of others. Clearly, some of these activities require us to rethink the boundaries of the economy to include work that is not assigned a monetary value. We can, however, go further still and ask whether there is a distinctively different feminist approach to the economy that would not just analyze it differently, but would re-imagine it in important ways.

We can identify several dimensions of a feminist vision of the economy. The first would be an openness to the role of difference and identity in economic life. As we have seen, this has been rooted in an understanding of how gender identities have affected the economic experiences of women, but it can also be extended to other forms of identity difference. For example, feminist debates have opened the way to thinking about how sexuality shapes workplaces and economic opportunities for gay men and lesbians. It is important to note that much of the discussion in this chapter has assumed a household arrangement that involves a heterosexual couple, often with children. This represents a rather narrow perspective on the range of sexual identities and household arrangements that might exist. Box 13.5 expands upon this point by examining how queer identities and queer theory have started to influence our understandings of economic geography.

A second contribution of feminist approaches is to highlight the linkages between productive jobs in the workforce and reproductive labor in the home. We saw in Section 13.5 that this has important implications for women's participation in the labor force, but it also opens a wide range of issues that are relevant to both men and women. Understanding how people's working lives are embedded in concerns such as childcare, commuting, and family relationships represents an important expansion of our understanding of economic processes. More broadly, then, feminism has focused attention on the social and cultural realities of everyday life that shape how both men and women participate in the economy. These are everyday realities that conventional economic analysis too often ignores.

A third influence of feminism is found in its ethical commitments. Rather than an assumption that economic life is about profit maximization, growth, and competition, feminism has fostered an alternative set of values based on "nurture, cooperation, sharing, giving, concern for the other, attentiveness to nature, and so on" (Cameron and Gibson-Graham, 2003, p. 153). It would be wrong to suggest that these values can only be arrived at through a feminist philosophy,

FURTHER THINKING

Box 13.5   A queer Economic Geography?

The study of sexuality in the field of Geography has developed significantly in the last 20 years, but its influence on how we think about economic processes is only slowly becoming apparent. There are, however, several ways in which queer identities have been important in Economic Geography. First, at the level of specific workplaces, sexuality represents a dimension of identity that may have a direct impact on workplace dynamics. A study in Australia, for example, shows that exclusionary heteronormative or homophobic places of work may present significant barriers to the labor market participation and advancement of young gay workers (Willis, 2009). A second area of interest is in the existence of a specifically gay economy in major urban centers. This might include a mainstream commercial gay scene featuring clubs, bars, other businesses, and commercialized Gay Pride parades. But it might also include alternative economic arrangements such as not-for-profit care-giving organizations, housing cooperatives, festivals, and other forms of shared or communal economies (Brown, 2009). A third, and more abstract, "queering" of Economic Geography is to be found in the role of queer theory in inspiring alternative ways of thinking about the economy. In particular, queer theory's core goal of destabilizing and questioning what is taken for granted as normal and natural has inspired some key thinkers to ask how we might rethink our relationship with capitalism. How might we "de-center" capitalist relations of work/production and market exchange to explore new ways of relating to each other economically? This line of thought has led to practical experiments in alternative/diverse/community economies. (See Gibson-Graham, 2006a and 2006b; and www.communityeconomies.org.)

but it is certainly true that feminism provides an approach that facilitates other motives in economic transactions. In particular, the attentiveness to otherness and difference that is a hallmark of feminism provides a basis for extending an "ethic of care." In the case of regional economic development, for example, feminist economic geographers have suggested that there is a need to "find innovative and creative ways of inviting ourselves and earth others [that is, the environment] into a different developmental relationship, one that denies domination and explores mutuality and interdependence" (Gibson-Graham, 2011, p. 15).

Finally, for some feminists, the big picture of an all-powerful capitalist system is a very masculine way of thinking about the world. Some have argued instead that capitalism should be seen as just one of many ways in which economic relationships are organized. Thinking in this way opens up the possibility of diverse economies. This was the key point behind the "iceberg" diagram in Figure 2.7, which suggests that there are many economic relations that exist outside of waged work for capitalist firms in a market economy. This point sits within a larger argument that a feminist approach to the economy looks beyond the overwhelming power of capitalism. This involves viewing women and men as having more diverse dimensions to their lives and identities than the way in which they participate in capitalist production, labor, and exchange relationships (Gibson-Graham, 2006a). In advocating alternative/diverse forms of economy, this project has a great deal in common with some of the ideas about alternative economies that we discussed in Chapter 2.

13.8   Summary

In this chapter we started by observing that unpaid work in the home is highly gendered, and this provided an important foundation for much of what followed. In exploring how women are increasingly participating in the waged labor force around the world, we saw that their work was often an extension of gendered roles within the household. Even where women do work that is similar to men's, it is often seen as unskilled, de-professionalized, and secondary to the "breadwinner" role of a male spouse. It is therefore paid less. We also saw that patriarchal relations may be reproduced in the workplace, thereby creating exclusionary spaces and barriers to women in the labor market. The increasingly interactive nature of service-sector employment means that employees have to *embody* the job and not just *do* the job. In this way, gender and sexuality are fundamental to whether someone is seen to "fit" with an occupation.

Women's participation in the labor force may be constrained in practical ways by their domestic roles. These constraints have a distinctly geographical dimension as they relate to the spatial integration of unpaid domestic work and paid non-domestic work. Some similar processes arise when entrepreneurship rather than employment is considered. Here, new businesses also rely on networks and support systems, and these tend to be gender-specific. Meanwhile, among the poorest in rural areas of the global south, women may sustain their families' livelihoods through noneconomic gathering and foraging of natural resources – activities that are vulnerable to environmental change and commodification.

Adopting a feminist perspective on economic processes allowed us to see the ways in which patriarchal power affects the economic lives of women and men. But such a perspective also opens up the possibility of rethinking how we view the economy – not just counting something not currently counted, but also envisioning an economic world that is based on different values and ethics as well.

Notes on the references

- In the last decade, several good overviews of the field of feminist geography have been published: Nelson and Seager (2005), Moss and Al-Hindi (2008), and Peake and Rieker (2012).
- For more on the devaluation of third world women, see the work of Wright (2006) and Cravey (1998) on Mexico and Silvey (2003) on Indonesia.
- At the other end of the spectrum, McDowell's (1997) study of banking in London is full of insights, and research by Mullings (2005) in the Caribbean examines gender and race in banking in an entirely different context. McDowell (2009) provides an excellent account of the performance of gender in a variety of service-sector settings.
- At various points in this chapter we have noted the variation in gender regimes across space. This is an idea developed in some detail by Massey (1984) through her *spatial divisions of labour* concept discussed in Chapter 3.
- On female entrepreneurship see Molloy and Larner (2013), Ekinsmyth (2011), and Hanson (2009).

Sample essay questions

- Does increasing access to the waged labor force represent liberation from patriarchal power structures for women?
- How might workplaces become masculine or feminine environments?
- How does the division of domestic labor within the household affect women's experiences of work in the waged labor force?
- Thinking about both advanced industrialized economies and developing countries in the world, do you believe that contemporary processes of economic restructuring favor men or women?

Resources for further learning

- Catalyst is a nonprofit organization committed to expanding opportunities for women in business. It is based in Washington, D.C., and has offices in Canada, Europe, and India. Its website, www.catalyst.org, provides a wealth of data and studies on gender in the workforce. Many of the figures cited in this chapter are derived from Catalyst reports.
- The Organization for Economic Cooperation and Development (OECD) focused its attention in 2011 on the issue of unpaid work. Its "Society at a Glance" publication provides extensive data and discussion: www.oecd.org/els/social/indicators/SAG.
- The Maquila Solidarity Network has campaigns and further information concerning women's work in export factories around the world: http://www.maquila solidarity.org.

- The United Nations Development Fund for Women (UNIFEM) has information on women's work around the world: http://www.unifem.org
- The Community Economies website provides thinking and examples of the diverse economies inspired by feminist and queer theorizing: www.community economies.org

References

Blake, M. (2006). Gendered lending: gender, context and the rules of business lending. *Venture Capital: An International Journal of Entrepreneurial Finance*, **8**, 183–201.

Brown, G. (2009). Thinking beyond homonormativity: performative explorations of diverse gay economies. *Environment and Planning A*, **41**, 1496–1510.

Cameron, J., and Gibson-Graham, J. K. (2003). Feminising the economy: metaphors, strategies, politics. *Gender, Place and Culture*, **10**, 145–57.

Cravey, A. (1998). *Women and Work in Mexico's Maquiladoras*. Oxford: Rowman and Littlefield.

Dowling, R., Gollner, A., and O'Dwyer, B. (1999). A gender perspective on urban car use: a qualitative case study. *Urban Policy and Research*, **17**, 101–10.

Ekinsmyth, C. (2011). Challenging the boundaries of entrepreneurship: The spatialities and practices of U.K. "Mumpreneurs." *Geoforum*, **42**, 104–14.

Gibson-Graham, J. K. (2006a). *The End of Capitalism (as We Knew It): A Feminist Critique of Political Economy*, Second Edition. Minneapolis: University of Minnesota Press.

Gibson-Graham, J. K. (2006b). *A Post-Capitalist Politics*. Minneapolis: University of Minnesota Press.

Gibson-Graham, J. K. (2011). A feminist project of belonging for the Anthropocene. *Gender, Place and Culture*, **18**, 1–21.

Gray, M., and James, A. (2007). Connecting gender and economic competitiveness: lessons from Cambridge's high-tech regional economy. *Environment and Planning A*, **39**, 417–36.

Hanson, S. (2009). Changing places through women's entrepreneurship. *Economic Geography*, **85**, 245–67.

Hanson, S. and Blake, M. (2009). Gender and entrepreneurial networks. *Regional Studies*, **43**, 135–49.

James, A. (2011). Work–life (im)"balance" and its consequences for everyday learning and innovation in the New Economy: evidence from the Irish IT sector. *Gender, Place and Culture*, **18**, 655–84.

Massey, D. (1984). *Spatial Divisions of Labour*. Basingstoke, U.K.: Macmillan.

McDowell, L. (1997). *Capital Culture: Gender at Work in the City*. Oxford: Blackwell.

McDowell, L. (1999). *Gender, Identity and Place: Understanding Feminist Geographies*. Minneapolis: University of Minnesota Press.

McDowell, L. (2003). *Redundant Masculinities? Employment Change and White Working Class Youth*. Oxford: Blackwell.

McDowell, L. (2009). *Working Bodies: Interactive Service Employment and Workplace Identities*. Oxford: Wiley-Blackwell.

Molloy, M., and Larner, W. (2013). *Fashioning Globalization: New Zealand Design, Working Women and the "New Economy"*. Oxford: Wiley-Blackwell.

Moss, P., and Al-Hindi, K. F. (2008). *Feminisms in Geography: Rethinking Space, Place, and Knowledges*. Lanham, MD: Rowman and Littlefield.

Mullings, B. (2005). Women rule? Globalization and the feminization of managerial and professional workspaces in the Caribbean. *Gender, Place and Culture*, **12**, 1–27.

Nelson, L., and Seager, J., eds. (2005). *Companion to Feminist Geography*. Oxford: Wiley-Blackwell.

OECD (2011). *Society at a Glance 201 – OECD Social Indicators*. www.oecd.org/els/social/indicators/SAG.

Oswin, N. (2010). The modern model family at home in Singapore: a queer geography. *Transactions of the Institute of British Geographers*, **35**, 256–68.

Peake, L., and Rieker, M. (2012). *Interrogating Feminist Understandings of the Urban*. London: Routledge.

Peck, J., and Theodore, N. (2010). Labor markets from the bottom up. In S. McGrath-Champ, A. Herod, and A. Rainnie, eds., *Handbook of Employment and Society: Working Space*, pp. 87–105. Cheltenham: Edward Elgar.

Preston, V., McLafferty, S., and Liu, X. F. (1998). Geographical barriers to employment for American-born and immigrant workers. *Urban Studies*, **35**, 529–45.

Rapino, M., and Cook, T. (2011). Commuting, gender roles, and entrapment: a national study utilizing spatial fixed effects and control groups. *The Professional Geographer*, **63**, 277–94.

Shearmur, R. (2006). Travel from home: an economic geography of commuting distances in Montreal. *Urban Geography* **27**, 330–59.

Silvey, R. (2003). Spaces of protest; gendered migration, social networks, and labor protest in West Java, Indonesia. *Political Geography*, **22**, 129–57.

Statistics Canada. (1995). *Households' unpaid work: measurement and valuation*. Ottawa: Statistics Canada.

Werner, M. (2012). Beyond upgrading: gendered labor and firm restructuring in the Dominican Republic. *Economic Geography* DOI: 10.1111/j.1944-8287.2012.01163.x.

Willis, P. (2009). From exclusion to inclusion: young queer workers' negotiations of sexually exclusive and inclusive spaces in Australian workplaces. *Journal of Youth Studies*, **12**, 629–51.

World Bank. (2011). World Development Indicators, http://data.worldbank.org/indicator.

Wright, M. (2006). *Disposable Women and Other Myths of Global Capitalism*. New York: Routledge.

Ye, J., and Kelly, P. F. (2011). Cosmopolitanism at work: labour market exclusion in Singapore's financial sector. *Journal of Ethnic and Migration Studies*, **37**, 691–707.

CHAPTER 14

ETHNIC ECONOMIES
Do cultures have economies?

Goals of this chapter

- To explore how finding a job can be shaped by one's ethnic identity
- To trace the emergence of ethnically-based economies in immigrant cities
- To examine the transnational dimensions of ethnic businesses
- To consider the implications of incorporating ethnic identity as a factor in economic life

14.1   Introduction

Toronto is, according to one of its own slogans, "the World in a City." Despite being an exercise in civic boosterism like those of so many other cities in North America and Western Europe, this image of global inclusiveness is one that the city can claim with some justification. With around 5.5 million people in the Greater Toronto Area in 2011, sprawling along the north shore of Lake Ontario and into the rolling farmland of southern Ontario, the city is Canada's financial, manufacturing, service, artistic, and academic hub. It is also home to immigrants and their descendants from almost every place on earth: few other cities combine the same proportion of foreign-born citizens (50% in 2006) with such a diversity to their places of origin (over 170 "home" countries), and few can claim as large a proportion of "visible minorities" (47% in 2006) among their population. Home to 8% of Canada's population, Toronto accounts for 20% of all immigrants to Canada in the 2000s so far (www.toronto.ca, accessed on January 3, 2012).

This is a feature of Toronto's social composition that a visitor would notice immediately upon landing at Pearson International Airport. Arriving passengers (and the relatives and friends waiting for them) would be as diverse as could be found anywhere, and so would the service-sector workers who help the visitor during a stay in Toronto. But among that diversity of humanity, the observant visitor would start to notice some patterns emerging. All of the airport taxi drivers seem to be South Asians, and a brief conversation might reveal that the driver has a master's degree in engineering and comes from Punjab in northern India. Arriving at the hotel, the housekeeping staff tends to be almost exclusively Filipina or black Caribbean women. Stepping out to pick up a newspaper, the visitor might discover that the store around the corner is being run by a South Korean couple, and so are the convenience stores a few blocks away in either direction. A shopping trip to a suburban mall reveals a linguistic environment where very little English is heard, all signage is in Chinese, and Cantonese (a southern Chinese dialect) is widely spoken, reflecting the ethnicity of both retailers and the overwhelming majority of shoppers. It seems, then, that the socio-spatial distribution of people among different jobs, sectors, and neighborhoods in this diverse metropolis is not entirely random. There is a visible pattern in which certain groups are either allocated to particular jobs in the labor force, or in which a particular entrepreneurial activity and neighborhood are dominated by a specific ethnic group. Toronto is not unusual in this respect: in every large and diverse city around the world, ethnicity appears to play a part in shaping people's experiences of the workplace, the commercial landscape, and more broadly the economy.

These are also fundamentally *geographical* processes. The sorting of various ethnic groups into different labor market niches often reflects the ways in which their places of origin or new residence are represented, or they reflect the geographical challenge of getting from ethnic residential neighborhoods to places of employment. The creation of ethnic commercial clusters within a city – Chinatown, Little Italy, Little India, for example – reflects distinct spatial patterns of economic activity that shape the economic lives of residents in both positive and negative ways. These ethnic neighborhoods are intentional attempts over time by various actors to reaffirm their ethnicity at the local scale in response to changing demographic and economic forces. In many urban economies, these attempts can start simply as residential congregations of similar ethnic groups that eventually translate into occupational segmentation and/or ethnic business clusters. Thus, we see a geographically dynamic relationship between ethnicity and urban economies. Finally, international migration flows into cities create social and economic ties with migrants' places of origin such that significant flows of capital now pass between households in different parts of the world, and in some cases transnational businesses and entrepreneurial activities are established with practices that appear to be distinctively ethnic forms of capitalism.

In all of these ways, we will argue that ethnicity matters a great deal in the geographies of economic life. In Section 14.2 we will start by examining the taken-for-granted argument that economic processes and logics are blind to ethnic difference. We then begin to refute that argument by looking at the role of ethnicity, both in the creation of a differentiated labor force (Section 14.3) and in the formation of distinctive ethnic economies (Section 14.4). In Section 14.5 we then look at a growing (but not necessarily new) phenomenon of transnationalism among migrant communities and explore some of its economic implications. Finally, we ask whether there is a specific form of business practice (even a distinctive form of capitalism) to be found in ethnic economies (Section 14.6).

14.2   "Color Blind" Economics

The above examples from Toronto allude to two ways in which ethnicity is important in shaping urban economies: first, the sorting of the workforce so that certain ethnic groups are disproportionately concentrated in particular kinds of jobs, and second, the spatial patterning of urban landscapes so that businesses owned by certain ethnic groups tend to cluster in particular neighborhoods and subsequently shape the economic lives of these ethnic populations and other residents in the same city. Before we elaborate upon a geographical interpretation of these processes, it is worth considering the notable *absence* of any role for ethnic identity in conventional economic analyses of these phenomena.

In Chapter 13, we noted the narrow analytical horizons of human capital theory in relation to the experience of women in the labor market. The same argument could be made for ethnic minorities. Conventional economic analysis has very little space for human diversity. Instead, a person's place in the labor market, meaning the job and wage they obtain, is seen as a reflection of their human capital: the collection of skills, qualifications, and experience they bring to the market for human labor. The labor market is, in theory, blind to the visible or cultural differences that an individual might exhibit. Every employee is simply a unit of labor input in the overall production function of a particular economic process, be it agriculture, manufacturing, or service. Everyday experience suggests, however, that this is far from true in reality. As we will show in this chapter, ethnicity is an essential variable in understanding who gets which jobs (and where) in the labor market.

Likewise, any social or cultural motivations for establishing or patronizing a business enterprise are obscured by conventional economic approaches, which focus on the incentive to profit from satisfying a market demand for particular goods and services that is driven by the price mechanism. Economic processes, in this form of technical analysis, become anonymous, universal, and rational. There is no room for personal ties, social relations, loyalty, discrimination, or culturally distinct practices. Indeed it is often assumed that modern capitalism sweeps aside these sentimental bases for economic action, and they are taken to be vestiges of "traditional" small-scale, artisanal, or even agrarian societies. In reality, we will see that the diverse enterprises established by ethnic minority groups are often a direct response to the marginalization they experience in relation to the mainstream economy and society.

The *location* of economic activities is also seen as the outcome of rational decision making. In the economics of retailing, as well as in Economic Geography's past focus on location theory, the decision to locate an enterprise was rooted in an unerring economic rationality and logic. This logic was one based upon profit maximization and cost minimization brought about, for example, by maximizing the consumer base available to a retailer, and minimizing the distance to be traveled by customers (and thus transport costs). This approach is captured by central place theory (see Section 11.2), which sought to model the ideal location for retailing and service activities in particular. In maximizing the available market area, a retailer would try to locate as far as possible from other similar retailers, as close as possible to customers, and consumers would minimize their travel costs by visiting the nearest supplier of a particular product. The urban landscape, however, tells another story that is much more complicated and "irrational." For any business, there may be advantages to be gained from clustering together with similar enterprises (as discussed in Chapter 12), but the social geography of many cities reveals a pattern in which it is not just similar enterprises that cluster together in space, but rather enterprises owned and patronized by people with similar ethnic backgrounds. Clearly, there are processes at work that go beyond either economic logic or the less tangible economic benefits of agglomeration.

Another issue that mainstream Economics tends to neglect is the strong set of economic ties that bind immigrants in global cities with other parts of the world. Local labor markets are usually treated as just that – local. Individual and household decisions are analyzed on the assumption that they are made in a local context of opportunities, barriers, skills, and qualifications. In reality, however, lived experiences are more complicated. Many immigrants in world cities have responsibilities to family members back in China, the Philippines, India, Ghana, Mexico, and so on. This will sometimes force them into taking on several jobs in order to earn enough to satisfy these distant needs. When looking for jobs, immigrants will often depend upon their preexisting social networks of family and co-ethnics, usually forged long before they arrived, rather than upon "objective" sources of information about opportunities and labor market conditions. Ethnic and family ties, transnational in their scale, are therefore an essential part of the economic lives of immigrants.

Clearly we need explanations for a variety of economic processes that go beyond economic logics to include cultural identities, emotional belongings, and social practices. While there are many dimensions to cultural identity and its intersection with economic processes (for example, one could explore the implications of youth culture or gay culture), here we will focus on ethnicity as the key variable. Box 14.1 explains the concept of ethnicity in more detail.

KEY CONCEPT

Box 14.1   What is Ethnicity?

Ethnicity is about difference, marked or coded in various ways. Thus when we talk about Chinese entrepreneurship, we are generally not talking about businesses in China itself, but about the business practices of the ethnic Chinese diaspora around the world. Chinese ethnicity only becomes significant when juxtaposed with different groups. In that sense, ethnicity is always a relational concept, and as a collective identity it includes *and* excludes at the same time. Ethnicity is based upon a common (real or imagined) historical experience, ancestry, or cultural commonality – for example, due to language or religion.

Ethnicity does not refer simply to minority groups, although it is often used in that way. To be ethnic is not to be somehow outside the mainstream or abnormal. In fact, everyone has an ethnicity. In that sense, a British pub in Bangkok owned by an expatriate is no less an ethnic enterprise than a London restaurant owned by a Thai.

Ethnic identities are not necessarily stable and unchanging. Different aspects of ethnic identity may be brought to the fore in different circumstances, and the strength of an ethnic identity may only emerge over time. For example, Italian immigrants may have arrived in North America in the 20th century with small-scale loyalties to villages and regions, but once there they developed an ethnic identity as Italians.

It is important to distinguish ethnicity from race. Race refers to the visible characteristics of human beings (hair, skin pigmentation, bone structure) but is a very unhelpful analytical category because most "races" are poorly defined and show as much internal diversity as they do distinctiveness. The notion of race tends to be used when groups with a particular appearance are racialized – that is, categorized and attributed certain characteristics by other groups. For this reason it is usual to recognize "race" as a social construction and place the word in quotation marks.

14.3   Ethnic Sorting in the Workforce

The market for labor is unlike any other market. Labor, as we saw in Chapter 6, cannot easily be bought and sold like other commodities. When employees are hired, fired, or promoted, the process of buying and selling is rather more complicated. Labor power comes in the form of complete and complex human beings who have many more attributes than simply their ability to perform a particular function in a workplace. They have gender, ethnic, class, regional, and a variety of other identities that often lead to an unfortunate predictability to the type of work they will end up doing.

To illustrate this point, we can look closely at the labor market experiences of different groups in a major metropolis. Table 14.1 provides data on employment in the Los Angeles metropolitan area, another ethnically diverse world city. It shows the distribution of four very broad ethno-racial groups (white, Asian-American, black, and Hispanic), as defined by the United States Equal Employment Opportunity Commission. The table lists a variety of occupational types and then the number of employees in each category. For each occupation category, a "concentration index" is also listed. This is simply the percentage of a particular group in a given occupation as a ratio of the percentage of all employees in a given occupation. Thus, where we see a concentration index of 1.6 for white employees in the "Officials and Managers" category, we know there is a significant overrepresentation of white employees in that occupation: 1.6 times as many as one would expect if all occupations were shared evenly among different groups.

By examining the shaded boxes, showing distinct levels of over-concentration (>1.0 in the index), we can see some patterns emerging. Most notably, white employees are far overrepresented in managerial and professional occupations, and Asian-Americans are over-concentrated in professional and technical jobs. Conversely, both groups are underrepresented in less desirable and less well-paid jobs, such as "laborers" and "service workers." For blacks and Hispanics, however, we see the reverse pattern, with underrepresentation in high-level managerial and professional jobs, and overrepresentation in these low-end occupations. These figures could be analyzed in much more depth. In some cases the figures conceal the kinds of gender disparities discussed in Chapter 13, and certainly there is great diversity within the categories used. "Asian-American," for example, bundles together a vast range of experiences from Taiwanese software engineers to Vietnamese refugees. But a broad pattern is clear – one in which blacks and Hispanics disproportionately occupy subordinate positions in the labor market.

A similar story, in which minority populations (in some, but not all cases, immigrants) are marginalized in the labor market, is replicated in many cities around the world (see Box 13.3 on labor market segmentation). Within these cities, ethnic minority workers from different origins are known to be active in different districts. In London, for example, low-paid migrant workers are sorted into different segments of the local labor markets in different boroughs (urban districts): Kurds in Hackney, South Asians (originating from the Indian subcontinent) in Ealing, and people of black Caribbean descent in Lambeth (Holgate et al., 2011). To Wills et al. (2010), this often fine-grained sorting of the migrant workforce into low-paid jobs in global cities represents a "new migrant division of labour."

Curiously, such marginalization does not happen only in cosmopolitan cities such as Los Angeles or London, but also in such "ordinary" cities as Urumqi (capital of China's Xinjiang autonomous region) and Marseille (a port city

Table 14.1   Distribution of ethno-racial groups in various occupations, Los Angeles, 2009

	All Employees	White	Asian-American	Black	Hispanic
Officials & Managers	228,052	142,320	29,977	11,817	39,930
Concentration Index		**1.64**	0.93	0.62	0.48
Professionals	344,699	172,769	97,972	20,223	44,874
Concentration Index		**1.32**	**2.01**	0.70	0.35
Technicians	96,896	35,037	23,553	8,198	27,010
Concentration Index		0.95	**1.71**	**1.00**	0.76
Sales Workers	233,809	104,325	19,083	18,740	84,668
Concentration Index		**1.17**	0.58	0.95	0.98
Office & Clerical Workers	282,559	98,918	39,908	32,149	102,748
Concentration Index		0.92	**1.00**	**1.35**	0.99
Craft Workers	96,991	30,705	10,366	5,193	48,848
Concentration Index		0.83	0.75	0.64	**1.37**
Operatives	157,616	25,162	16,696	12,708	100,723
Concentration Index		0.42	0.75	0.96	**1.73**
Laborers	133,017	15,798	7,059	7,760	100,239
Concentration Index		0.31	0.37	0.69	**2.05**
Service Workers	342,177	103,186	26,873	44,613	156,825
Concentration Index		0.79	0.55	**1.55**	**1.24**
Total Employment	1,915,816	728,220	271,487	161,401	705,865

Note: This relates to private-sector employment only. Public-sector employees were not included. The ethno-racial categories used are those employed by the U.S. Equal Employment Opportunity Commission.
Source: Calculated from U.S. Equal Employment Opportunity Commission data for 2009 (www1.eeoc.gov, accessed on January 3, 2012).

in southern France). As noted in Figure 14.1, Urumqi is the capital city of an ethnically diverse region bordering Russia, Mongolia, Kazakhstan, Kyrgyzstan, Tajikistan, Afghanistan, Pakistan, and India. The recently enacted "Open Up the West" policy by the Chinese government in 2001 has helped to create a vibrant labor market in this city (Howell, 2011). The "frontier" status of the region means that Xinjiang's almost 22 million population is today constituted by nine or more diverse ethnic groups. The two dominant groups are Uyghur (45%) and Han (41%). And yet of Urumqi's 2.4 million population, over 72% of the population belong to ethnic Han Chinese, who constitute about 92% of China's total population. Ethnic Uyghur, Hui, Kazak, and others are the minorities in the capital city, but they tend to concentrate in several city districts, particularly Tianshan (40% Uyghur and 24% Kazak) and Shayibak.

Howell (2011) found that Han Chinese tend to control local natural resources (e.g., oil and gas) and high-skill technical, administrative, and professional jobs. Uyghurs are typically excluded from the industrial job market and the energy service sector. Uyghur migrants from the less developed southern Xinjiang have little choice but to work in low-paying service jobs (e.g., petty vendors and the informal sector), a phenomenon not unlike that of the Hispanics in Los Angeles shown in Table 14.1. Even in such informal sector jobs, further ethnic segmentation occurs in Urumqi. Han migrants tend to be engaged in a larger range of service work, including selling specialty foods, fruits, appliances, clothing, jewelry, and shoe vendors, key makers, truck drivers, painters, construction, bike and electronics repair, and shoe shiners. Uyghur migrants in the informal sector tend to concentrate in selling foods (e.g., flat bread and roast lamb meat) or goods of higher value (e.g., jewelry and electronics) on the streets. The case of Urumqi shows that ethnicity is an important factor in channeling residents and migrants from within the *same* country (China) into particular segments of the labor market.

Key questions are: Why does the workforce get sorted in this way? Why are certain groups drawn into certain jobs? Why do employers take on disproportionate numbers of a certain ethnic group, thereby apparently excluding others? Why does an ethnic group continue to congregate in what is often an unappealing occupation? A wide range of processes can be identified, each with an important *geographical* dimension:

- *Qualifications and skills*: The standard economic argument is that workers are allocated to jobs based on their qualifications and skills. Thus a subordinate position in the labor market would be expected if a group, as a whole, has relatively low levels of educational achievement. This assumes, however, that educational attainment is a reflection of innate abilities. In reality, it is often a reflection of unequal distribution of wealth and income within and between different ethnic groups that in turn affects the ability of parents to support their children's studies both financially and intellectually; there is therefore a strong process of intergenerational reproduction of low or high educational

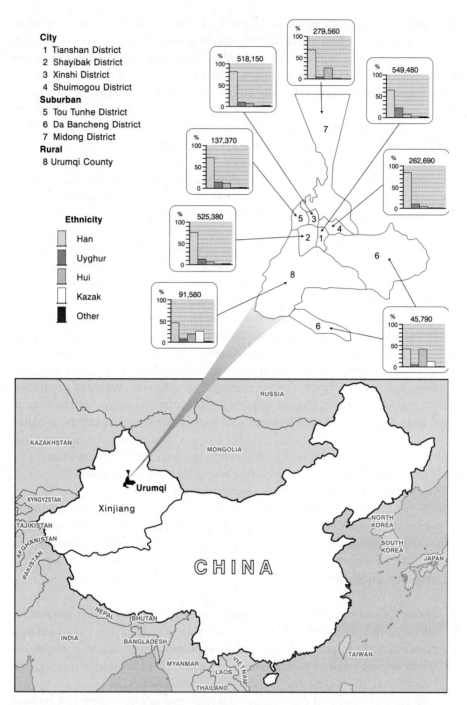

Figure 14.1 Population of Urumqi in China's Xinjiang autonomous region by ethnicity and district, 2009

Source: Data from Howell (2011), Table 1.

achievement. Different education attainment is a product of social relations. Furthermore, this difference is powerfully *place-based*: if you grow up in a working class community where very few of the youth go off to college, the chances that you will do so are equally slim, even if another context might have made this a possibility. But even if we accept that education is a reflection of innate abilities and skills, the connection between this and employment opportunities is far from clear. Immigrants arriving in cities such as Vancouver, Los Angeles, London, Hong Kong, and Sydney are often more educated than the native-born population. And yet they find themselves disproportionately concentrated in low-status, low-paid, and insecure jobs because their qualifications may not be recognized in a new context, or because they have developed skills that are specific to a particular sociocultural context. The skills of a civil servant, teacher, journalist, architect, graphic designer, or marketer may not "travel" well. In this way, the human capital associated with education or professional experience can be substantially *devalued* across international space. This points, in part, to cultural differences across space, but it also points to institutional barriers.

- *Institutional barriers*: We have noted that many ethnic minorities who arrive as immigrants may have qualifications, but these are not treated as equivalent to the local credentials required by employers. In many labor markets, professions are regulated by institutions that decide who will be licensed to practice within their jurisdiction: colleges of physicians, bar associations, professional engineering societies, and so on. Where they do not recognize foreign qualifications, or demand costly upgrading courses, an immigrant may be destined for de-professionalization. In other cases, certain immigration programs may place restrictions that limit an individual's ability to upgrade their qualifications or access their profession. In Canada, for example, many Filipina women have arrived under a program that permits full immigration after two years of work as a domestic caregiver in the house of an employer. Its restrictive terms mean that while nurses, teachers, and accountants may enter the program, those who graduate from the program are usually destined for low-paid work in healthcare or personal service occupations. In the United States, migrants from countries such as Mexico and El Salvador who have arrived without legal status often find themselves without any choice but to accept insecure, low-paid, and sometimes dangerous work. In China, as demonstrated in the case of Urumqi, minority Uyghur migrants can be disadvantaged due to institutionalized systems of political control and preferential policies in favor of ethnic Han Chinese. In all of these instances, whether professional regulatory bodies, immigration programs, or undocumented migrants, it is the institutions that govern labor markets in particular territories that create barriers to *upward* mobility for *geographically* mobile people. Understanding the place of immigrants in labor markets, then, requires a close examination of localized institutional structures.

- *Discrimination and stereotypes*: The major immigrant-receiving countries around the world today (the U.S., U.K., Canada, and Australia, for example) present themselves as multicultural and tolerant societies and have laws that prevent discrimination or unfair treatment on the basis of ethnic identity. Nevertheless, when decisions related to hiring or promotion are made, employers and managers inevitably work within a set of cultural codes and understandings that are difficult to regulate. Over the last few decades, for example, hundreds of thousands of Filipina women have traveled overseas to work as domestic helpers, nannies, and nurses. In many countries, both in Asia and beyond, stereotypes have developed that equate being Filipino with aptitudes for nurturing, caring, or domestic work. Furthermore, within workplaces such as hospitals or nursing homes it is the "front-line" or "bedside" jobs that Filipinas tend to be doing, rather than managerial, supervisory, or instructional work. Rather than any innate ethnic aptitude, or lack of it, it would seem that this is due, at least in part, to the stereotypes of Filipina identity that exist in the labor market. We therefore see the stereotyped association between a particular *ethno-national space* and certain labor market outcomes. Such stereotyping may also take other more localized forms with certain neighborhoods becoming associated with certain kinds of (inferior and unreliable) labor. There is, then, the possibility of a localized neighborhood effect on labor market outcomes, and the more segregated and ghettoized an ethnic minority population becomes in urban social space, the greater the possibility of this occurring.

- *Home–work linkages*: In some cities, residential segregation on the basis of ethnicity is quite pronounced. Certain neighborhoods are strongly associated with particular ethnic groups. Where this is the case, for example in many large American cities, there is often a geographical problem to be resolved in getting from home to work. It may mean that many visible minority immigrants have to travel much further to work than their white and non-immigrant counterparts, a phenomenon known as *spatial mismatch*. This process is contingent on the geography of immigrant residences and workplaces (Wang, 2010). As depicted in Figure 14.2, mapping Chinese immigrants in the San Francisco metropolitan area points to three areas of high Chinese concentration of residence and work: San Francisco county (Chinatown), the Silicon Valley region, and the Oakland area. This residential neighborhood effect seems to be more important for women. Chinese women living in majority Chinese areas have a significantly greater chance of working in a job that accommodates their home–work constraints. There are, however, limits to this effect. In large metropolitan areas, ethnic residential enclaves may be sufficiently far from downtown or suburban concentrations of "good" jobs. For at least one adult in a family with children to pick up from school and other errands to run, being close to home may be necessary (see Chapter 13 on the gendering of

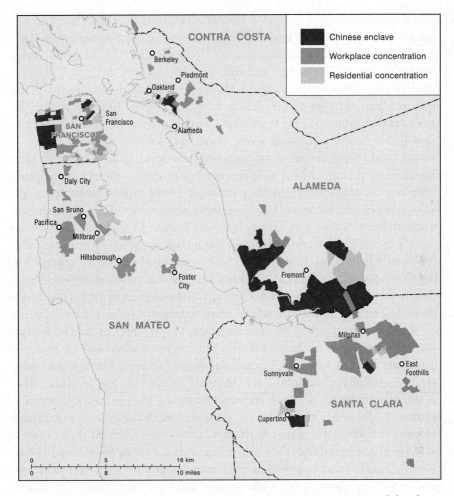

Figure 14.2 Geography of Chinese immigrants in the San Francisco Consolidated Metropolitan Statistical Area, by type of concentration pattern
Source: Adapted from Wang, Q. F. (2010) How does geography matter in the ethnic labor market segmentation process? A case study of Chinese immigrants in the San Francisco CMSA Annals of the Association of American Geographers, 100(1), 182–201, Figure 1.

such roles). Hence ethnic residential *segregation* can translate into occupational *segmentation* as the geographical scope of possible jobs is limited by the need to minimize commuting time and cost on public transport. Jobs may also themselves be on the move: where manufacturing or service industries have relocated to suburban areas, especially in North American cities, they have often left behind ethnic minority neighborhoods in downtown areas. In San Francisco, living in a residentially concentrated area lowers the propensity of Chinese male immigrants to work in jobs suitable to their circumstances, compared to those men living in non-Chinese-immigrant-concentrated areas (Wang, 2010).

- *Social capital and the search for a job*: The final feature of labor markets that needs to be incorporated into our explanation of ethnic segmentation is the role of social networks in finding work opportunities. The idea of a "labor market" assumes that buyers and sellers of labor power (i.e., employers and their employees) will know about all possible alternatives in terms of the jobs available, current market trends in wages, and so forth. In reality, the labor market is far less easy to navigate – many people will find jobs because they had contacts who hired them, or actively lobbied for their hiring, or simply made them aware of an imminent job opening. In particular, these social networks are critical to labor market processes for new immigrants who may otherwise have difficulty knowing where to start in finding a job. Such networks do, however, have distinctive geographies. They are frequently *neighborhood based*, and as we saw in Chapter 13 and the case of San Francisco, evidence suggests that this is more often the case for women than for men. Where ethnic minorities exhibit patterns of residential segregation, the likelihood of a neighborhood-based and co-ethnic social network being used is very high. There are two important consequences of this geographical process. The first is that an ethnic group concentrated in a particular line of work will perpetuate its concentration in that line of work. Second, such co-ethnic-network-based referrals tend to work when seeking low-level positions, such as jobs in kitchens, hotels, or factories, but a whole other set of contacts and networks are required to get managerial work. Thus a reliance on social ties and networks tends to keep ethnic minorities in certain sectors and in certain jobs in those sectors. Breaking out of particular sectors, and particular places in the city, can be very difficult. For example, a recent study found that different ethnic immigrants in Swedish cities may initially benefit from help within intra-ethnic concentrations for job seeking and income generation, but such benefits can turn into disadvantages over time as these minority immigrant residents become locked into certain labor market segments (Musterd et al., 2008).

There are, then, multiple processes operating to channel certain groups into particular occupations or parts of the labor market. In all cases, though, there is an important *spatial* dimension to these processes. First, place-based processes are often important in determining what kinds of qualifications a person will bring to the labor market, and the place where they come from may be stigmatized by employers. Second, this spatial bias extends to particular *ethno-national* identities, such as Filipinas, who may find themselves "coded" or represented in certain ways in international labor markets. Third, the construction of institutional barriers to entering certain kinds of professions is based on administrative territories and jurisdictions, and the migrant populations that move between them are often de-professionalized in the process. Fourth, the geographical balancing act that workers have to engage in between work places and home places can lead to limited employment opportunities, especially when ethnic groups are residentially concentrated in urban space. Finally, the social networks that are so important in finding jobs are usually local and place-based in nature and lead to certain kinds of (low-paying) work.

14.4   Ethnic Businesses and Clusters

We have now seen the ways in which ethnic groups (many of them immigrants) are differentially sorted into various kinds of jobs in urban labor markets. In some cases, ethnic minorities have responded to marginalization in the labor market by establishing their own businesses and, as we shall see, their own commercial neighborhoods. This is an *entrepreneurial process* with a long history. For example, some Chinese and Jewish neighborhoods in major cities around the world can be traced back several centuries. It is, however, a phenomenon that has become even more prevalent with increasing flows of migration since the last quarter of the 20th century. Major immigrant cities may now have a distinctive Chinatown, Koreatown, Little India, Little Tokyo, Greektown, Little Italy, Portuguese Village, and so on (part of Singapore's Little India is shown in Figure 14.3). In this section we will address two issues: the reasons for the emergence of ethnic entrepreneurship, and the processes that sometimes lead to the spatial clustering of such enterprises in the urban landscape.

Figure 14.3   Singapore's Little India

Ethnic Entrepreneurship

Many ethnic minority groups, especially immigrants, have turned to self-employment as a means of livelihood. Indeed, immigrants from minority ethnic groups often show a higher tendency to establish their own businesses than members of the host societies in which they have settled. The most visible, and identifiably ethnic, of these businesses would be restaurants, grocery stores, laundrettes, video rental shops, telecommunications shops, and travel agencies – those enterprises that are observable on the urban landscape. As Figure 14.4 illustrates, some businesses may be established to cater to multiple needs of minority immigrants – in this case San Francisco Street in Bilbao, Spain, houses an ethnic restaurant next to a telephone and internet shop that cater to immigrants from Africa (e.g., Morocco) and Spanish-speaking countries in South America (e.g., Bolivia and Colombia). Beyond these activities, ethnic enterprises may also extend into a full range of less visible activities, including banking, real estate, construction, and manufacturing.

Different ethnic groups have varied in their tendency to establish businesses in their host societies. Where immigrant groups have been quickly assimilated into the waged labor force (for example, British, Jamaican, or Filipino immigrants to Canada arriving with the necessary linguistic skills) then levels of self-employment are generally low. But where the cultural barriers to employment are difficult to surmount, self-employment has been more common. This has been especially true where immigrant communities have also arrived with sufficient capital – for

Figure 14.4   Ethnic businesses on San Francisco Street, Bilbao, Spain

example, in the case of recent South Korean or Taiwanese immigrants to the United States. Indeed, it is with resources from within the ethnic community that enterprises are often established. Box 14.2 relates the story of Korean convenience store owners in Ontario, Canada, and highlights the importance of these *ethnic resources*.

CASE STUDY

Box 14.2   Ontario's South Korean convenience stores

South Korean immigration to Canada was minimal until new immigration laws in 1967 made the process less biased toward white European immigrants. Immigrants from South Korea started to arrive in the early 1970s, but it was in the 1990s that a significant immigration stream emerged. Of almost 100,000 Korean immigrants living in Canada by 2006, 60% had arrived in the previous 10 years, reflecting a huge surge in immigration following the financial crisis in South Korea in the late 1990s. A remarkable 29% of those in the labor force were self-employed (compared with 12% for native-born Canadians), reflecting the fact that nearly half of all Korean immigrants in the 1990s had entered Canada through the business immigrant program. Many of these Korean entrepreneurs established video rental shops, restaurants, dry cleaners, and, in particular, convenience stores. By 2005, there were 3,200 convenience stores operated by Korean immigrants in the province of Ontario alone.

Why did so many Korean immigrants establish convenience stores? On one side of the equation, from the late 1980s onward a restructuring in the retail sector led several corporate owners of chains of convenience stores to sell off their retail outlets – hence a business opportunity presented itself at precisely the time when Korean business immigrants were arriving. It was also a business on a scale that could be operated by a family, required no specific technical know-how, supplied a modest but reliable daily cash income, and could be acquired for about the required investment under Canada's business immigration program (Canadian $300,000–400,000 depending on the place of settlement). But why Koreans in particular? In part this reflects the relatively low levels of English language fluency among Korean immigrants, which made only certain kinds of business possible. But a large part of the explanation relates to the co-ethnic support networks created by the Korean immigrant community. As noted earlier in this chapter, when a particular group gains a foothold in a particular sector of the economy or the labor market, it can lead to deepening concentration as new arrivals are guided by previous immigrants into the field that they themselves have entered. In the Korean case, this was also institutionalized through the Ontario Korean Businessmen's Association (OKBA). The OKBA was established in 1973 and emerged to become a major factor in channeling new Korean immigrants into retail enterprises, providing information, credit, and wholesale services to convenience store owners. Less tangibly, the OKBA also created a social environment in which members could find mutual respect and solidarity – a reinforcement of self-esteem in a Canadian context of de-professionalization and cultural difference (for more, see Kwak, 2002).

Co-ethnic networks may provide capital for the initial establishment of a business, and credit during its operation, but they may also offer important information on how to navigate the bureaucratic intricacies of establishing a new business, market intelligence on where opportunities are to be found, as well as the suppliers, employees, and customers that a new business needs. Such ethnic resources might amount to nothing, however, if the *opportunity structure* for establishing a business is not conducive. This relates to the opportunities and barriers that a host society presents. There may, for example, be government policies concerning the establishment of businesses in a particular sector. The general economic climate will also affect whether small businesses are able to take root and grow. Also, the size of an ethnic community and its spatial distribution might determine whether businesses can be established that will serve co-ethnics alone. In U.S. cities, the presence of larger co-ethnic consumer markets among blacks and Hispanics can create a more favorable social and political environment for business opportunities and ethnic entrepreneurship (Wang, 2009).

In some cases, ethnic minority groups may be forced into certain kinds of businesses or sectors. In the early part of the 20th century, Chinese immigrants in North America, for example, established restaurants and laundry businesses because they were largely excluded from all but the most menial of tasks in the waged labor force. This motivation for self-employment is often referred to as the *blocked mobility thesis*. In many Southeast Asian countries, such as Indonesia and the Philippines, where Chinese traders had settled for hundreds of years, citizenship was often denied and legislation in the 1950s explicitly excluded noncitizens from land ownership or any form of retail business. This forced many ethnic Chinese into wholesaling, banking, and manufacturing industries, which, ironically, proved to be far more profitable and dynamic sectors over the subsequent decades (Yeung, 2008). The same has been true for Jewish immigrants in North America, and elsewhere – exclusion from land ownership and skilled worker guilds in Europe forced many Jewish entrepreneurs to take on roles as "middleman" traders, shopkeepers, and moneylenders.

By the late 20th century, such institutionalized racism was less common, but ethnic differences (including, for example, linguistic and religious differences) were still a basis for the marginalization of some groups in immigrant-receiving societies. In addition, several major destination countries had put in place "business immigrant" programs that granted residency or citizenship to individuals with entrepreneurial experience and large amounts of financial capital to invest. In Canada, Australia, and New Zealand, for example, many ethnic Chinese from Greater China (China, Hong Kong, and Taiwan) took advantage of these programs and were thereby obliged to establish businesses. While these schemes have brought a huge amount of capital into these host countries, the enterprises themselves have often been less successful (Ley, 2010). Although such business immigrants had to demonstrate previous aptitude and success in entrepreneurial activity, the reality is that many fail to succeed in a new cultural and institutional environment. In some cases this is because of linguistic limitations, but in others it reflects unfamiliarity with labor practices, tax codes, as well as the less definable cultural dimensions of doing business. Business acumen is, therefore, to some extent *place-bound*, and success as an entrepreneur does not guarantee a universal ability to make enterprises work in a different context (a phenomenon that echoes the points made about national business cultures in Chapter 10).

Ethnic Business Clusters

We have noted some of the reasons that ethnic minorities have turned to self-employment and entrepreneurship, especially when arriving in a host country as new immigrants. In many instances, a further feature of these businesses is their congregation in *spatial clusters*, forming identifiable commercial districts such as a Chinatown or a Little India. It would seem that space therefore plays an important part in the emergence of an ethnic economy. As illustrated in Figure 14.5, these connections can work in several different ways (here we draw on Kaplan and Li, 2006; Wang, 2009).

First, ethnic business clusters are usually located in the heart of residential neighborhoods associated with the same ethnic group. The neighborhood is likely a source for employees, who may have been attracted to the area in the first place because of employment opportunities in ethnic enterprises. The advantage of such employment for new immigrants is that barriers such as nonrecognition of credentials, linguistic limitations, and discrimination are much less likely to present problems than in the mainstream labor market (as noted in Section 14.3). Where a business is specifically targeting co-ethnics (for example, a video rental or grocery store) then it also clearly benefits from this concentration of potential customers. Even where ethnic business clusters are not located in the midst of the ethnic residential concentrations, an identifiable neighborhood becomes a natural destination for members of that group when seeking goods and services, or simply when they want to operate in their native language.

Second, there may be *agglomeration economies* (described in Chapter 12) that generate benefits for businesses locating close to their competitors. Restaurants,

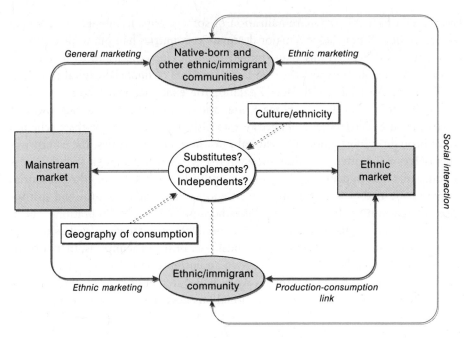

Figure 14.5 How does ethnicity matter in the formation of business clusters and markets?

Source: Adapted from Lo, L. (2009) The role of ethnicity in the geography of consumption. Urban Geography, 30(4), 391–415, Figure 1.

for example, might collectively benefit from proximity to grocery stores selling the types of ingredients that they require. They might also benefit from customers (especially those from other ethnic groups) who are attracted to the neighborhood as a whole as much as to an individual shop or restaurant within it. Large Chinatowns in New York, San Francisco, London, Manchester, and Sydney, for example, are tourist attractions in their own right and locating within them offers access to a wealthy tourist clientele, as well as ethnic Chinese clients living, working, or studying in these cities. In the United Kingdom, the Indian and Pakistani restaurants of London's Brick Lane or Manchester's "Curry Mile" in Rusholme similarly benefit from the agglomerations they form. In Chicago's Andersonville (see Figure 14.6), local business actors actively use a Swedish identity at the neighborhood level to create a sense of "Swedishness" to brand the locality. To preserve its urban village character, local communities and residents seek to direct visitors toward restored buildings, museums, and other attractions (Johansson and Cornebise, 2010).

Third, ethnic business clusters enable enterprises to access the ethnic resources mentioned earlier. Networks that provide capital, know-how, and market information are often based on face-to-face interaction, which requires spatial proximity. These networks have been shown to be critical in a variety of settings where ethnic minorities have successfully established business, and for

Figure 14.6   Chicago's Andersonville

them to work effectively, the clustering of related businesses is often a necessary prerequisite.

Finally, a spatial enclave collectively forms a tangible basis for the nurturing and maintenance of an ethnic identity. The architecture, languages spoken, products sold, and even faces seen in an ethnic business cluster ensure that identities are reinforced and positively affirmed. In other words, to be Chinese in Chinatown is to be reminded of one's "Chinese-ness," which in turn increases the likelihood that one will continue to demand Chinese goods and services. To be in Chicago's Andersonville is to be immersed into a "Swedish" environment. In this way, ethnic enterprises collectively benefit from their visible presence on the urban landscape in ethnic places.

Through a variety of processes, then, spatial proximity and the creation of ethnic places are important elements of an ethnic economy (see Figures 14.3, 14.4, and 14.6 again). A further effect of spatial clustering may be the increasing *closure* of an ethnic economy. This refers to the extent to which an ethnic economy is conducted exclusively by, and for, members of the same ethnic group. At the furthest extreme, an *ethnic enclave economy* involves a collection of entirely *intra-ethnic* economic relations, with ownership, management, employees, customers, and suppliers across a range of economic sectors all sharing the same ethnic identity. In reality, this situation is very rare. In most cases, ethnic businesses are made up of some combination of ethnic and non-ethnic participants in each of these roles. Indeed, the line between an ethnic and a mainstream enterprise can be a blurred one. However, where the clustering of ethnic businesses coincides with a distinct pattern of residential segregation, so that an ethnic community can live, work, and shop in the same neighborhood, then the possibilities of enclave formation increase (e.g., San Francisco in Figure 14.2).

There has been much debate concerning the desirability of spatial clustering and enclave formation among ethnic groups in major immigrant-receiving cities. On the positive side, the ethnic economy is seen as providing opportunities for new immigrants, access to particular market segments, and an array of mutual support mechanisms. On the negative side, however, ethnic economies are seen as marginalized, providing poor wages and working conditions, and holding limited opportunities for expansion. These various positions are summarized in Table 14.2. It is, however, important not to assume that the movie-induced stereotypes of an ethnic economy hold true in all cases. Informal contract arrangements, small-scale businesses, and low-skilled, low-paid work might still hold true in some places and for some groups, but a more complex picture must now be acknowledged. Many ethnic communities in the largest immigrant gateway cities now have sophisticated and formalized forms of business.

Table 14.2   The two sides of ethnic enterprise

Dimension	Positive	Negative
Causes for entrepreneurship	• Market opportunities • Entrepreneurial initiative	• Discrimination • Blocked mobility
Networks and organizational Structure	• Ethnic resources • Organizing capacity of ethnic groups • Cooperation in raising capital	• Excessive internal competition • Dependence on networks • Commodification of ethnicity
Networks and labor	• Job opportunities for immigrants • Efficient labor recruitment • Advantages due to low-cost labor • Returns to human capital	• Containment and segregation of ethnic groups • Poor pay rates and exploitative working conditions • Low returns to human capital
Networks and markets	• Access to protected markets • Successful business transactions	• Limitations of a closed ethnic market
General outcomes	• Internal support mechanisms and social mobility	• Exploitative labor relations and broader structural marginalization • Spousal exploitation

Source: Adapted from Walton-Roberts and Hiebert (1997), Table 1.

We should also note that ethnic economies have proven to be dynamic and mobile phenomena. In recent years, with a new generation of wealthier immigrants, older downtown ethnic neighborhoods in some cities have been joined by new ethnic residential and commercial concentrations in the suburbs. Dubbed *ethnoburbs* by some researchers, these are often not marginalized spaces of ghettoization, but are thriving concentrations of ethnic business that depend to a large extent on the ability of ethnic banks to finance their development. In Los Angeles, the ethnic Chinese banks mentioned above have, through both mortgage lending and commercial loans, brought about an increased concentration of ethnic Chinese in particular neighborhoods, including both the downtown Chinatown and the Chinese ethnoburb of Monterey Park. Similar dynamic processes are occurring in the San Francisco Bay area (see Figure 14.2). In Toronto (as well as in Vancouver's Richmond and Sydney's Chatswood), suburban neighborhoods have become the major destinations of new, and often wealthy, immigrants from Hong Kong, Taiwan, and China. Toronto's two downtown Chinatowns are now relatively minor residential and business concentrations for the Chinese population, and the presence of other groups such as the Vietnamese has increased in these areas. The primary centers of Chinese commercial activity are now the city's northeastern suburbs, where numerous Chinese shopping malls have been developed.

In all of these cases, we see ethnicity affecting the location of businesses in a way that is entirely unpredictable when economic analysis is blind to the importance of culture and ethnic difference. Ethnic bonds motivate businesses to cluster together, in part because ethnic bonds result in consumers and clients gravitating toward businesses that provide a familiar environment. But the very act of creating a cluster also strengthens those bonds as a place-based ethnic economy reinforces the sense of identity and difference among ethnic minorities.

14.5   The Economic Geographies of Transnationalism

So far we have seen the diverse ways in which ethnic enterprises are related to changing geographies at the scale of the city. Until recently, this was the spatial scale at which most geographical analyses of ethnic economies would stop. Ethnic entrepreneurs and ethnic economies were understood as a localized part of the integration and settlement of immigrants in a host city/society. Over the last two decades, however, it has become increasingly evident that immigrants (of any ethnicity) do not simply arrive, settle, and start anew. Instead, they maintain close ties with their country of origin. These ties are global in their reach and have been collectively labeled *transnationalism* (see Box 14.3). In this section we will consider two forms of transnational ties that are commonly established by

KEY CONCEPT

Box 14.3   Transnationalism

Since the early 1990s, the notion of transnationalism has been used to describe the multiple linkages that immigrants maintain with their places of origin. These linkages are not necessarily new. Immigrants have always maintained social, cultural, economic, and political linkages with their homelands. But in recent years several developments have intensified these linkages. First, the sheer growth in temporary and permanent migration, and the diversity of countries involved, has increased the total numbers of people living outside the country of their birth, hence the scale and scope of transnational linkages has multiplied. Second, technological developments have greatly facilitated the maintenance of strong ties: inexpensive telephone calls, email and text messaging, and even discount airfares have all enabled relationships to be continued across global space with an intensity that was not previously possible. Third, institutional structures increasingly recognize and even encourage transnational activities – overseas investment by diasporas, for instance, is positively encouraged by many governments, and is increasingly backed up with provisions for dual citizenship, absentee voting, and state support to maintain the bonds of overseas citizens. The result is an increasingly dense network of ties based on migration, flows of financial remittances and commodities, and flows of information. In all of these ways, then, contemporary transnationalism is not a new process, but a phenomenon that has increased dramatically in both its extent and its intensity.

migrants between home and host countries. The first relates to the sending of *financial remittances* to family members left behind in their home country. These funds often have significant economic effects at the national scale in the receiving country, but they may also transform very significantly the lives of households who receive them. The second relates to the development of *transnational business practices* that are rooted in ethnic bonds, personalized trust, and cultural practices.

Transnational Remittances

All migrants maintain ties of one sort or another with their home country, and for many there will be a *financial* dimension to this linkage. Temporary contract workers in particular maintain strong economic ties with their place of origin by sending money back home. Globally, officially recorded remittances to developing countries in 2010 amounted to US$440 billion, equivalent to

Table 14.3   Top twenty remittance-receiving countries in 2010

Country	Remittances in 2010 (US$ millions)	Remittances as Share of 2009 GDP (%)
Tajikistan	2,032	35.1
Tonga	111	30.3
Samoa	143	26.5
Lesotho	490	26.2
Nepal	3,507	23.8
Moldova	1,306	22.4
Lebanon	8,409	21.9
Haiti	1,499	21.2
Kyrgyz	1,160	21.7
Honduras	2,649	17.6
El Salvador	3,557	16.5
Jamaica	2,011	15.8
Jordan	3,813	14.3
Guyana	308	13.7
Serbia	4,896	12.6
Nicaragua	820	12.5
Philippines	21,373	12.3
Bosnia and Herzegovina	1,913	12.2
Bangladesh	10,804	11.8
Albania	1,296	11.0

Source: Data from Migration and Remittances Factbook 2011, Worldbank.org, accessed on January 3, 2012.

about three-quarters of total foreign direct investment flows (US$573 billion) (www.migrationinformation.org, accessed on October 17, 2011). But if we include unrecorded formal and informal flows, the actual size of remittances is likely to be much larger. Remittances are increasingly becoming the most important source of income in many developing countries, as Table 14.3 indicates. While the absolute levels of remittance income are highest in India (US$53 billion in 2010), China (US$51 billion), and Mexico (US$22 billion), remittances are very significant economically for many smaller nations. In Mexico, remittances are larger than foreign direct investment inflows. They exceed tourism receipts in Morocco, revenue from tea exports in Sri Lanka, and the revenue from the Suez Canal in Egypt. In all 20 countries listed in Table 14.3, remittances account for more than 10% of the home country's economy, with Tajikistan and Tonga at the extreme end, having over 30% of their GDP made up of remittances.

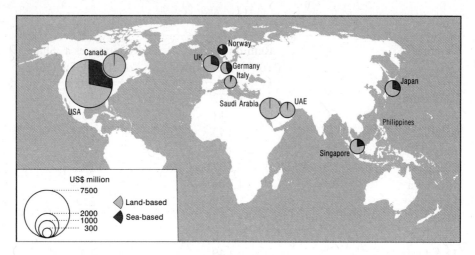

Figure 14.7   Remittances into the Philippines by top 10 source countries,
January–November 2011
Source: Data from http://www.bsp.gov.ph/statistics/keystat/ofw.htm, January 20, 2012.

Across the developing world, remittances are now worth more than double the
amount provided by official development assistance (i.e., aid) from all sources.
Figure 14.7 shows the top ten source countries of remittances for the Philippines. It
is interesting to note that the top five source countries are either highly developed
countries (the U.S., Canada, and the U.K.) or oil rich kingdoms (Saudi Arabia
and UAE). There are also significant differences in remittances emanating from
land-based activities and sea-based work (e.g., ship workers).

Clearly remittances have an effect in reducing poverty in developing countries.
Moreover, remittance flows are in most cases much more immune to the vagaries
of economic and political uncertainty than investment flows, especially in countries such as Bangladesh and the Philippines where money is being sent from
all around the world. For example, remittances were very resilient during the
2008–09 global financial crisis. While foreign direct investment flows to developing countries dropped 22.4% from US$658 billion in 2008 to US$510 billion
in 2009, remittances fell by only 6% during the same period. But the benefits of
remittances are the subject of some debate. Critics point out that it is never the
poorest who are able to migrate, and so remittances may increase inequality in
migrant-sending societies. Remittances are also often used for consumption (and
often imported) goods, or for services such as healthcare and education, not for
productive investment. They may push up the cost of land, housing, good quality education, and other commodities, such that migration becomes a necessity
in order to enjoy a comfortable quality of life. A high volume of remittances

may also push up the exchange rate of the country's currency, therefore making its exports more expensive and further deepening its dependence on migration. These, of course, are only the economic costs; to generate remittances, especially by temporary workers, families must often be oceans apart, so that children are separated from parents and spouses from each other, creating a very human cost (Pratt, 2009).

There is, however, a counterargument. Where remittances are used for the education and healthcare of children, they are fostering the long-term well-being (and productivity) of the workforce (quite apart from the less economic and more humane benefits). When remittances are used for immediate consumption needs, they are circulated among local businesses and can therefore have multiplier effects on the local economy in the country of origin (Kelly, 2009). In some cases, collective remittances by hometown associations will develop public infrastructure such as potable water projects, health clinics, orphanages, and schools. Remittances may also have positive cultural impacts: financing religious and ritual activities and therefore keeping traditions alive; or enhancing the power of women in households where they become the prime breadwinner due to their overseas work.

Transnational Ethnic Enterprises

Aside from household expenditures, migration also gives rise to a transnational variety of ethnic entrepreneurship. These are not the powerful transnational corporations discussed in Chapter 10, but generally small enterprises that nevertheless transcend national borders. Four specific types of transnational migrant enterprises have been identified (Landolt et al., 1999):

- *Circuit firms*: provide services transporting or transmitting goods and remittances, or in facilitating the migration process itself, for example as recruitment agencies.
- *Cultural enterprises*: import and sell cultural goods such as magazines, music, and videos from the home country.
- *Ethnic enterprises*: supply goods such as food and clothing from "home" to both immigrant and non-immigrant urban markets.
- *Return migrant micro-enterprises*: are established by returnees to their home country and utilize contacts made while overseas.

In Germany in recent years, Turkish enterprises of all of these kinds have expanded prolifically, changing the relationship between Turks and Germans, and between Turks and their places of origin in Turkey. Box 14.4 examines this phenomenon in more detail.

CASE STUDY

Box 14.4   Ethnic Turks in Germany

Turks first migrated to Germany in the early 1960s as contract labor migrants, or "guest workers" (*Gastarbeiter*). Many stayed, and they now constitute by far the largest ethnic minority population in Germany (and one of the most significant in any European country), numbering over 3 million in 2011. Turkish businesses are a common sight in the landscapes of German cities. In Berlin alone, it is estimated that there are 5,000–6,000 German-Turkish enterprises. Initially, businesses were established in the 1960s to cater to the demand for Turkish services and products among guest workers – circuit firms and cultural enterprises such as travel agencies, translation firms, restaurants, and cafes. By the 1970s, cultural enterprises had expanded to provide clothing, music, groceries – food and grocery stores still constituted over 60% of all German-Turkish enterprises.

By the 1990s, however, the vast majority of businesses were not just drawing upon the Turkish community but also had ethnic Germans among their suppliers and customers. Almost one-third had non-Turkish employees. Thus, Turkish entrepreneurship had shifted from solely circuit or cultural enterprises and into ethnic enterprises, catering to increasing demand by Germans for ethnic food and other products. We also now see the emergence of transnational return migrant enterprises as German-Turks invest in businesses in Turkey. These have included clothing manufacturers established in Turkey to produce for the German market, and German-Turkish involvement in the tourism business in Turkey. In more recent years, the German government has intensified its quest to encourage more Turkish entrepreneurs and business associations to provide vocational training for Turkish trainees so that they can be better integrated into the German job market. This intensification came after a change in German law in 2004 that allowed Turkish companies to give vocational training (www.todayszaman.com, accessed on January 3, 2012).

In recent years, the business practices associated with transnational ethnic firms have received increasing attention. The economic practices of ethnic Chinese entrepreneurs have been a particular source of fascination, largely because the rise of China itself as an economic force has been paralleled by the great success of ethnic Chinese communities around the Pacific Rim and elsewhere across the world. From Sydney, to Singapore, to San Francisco, ethnic Chinese immigrant communities (or, more accurately, some segments within them) have thrived economically, in part based on the global networks forged by bonds of co-ethnicity (Yeung, 2008). It is useful, however, to note that such transnational entrepreneurship is not unique to the ethnic Chinese. In southern France's port city Marseille, Mitchell (2011) shows that the city's trading networks converge toward Belsunce in the city center. This ethnic neighborhood is formed primarily by Algerian vendors and Sephardic wholesalers and continues to serve today as the most important landing site for contemporary Maghrebi and Sub-Saharan immigrants from former French colonies (e.g., Tunisia). Operating through a highly personalized, trust-based system, Belsunce now offers an excellent selection of specialty food items and other ethnic supplies to second- and third-generation immigrants in the local and regional markets in Southern France and even weekend customers from North Africa. This city center–based trading node has deep and tightly linked transnational ethnic networks that connect it with the entire Mediterranean region and beyond.

There are a number of distinctive business practices that seem to prevail in such global networks that are not always found in the mainstream market economy and that appear to have been especially well adapted to competition in the contemporary global economy. These practices emphasize the importance of personalized trust and sociocultural connections in initiating and cementing what is otherwise a business relationship. In the case of ethnic Chinese business, this has become known as Chinese, or Confucian, or *guanxi* capitalism, referring to the Mandarin Chinese word for "relationship." Five characteristics in particular are seen to be competitive advantages of ethnic Chinese firms:

- Family ties are a significant component of business management. Even some very large multi-billion-dollar ethnic Chinese businesses are controlled by sons, sons-in-law, brothers, and nephews of a founding father figure (and such managerial posts are, on the whole, the domain of male relatives alone). The advantage of family ownership is that it allows quick decision making even across a transnational business organization.

- Business transactions outside of the family circle will often be based on other forms of cultural affinity: membership of the same clan, place-based identities, and linguistic commonality. There is, then, a social basis to business ties that cements the business relationship into one of mutual trust and reciprocity that goes well beyond a purely contractual relationship. Where joint business ventures with foreign partners are involved, this creates a reliable relationship where legal and contractual agreements may not be enforceable. This has been the basis for a great deal of the investment by Taiwanese and Hong Kong manufacturers in mainland China since the early 1990s.

- Both familial and cultural relationships point to the importance of reputation and trust in business practices. These might apply to relationships between competitors, between contractors and suppliers, and between employers and employees. The bond of trust in business relationships is especially effective in ensuring that contracts are honored even when partners are oceans apart – a relationship based on trust is far more easily stretched across global space than one based on legal contracts.

- The ability to raise capital for new business ventures through informal networks creates a great deal of flexibility. Instead of going through formal banking or stock market channels to raise money, a new venture can be financed through loans from members of the same co-ethnic network. In ethnic Chinese communities such mutual aid has been a long tradition, largely on account of the discriminatory environments in which they found themselves initially and the lack of access to formal credit channels. In a contemporary transnational business operation, such sources of credit make the raising of capital for new ventures both relatively easy and inexpensive.

- In the exchange of market intelligence, relationships in networks can be cemented not just through business partnerships but also through binding indebtedness – not of a financial kind, but through the exchange of gifts and favors, including information. A cultural understanding of the "gift economy" then dictates that a gift can never be exactly repaid, thus establishing an ongoing relationship.

In this section we have seen how familial or co-ethnic bonds can operate at scales much greater than the neighborhood scale of the ethnic economy in a particular city. Migrant remittances represent one form of transnational linkage with households in migrant-sending countries, often with profound implications for both individual recipients and national economies. Transnational ethnic enterprises, on the other hand, represent a stretching of the same bonds and ethnic resources that create urban business clusters. In some cases, particularly the example of ethnic Chinese business networks, these bonds have proven to be well adapted to the need for flexible and reliable business partnerships in the contemporary global economy.

14.6   The Limits to Ethnicity

We have sought in this chapter to highlight the importance of ethnicity in economic life at multiple geographical scales. In closing, however, it is important to note some of the inherent limitations that exist in using ethnicity as a lens to understand various dimensions of the economy. In discussing ethnicity there is always an inclination to create groups and categories and to assume that they are both meaningful and homogeneous. In other words, it is often assumed that there is something *essential* to ethnicity. There is also sometimes a tendency to see culture or ethnicity as something fixed – something that defines how we behave because of some innate characteristics, hence the notions of "youth culture," or "Chinese-ness," or "American culture." In each case, the behavior of an individual is assumed based on their identity as a youth, Chinese, or American.

In reality, however, ethnicity is a complex and contradictory phenomenon. In this book, we have steered away from a fixed notion of ethnicity and have instead used the concept to refer to shared practices, norms, and conventions, rather than innate characteristics. Ethnicity does not determine practices; instead, practices are where ethnicity is to be found. This is a subtle distinction, but an important one. It allows us to distinguish an argument that "Germans do business in a certain way," from one that says "we find certain kinds of business practices in Germany." While the first statement implies that things are being done a certain way *because* they are being done by Germans, the second statement implies only that we find a certain commonality of practices in Germany and leaves open the reasons why this might be the case. This liberates us from dead-end discussions that imply a person with a certain identity *must* belong to a certain ethnic group and therefore behave in a certain manner. In short, ethnicity can operate differently in places and at scales that make it really hard to generalize. We should not assume that ethnicity predicts any particular forms of economic practice or exhibits any essential characteristics.

This caveat is needed when we look at statistical data on the employment experiences of particular ethnic groups – ultimately, there is no "average" experience for Filipinos, Koreans, Ghanaians, and so on in the urban labor market. Such groups often exhibit as much diversity as distinctiveness. In 2009, for example, just over 28,000 immigrants arrived in Canada from the Philippines. Around 4,500 arrived as "skilled workers," while over 11,000 were granted permanent resident status through the Live-In Caregiver Program mentioned earlier in this chapter. Another 69 arrived under the business immigrant program. Meanwhile, just over 1,000 were under the age of 5 and would therefore receive all of their formal schooling in Canada. All are ethnically Filipino immigrants, and yet the circumstances of their arrival alone will have huge impacts on their experiences of economic life in Canada. There is, in short, no such thing as an average Filipino immigrant. There is a danger that using this ethnic category to group or essentialize diverse individuals will obscure as much as it reveals.

Ethnic Chinese businesses in East and Southeast Asian societies provide another example of the changes that can limit intra-ethnic cooperation (Yeung, 2008). As ethnic Chinese firms from these regions have expanded around the globe, and as foreign firms also increase their presence in these areas, changes have occurred in their business practices. In particular, we see the greater inclusion of nonfamily members and professional managers, the broadening of "Chinese" business networks to include non-Chinese actors, and the increasing participation of Chinese elites in non-Chinese business networks. These processes represent significant steps toward the *hybridization* of ethnic Chinese business in a context of globalization. The resultant form of Chinese business system resembles neither the earlier form of family-centered business organization nor the Anglo-American form of market culture driven by anonymous professional managers. Instead, a new hybrid culture of business practice is becoming apparent in ethnic Chinese business. This new culture comprises some elements of earlier Chinese business practices (the continued role of family members in ownership and management) *and* new elements of professionalism and knowledge-based decision-making processes developed through cross-border learning in different international business environments.

A further point of caution is that ultimately other factors can trump ethnicity in the conduct of business transactions. For example, there exist extensive linkages between the Hsinchu high-technology region in Taiwan and Silicon Valley in California, based on co-ethnic Chinese networks. But while co-ethnic Chinese bonds may facilitate linkages such as subcontracting and sharing of information, many entrepreneurs remain cautious with respect to bonds of co-ethnicity (Saxenian and Sabel, 2008). More important are the technological and economic logics for collaboration, rather than the cultural logics that supposedly shape "*guanxi* capitalism." *Guanxi* is not the dominant organizational principle behind such relationships; *guanxi* is the lubricant in the system, rather than the system itself. Ultimately, technological competencies and a market-based rationality are the determining factors in shaping these high-tech industries. This reminds us that we must not replace a simple economic logic with one that overemphasizes cultural processes. Cultural and social processes are integral to how economic processes work, but they are neither necessarily determining of, nor substitutes for, economic logics.

14.7 Summary

This chapter has explored the diverse ways in which ethnic identity intersects with economic practice, and how space is integral to this intersection. We should now recap some of these geographical arguments. In the case of ethnically diverse *labor markets*, we saw the sorting through which certain ethnic groups get matched to certain kinds of work. This phenomenon of segmentation was explained on the basis of human capital, institutions, discrimination, and home–work linkages. In each case, the spatiality of the process was critical. Qualifications and professional experience are frequently devalued as they travel across jurisdictional boundaries, while skills may be specific to particular cultural contexts. The localized institutional and cultural specificity of labor markets is an important basis for understanding segmentation, especially among those such as migrants who move between them. Equally, we cannot dismiss the role that stigmatization and discrimination play in segmentation, and this is frequently spatialized, with particular places seen as producing stereotyped workers. An explicitly spatial argument exists in relation to home–work linkages, where residential segregation can lead to occupational segmentation.

In exploring *ethnic businesses* we saw the significance of geographical processes. For many businesses, the establishment of ethnic concentrations in certain parts of major cities allows an economy to develop that draws upon co-ethnic financing, customers, and employees. The emergence of ethnic places in the city also reinforces a sense of ethnic identity. But ethnic neighbourhoods are not static. In many cases, new ethnic business clusters have emerged, often reflecting the greater wealth of more recent immigrants. Driving these new suburban developments have been ethnic financial institutions, highlighting the fact that ethnic businesses are far more diverse and sophisticated than the popular image of grocery stores and restaurants.

We then examined the *transnational dimensions* of the economic connections established by migrants. Remittances between households, which have multiplied dramatically in recent years, can have significant effects on economies and households where they are received. We also saw how transnational business enterprises can bind together different local economies in ways that are rather distinct from the patterns of transnational corporations described in Chapter 10. A celebrated feature of these enterprises has been their supposedly ethnic quality, making them distinct from mainstream corporate models. In closing, however, we pointed out the pitfalls in placing too much emphasis on ethnicity as an explanatory factor in understanding economic experiences and practices.

Notes on the references

- For recent geographical analysis of ethnic differentiation in urban labor markets, see Wang (2009, 2010) for North American cases, Mitchell (2011) for French cases, and Wills et al. (2010) for the case of London.
- Much of the research on ethnic entrepreneurship and ethnic economies has been conducted by American sociologists: Roger Waldinger, Alex Portes, Ivan Light, Edna Bonacich, and their various collaborators have been especially influential. See Dana (2007) for a recent collection of essays. More recently, geographers have also paid increasing attention to this topic. See, for example, Kaplan and Li (2006), Johansson and Cornebise (2010), and Ley (2010). On entrepreneurship in general, see Hanson and Blake (2009) and Yeung (2009).
- On remittance economies, see Conway and Cohen (1998) on Mexico, Walton-Roberts and Hiebert (1997) on India, and Kelly (2009) on the Philippines.
- For a feminist take on transnational migration, see Pratt (2009).
- For recent geographical work on the ethnic Chinese diaspora and their transnational business networks, see Ma and Cartier (2003) and Yeung (2008).

Sample essay questions

- How does ethnicity influence the spatial organization of labor markets?
- What are the main factors accounting for the clustering of ethnic business in major gateway cities?
- Why is ethnicity a relevant explanation for transnational economic activities?
- What are the dynamic changes that ethnic economies experience in today's globalizing world economy?
- Describe and explain how ethnicity works in your own economic life.

Resources for further learning

- Jan Rath, a researcher in Amsterdam, has an extensive website relating to ethnic entrepreneurship: http://www.janrath.com/?tag=ethnic-entrepreneurship.
- The European Commission also maintains a useful website on ethnic entrepreneurship: http://ec.europa.eu/enterprise/policies/sme/promoting-entrepreneurship/migrants-ethnic-minorities/index_en.htm.
- The U.S. Equal Employment Opportunities Commission has statistics for all U.S. cities and states along the lines discussed for Los Angeles in Section 14.3: http://www.eeoc.gov/eeoc/statistics/employment/index.cfm. Data on Britain are available from the Equality and Human Rights Commission: http://www.equalityhumanrights.com/publications/our-research. Statistics Canada's E-stat system provide extensive data on migration, ethnicity, and labor markets: http://www.statcan.gc.ca/estat/licence-eng.htm.
- The Migration Policy Institute and its Migration Information Source have extensive information on global migration, transnationalism, and remittances: http://www.migrationinformation.org.
- The website of the 2011 meeting of the biennial World Chinese Entrepreneurs Convention (WCEC) in Singapore provides an interesting window onto the activities of the international Chinese business community (http://www.11thwcec.com.sg/en/index.html).

References

Conway, D., and Cohen, J. (1998). Consequences of migration and remittances for Mexican transnational communities. *Economic Geography*, 74, 26–44.
Dana, L. P., ed. (2007). *Handbook of Research on Ethnic Minority Entrepreneurship*. Cheltenham: Edward Elgar.
Hanson, S., and Blake, M. (2009). Gender and entrepreneurial networks. *Regional Studies*, 43, 135–49.

Holgate, J., Pollert, A., Keles, J., and Kumarappan, L. (2011). Geographies of isolation: how workers (don't) access support for problems at work. *Antipode*, **43**, 1078–1101.

Howell, A. (2011). Labor market segmentation in Urumqi, Xinjiang: exposing labor market segments and testing the relationship between migration and segmentation. *Growth and Change*, **42**, 200–26.

Johansson, O., and Cornebise, M. (2010). Place branding goes to the neighbourhood: the case of pseudo-Swedish Andersonville. *Geografiska Annaler*, **92B**, 187–204.

Kaplan, D., and Li, W., eds. (2006). *Landscapes of the Ethnic Economy*. Lanham: Rowman and Littlefield.

Kelly, P. F. (2009). From global production networks to global reproduction networks: households, migration and regional development in Cavite, Philippines. *Regional Studies*, **43**, 449–62.

Kwak, M. J. (2002). *Work in Family Businesses and Gender Relations: A Case Study of Recent Korean Immigrant Women*. Master's thesis, Department of Geography, York University, Canada.

Landolt, P., Autler, L. and Baires, S. (1999). From hermano lejano to hermano mayor: the dialectics of Salvadoran transnationalism. *Ethnic and Racial Studies*, **22**, 290–315.

Ley, D. (2010). *Millionaire Migrants: Trans-Pacific Life Lines*. Oxford: Wiley-Blackwell.

Lo, L. (2009). The role of ethnicity in the geography of consumption. *Urban Geography*, **30**, 391–415.

Ma, L. J. C., and Cartier, C., eds. (2003). *The Chinese Diaspora: Space, Place, Mobility and Identity*. Boulder, Colo.: Rowman and Littlefield.

Mitchell, K. (2011). Marseille's not for burning: comparative networks of integration and exclusion in two French cities. *Annals of the Association of American Geographers*, **101**, 404–23.

Musterd, S., Andersson, R., Galster, G., and Kauppinen, T. M. (2008). Are immigrants' earnings influenced by the characteristics of their neighbours? *Environment and Planning A*, **40**, 785–805.

Pratt, G. (2009). Circulating sadness: witnessing Filipino mothers' stories of family separation. *Gender Place and Culture*, **16**, 3–22.

Saxenian, A. L., and Sabel, C. (2008). Venture capital in the "periphery": the new Argonauts, global search, and local institution building. *Economic Geography*, **84**, 379–94.

Walton-Roberts, M., and Hiebert, D. (1997). Immigration, entrepreneurship, and the family: Indo-Canadian enterprise in the construction industry of Greater Vancouver. *Canadian Journal of Regional Science*, **20**, 119–47.

Wang, Q. F. (2009). Gender, ethnicity, and self-employment: a multilevel analysis across U.S. metropolitan areas. *Environment and Planning A*, **41**, 1979–96.

Wang, Q. F. (2010). How does geography matter in the ethnic labor market segmentation process? A case study of Chinese immigrants in the San Francisco CMSA. *Annals of the Association of American Geographers*, **100**, 182–201.

Wills, J., Datta, K., Evans, Y., Herbert, J., May, J., and McIlwaine, C. (2010). *Global Cities At Work: New Migrant Divisions of Labour*. London: Pluto.

Yeung, H. W. C. (2008). Ethnic entrepreneurship and the internationalization of Chinese capitalism in Asia. In L. Dana, ed., *Handbook of Research on Ethnic Minority Entrepreneurship* pp. 757–98. Cheltenham: Edward Elgar.

Yeung, H. W. C. (2009). Transnationalizing entrepreneurship: a critical agenda for economic geography. *Progress in Human Geography*, **33**, 210–35.

CHAPTER 15

CONSUMPTION
You are what you buy

Goals of this chapter

- To recognize the significance of consumption processes within the capitalist system
- To appreciate the changing global landscape of consumption
- To reflect on the ways in which consumption, place, and identity are interrelated
- To think about the ways in which it is possible to develop consumption politics
- To consider how places themselves may be consumed through tourism

15.1   Introduction

Certain commodities are seen as emblematic of globalization and homogenization. Think for example, of McDonald's fast food, Starbuck's coffee, or Nike sports shoes. But how "global" are these products really? To what extent are they interpreted differently, and given different meanings, in different places and by different groups of people? Coca-Cola is another iconic and seemingly ubiquitous brand within the global economy. However, while Coca-Cola is undeniably a global commodity in terms of its wide availability, it is always consumed and experienced within particular localities. In Trinidad in the Caribbean, for example, it is possible to argue that rather than representing the dominance of Western consumer culture or American cultural imperialism, Coca-Cola has become an essentially Trinidadian product in the way it is consumed (Miller, 1998). Coke is seen as an everyday necessity and is widely available. The national drink of Trinidad is rum (a reflection of the history of sugar cane production in the country), which is always consumed with a mixer drink, usually Coca-Cola. In this way, Coke is part of the beverage that is most directly associated with what it is to be Trinidadian. Even without that link, however, Coke can still be seen as a specifically Trinidadian drink. Soft drinks in Trinidad tend to be seen as either "red" or "black," with both having historically embedded connotations. Red drinks are particularly sweet and are commonly associated with roti (flatbread), a staple food among the East Indian population. Black drinks have more urban and sophisticated connotations and, partly through the rum and coke connection, are linked more strongly to the African and white populations. These meanings are far from fixed, however – reflecting the hybrid nature of Trinidadian identity – with Indians being strong consumers of Coke and with a "red and a roti" often being consumed by Trinidadians of African origin.

This example introduces several of the main themes that underpin this chapter. Most centrally, consumption dynamics are highly geographically variable because they reflect place-specific social and cultural practices, or "local consumption cultures" (Jackson, 2004). Even global products such as Coca-Cola are reinterpreted in different contexts and given specific meanings. These meanings may run counter to, and challenge, the dominant discourses associated with particular commodities (e.g., westernization). They may have a national dimension – for example, the links between Coke and what it is to be "Trinidadian" – but may also have more localized and individual dimensions as they intersect with other facets of identity such as ethnicity, gender, and age. In turn, the urban and sophisticated dimensions of the black drinks remind us that products are not simply purchased for their basic use value, but are also selected for the more symbolic value that is associated with their consumption. Put another way, as consumers we actively select the commodities that we eat, drink, wear, and use as part of our identity formation. Such processes always take place within, and are shaped by, the attributes of particular places. And finally, these place-specific associations should not be seen as static, as they ebb and flow with shifting fashions and cultural practices. Overall, consumption processes demand that we broaden our perspective beyond the economic transactions in which commodities are purchased to explore the wider sociocultural terrain in which such purchases are embedded.

The analysis in this chapter unfolds in five stages. First, we introduce consumption and consider different views on how best to interpret consumption processes within wider capitalist dynamics (Section 15.2). This chapter seeks to move beyond the limitations of simplistic economistic interpretations to develop an understanding of consumption as a sociocultural phenomenon. Second, we consider the changing landscape of consumption at the global scale, and in particular, the rapidly expanding numbers of middle-class consumers in countries such as China and India (Section 15.3). Third, we look at how consumers use the commodities they buy as part of place-specific processes of identity construction (Section 15.4). We also consider how consumers actively interpret and rework consumption spaces. Fourth, we reflect on how our individual and collective consumption decisions matter, by looking at the extent to which they are informed by different notions of ethical consumption (Section 15.5). Finally, we explore how consumption is not only heavily shaped by the places in which it occurs; places themselves are also actively consumed through tourism (Section 15.6).

15.2   Interpreting the Consumption Process

What exactly do we mean when we talk of "consumption"? As we saw in Chapter 8, the commodity chain concept allows us to see how consumption is the final step in a series of material transformations and value-adding activities. Importantly, consumption refers to the sale, purchase, *and use* of commodities. Hence it is not only about the interface between those offering products/services for sale, and those making decisions about which products/services to buy (as discussed in Chapter 11), but it is also about what people do with commodities after buying them. Consumption is thus not just about simple economic transactions at the point of sale. Rather, it is a *process* that includes sociocultural aspects of commodities and their use, encompassing a range of activities: purchasing, shopping, using, discarding, recycling, reusing, wearing, washing, eating, leisure, tourism, and home provisioning and renovation among others. Consumption is therefore underpinned by a wide range of activities such as shops, restaurants, and hotels, and repair, servicing, cleaning, and recycling operations, many of which are significant economic sectors in their own right.

The everyday nature of many consumption spaces means that they necessarily have a more extensive and disparate geography than other economic activities. Put simply, while not every large city or region will have a major manufacturing plant, all will have supermarkets, cinemas, laundrettes, restaurants, libraries, recycling plants, and the like. Indeed in some places, whole city economies can be based primarily around such activities (see, for example, Box 15.1 on the shifting fortunes of Las Vegas). Equally, however, we need to comprehend the ways in which the *nature* of consumption varies geographically, due in large part to its sociocultural dimensions. The spaces of consumption not only overlap with the retail spaces described in Chapter 11 – shops are spaces of both retail and consumption – but extend well beyond them into a wide range of other business and leisure spaces. They also incorporate more personal spaces such as the home. While most commodities that we buy end up at home, home furnishings (e.g., furniture, curtains, bed sheets, paint, flooring) are explicitly deployed to create particular environments and spaces within the home. This material process of decoration and maintenance contributes to identity formation as the home becomes a micro-space of creativity and personal expression (Leslie and Reimer, 1999).

Before going any further, however, we need to consider the wider significance of consumption processes within capitalism. Clearly, production and consumption

CASE STUDY

Box 15.1   Viva Las Vegas!

It is hard to think of a more iconic example of a consumption space than the city of Las Vegas, in Nevada, United States. The city, famously founded on the site of a small desert railroad and agricultural town in 1905, was by 2010 a city of 2 million people. As is well known, the city has now grown into a massive center of gambling, leisure, entertainment, and retailing, attracting 37 million visitors in 2010 who spent US$5.8 billion on gambling alone. The employment structure of the city clearly reflects its reliance on consumption activities: in August 2011, 263,000 workers (33% of the total) were employed directly in leisure and tourism activities, with many others employed in related and support industries such as government, health, education, business services, and construction. Manufacturing, on the other hand, employed only 18,400 workers (2.3% of the total). This, clearly, is a massive urban economy founded on, and sustained by, consumption. Over the period from the mid-1970s until the mid-1990s, Las Vegas transformed itself from an American casino resort to a tourist attraction with global appeal, a transformation supported by national and local regulatory conditions that facilitated quick decision making on planning and development requests. The leading theme park, retail, and casino attractions are clustered along the city's famous "Strip" (see Figure 15.1), with the 12 leading casino-hotels employing over 3,000 workers each. Many of the sites along the Strip are themed in order to increase their attractiveness to visitors by "transporting" them to alternative times and places. Places as far away as France, Venice, Egypt, and Polynesia are evoked, in addition to fantasy (e.g., Treasure Island) and historical (e.g., Greco-Roman) references. In the case of the MGM Grand, a 7,000-room hotel, casino, and entertainment complex that includes a 17,000 seat auditorium and 16,000 square meters of gaming space, the theme is actually MGM's own Hollywood studio complex! The example of Las Vegas speaks both to the importance of consumption for the economy of particular places, and also to the way in which attractions are themed as a strategy for inducing visitors to spend money, and in this instance, gamble. The Las Vegas economy has struggled since recession hit in the late 2000s, however. By October 2011 the unemployment rate was 14.3%, property prices and real wages had tumbled, poverty and homelessness rates were rising sharply, and several uncompleted construction projects stood empty at the north end of the Strip, including the US$2.9 billion, 3,889-room Fontainebleau development (*The Guardian*, 2011).

Figure 15.1   A consumption landscape – The Strip, Las Vegas, Nevada

are two mutually dependent spheres of activity, but which drives which? Here we can look at three important perspectives on this question. The first can be characterized as the *consumers-as-dupes* viewpoint, which emphasizes the primacy of production within the economic system as a whole. From such a perspective, consumption is read as the outcome of changes in the nature of the production process. It is a pleasure-seeking but ultimately regressive activity in which passive consumers are enticed into parting with their money, thereby delivering profits for producers and refueling the basic capitalist model. As we saw in Chapter 11, this argument can be extended to explore how particular retail spaces – for example, a shopping mall – are actively designed to induce consumers to spend as much money as possible during each visit. Most importantly, however, the consumers-as-dupes perspective places capitalist corporations rather than consumers in control of consumption processes and spaces. While there are strong elements of truth to this interpretation, as we shall see in this chapter it underplays the extent to which individuals and groups of consumers demonstrate their agency through their consumption practices.

This can be compared with the *consumer sovereignty* view, which emphasizes the free will of individual consumers. Deriving from neoclassical economics, consumption is seen here as an economic transaction dependent on price-based decisions. Rational consumers, acting on full information about different products and prices, will make informed decisions about which products to buy. In turn, these actions will have an impact on the production process, as market information about which products are popular feeds back through retailers to manufacturers. In short, consumers can choose whether to buy the products on offer or not, and as such they have considerable sovereignty, or control, over the economic system as a whole. Again, as with the consumer-as-dupes perspective, this approach is not entirely without merit. Capitalist firms – whether manufacturers, retailers, or other service providers – do indeed closely monitor their markets and respond rapidly to signals they receive about the popularity (or not) of the commodities they produce. The atomistic, rational view of consumers inherent in the sovereignty view has clear limitations, however, particularly in the way that it represents consumption decisions as being taken by autonomous consumers unaffected by wider society and its various norms and expectations.

By contrast, a *sociocultural* perspective emphasizes the way in which consumers actively construct their own identity through their consumption practices. Such an approach draws attention to the other intersecting facets of identity – such as gender, ethnicity, age, and sexuality (see Chapters 13 and 14) – and looks at how consumers knowingly purchase certain commodities and use them in specific ways as part of an active process of identity construction. These processes of identity construction, however, are not individualized as in the consumer sovereignty view, but are instead part of larger social forces and trends within society (e.g., teenage fashion or ethical purchases). Here the interactions between consumers and producers are complex and two-way. Consumption processes will clearly be influenced by manufacturing and retail corporations and the commodities they offer, but consumers purchase such goods selectively and knowingly; firms, in turn, will respond to emerging consumption dynamics within society. This approach, which underpins the remainder of this chapter, also allows the geographies of consumption to be laid bare, as the processes of identity construction are very much place-specific. It also has implications for how we might read particular retail and consumption spaces, suggesting that consumers are actively aware of the way in which spaces are designed to induce consumption.

15.3   The Changing Global Consumption Landscape

The global geographies of consumption are changing as economic power steadily gravitates toward a cohort of emerging economies. More specifically, profound shifts in wealth and spending power are seeing Asia emerge as the key arena for consumption within the global system. One way of charting this rise is to look at the global distribution of the middle class. While the middle class has traditionally been defined by its position in wider society (i.e., between the upper and working classes), it is increasingly being interpreted through the lens of consumption practices. The essential characteristic is the possession of a significant amount of disposable income after providing for basic food and shelter, allowing middle-class consumers to buy products like televisions and cars but also to invest in services such as healthcare and education. The middle class is hence associated with a particular form of consumerism that signals a shift away from subsistence living and toward the purchase of a range of more conspicuous, status goods.

While the global middle class was historically dominated by the triad of Europe, North America, and Japan, the 1970s and 1980s saw significant middle classes emerge in the newly industrializing economies of South Korea, Brazil, Mexico, and Argentina. Since the 1990s, however, there has been massive expansion in the middle classes in China, India, and other Asian emerging economies, and the rate of growth is increasing. The key driver is economic growth and associated shifts in income levels. The World Bank estimates that the global middle class (defined as those making between US$10 and $20 per day) will expand from 430 million in 2000 to 1.15 billion by 2030. While 56% of the global middle class were living in developing countries in 2000, the proportion is projected to increase to 93% by 2030. Others define the middle class more broadly as those on an income of between US$10 and US$100 per day (an annual income range of US$3,650–US$36,500). Projecting forward to 2030, the locus of the global middle class will almost certainly continue to shift decisively toward Asia (see Table 15.1). It is estimated that the global middle class, in this broader definition, will expand from 1.8 billion to 4.9 billion people over the period 2009–2030, with 85% of this growth being accounted for by Asia. The share of middle-class consumers in North America and Europe is projected to fall from 54% of the world total in 2009 to 21% in 2030, while Asia-Pacific's share will expand from 28% to 66%. In terms of actual spending power, it is estimated that North America and Europe will fall from 64% to 30% of the global total, with the

Table 15.1 Asia's burgeoning middle class?

Region	2009		2030	
	Number (millions)	Global share	Number (millions)	Global share
North America	338	18%	322	7%
Europe	664	36%	680	14%
Central and South America	181	10%	313	6%
Asia Pacific	525	28%	3,228	66%
Sub-Saharan Africa	32	2%	107	2%
Middle East and North Africa	105	6%	234	5%
World	**1,845**	**100%**	**4,884**	**100%**

Source: Based on data from table 2 Numbers (millions) and Share (percent) of the Global Middle Class from Kharas, H. (2010), "The Emerging Middle Class in Developing Countries", OECD Development Centre Working Papers, No. 285, OECD Publishing. doi: 10.1787/5kmmp8lncrns-en

Asia-Pacific region rising from 23% to 59%. Within these shifts, two countries dominate, with China and India expected to account for almost 40% of the global middle class by 2030 (data from Kharas, 2010).

The precise methodology behind, and accuracy of, these predictions is less important than the central observation that the geography of consumer demand is shifting inexorably toward Asia. There are two important implications of these global market shifts. First, there has been steady growth in so-called "South–South" trade, as flows of trade, investment, and aid between emerging economies have intensified. As part of this dynamic, leading emerging economy businesses are increasingly turning their attention to the domestic market as opposed to exports. By the late 2000s for example, over half the apparel produced in China was destined for the domestic market. Second, transnational corporations from the mature economies of North America, Western Europe, and Japan are no longer seeing leading Asian markets simply as sources of labor and bases for export production, but rather as populations able to afford a wide range of consumer goods. By 2011, for example, China had already become the world's single largest market for televisions, mobile phones, personal computers, and automobiles (13.5 million cars were sold there in 2010).

Importantly, there appear to be different consumer preferences in emerging markets, where price is an overwhelming consideration, and product variation matters less. Even though millions are reaching middle-class status, they do not yet have the same levels of income as their counterparts in mature economies, and price sensitivity is therefore of paramount importance. In such contexts, the product differentiation (i.e., highly priced products based on variety and quality) characteristic of mature markets and post-Fordist production systems (see Chapter 9) gives way to Fordist-style product commodification (i.e., standardized, low-cost products based on mass production) (Kaplinsky and Farooki, 2010). The potential of tapping the consumer power of the newly emerging consumer class has been popularized through the notion of the "fortune at the bottom of the pyramid" (see Box 15.2). Another characteristic of consumer markets in emerging economies may be relatively poorly developed and weakly implemented product standards, as witnessed in incidents involving contaminated baby milk in China (2008) and pesticide levels in soft drinks in India (2003).

The commodification and standardization of existing goods to push down prices is often led by developing country firms. In India in January 2008, for example, Tata famously launched the Nano, a stripped-down 100,000 rupee small car (approximately US$2,100) designed to appeal to the country's millions of scooters riders (see Figure 15.2). Another example is the low-cost washing machine produced by the Indian consumer electronics company Videocon. These are not only cheap but are also attuned to the needs of the Indian consumer; for example, they automatically finish the wash cycle after a power outage, and they do not have a drying option as it is not required in India's climate. Cheap

KEY CONCEPT

Box 15.2   Bottom of Pyramid markets

The notion of the market at the bottom (or base) of the pyramid (BoP) was first put forward by business scholars C. K. Prahalad and Stuart Hart in the late 1990s and was popularized in the early-to-mid 2000s. In their initial formulation, they made the simple yet powerful observation that 4 billion people were living outside the global market system (the figure is undoubtedly much lower now) and that they offered huge untapped market potential for global businesses. Investing in the specific requirements of these markets was argued to offer different sources of opportunity for TNCs: some BoP markets are huge in their own right (e.g., India); local innovations could be transferred to other BoP markets; some BoP innovations could be transferrable to developed markets; and there might be management lessons that benefit the firm as a whole. Prahalad (2005) argued that the key to success lay not simply in "tweaking" global products and services, but in really understanding the needs of low-income consumers. One often-quoted example is how Hindustan Unilever Limited (Unilever's Indian subsidiary) responded to local competitor Nirma in developing an eco-friendly washing detergent (called Wheel) that suited an Indian market with low water availability. Importantly, this process involved not just reworking the product itself, but also adopting a more decentralized and labor-intensive production and distribution system. The lessons from India were in turn transferred to other markets – for example, sales of Ala detergent in Brazil. While the theory originated in the domain of business strategy, it has subsequently become part of a wider debate about the extent to which market mechanisms can be used to lift people out of low-income and poverty situations. A central element in this debate is whether TNCs are the appropriate actor for driving such changes, or whether there should be also be roles for local businesses, social entrepreneurs, civil society organizations, and state agencies.

netbook computers are another emerging market innovation. The first mass-produced netbook, the Eee PC 700, was introduced by the Taiwanese firm ASUS in 2007 at a price of just US$260. These market conditions can be challenging for foreign transnational corporations. While some firms, such as KFC in China, may outdo the performance in their home market, many others struggle to adapt.

Figure 15.2 Tata's Nano car

The U.S. retailer Home Depot, for example, has struggled in the Chinese context where labor is cheap and hence do-it-yourself home improvements are not the norm. In China's mobile phone sector, a lack of variety of low-end product lines, higher prices, incompatible standards, and restrictive regulations have all driven a shift from foreign to local handsets among consumers. The prevalence of fake handsets, known as *shanzai* or mountain villages – in a context of weak intellectual property laws and enforcement – is another important barrier for foreign providers (see Figure 15.3). In short, incoming transnationals need to be able to adapt their business to meet the specific needs of emerging-market consumers. Coca-Cola, for example, has developed a "layered" strategy in China, where Coke is sold at a slightly lower price than Western markets in urban areas but less in the countryside, with customers being asked to return their bottle to the vendor immediately to save costs.

Persistent regulatory barriers prevent the expansion of foreign service industries in China and India, and so consumer goods are currently central to the consumption landscape. For Chinese consumers, expensive consumer goods have become emblematic of success within wider society. Shopping is the leading leisure activity, with consumers spending over twice as much time shopping each week as their American counterparts in the mid-2000s (Kharas, 2010). Growth has been given added impetus by the Chinese government's prioritization of growing domestic consumption in its 2011–2015 five-year plan. In India, it is predicted that middle-class consumers will expand from 50 to 583 million by 2025 as it grows from the twelfth to fifth largest consumer market in the world. The

Figure 15.3   Real (on the left) and fake (on the right) iPhones available in the Chinese market

consumption choices of India's burgeoning middle class (comprising government officials, college graduates, rich farmers, traders, business people, and professionals) are creating demand for new kinds of products, such as packaged and prepared foods. They are also fueling consumer credit, as seen in the growing use of credit cards.

As this section has shown, one geographical approach to consumption is to look at shifting patterns of activity at the global scale. Another, however, involves exploring how the *nature* of consumption processes is geographically variable, with the meanings attached to commodities differing from place to place. We turn to this topic in the next section.

15.4   Cultures of Consumption, Place, and Identity

The nature of consumption has unarguably changed over the last few decades. In line with our discussion of post-Fordist production systems in Chapter 9, it is possible to identify the key attributes of what might be called post-Fordist consumption (see Table 15.2). In general, the Fordist era was characterized by the large-scale, mass consumption of a relatively limited range of standardized commodities (originating, most famously, with the Model T car, available as Henry Ford famously declared, in "any color so long as it's black"). The system

Table 15.2   Mass consumption and post-Fordist consumption compared

Characteristics of Mass Consumption	Characteristics of Post-Fordist Consumption
Collective consumption	Increased market segmentation
Demands for familiarity from consumers	Greater volatility of consumer preferences
Undifferentiated products/services	Highly differentiated products/services
Large-scale standardized production	Increased preference for non-mass-produced commodities
Low prices	Price one of many purchasing considerations, alongside quality, design, etc.
Stable products with long lives	Rapid turnover of new products with shorter lives
Large numbers of consumers	Multiple small-niche markets
"Functional" consumption	Consumption less "functional" and more about aesthetics
	Growth of consumer movements, alternative and ethical consumption

was driven by economies of scale and low prices that enabled the widespread consumption of a broad range of household and personal goods. The post-Fordist era, by comparison, seems to have engendered a much more fragmented consumption pattern in which many highly differentiated products are offered to a wide range of consumer groups or *niches*. This mode of consumption is driven less by the price and functionality of commodities, and more by the aesthetic and symbolic value they bring to consumers. It is shaped by knowledgeable end-consumers making strategic decisions about the commodities that they buy. Another notable feature of the post-Fordist era is the growing importance of consumption-related employment (see Box 15.3). Persuasive though such arguments are, we need to keep them in check, as the shifts toward emerging market middle-class consumption described in the previous section are arguably invigorating some of the dimensions of mass consumption in those contexts (e.g., mass produced, affordable goods). The precise balance of mass and post-Fordist consumption processes will play out in place-specific cultures of consumption and will also shift dynamically over time.

FURTHER THINKING

Box 15.3   Consumption work

An important characteristic of what we are calling the post-Fordist era in developed economies is that a growing proportion of the workforce is employed in sectors such as retailing, restaurants, tourism, and entertainment. At the same time, the rapid growth in these kinds of jobs has raised questions about their quality and desirability. These doubts have prompted economic geographers (and other social scientists) to look within workplaces and explore the nature of work in the consumption sphere. This research has revealed that consumption-related jobs tend to be characterized by some, or all, of the following features:

- Socially constructed as low-status jobs within society (e.g., the waitress, the flight attendant, the checkout assistant) and receiving relatively low wage rates.
- Low levels, or indeed the complete absence of, collective representation and unionization.
- Dominated by part-time and temporary contracts rather than full-time positions. In some instances, such as tourism, there may be a marked seasonality to availability of employment.
- Technology-based surveillance of the workforce to both monitor workers (e.g., listening in to telephone sales calls, measuring productivity of fast-food operatives) and deploy them efficiently (e.g., gauging the appropriate number of checkout staff in a supermarket).
- Based around labor-intensive, repetitive tasks that cannot be replaced by technology (e.g., waiting on tables or cutting hair).
- Having a "performative" component in which workers are required to take on a particular role or personality. This can occur in both face-to-face (e.g., the scripts given to fast-food workers) and technologically-mediated situations (e.g., Indian call center workers taking on English names as part of scripted encounters with customers). These performances may challenge workers' true senses of identity and lead to substantial stress levels.
- The social construction of many of these jobs as essentially "female" (e.g., flight attendants), reinforced through recruitment practices.
- The social construction of many of these jobs as being for young people (e.g., fast-food workers, bar and nightclub staff), reinforced through recruitment practices.

- Selective recruitment on the basis of a wide range of other personal attributes – ethnicity, bodily appearance, weight, bodily hygiene, dress and style, interpersonal skills – to fulfill the requirements of the performative encounter with customers.

For more on the subject, see McDowell (2009).

Nonetheless, consumers can be usefully thought of as active agents, constantly making informed decisions about which commodities to buy in order to construct particular identities, or senses of self. This is particularly true of certain kinds of highly visible symbolic goods – such as clothing, jewelry, and cars – that are often taken to signify a certain standing or position within society. Other consumption activities, including the choice of restaurants, bars, and holiday destinations, may be part of the same process. Equally, purchases may be used as markers of difference, to indicate a position outside, or toward the margins of, the mainstream of society (e.g., buying a car that runs on vegetable oil or wearing distinctive forms of clothing). In terms of clothing, jewelry, make-up, and the like, the body itself becomes a site of inscription through which key elements of individual identity are communicated (see Box 15.4). It is important to recognize, however, that the ability of consumers to engage in these forms of consumption varies both socially and spatially. Socially, it may only be certain groups within the population, and particularly middle-class consumers with the necessary disposable income, that choose to take part in symbolic consumption. Spatially, such consumption will be more widespread in wealthy countries than in poorer countries, where simply meeting basic needs requirements (water, food, shelter, etc.) may be the only priority. Even within wealthier countries, poor urban or rural dwellers with low income levels, limited local retail provision, and poor access to transportation may effectively be excluded from the consumption of symbolic goods. In short, we need to recognize the profound social and spatial variation in the degree of consumer agency with respect to identity formation.

Moreover, the *nature* of the relationship between consumption and identity works out differently in different places – in other words, it is *place-specific*. This is because of the social and cultural specificity of consumption practices such as shopping and eating. This variation may reflect a consumer's unique individuality, their membership of a particular group in society (e.g., youth or ethnic cultures), or their position within a particular regional or national culture. Think, for example, of a basic cosmetic product such as lipstick. What it means to use lipstick is clearly associated with certain gender and age norms but also

KEY CONCEPT

Box 15.4   Bourdieu's cultural capital

In his now famous 1984 book *Distinction: A Social Critique of the Judgment of Taste*, the French sociologist Pierre Bourdieu outlined a theory of cultural capital. His analysis focused on the symbolic value of commodities and how social differences were produced through the consumption of goods. Based on a study of class-based distinction in 1960s French society, he argued that an individual's position in society was linked to certain forms of consumption practices and processes of self-identification relating, for example, to the arts, education, cuisine, and fashion. From this perspective, social difference or "distinction" was based less on wealth itself, and more on the ability of different groups to *display* wealth through investing in symbolic goods. Hence dominant fractions of society derived their status and prestige from their ownership of cultural or symbolic capital. Cultural capital, in turn, was conferred through the ability to implement judgment or taste with relation to the consumption of commodities (e.g., what is fashionable or what constitutes a high-quality bottle of wine). For Bourdieu, education was critical to the acquisition of cultural capital and the distinctions between different social groupings. These distinctions were seen to be reproduced through routinized and often unconscious social practices that Bourdieu termed "habitus." Overall, his work has made a crucial contribution to consumption research by linking the societal positions of groups and individuals to the symbolic meanings of the various commodities that they consume. Moreover, Bourdieu's emphasis on the importance of local social context has made his work particularly appealing to geographers.

varies geographically as different forms of "youthful feminities" are created in different societal contexts. In the West, for example, lipstick resonates with a well-established beauty myth, propagated through media and advertising circuits, that sustains a huge and highly profitable cosmetics industry. In central and eastern Europe, lipstick may indicate different forms of female participation in capitalist labor markets after the collapse of state socialism. In the Middle East, wearing lipstick may be a marker of rebellion against prevailing social norms. In cities across China, lipstick may signify participation in a globalized form of urban modernity associated with a broad range of conspicuous consumption practices. In contexts like the Philippines, it may infer a post-colonial desire to achieve certain attributes of "whiteness." What these examples suggest is that even if the lipstick is produced by the same cosmetics firm, its use as a bodily marker carries highly variable meanings across different social contexts as "young women appropriate, adapt and subvert globally marketed versions of femininity" (Kehily and Nayak, 2008, p. 339).

At the same time, while processes of identity construction predominantly take place within local consumption cultures, we must not fall into the trap of reading those cultures in static terms. Local consumption cultures are increasingly open to changes brought about by transnational connections. Television, film, and new media technologies, for example, form part of global circuits of culture that circulate a wide range of images about norms and aspirations of consumption. Consumption processes are also shaped, however, by links forged through migration and diasporic connections. London and Mumbai, for example, are part of a transnational British-Asian domain of fashion that is constituted by intense exchanges of people, information, images, commodities, and capital and that has been shaped by a long history of colonial and post-colonial relations (drawing on Jackson et al., 2007). These interconnections serve to blur what is understood as authentic British or Indian fashion as a variety of *hybrid* forms of clothing emerge. These hybrid fashions are interpreted locally in both London and Mumbai, and those interpretations vary across consumers of different ages, genders, occupations, and educational level. The transnational domain should not be read, therefore, through simplistic notions about the Westernization of Asian dress or the imposition of a Western modernity. Instead, processes of change are spatially and socially uneven in both London and Mumbai. Indian consumers in Mumbai can resist or even reverse the commonly assumed contrast between Western modernity and Eastern tradition, with Mumbai being seen as a fast-paced consumption arena open to a broad range of global influences and with wide-ranging consumer choice. University students in Mumbai may feel that their overseas cousins are behind in terms of dress because of the pace of change in Mumbai's fashion circles. Equally, British-Asian students living in the United Kingdom may perceive themselves to be lagging the trends being set in Mumbai through, for example, the Bollywood film industry, itself an increasingly global cultural form (see Figure 15.4). Overall, the persistent "localness" of consumption cultures should not obscure the ways in which those local cultures are increasingly embedded in complex and overlapping transnational arenas of consumption.

Further mobilizing this sociocultural perspective on consumption, we can start to reinterpret shopping, a seemingly routine and everyday activity that was implicitly an integral part of our earlier discussion of retailing (Chapter 11). Shopping is clearly much more than a robotic response to the stimuli provided by retailers and shopping mall designers. Instead, it can be thought of as a culturally and societally specific set of practices (Miller et al., 1998). Shopping is a social activity as well as a simple exchange of commodities, entailing, for example, interaction with shop workers and discussions with friends about the relative merits of products both before and after the act of purchase. "The activity of shopping keeps modern economies afloat ... but shopping is also a cultural activity. Because it

Figure 15.4   Hybrid cultures? Bollywood dancing outside the National Gallery, London

is usually what we do when we 'go out,' shopping is how we satisfy our need to socialize – to feel we are a part of public life" (Zukin, 2004, p. 7). Social relations within the family are particularly important, as many purchases are made for other members of the family group. Consumers may also make decisions in the context of wider social debates and dilemmas concerning, for example, what might constitute ethical or environmentally responsible consumption (we will return to this topic in Section 15.5).

   With this more nuanced understanding of shopping practices, we can revisit our discussion of shopping malls from Chapter 11. Three observations can be made (Miller et al., 1998). First, while it is important to recognize the general principles that underpin the design of shopping malls, this does not mean that they are all the same. Shopping malls will each have their own distinctive histories, mixture of attractions, and senses of place, and consumers will make choices between different malls accordingly. Second, shopping malls are designed for an average, universal customer. In reality, different kinds of shoppers will use the mall in a wide range of ways. Not all shoppers, for example, will browse the space in a relaxed way as a form of leisure. Many will not be enjoying the experience, and will want to minimize the time spent in making a series of predetermined purchases. Hence there is huge variety in the ways that consumers choose to interact with such spaces. Third, shopping malls have to be seen in the context of the other retail formats on offer in a particular society (online shopping, second-hand markets, city center retailing, etc.). Again, different shoppers will use a mall as part of different combinations of retail options. Far from being homogeneous corporate spaces, therefore, shopping malls are in reality highly varied and distinctive environments that are strategically and knowingly used by consumers as one of the many spaces through which they undertake their ongoing consumption practices. Similarly, seemingly standardized fast-food spaces are in fact used and interpreted differently by consumers in different contexts (see Box 15.5).

CASE STUDY

Box 15.5   McDonald's in Asia

The expansion of McDonald's fast-food restaurants in East Asia since the early 1990s provides a useful case study of how consumers respond to and use consumption spaces in different ways (here we draw on Watson, 1997). Far from being a simple imposition of the American fast-food model onto the Asian context, what has in fact happened can be characterized as a two-way localization process requiring adaptation and accommodation on the part of both the company and consumers. On the one hand, McDonald's has had some effect on local diets, particularly among younger people. What was initially seen as an exotic food choice – a taste of Americana – has gradually gained acceptance as an "ordinary" food for certain sections of the population. Once broadly accepted socially, the retail chain can then expand more rapidly. In South Korea, for example, the first McDonald's restaurant opened in 1988: by 1990, there were 22 restaurants, by 1995 there were 58, and by 2011 the number had grown to 243. To facilitate this process, the company has altered its offerings and developed particular products for different national contexts (e.g., the McSpaghetti product, which originated in the Philippines, or the shrimp, kimchi, and bulgogi (marinated beef) burgers available in South Korea). On the other hand, consumers have transformed McDonald's into local institutions through the way in which they consume McDonald's products and restaurants. While some aspects of the McDonald's system have become accepted – queuing and self-service, for example – the notion of "fast" food has been adapted and altered. While food delivery is fast, its consumption may not be, as young consumers have turned their local McDonald's into a kind of leisure center or after-school club, a place for socialization and hanging-out rather than a brief visit. In this way, a global icon is given local meaning through place-specific consumption practices that reinterpret corporate business models: "East Asian consumers have quietly, and in some cases stubbornly, transformed their local McDonald's into local institutions" (Watson, 1997, p. 6).

15.5   Toward an Ethical Consumption Politics?

Consumption, of course, is not only about the selfish motives of looking and feeling good, and visibly demonstrating our position in wider society. It also has moral and ethical dimensions (Barnett et al., 2005). This is true of both the ordinary practices of consumption that often involve choices involving a concern for various others (e.g., family members and friends) but also to the purchase of the rapidly expanding variety of goods and services that are explicitly labeled as "ethical." It is these ethical products that form our primary focus here: "These include the consumption of organic foods, fair trade goods, and environmentally friendly products. Ethical consumerism is also seen to encapsulate forms of green and socially responsible tourism, the purchase and use of second-hand goods, the procurement of locally produced goods, ethical banking and consumer boycotts of specific commodities and brands" (Hughes et al., 2008, p. 350). Through the choices we make about the commodities we buy and consume, we demonstrate our concern (or not) for the economic, social, and environmental conditions under which those goods were produced. Thus consumption is unavoidably a political act in the sense that it implicates and involves us in the wider commodity chains that underpin all commodities. This realm of consumption politics is a challenging one to negotiate. As described in Chapter 8, capitalist production, and the work of intermediaries such as advertisers, serves to obscure the origins and production conditions of commodities. Ethical consumption entails exercising a "geography of responsibility" that extends beyond the everyday and the local to incorporate notions of "caring at a distance" for less fortunate others (Massey, 2004). Moreover, consumers must decide which dimension of ethicality – for example, enhancing worker conditions or limiting environmental degradation – they are most concerned with, and the extent to which they are prepared to follow those principles in their consumption practices (e.g., buying a certain brand of coffee as compared to making sure all their money is invested ethically). Ethical consumption will also usually entail paying a premium over "non-ethical" goods.

Ethical consumer action can take different forms. Perhaps the simplest form of consumer campaign is to *boycott* – that is, not purchase – the products of a particular company. For example, a boycott launched in 1977 against the giant Swiss foodstuffs company Nestlé, prompted by concerns over the company's promotion of baby milk formula in the developing world, is still in operation through the actions of the International Baby Food Action Network (IBFAN), which links 200 citizens groups across 100 countries. An extreme version of this approach is "Buy Nothing Day," an international event each November in which consumers are encouraged, under the slogan "shop less, live more," to not shop for a 24-hour period as part of a wider critique of modern consumerism (see Figure 15.5). The organizers describe how "Buy Nothing Day is an opportunity for you to make a commitment to consuming less, recycling more and challenging corporations to clean up and be fair. Modern consumerism

Figure 15.5   Poster for "Buy Nothing Day"

might offer great choice, but this shouldn't be at the cost of the environment or developing countries. Buy Nothing Day isn't about changing your lifestyle for just one day – it's a lasting relationship – maybe a life changing experience!" (http://www.buynothingday.co.uk/faq.html, accessed October 31, 2011). We can also identify *corporate campaigns* that seek to target highly visible corporations by mobilizing information regarding violations against workers or the environment. The United Students Against Sweatshops (USAS) campaign in the United States, for example, used a series of university campus protests and sit-ins in 1999 to force university administrators and firms such as Nike to be far more transparent about their licensing arrangements and production systems in the multi-billion-dollar college apparel sector (for more, see *Antipode*, 2004).

While some interventions are concerned with trying to prevent consumption of certain goods and services, increasingly consumer campaigns seek to encourage the consumption of commodities that have met certain measurable ethical standards. They are thus tied to pushing for corporate adherence to the growing range of labels, standards, and codes of responsibility that were introduced in Chapter 8. In this way consumers are actively enrolled in reworking consumer–producer links within broader production systems in favor of the latter. Ethical consumption has grown strongly in developed markets over the past 10–15 years. In the United Kingdom, for example, the ethical market grew from £13.5 billion in 1999 (3% of total household spending) to £46.8 billion (9%) in 2010.

The average ethical spend per household increased from £291 per year to £868 per year over the same time period. Food and drink was the largest ethical expenditure on average (£256), but ethical consumption was also to be found in the areas of energy efficiency (£176), travel (£127), sustainable home products such as long-life light bulbs (£99), local shopping (£91), clothing (£55), renewable energy (£24), cosmetics (£21), and "green" mortgages (£19) (*Financial Times*, 2011). Ethical consumption has thus become increasingly mainstream and is no longer a marginal concern to producers and consumers. A key component of growth has been the conversion of many large brands and retailers to Fairtrade, a labeling and certification scheme that guarantees a minimum price to producers (see Section 8.4). What is clear is that consumers' purchasing decisions are increasingly shaped by consideration of the ethical reputations of retailers and producers.

Ethical consumption decisions are not simply individual choices about purchasing priorities, but rather are embedded in wider societal debates. As a result, ethical consumption practices are geographically uneven in their prevalence. The uneven geographies of ethical consumerism can be considered at three spatial scales. Firstly, at the global scale, ethical consumption is largely a rich-country phenomenon. It is perhaps understandable that "caring at a distance" becomes more feasible once a certain level of income and financial security has been achieved. The Fairtrade movement, for example, has a distinctive global geography. It encompasses 19 different labeling initiatives across 26 countries in Central and Western Europe and North America, along with Australia, Japan, New Zealand, and South Africa. On the producer side, the Fairtrade Labeling Organization has three regional associations representing certified producer organizations in Africa, Latin America and the Caribbean, and Asia. Fairtrade, then, remains underpinned by a traditional developed/developing country, consumer/producer divide, and it is important to remember that ethical consumption is currently largely associated with high-income markets. It remains to be seen how long it will be before the macro-shifts in consumption at the global scale described in Section 15.3 feed through into new global geographies of ethical consumption.

Second, national context remains important, with ethical consumer campaigns taking on different forms in different countries. For example, the anti-sweatshop campaign described above was a distinctively American campaign, and clothing and toy retail and brands have been the most frequent targets of ethical consumers in the United States. In the United Kingdom, where food retailing is much more concentrated in the hands of a few firms, it has been the leading food and general merchandise retailers that have primarily been the focus of campaigns (Hughes et al., 2008).

Third, and important given the focus on local consumption cultures in this chapter, there are often strong local dimensions to ethical consumerism. Fairtrade Towns provide one interesting example. The status is awarded by the Fairtrade Foundation once five goals have been met. These relate to: the local council passing a resolution supporting Fairtrade; the availability of Fairtrade products in local retail and catering outlets; the demonstrable support of local workplaces and community organizations such as churches, schools, universities, and colleges; media coverage and events raising awareness and understanding of Fairtrade across the community; and the establishment of a local Fairtrade steering group to ensure that the campaign maintains momentum. The concept spread from the small Lancashire market town of Garstang in the United Kingdom, which became the first Fairtrade Town in April 2000; there are now around 500 Fairtrade Towns in the United Kingdom and 1,000 in total across 19 countries in Europe and North America. Another example is the Slow Food movement, an international nonprofit association and network of some 100,000 members across 153 countries concerned with promoting "good, clean and fair food for all" and improving how food is produced and distributed. The idea originated in Piedmont in northern Italy in the mid-1980s, and the international Slow Food Manifesto was signed in 1989. While coordinated globally and nationally, the key units are the 1,300 "convivia" or local chapters, which work largely autonomously to preserve distinctive and sustainable local food cultures. The convivia organize shared meals and tastings, visits to local farms and producers, film screenings/festivals, and the promotion of local produce markets.

Overall, the analysis in this section has served to show how the rise of ethical consumerism is an inherently geographical phenomenon in two interrelated ways. First, ethical campaigns are often multiscalar, with local, place-based action being coordinated through, and informed by, national and/or international initiatives. Second, ethical consumption brings into play new geographies of responsibility, with the local consumption practices of groups and individuals being explicitly connected to the livelihoods and fortunes of distant "others."

15.6   Consuming Places: Travel and Tourism

So far in this chapter we have looked at the ways in which consumption processes are given a distinctive flavor by the places in which they occur. But there is another angle we can take here – namely, that a significant proportion of consumption activity is actually concerned with the consumption of places themselves. Tourism is centrally about the generation of different kinds of pleasurable and/or memorable in-place experiences. Tourist experiences are often about the combination of tangible elements (e.g., hotel, flight) and intangible experiental elements (e.g., sunset, mood). Tourism is a huge industry, perhaps the world's largest (see Box 15.6), and is increasingly global in scope. International tourist arrivals have shown virtually uninterrupted growth in recent decades, rising from 25 million in 1950 to 277 million in 1980, 435 million in 1990, 675 million in 2000, and 940 million by 2009. Growth has been particularly fast in the developing world, where international tourist arrivals have risen from 31% of the global total in 1990 to 47% in 2010. Emerging economies are also increasingly important sources of international tourists, most notably China. Table 15.3 charts the top ten destination and source countries for international

CASE STUDY

Box 15.6   The world's biggest industry?

Tourism is sometimes described as the world's biggest industry, although it is perhaps best characterized as a complex of interlinked industries offering a broad range of accommodation, transport, and entertainment services. Globally, travel and tourism directly generated an estimated US$1,850 billion in revenues in 2011, equivalent to 2.8% of world economic output. This primarily reflects the economic activity generated by industries such as hotels, travel agents, airlines, and other passenger transportation services but also includes, for example, the activities of the restaurant and leisure industries directly supported, in part, by tourists. The total economic contribution of travel and tourism to the world economy, however – when the multiplier, supply chain and induced government expenditures (e.g., in transport infrastructure) are also taken into account – was US$5,992 billion, equivalent to 9.1% of total output. In employment terms, travel and tourism generated an estimated 99 million direct jobs (3.4% of total employment) and 256 million in total (8.8% of total employment). Roughly three-quarters of the revenue (76.7%) generated by the sector comes from leisure travel spending with the remaining quarter (23.3%) deriving from business travel spending (all data from WTTC, 2011). Many segments of the tourism industry are dominated by transnational corporations from developed countries. In terms of hotels, for example, the leading providers include the InterContinental, Wyndham, Marriot, Hilton, Accor, and Hyatt groups. As of September 2011, the U.K.-based InterContinental Group had 4,520 hotels offering 666,476 rooms in more than 100 countries across brands such as Crowne Plaza, Hotel Indigo, and Holiday Inn.

Tourism is also hugely significant as an internationally traded service. International tourism has become one of the world's major trade categories, with inbound tourists effectively counting as an export category. The overall export income generated by international tourism, including passenger transport, exceeded US$1 trillion in 2010, or close to US$3 billion a day. Tourism exports account for around 30% of the world's exports of commercial services and 6% of overall exports of goods and services. Globally, as an export category, tourism ranks fourth after fuels, chemicals, and automotive products (UNWTO, 2011).

Table 15.3 International tourism receipts and expenditure – top 10 countries in 2010

Rank	Country of Destination	2010 Receipts (US$ billion)	Country of Origin	2010 Expenditure (US$ billion)
1	United States	103.5	Germany	77.7
2	Spain	52.5	United States	75.5
3	France	46.3	China	54.9
4	China	45.8	United Kingdom	48.6
5	Italy	38.8	France	39.4
6	Germany	34.7	Canada	29.5
7	United Kingdom	30.4	Japan	27.9
8	Australia	30.1	Italy	27.1
9	Hong Kong	23.0	Russian Federation	26.5
10	Turkey	20.8	Australia	22.5

Source: Adapted from UNWTO (2011).

tourists in revenue terms, with the United States, Spain, and France being the biggest recipients of international tourist spending, while Germany, the United States, and China are the biggest sources. Figure 15.6 shows the dominance of flows within and between North America, Western Europe, and East and Southeast Asia.

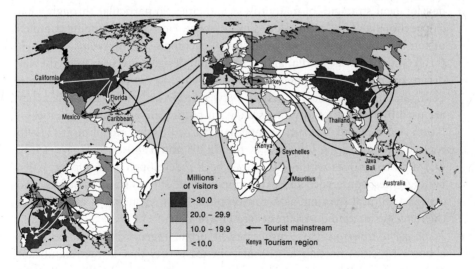

Figure 15.6 World tourist flows, 2010
Source: Updated and adapted from
http://www.seos-project.eu/modules/landuse/landuse-c04-p09.html.

How have tourist spaces evolved over time? Large-scale tourism dates to the second half of the 19th century, when first the middle classes and then the working classes in countries such as the United Kingdom took advantage of social change and improvements in transport to travel to coastal resorts such as Blackpool and Southend for rest and relaxation. International tourism flourished from the 1950s onward, driven by increased disposable incomes and leisure time in developed countries, the rise of airline travel, and the entry of transnational corporations into the industry in areas such as hotels and travel agencies, thereby increasing the availability and affordability of so-called "package" holidays combining travel, accommodation, and other tourist services. The 1960s and 1970s were the golden era of international mass tourism in Europe in particular, with large numbers of consumers purchasing standardized products. This period of strong growth was associated with the development of standardized tourist resorts along the Mediterranean coast but also on the Black Sea and in the Alps.

As with consumption trends more generally (see Table 15.2), the tourist industry has changed significantly in recent decades, becoming more post-Fordist in nature. More individualized consumption patterns have become apparent in tourism as they have in many other consumer-oriented industries. Most importantly, there has been something of a rejection of mass tourism from middle-class consumers who now seek a wider range of tourism experiences, driving growth in a number of areas, including:

- *Urban and heritage tourism*: Urban tourism, often undertaken in short trips, has grown rapidly. As a result, tourism has become central to the economic development strategies of many towns and cities, in particular through what can be thought of as "culture-led" development. Here an overlapping complex of tourist, cultural, and creative sectors – encompassing museums, galleries, theaters, the arts, the media, architecture, and design – are seen to be the key drivers of growth. These amenities, and associated services such as restaurants, clubs, and bars, blur the distinctions between tourism and leisure in that they are used by residents and visitors alike. In particular, postindustrial cities in North America and Western Europe have turned to this model of growth to replace manufacturing jobs lost in the 1970s and 1980s (see Figure 15.7). Old industrial zones within cities have been redeveloped as shopping malls, heritage sites, conference and exhibition centers, entertainment districts, and arts and cultural quarters as a consumption-based economy is carved from the remnants of an industrial one (see Figure 15.8).
- *Mega-event tourism*: These processes of urban regeneration can be given a boost by cities hosting large-scale one-off events that attract significant numbers of visitors, such as a World Expo (e.g., Shanghai, 2010) or Olympic Games (e.g., London, 2012). The Barcelona Olympics of 1992, and the associated revitalization of its built environment, are widely seen to have played a central role in Barcelona becoming one of the leading tourist cities

The authors

Figure 15.7   Culture-led redevelopment – the waterfront in Bilbao, Spain

The authors

Figure 15.8   Reclaimed industrial heritage – Legoland in Duisburg, Germany

in Europe. The Olympics gave planners and politicians the opportunity to undertake large-scale public works, for example opening up the city's waterfront as a consumption space. The 2008 Olympics were similarly pivotal in presenting modern urban China on the world stage.

- *Theme park tourism*: Here visitors pay for admission to an entirely themed complex, placing particular emphasis on the architecture and its symbolism. At the heart of the theme park model is the consumption of multisensory experiences that may combine simulated environments (e.g., natural, cultural, historical, or technological); the humanizing of those environments by live interpretations, performances, and commentaries (e.g., reenactments of historical events); state-of-the-art technological devices (e.g., films, rides, and games); and themed exhibits and eating places. The focus of the global theme park industry is increasingly turning to Asia, as developers seek to benefit from the rapidly growing middle classes in many Asian markets (*Bangkok Post*, 2011).

- *Ecotourism*: This has been defined by the International Ecotourism Society as "responsible travel to natural areas that conserves the environment and improves the well-being of local people." Ecotourism is concerned with enjoying the natural environment through low-impact and sustainable activities such as trekking and animal/bird watching. It is associated with certain kinds of relatively wealthy tourists and also with certain destinations that are rich in natural amenities, such as New Zealand.

Overall, while mass tourism is still an important component of the industry, it is now supplemented by a wide range of other specialized niche markets offering more distinctive, individualized experiences. One outcome has been a significant blurring of the boundaries of tourism with other activities such as leisure, recreation, shopping, entertainment, education, and sport. Another has been the proliferation of places and local economies that now depend, to a greater or lesser extent, on tourism-cum-leisure revenues.

Accordingly, state agencies and businesses involved in promoting and selling tourism have sought to shape distinctive images of their localities through place-marketing campaigns. In these campaigns, places are presented to appeal to particular types of tourists who may wish to consume different aspects of those places, such as "culture and tradition, ecology and nature, or more prosaically, sun, sea and sand" (Hudson, 2005, p. 172). These place-marketing campaigns can operate at a number of different spatial scales including an individual street (e.g., Chicago's "Magnificent Mile" of shops), a district (e.g., "The Rocks" district of Sydney), a city (e.g., "Marketing Manchester"), a region (e.g., the Lord of the Rings "Trilogy Trail" in southern New Zealand) or a national economy. An example of the latter is the 'Magical Kenya' campaign, designed to promote Kenya's undoubted attractions to international tourists (see Figure 15.9). Each poster in the series bears the greeting "Jambo! Welcome to magical Kenya" (Jambo means hello in Swahili). The logo uses the colours of the Kenyan flag and the arch in the middle represents a necklace from the Masai people, one of Kenya's most famous tribal groups. Other images in the series show a tropical beach, a fisherman brandishing a huge catch and dancing tribespeople. As with all place marketing campaigns, however, it is a particular set of images of Kenya that are presented to the wider world. This is a Kenya of big landscapes, pristine

Figure 15.9   Magical Kenya

beaches, wild animals, stunning vistas, and traditional tribespeople. There is understandably no reference to poverty or the sprawling Kibera slums of Nairobi. The wider point here is that crafting and disseminating a place image is a crucial component of the economic development strategies of localities that need to attract visitors in order to fuel their tourism and recreation sectors.

15.7   Summary

This chapter has explored different aspects of the geographies of consumption. We started by advocating a sociocultural perspective on consumption dynamics that affords consumers a significant degree of agency as they go about constructing their identities through the purchase and subsequent use of commodities. Four arguments were then developed in the subsequent analysis. First, we looked at the changing geographies of consumption at the global scale. The rise of the middle class across a range of emerging economies, and China and India in particular, is already starting to rework the assumption that consumption is dominated by the economies of North America, Western Europe, and advanced Asian economies such as Japan and South Korea. Second, we profiled the way in which consumption is integral to identity formation for groups and individuals, and how those processes of identity formation are unavoidably place-specific, with commodities assuming different meanings in different sociocultural contexts. These insights allow us to reveal the agency of consumers as they seek to reinterpret and even subvert the intended meaning of both commodities and consumption spaces in geographically variable ways.

Third, we considered the extent to which individual and collective consumption decisions are increasingly being shaped by ethical concerns for the economic, social, and ecological environments in which commodities are produced. Linking back to the discussion of commodity chains in Chapter 8, it seems that consumption politics have become an increasingly important, though not unproblematic, arena for debating and seeking to ameliorate some of the significant downsides to global capitalism. Finally, we looked at tourism and leisure, a wide-ranging set of economic activities concerned with enabling consumers to experience places of different kinds. Changes in the tourism industry mean that an ever widening variety of places are now involved in such activities, and place marketing has become an integral part of the economic development strategies of many cities and regions.

Notes on the references

- Mansvelt (2005), Jayne (2006), and Miles (2010) deliver student-friendly reviews of the best geographical work on consumption. See also Goodman et al. (2010) for a range of geographical perspectives on consumption processes.
- Miller et al. (1998) and Zukin (2004) offer compelling accounts of the varied and everyday practices of shopping.
- For more on the case studies of Coca-Cola in Trinidad, McDonald's in Asia, and fashion and food cultures in London and Mumbai, see Miller (1998), Watson (1997), and Jackson et al. (2007), respectively.
- Barnett et al. (2005) and Hughes et al. (2008) offer excellent accounts of the rise of ethical consumption.
- Shaw and Williams (2004), Hall and Page (2006), and Williams (2009) provide comprehensive accounts of the changing nature of tourism and leisure spaces. For an explicitly economic geography take on tourism, see Ioannides and Debbage (1998).

Sample essay questions

- Why is consumption such a geographically variable process?
- How are different consumption spaces manipulated in order to induce consumer purchases?
- In what ways can consumers actively contest and subvert the intended meanings of consumption spaces?
- How do individuals use consumption to develop place-specific identities?
- In what ways can places themselves be consumed through travel and tourism?

Resources for further learning

- http://www.consume.bbk.ac.uk/index.html: The Cultures of Consumption site hosted by Birkbeck College, University of London, hosts a range of cutting-edge research materials on consumption dynamics worldwide.
- http://www.fairtradetowns.org/ and http://www.slowfood.com/: For more on Fairtrade towns and the Slow Food movement.
- http://www.buynothingday.co.uk/: Offers an insight into the anti-consumption protest movement.
- www.unwto.org: The website of the World Tourism Organization hosts a huge range of materials about the global tourism industry. See also the World Travel & Tourism Council (WTTC): http://www.wttc.org/.

References

Antipode. (2004). Intervention symposium: geographies of anti-sweatshop activism. *Antipode* **36**, 191–226.

Bangkok Post. (2011, August 15). Asia theme park boom is big business for designers, p. B9.

Barnett, C., Cloke, P., Clarke, N., and Malpass, A. (2005). Consuming ethics: articulating the subjects and spaces of ethical consumption. *Antipode*, **37**, 23–45.

Bourdieu, P. (1984). *Distinction: A Social Critique of the Judgment of Taste*. Cambridge, MA.: Harvard University Press.

Goodman, M. K., Goodman, D., and Redclift, M., eds. (2010). *Consuming Space: Placing Consumption in Perspective*. Farnham, U.K.: Ashgate.

Guardian, The. (2011, October 18). Vegas: city of glitter and gloom, p. 29.

Hall, C. M., and Page, S. J. (2006). *The Geography of Tourism and Recreation*, Third Edition. London: Routledge.

Hudson, R. (2005). *Economic Geographies*. London: Sage.

Hughes, A., Buttle, M., and Wrigley, N. (2008). Global production networks, ethical campaigning, and the embeddedness of responsible governance. *Journal of Economic Geography*, **8**, 345–67.

Ioannides, D., and Debbage, K. G., eds. (1998). *The Economic Geography of the Tourist Industry: A Supply Side Analysis*. London: Routledge.

Jackson, P. (2004). Local consumption cultures in a globalizing world. *Transactions of the Institute of British Geographers*, **29**, 165–78.

Jackson, P., Thomas, N., and Dwyer, C. (2007). Consuming transnational fashion in London and Mumbai. *Geoforum*, **38**, 908–24.

Jayne, M. (2006). *Cities and Consumption*. London: Routledge.

Kaplinsky, R., and Farooki, M. (2010). Global value chains, the crisis, and the shift of markets from North to South. In O. Cattaneo, G. Gereffi, and C. Staritz, eds., *Global Value Chains in a Postcrisis World*, pp. 125–53. Washington, D.C.: World Bank.

Kehily, M. J., and Nayak, A. (2008). Global femininities: consumption, culture and the significance of place. *Discourse: Studies in the Cultural Politics of Education*, **29**, 325–42.

Kharas, H. (2010). The emerging middle class in developing countries. *OECD Development Centre Working Paper No. 285*, OECD, Paris.

Leslie, D., and Reimer, S. (1999). Spatializing commodity chains. *Progress in Human Geography*, **23**, 401–20.

Mansvelt, J. (2005). *Geographies of Consumption*. London: Sage.

Massey, D. (2004). Geographies of responsibility. *Geografiska Annaler B*, **86**, 5–18.

McDowell, L. (2009). *Working Bodies: Interactive Service Employment and Workplace Identities*. Chichester, UK: Wiley-Blackwell.

Miles, S. (2010). *Spaces for Consumption*. London: Sage.

Miller, D. (1998). Coca-Cola: a black sweet drink from Trinidad. In D. Miller, ed., *Material Cultures: Why Some Things Matter*, pp. 169–87. London: UCL Press.

Miller, D., Jackson, P., Thrift, N., Holbrook, B., and Rowlands, M. (1998). *Shopping, Place and Identity*. London: Routledge.

Prahalad, C. K. (2005). *The Fortune at the Bottom of the Pyramid: Eradicating Poverty through Profits*. Upper Saddle River, NJ: Pearson.

Shaw, G., and Williams, A.M. (2004). *Tourism and Tourist Spaces*,. London: Sage.

UNWTO (World Tourism Organization). (2011). *UNWTO Tourism Highlights: 2011 Edition*. Available from www.unwto.org.

Watson, J. L., ed. (1997). *Golden Arches East: McDonald's in East Asia*. Stanford, CA: Stanford University Press.

Williams, S. (2009). *Tourism Geography: A New Synthesis*, Second Edition. London: Routledge.

WTTC (World Travel and Tourism Council). (2011). *The Economic Impact of Travel and Tourism*. London: WTTC.

Zukin, S. (2004). *Point of Purchase: How Shopping Changed American Culture*. New York: Routledge.

PART V

CONCLUSION

CHAPTER 16

ECONOMIC GEOGRAPHY
Intellectual journeys and future horizons

Goals of this chapter

- To trace the development of ideas in the field of Economic Geography
- To place these ideas in the social and intellectual context of their development
- To think about how new contexts and circumstances will change Economic Geography in the future

16.1 Introduction

In early 2009, the Australian government announced that it would inject AU$27 billion (about US$18 billion at the time) of new spending into its economy through a program called the Nation Building–Economic Stimulus Plan. The money would pay for almost 50,000 construction projects across the country, including new school buildings, community infrastructure, road and rail improvements, and social housing. Most of these projects were completed by the middle of 2011. The purpose of this spending was to stimulate economic growth and to avert a deep recession. The investment would also boost future competitiveness by developing transportation facilities, education, and other productivity-enhancing infrastructure.

Many other countries did the same thing (see Section 4.1). Meeting in November 2008, the Group of 20 leading industrialized and emerging economies (G20) vowed to "use fiscal measures to stimulate domestic demand to rapid effect." The level of international cooperation and agreement concerning these stimulus packages was remarkable. Within a few months at the end of 2008 and beginning of 2009, dozens of countries had created spending programs that collectively pumped trillions of dollars into the economic system. Even more remarkable was the fact that these countries were collectively subscribing to a set of economic ideas that had been out of favor for several decades. In fact, since the mid-1970s governments in the West had generally adhered to the idea that in lean times governments must cut back their spending, otherwise government dollars would crowd out productive spending and investment by the private sector.

The policies of 2008–09 represented a return to ideas that had held sway through most of the mid-20th century, having first been proposed in the 1930s by the British economist John Maynard Keynes. As we noted in Chapter 2, it was his work that provided comprehensive evidence that economies could be managed. A part of this management regime would be exactly the kind of stimulus spending to counteract economic cycles that was implemented in 2008–09. Keynes would doubtless have taken great satisfaction in seeing the rejuvenation of his ideas. After all, he had written in 1936 that "[p]ractical men, who believe themselves to be quite exempt from any intellectual influence, are usually the slaves of some defunct economist" (p. 383). In this case, Keynes himself was the "defunct economist" whose ideas were influencing world leaders. A headline in the *Sydney Morning Herald* in 2009 announced that "Keynes is back," and books appeared with titles such as *Keynes: The Return of the Master*.

An important lesson we can learn from this resurgence of interest in Keynes is that ideas about how the economy works are very much a product of the time and place in which they are developed. Their popularity and influence are a product of circumstances (and, it should be added, the power of different interest groups). While Keynes developed his ideas in an inhospitable intellectual climate in the 1930s, they were quickly adopted in the context of postwar reconstruction from the late 1940s to the mid-1970s. In the face of difficult economic times in 2008–09, they were again deemed to be relevant.

The ideas of economic geographers have never been as influential as those of Keynes, but we can certainly see the ways in which the development of concepts has reflected the times in which scholars were working and the challenges to which they applied themselves. For the most part, this book has explored the work of economic geographers without making specific reference to the debates and ideas that have driven the intellectual journey of the discipline. But it is possible to see many of the ideas presented in this book as part of an intellectual journey that the field of Economic Geography has taken over the last half century. In this concluding chapter, we will review the ways in the field has changed, to reflect the different circumstances and contexts (both intellectual and worldly) in which it was working. We will then look to the future to see what challenges lying on the horizon will likely shape the future of Economic Geography.

16.2   A Changing Field

During the second half of the 20th century, the field of Economic Geography saw the emergence of several different styles of inquiry or intellectual traditions (Scott, 2000). Each of them asked distinctive questions and developed particular ways of uncovering answers. Telling the story of these different traditions is difficult for several reasons. First, it is all too easy to place tidy labels on schools of thought that were internally diverse and overlapping with each other. The reality is that intellectual history is always messier than the categories that we impose with hindsight. Second, it is tempting to construct an intellectual lineage that implies the redundancy of "old" ways of thinking and the enlightenment of current approaches (especially those favored by the author!). In fact, we can usually learn a lot from approaches that were once the cutting edge of scholarship, and in any case many do not disappear, but instead become integrated into the mainstream practices of the discipline. Third, any disciplinary history will tend to create "periods" that are all too neat, with the implication that scholars moved forward collectively and in unison. In reality, the new frontiers of inquiry at any given time likely do not actually reflect what the majority of Economic Geography students are learning about in their classrooms.

Notwithstanding these caveats, it is worthwhile to give a sense of how Economic Geography has changed over the years, and to note how this has reflected the times and places in which it was being practiced. The influence of the wider social context has been important since the very beginning, both for Economic Geography and for the discipline of Geography as a whole. The earliest institutions of Geography in the English-speaking world emerged in the late 19th century as a part of the British colonial project. Scholarly societies such as the Royal Geographical Society in the United Kingdom, the journals they published, and the meetings they organized were all focused on supporting colonialism in intellectual and practical ways. For economic geographers, this consisted of assessing the resource potential of colonial territories and analyzing trade patterns within the empire. Continuing through the first half of the 20th century, a tradition of regional description held sway. Although the linkage with colonial knowledge production became less direct, the purpose was still largely one of integrated analysis of resources and resource use in a given region. It was in the post-World War II years that this tradition was challenged and the story of modern Economic Geography begins.

In the following three sections we organize the intellectual twists and turns of Economic Geography of the last 60 years into three broad categories: positivism, structuralism, and post-structuralism (based in part on Sheppard, 2006). These categories reflect different philosophies that have shaped the types of questions that were asked by economic geographers and how answers were sought through

KEY CONCEPT

Box 16.1 *Ontology, Epistemology, and Methodology*

The philosophy of knowledge consists of several distinct concepts. *Ontology* refers to a set of philosophical beliefs about what exists in the world and therefore what it is possible to know. When we use categories such as economy, nature, and culture, we are making ontological claims that these are objects that exist in the world and are therefore knowable. When we debate the existence of these objects (as we have at several points in this book) then we are engaging in ontological discussions.

Epistemology is a set of philosophical beliefs about what constitutes knowledge and how it can be obtained. Some economic geographers, for example, might argue that knowledge is derived from objectively constructed data that can be replicated by other researchers, while others would insist that subjective interpretations provide important means of understanding the economic world. Obviously an epistemological position is closely linked to an ontology, but they are not necessarily the same thing. One might believe, for example, that something exists without believing that it is knowable.

A *methodology* is a framework for evaluating knowledge claims made about the world; in particular, it provides the principles by which specific techniques of data collection are chosen and evaluated. For example, a feminist economic geographer would likely want to employ data collection techniques that empowered the women involved, and that allowed careful reflection on the relationships created in the research process.

research. In order to see how they are different it is important to understand something about how a philosophy of knowledge is constructed through the concepts of ontology, epistemology, and methodology (see Box 16.1).

Positivism: Science and Quantitative Economic Geography

The decades after World War II provided a very different social context for Economic Geography as a field of inquiry. It was a period in which European empires were giving way to independent states in the developing world, while the developed countries of North America, Europe, and Australia/New Zealand were experiencing a prolonged boom of economic growth, urban expansion, and (as noted earlier) a Keynesian style of national economic management.

In these circumstances, economic geographers turned their attention to questions that addressed the times and places in which they lived and worked (just as they had in the era of colonialism). Issues of industrial location, patterns of urban growth, changes in land use, the development of transportation networks, and the dynamics of trade were all primary concerns in a context of rapid growth. Furthermore, given the Keynesian strategy of economic management and the deep involvement of governments in planning and reconstruction after the war, there was also a strong sense that these were processes that could be shaped and directed through state interventions informed by academic research.

At the same time, there was an increasing desire for systematic and "scientific" modes of knowledge production in Economic Geography. This took the form of: (a) seeking universal "laws" or principles that underpinned spatial patterns of economic activity; (b) using quantitative data to identify such patterns; and (c) applying statistical techniques to derive mathematically rigorous proofs of the patterns being identified. These practices were applied to patterns of urbanization, regional growth, and industrial location, and flows and interactions through space.

The quantitative tradition in Economic Geography had two strands (Scott, 2000). One strand focused on spatial analysis using mathematical models, and utilized early computer technology for analytical purposes. Those who lived through those times will tell stories of punch cards and computer analysis that could take several days to perform. While advances in computational technology formed one context for these approaches, government funding for scientific and military research was also a key factor (Barnes, 2008).

The other strand of the quantitative tradition sought ways of integrating space and location into neoclassical models of economic theory. This line of work became known as regional science and forms the basis for the contemporary work of economists who analyze geographical economics. The most famous of these is Paul Krugman, whose Nobel Prize in Economics in 2008 recognized his contributions to understanding how space, location, and distance are fundamental for trade and the agglomeration of economic activities (see Box 1.2).

In both spatial analysis and regional science, earlier classics in the German tradition were rediscovered. The agricultural land use theories of Johan von Thünen, originally published in 1826, were used to create models of optimal land use patterns (see Chapter 1). Walter Christaller's central place theory (from 1933) inspired models of urban system formation and consumer behavior (see Chapter 11). Alfred Weber's industrial location theory (published in 1909) informed approaches to understanding how manufacturing facilities were optimally sited (see Chapter 12). And while these classics provided an important foundation, the rise of quantitative Economic Geography generated its own classics. The intellectual excitement of a new style of knowledge that seemed to sweep away the fuzzy vagueness of a previous generation was very appealing to a new generation of bright young scholars. Among them, charismatic and extremely capable leaders emerged who defined the field and in turn attracted the best young graduate students. In a relatively small discipline such as Geography, internal sociologies of this kind can have a significant influence (Barnes, 2001, 2012). By the late 1960s and early 1970s, a quantitative, model-building, 'scientific" approach to Economic Geography may not have been universal, but it was certainly dominant in the leading centers of research.

What bound together these various approaches was a positivist approach to knowledge production. This implied that a truth about the world could be uncovered by directly observing and measuring phenomena. Moreover, the things being measured were believed to display a consistency across time and space. Such consistency meant that it was possible to develop and test general principles/laws/models from empirical observations.

It would be a mistake to believe that positivist Economic Geography is a thing of the past. As noted above, the science of spatial relations has energized a new generation of economists, and the connections between Economics and Economic Geography are perhaps stronger now than they have been in several decades (for example, the *Journal of Economic Geography* was created in 2000 to publish the research of both economists and geographers). Economic geographers continue to find mathematical and model-building approaches relevant, especially where large data sets provide answers to research questions on issues such as international trade patterns, regional development, and housing or labor market trends.

Structural Approaches

A second set of approaches takes a quite different view of the world than the positivist/scientific philosophy. Rather than understanding the world through observation and measurement, they see underlying structures that are not necessarily directly observable. Furthermore, these structures shape and constrain people's actions in the world and create differences of power. The structures and inequalities created by capitalism/class, gender, and race have been especially important in Economic Geography (a structural approach to economic power was discussed in some detail in Chapter 3). In addressing these issues, economic geographers have been influenced (but also, inspired, angered, moved, and stimulated) by the pressing social issues of the world around them. They have also sought to develop normative agendas, meaning that they do not simply wish to describe and analyze the economic world, but also to shape it. For the overwhelming majority of economic geographers, this has meant a concern with social justice, economic inequality, and environmental sustainability.

The reason for this progressive normative agenda becomes evident when we examine the roots of structural approaches to economic questions. While quantitative geography and locational analysis emerged to address the needs of growing economies in the mid 20th century, various social movements took hold in the late 1960s and economic growth began to slow down in the early 1970s. In the United States in particular, civil rights protests, feminist and environmental movements, and the anti–Vietnam War campaign all gave rise to alternative and anti-establishment thinking. In Economic Geography, the scientific certainties of the quantitative approach seemed less capable of addressing the clear social problems that these movements highlighted. In North America and Western Europe, urban poverty and segregation, gender inequalities, uneven international development, and deindustrialization were all pressing issues. They demanded an understanding of the structural processes that underpinned observable economic phenomena, and a political agenda to change them.

The most celebrated conversion to a politicized and radical approach to economic questions was found in David Harvey. Harvey was trained in Britain in the descriptive tradition of historical geography, but made his name through contributions to the quantitative revolution in the 1960s. His book, *Explanation in Geography* (1969), was seen by many as the high point of quantitative geography, as it laid out the philosophical fundamentals of scientific method for geographers. Within just a few years, though, Harvey's approach had changed dramatically. Moving from Britain to Baltimore in 1969, he came face-to-face with urban poverty, race-based segregation, and deindustrialization and turned to Marxist theory to understand how the economy was fundamentally driven by class relationships.

Over the following decades, Harvey's influence through his students, and his students' students, has been significant. Seeing the economy as tending toward crisis rather than equilibrium and as driven by class antagonism not harmonious divisions of labor, he and others provided a perspective on structural power that was not available through positivist Economic Geography (see Chapter 3). While the approach was controversial initially, many aspects have gradually become mainstream even among scholars who do not subscribe to the entire framework provided by Marxian theory. Our ability to think about the structural logics of capitalism, to identify class-based power relations in economic processes, and to conceptualize uneven development are all dependent on the legacies of Marxist theory in the field of Geography.

While Marxist thought sees crises as the inevitable outcome of capitalism's fundamental contradictions, one branch of Marxian thought in Economic Geography asked instead how capitalism has been able to avert crises. Known as *regulation theory*, and emerging in the 1970s, this line of inquiry sought to find out how institutions have been created to divert or delay crises effectively. Of particular interest to regulation theorists were the three decades following World War II, when a period of sustained economic growth was maintained in Western Europe and North America under the regime of Fordism (see Boxes 3.1 and 9.3). What came after Fordism – post-Fordism or flexible specialization – was also a subject of close analysis by economic geographers. The new geographies of production in clusters or high-growth regions (see Chapter 12), and the local social and institutional foundations for these success stories, have energized economic geographers for the past 20 years or more. An understanding of the role of the state and other key institutions in economic processes at multiple scales has been especially important (see Chapter 4).

The analysis of global commodity chains or global production networks is another strand of inquiry that has been built on the careful analysis of the institutional foundations of capitalist growth (see Chapter 8). By understanding corporate and noncorporate players and the ways in which they interact, such analyses have tried to explain how specific sectors organize their activities across global economic space. The institutional forms of corporate organizations (see Chapter 10), and their relations with the state and other regulatory institutions, have been central to this line of inquiry. Although rooted in an attempt to understand capitalism, research on global production networks has moved a long way from Marxian theories of class relations and crises. Like other institutional approaches, it falls within the broadly political-economic approach that Marxist geography inspired, while downplaying the centrality of capitalist fundamentals in explaining all observable phenomena.

Another spin-off from Marxist geography was the emergence of a subfield focused on labor. This field took two forms (as described in Chapter 6). The first examined the geographical strategies of organized labor in the trade union movement. It came (in the 1990s) at a time when the trade union movement had seen its power significantly reduced by deindustrialization and antagonistic government policies (especially in the U.S. and U.K.). It was therefore a time when strategizing about new ways to organize labor and to reach out to "non-traditional" sectors (i.e., immigrants, women, ethnic minorities) was a growing imperative. As corporate entities got bigger, it also became important to think about the scale at which labor organizing would occur. A second strand in labor geography took Marxian analyses of institutions to the local scale and sought to understand how labor markets were shaped and regulated. This line of inquiry took on particular relevance in a time when governments across the developed world were adopting neoliberal policies (see Box 4.4).

One of the strongest lines of critical inquiry in Economic Geography in recent decades has come from feminist geography (see Chapter 13). The wider social context for this development was in the growth of a feminist movement, the increasing participation by women in the labor force and, not coincidentally, the greater representation of women within academia. These circumstances meant that questions around gendered structures of power in economic life came to the fore – questions, it should be noted, that had always been relevant but were very seldom asked.

If class and gender are two fundamental structures of uneven power, then race represents a third. As Chapter 14 showed, ethno-racial difference has come to be recognized as a major line of division that has profound economic manifestations. A critical Economic Geography that has been explicitly anti-racist has therefore been an important thread of scholarship in recent years, examining labor markets, workplaces, and housing markets. While attention to race has reflected the small but increasing number of non-white scholars teaching and researching in anglophone Economic Geography, it has also reflected the extent to which ethnic difference now plays a part in everyday economic life. Some cities have long hosted diverse populations, but intensifying migration flows are creating more and more settings in which ethno-racial differences are juxtaposed and play a role in defining individual experiences of economic processes. As we will note later in this chapter, the growing use of international migrant labor around the world also means that citizenship regimes are closely tied to different levels of access to economic prosperity.

The approaches mentioned here – from Marxism, to institutionalism, to feminism and anti-racism – are certainly diverse and have occasionally been in profound disagreement with each other. What their manifestations in Economic Geography have in common, however, is that they have generally shared a belief in the existence of underlying structures of power that are described in theoretical writing and revealed through empirical research. The flourishing of these diverse approaches has also coincided with a turn toward greater concern for the intersection of economic and cultural processes – a so-called "cultural turn" in Economic Geography (discussed further in Box 16.2).

FURTHER THINKING

Box 16.2   The "Cultural Turn" in Economic Geography

This chapter distinguishes structural and post-structural approaches to Economic Geography because of their different approaches to the production of knowledge. There has, however, also been a shift in the topics of research addressed by Economic Geography under both structural and post-structural frameworks. In particular, a concern with the implications of culture for economic processes has been rising since the 1990s. This has manifested itself in numerous ways (many of which have been addressed in this book).

- First, the role of culturally-coded identities, such as gender or ethnicity, in workplaces and labor markets has been an important way of understanding how people experience economic life as embodied individuals, and not just as disembodied economic actors. This has become especially important as service-sector jobs, which involve an interactive "performance," have become more and more common.
- Second, culturally prescribed ways of interacting have been shown to be essential in understanding how knowledge exchange and innovation occur in industrial clusters. Cultures of work are also important for understanding why certain technologies and work practices might be successful in some contexts but not others. More broadly, there has been much discussion of different "cultures of capitalism" as researchers have realized that both business and government regulation are done quite differently across the world.
- Third, it has been clearly shown that firms themselves have a corporate culture. Employees, managers, and owners do not behave in a rational

and objective manner, but are guided by their habits, assumptions, and prejudices – a set of cultural practices that are often cultivated and perpetuated within firms.

- Fourth, consumers do not buy products and services purely because of their utility, but also because of the symbolic meanings that they embody. This kind of "cultural capital" increasingly shapes economic decisions. Furthermore, those sectors of the economy that produce purely cultural or symbolic outputs (such as movies, music, design, or advertising) have been a growing segment of most economies and an increasing focus of economic geographical research.

- Finally, there has been a growing realization that the "texts" or representations that shape understandings of the economy need as much attention as the flows of capital, commodities, and labor that form the material reality of the economy. The role of think tanks, business schools, and even textbooks, all bear scrutiny for the realities that they construct.

Together, all of these factors have led to a greater concern on the part of economic geographers with the cultural context for economic life – and hence a "cultural turn."

Post-structural Economic Geography

Approaches that trace underlying structures of power have taken geographers a long way from the positivist study of quantitative patterns in economic life. But structural approaches have weaknesses of their own, and a contrasting set of ideas have emerged that we will bundle together under the banner of post-structuralism. While positivism finds truth in measurable phenomena, and structuralism finds truth in the structures that underlie social processes, post-structuralism is an approach to knowledge that denies that an absolute truth is discoverable at all. Instead, there may be multiple "truths" to tell about a given phenomenon. In this sense, any of the preceding intellectual approaches in Economic Geography has a truth to convey, but it will be partial and reflective of the thinker's circumstances and perspectives. The argument that knowledge is partial, situated, and contingent is central to post-structuralism and has many implications.

This philosophical position started to gain traction in Economic Geography in the 1990s and was manifested in several ways. First, economic geographers began to pay attention to the powerful implications of how we understand and represent economic processes. Julie Graham and Katherine Gibson, for example, showed that the ways in which we conceptualize capitalism shapes how we respond to it. They argued that instead of seeing capitalism as an all-encompassing and powerful

KEY CONCEPT

Box 16.3   What is discourse?

A discourse is concerned with the entire package of techniques that we employ to conceptualize "things," to order things and make them comprehensible to ourselves and others. *The Order of Things* is, in fact, the title of a key text by the French philosopher Michel Foucault, who contributed a great deal to our understanding of how language affects our conception of the world and thereby affects how we act in the world. Foucault was especially interested in how people can be represented in certain ways (for example, as criminals, as insane, as abnormal) using a range of language, technologies, and institutions.

To analyze a discourse, then, is to consider how our thinking on a particular subject is shaped by the accepted vocabulary used, the ways in which expertise in the subject is constructed, and how the analysis of a particular phenomenon is institutionalized (that is, how ideas get embedded in the institutions that organize our lives: governments, laws, customs, religions, academic disciplines, etc.). The institutionalization of discourses is important because it highlights the fact that discourses are not the conscious creations of a single author, but are collective understandings that must be constantly recreated (or "performed") in order to continue. Not just any discourse, however, succeeds in becoming dominant. Discourses reflect, and recreate, configurations of power. They reflect them, as only the powerful can construct inclusions and exclusions, the normal and the abnormal, according to their own requirements; but they also recreate power, as a discourse powerfully constitutes and naturalizes the objects that it describes.

structure, we could view it as vulnerable to being undermined and subverted. With our understanding liberated in this way, they argued that we could start to imagine alternative and diverse forms of economic life (Gibson-Graham, 2006). This approach, then, is less a matter of analyzing capitalism as a structure, and more a questioning of the constraints created by our modes of thinking (such as the metaphors discussed in Box 2.1).

A second strand of post-structural thinking in Economic Geography is found in studies that explore the power of representation and discourse in economic life. The concept of discourse is explained further in Box 16.3 and its implications are widespread. Paying attention to discourses means being concerned less with material processes such as class relations, poverty, or state regulation, and more with the ways these processes are rendered understandable. Economic geographers have taken this concept in many different directions:

- Some argue that there are "cultural circuits" in contemporary capitalism that shape the world in important ways – for example, the representations and understandings developed and spread in business schools (Thrift, 2005).
- Studies of globalization have gone beyond understanding what is happening in the global economy to ask instead how representing economic processes in this way leads to certain kinds of policy responses that demand a disciplined competitiveness (Cameron and Palan, 2004).
- Many other dimensions of economic life are open to examination based on how they are represented and understood. Nothing is simply taken at its face value – even economic categories that are apparently fundamental and objective, such as the wage (Mann, 2007).

A third strand of post-structuralist perspectives serves to question individual identities in economic processes. Rather than seeing a person's identity as categorical (for example, as a worker, woman, black, gay, or immigrant) and bound up in powerful structures, post-structuralism sees identities as much less fixed. These identity categories are constructed in particular circumstances and in the context of relationships with other people – they are therefore contingent and relational. Furthermore, identities are changeable because they amount to nothing more than the repeated performance of a particular role. Thus in Chapter 13 we saw that women joining the workforce have performed a different version of femininity, while the masculinities described in Box 13.5 constrain the economic opportunities of young men.

A final strand of post-structuralism questions who has the power to produce knowledge. In particular, a set of approaches termed post-colonialism examine the role of knowledge production in supporting the political, cultural, and economic dominance of the developed world over the developing (and formerly colonized) world. In opening up this issue, post-colonialism also highlights the link between where knowledge is produced and what kinds of understandings are created, paying particular attention to who benefits in the process. The study of economic development and poverty has been particularly affected by this line of thinking. The idea of development itself is a conceptual construct that has been used to impose a Western model on other parts of the world, thereby both diagnosing problems and prescribing solutions. Beyond the critique of existing models, post-colonialism implies a need for economic theories that are sensitive to culture and context, rather than being imported from the West. An example is provided by Vinay Gidwani (2008), who attempts to rethink how fundamental categories such as work, value, capital, and production are constructed in rural Gujarat, India.

The emergence of post-structural approaches in Economic Geography is perhaps too recent to identify the broader historical circumstances that led to this intellectual direction, but a few guesses are possible. First, in an era where the internet and other forms of communication have made a deluge of information so widespread and so accessible, it is perhaps not surprising that we increasingly ask ourselves where an idea or representation comes from, whose interests it serves, and how it shapes our thinking. With so many perspectives available, the idea of a fixed and absolute truth is understandably less convincing. Second, the restructuring of the world economy means that the idea of a wealthy, developed West exporting knowledge and economic models to the poorer developing world of the South is less and less tenable. The world's most dynamic economies now are those that were once subjected to the diagnoses and "assistance" of foreign development experts. It is perhaps to be expected that theories developed in the West are no longer seen as quite so infallible. In the words of one post-colonial scholar, we are now seeing a process of "provincializing Europe" (Chakrabarty, 2007).

What binds all of these post-structural approaches together is a view that scholarship is not about seeking the truth. While positivist approaches view truth as coming from empirical data, and structuralist approaches see it in the form of underlying structures of power, post-structuralism sees truth as altogether more difficult to pin down. Instead, post-structural approaches to Economic Geography ask how we are constructing our knowledge about the economic world and what are the consequences of understanding things in that way.

As we noted earlier, it would be wrong to suggest that the three sets of approaches discussed here (positivism, structuralism, and post-structuralism) represent a sequence of advancing knowledge in the field of Economic Geography. But it is also important to acknowledge that the account given here is very much focused on the anglophone world and that the story of intellectual lineages in other linguistic contexts could look quite different. Box 16.4 emphasizes this point by highlighting how Economic Geography in the wider non-anglophone world has moved to a different beat. In France, Germany, China, and elsewhere, the intellectual currents described above would be only partially recognizable.

FURTHER THINKING

Box 16.4 Economic Geography beyond the Anglosphere

While this chapter focuses almost exclusively on the English-language tradition in Economic Geography, the field has developed elsewhere as well, often in quite different ways. A few examples serve to illustrate these divergent approaches. The French tradition of geographical inquiry has been relatively influential in the English-speaking world at various times, but the passage of the "quantitative revolution" passed largely unnoticed in French Geography, which retained a strong descriptive style. In Germany, a descriptive style of inquiry was evident in the early 20th century (and influential in North America) and a strong quantitative tradition did also take off, but the more radical structural approaches were largely ignored. This might be partly attributed to fairly hierarchical academic institutions in which it was more difficult for young scholars to break into the mainstream than it was in North America and the United Kingdom. Interestingly, though, elements of German and French thought have been highly influential in the development of English-speaking Economic Geography. German location theorists of the 19th and early 20th centuries were rediscovered by English readers in the 1950s and 1960s. As we have seen elsewhere in this book, the work of Johan von Thünen, Alfred Weber, and Walter Christaller were all key texts for the "scientific" economic geographers of the 1950s and 1960s. French scholarship has also been influential in a quite different way, as the nongeographical work of Michel Foucault, Jacques Derrida, Pierre Bourdieu, and others has informed the development of theoretical approaches since the 1990s, some of which have touched Economic Geography.

Outside of Western Europe, the Chinese tradition of Economic Geography highlights the importance of the broader social and political context for the type of scholarship being conducted. From the 1950s until reforms in the late 1970s, the Chinese government emphasized the development of heavy industry and physical infrastructure. Academic geographers were expected to work uncritically in the service of the state, and so their attention was focused on surveys of natural resources, the selection of sites for industrial plants and railways, land use planning in agriculture, and the integrated planning of industrial sectors. Following reforms that opened the country up to trade and investment in the 1980s, economic geographers started to focus on the spatial distribution of industry and the restructuring of preexisting patterns of industrial and urban growth into areas of rapid growth in the coastal provinces of Guangdong, Fujian, Zhejiang, Shanghai, and Jiangsu. As the environmental consequences of rapid growth have become apparent, the costing and amelioration of these impacts have also become a focus for economic geographers. In all of these endeavors, though, Chinese economic geographers have tended to play the role of planners and consultants to state authorities rather than critical social scientists. But as with scholars everywhere, this reflected the immediate institutional environments in which they worked, and the wider social and political context for their scholarship.

16.3   A Changing World

Just as the field of Economic Geography has evolved to reflect its wider social context in the past, it will no doubt continue to do so in the future. It is hard to say what changing circumstances will shape this future evolution, but some present trends suggest important changes occurring that will demand attention and adaption by economic geographers. Here we identify four such changes: a shift in global economic power, new forms of global integration, continued dynamism in the development of new technology, and the need to address environmental challenges.

A New Geography of Global Economic Power

In the late 1980s, the U.S. economy generated just over 25% of global wealth (when GDP values are adjusted according to purchasing power in each country). This was despite having slightly less than 5% of the world's population. By 2009, the U.S. share of global output had fallen to 20.1%, and by 2016 it is projected to be just 17.7% (all data from IMF, 2011). This trend is part of a wider restructuring in the global economy. The seven large industrialized economies that formed the G7 grouping, for example, accounted for 56% of global output in 1990 but only 40% by 2009. In 2016 their share is projected to decline to just under 35%. Although this still indicates a hugely uneven distribution of global wealth, clearly there is a reorganization of the global economic map under way.

China, with a population that was 21.6% of the world's total in the late 1980s, claimed just under 4% of global output. But by 2009, China had surged ahead to generate 12.9% of global wealth. By 2016 this share will likely have risen to 18% – the first year in which it will be larger than that of the United States. India's growth has come later and slower, but with around 17% of global population, it will see its share of global wealth rise from 3.8% in 2000 to 6.6% by 2016. All together, the developing countries of Asia will see their share of global wealth rise from 11% in 1990 to over 30% by 2016.

Clearly, the global economy is undergoing a major geographical shift (as noted in relation to consumption in Chapter 15). The work of economic geographers will no doubt reflect this change. In part this will mean understanding how growth and decline are affecting different places around the world. But it will also mean understanding *how* economic growth occurs in new kinds of contexts. When wealth and economic growth were centered on the advanced developed countries, it was easy to assume that their ways of doing things would become universal. The restructuring of the global economy means that economic geographers in the English-speaking world will have to take different forms of organization and behavior increasingly seriously. For example, different forms of corporate organization, consumer behavior, and state regulation found in China, India, and elsewhere will need to be understood. As we noted earlier, knowledge and understanding are time and place specific – existing ideas will no doubt be tested in new settings, and new ideas will emerge.

New Forms of Integration

As the economic world becomes multi-polar, new forms of integration and interdependency are becoming apparent. Increasing flows of migration, both permanent and temporary, have become a reality in many parts of the world. These migrations are both intra- and international. They will likely heighten issues of unequal access to economic resources on the part of minority populations who are marginalized because of the immigration/citizenship status, their ethnic difference, or their racialized identities. We noted in Chapter 6 that some places are massively dependent on migrant labor (for example, Dubai and Singapore), but this would seem to be only the beginning of a global trend toward the use of disenfranchised sources of labor. Economic geographers interested in labor issues will therefore be increasingly attentive to regimes of citizenship. The role of cultural difference will also be a significant issue as people of different ethnic, linguistic, racial, and religious origins coexist in the same spaces. When uneven access to employment and economic resources becomes aligned with other identities then the possibility for social unrest is also heightened.

Even more so than labor, financial capital has also become increasingly mobile in recent years (see Chapter 7). This has made national economies vulnerable to the exodus of investment, but the degree of debt ownership across borders means that one country is not immune to another's misfortune. If the banking system in one country enters a period of crisis because borrowers are defaulting on debts (as happened in the U.S. in 2007–08), then investors elsewhere are affected too. Similarly, if a country's national government finds itself potentially unable to pay its bills (as happened with Greece, Italy, and Ireland in 2008–11) then the impacts on its currency, creditworthiness, and creditors will be felt far and wide. The result is that we increasingly see coordinated action in global financial matters, from the coordinated stimulus packages of 2008–09 described at the start of this chapter, to the intense negotiations over Greek and Italian government finances in 2011. It seems possible that the future will witness attempts to regulate the workings of financial markets, to implement collective oversight of national government fiscal policies (especially in places where structures exist for such oversight, such as Europe), and to intervene when crises arise in national banking systems.

New Technologies

The role of new technologies in everyday life has changed dramatically in the last 20 years. When many student readers of this book were born, email and the web were just emerging, Twitter and Facebook were nonexistent, personal computers and cell phones were much less common, and file downloads or online shopping had not emerged as everyday activities. In many parts of the world, all of these information and communication technologies have profoundly changed how we consume, what we consume, and how we work.

Few of these technologies were imaginable before they emerged, and so it is very difficult to foresee how new technologies may change economic processes in the future. Nevertheless, it seems inevitable that the ease of communications that existing technologies permit will be intensified still further. It also seems likely that their diffusion will become wider and wider – already many parts of the developing world are leap-frogging over land-line telephone technology and moving straight to wireless phone connections (as noted in Chapter 9). But these developments will also mean that the gulf between those with access and those without will deepen still further.

Alongside accelerating communications technologies, the digitization of every-day life is also intensifying. Economic transactions, intellectual property, products, and services are all increasingly created and delivered in electronic form. This creates new questions about the role of geographical boundaries as flows of infor-mation are technologically easier but still regulated by national legal regimes. Data itself will increasingly become a valuable commodity. Forms of work are also changing as the world digitizes – from data entry, to computer programming, to the creation of mathematical algorithms to analyze financial markets.

Other technologies are even harder to predict, but there can be little doubt, for example, that medical and pharmaceutical technology will continue to develop and will likely keep people (who have access to it) alive and healthier for longer. This will have implications for pension liabilities and public spending on healthcare. We will likely see a gradual increase in the retirement age and a tendency to work later in life (which already exists in those places where things such as retirement are not provided for). This will perhaps have implications for the geographies of labor markets and workplaces as much older co-workers are increasingly common. It may also exacerbate the problem of youth unemployment that exists in many countries.

New Environmental Challenges

Another dimension of our global interdependence is found in the natural envi-ronment upon which all economic activity depends. As governments have come to terms with this issue and have tried to find ways of addressing it, some major stumbling blocks have emerged. In particular, there is the vexing question (noted in Chapter 5) of whether emissions reductions should be shouldered by those countries with the highest per capita carbon emissions, those with the largest total emissions, or those with the fastest growing emissions. There is a good argument for demanding that the greatest cuts should come from those populations whose high standards of living contribute the most carbon emissions on a per capita basis. But at a time when economic growth in the developed world is fragile, many governments are unwilling to sacrifice growth for the sake of reducing carbon emissions. Some countries also resist taking action unless all countries have signed on to an emissions reduction plan. This is, then, fundamentally a geographical problem of assigning responsibility for a problem that is unevenly caused and unevenly experienced.

There is little doubt, however, that the economic costs of a changing climate will be significant. One of the most comprehensive studies ever undertaken of the costs of climate change was published in the United Kingdom in 2006. The report, known as the Stern Review of the Economics of Climate Change, laid out a "business-as-usual" scenario in which carbon emissions continued at the present rate of growth. The result would be a profound impact on climate and various environments around the world:

> Such changes would transform the physical geography of the world.
> A radical change in the physical geography of the world must have
> powerful implications for the human geography – where people live,
> and how they live their lives. (Stern, 2007: iv)

The cost of such changes, according to the Stern Review, would amount to about 5% of global GDP every year, but such outcomes are inherently uncertain and costs could be as high as 20% of GDP. Even if carbon emissions are stabilized there will still be ongoing costs associated with an already changing climate as extreme weather events, shifting zones of cultivable land, and coastal inundation affect certain areas of the globe. On the other hand, the costs of acting to reduce carbon emissions and stabilize atmospheric concentrations by 2050 were estimated by the Stern Review to be around 1% of global GDP per year (although in 2008 this estimate was revised upward to 2%). Who (and where) should shoulder these costs is a matter of contention and a fundamentally geographical problem.

New economies are also being created by the imperative to tackle climate change and other forms of environmental disruption (as noted in Chapter 5). So-called green technologies, such as wind power, fluorescent lightbulbs, electric/hybrid fuel cars, and solar panels, are all growth sectors. So too is the development of carbon trading systems, whereby carbon emission credits are traded or offset credits are bought. If the market in emissions offsetting expands still further it seems possible that it could have an extensive impact upon land use in the developing world in particular.

More broadly, environmental change presents some familiar questions with a new urgency. In particular, it requires us to think about what impacts we account for when measuring the economy, whether economic growth is prioritized in public policy making (and at what cost), and what exactly we are looking for when striving individually and collectively for quality of life.

16.4 Summary

For the most part, this book has avoided representing Economic Geography as a field, and has instead focused on the ideas that the field has contributed to our understanding of economic processes. This has meant that the internal debates within the field, and their development over time, have not been given much attention. In this chapter, we have pulled back the curtain to present a brief and simplified picture of the approaches that have held sway at various times, and continue to inform research activities today. By dividing the field into positivist, structural, and post-structural approaches to knowledge, we have highlighted the different strategies used by scholars to understand the world. In various ways, all of these approaches have been represented in this book. A key point of the discussion in this chapter, however, is that knowledge production in Economic Geography itself has a geography. In other words, it is only possible to understand why intellectual approaches have changed by also looking at the varied contexts (of time and place) in which they were developed. Given this argument, it is equally likely that Economic Geography in the future will change to reflect the times and places in which it will be practiced. We have suggested that new patterns of global economic power, new forms of regulation, continued technological change, and environmental issues will likely shape the field's future.

Notes on the references

- Barnes (1996) provides the most thorough account of the philosophical underpinnings of modern Economic Geography available, and an engagement with positivist, structural, and post-structural approaches. Excellent overviews of changing trends in Economic Geography over time, and the distinctions between different philosophical approaches, are provided by Scott (2000), Sheppard (2006), and Walker (2012).
- For more on different national traditions of geography see various entries in the *International Encyclopedia of Human Geography*, for example Hess (2009) on Germany, Benko and Desbiens (2009) on Francophone Geography, and Tang (2009) and Liu and Lu (2002) on Chinese language Geography.

- There are now several volumes that provide overviews of Economic Geography as a field, reflecting on its disciplinary history and various subfields. See, for example, Leyshon et al. (2011) and Barnes et al. (2012).
- The idea of a post-colonial Economic Geography is developed by Pollard et al. (2010).
- Nasar (2011) provides a readable and entertaining history of economic thought with particular attention to the ideas and life of John Maynard Keynes.

Sample essay questions

- How have the ideas and frameworks of Economic Geography reflected the social and political circumstances of the time?
- Should Cultural Geography and Economic Geography still be seen as separate fields?
- What role do quantitative approaches play in contemporary Economic Geography?
- What do you think will be the major subject headings in an Economic Geography textbook published in 2040?

Resources for further learning

- The work of David Harvey, including Harvey's audio and video presentations, are featured on the blog davidharvey.org.
- The Economic Geography Specialty Group of the Association of American Geographers provides resources and links to the world of Economic Geography. See http://www.geography.uconn.edu/aag-econ/index.html. The equivalent group in the U.K. is the Economic Geography Research Group of the Institute of British Geographers: http://www.egrg.org.uk/about.html.
- Geography departments around the world practice Economic Geography in quite different ways, and this can often be gleaned through information on their website. The Geography Departments Worldwide web portal is a useful tool for exploring the world of geography: http://univ.cc/geolinks/.

References

Barnes, T. (1996). *Logics of Dislocation: Models, Metaphors and Meanings of Economic Space*. New York: Guilford Press.

Barnes, T. (2001). Lives lived and lives told: biographies of geography's quantitative revolution. *Environment and Planning D: Society and Space*, 19, 409–29.

Barnes, T. (2008). Geography's underworld: the military-industrial complex, mathematical modelling and the quantitative revolution. *Geoforum*, 39, 3–16.

Barnes, T. (2012). Roepke lecture in economic geography – notes from the underground: why the history of economic geography matters: the case of central place theory. *Economic Geography*, **88**, 1–26.

Barnes, T., Peck, J., and Sheppard, E., eds. (2012). *The Wiley-Blackwell Companion to Economic Geography*. New York: Wiley-Blackwell.

Benko, G., and Desbiens, C. (2009). Francophone geography. In R. Kitchin and N. Thrift, eds., *International Encyclopedia of Human Geography*. Amsterdam: Elsevier.

Cameron, A., and Palan, R. (2004). *The Imagined Economies of Globalization*. London: Sage.

Chakrabarty, D. (2007). *Provincializing Europe: Postcolonial Thought and Historical Difference* (New Edition). Princeton, NJ: Princeton University Press.

Gibson-Graham, J. K. (2006). *The End of Capitalism: A Feminist Critique of Political Economy*, Second Edition, Oxford: Blackwell.

Gidwani, V. (2008). *Capital Interrupted: Agrarian Development and the Politics of Work in India*. Minneapolis: University of Minnesota Press.

Harvey, D. (1969) *Explanation in Geography*. London: Edward Arnold.

Hess, M. (2009). German-language geography. In R. Kitchin and N. Thrift, eds., *International Encyclopedia of Human Geography*. Amsterdam: Elsevier.

IMF (International Monetary Fund). (2011). *World Economic Outlook Database*. http://www.imf.org/external/data.htm (accessed April 2011).

Keynes, J. M. (1936). *The General Theory of Employment, Interest and Money*. London: Macmillan.

Leyshon, A., Lee, R., Sunley, P., and McDowell, L., eds. (2011). *The Sage Handbook of Economic Geography*. London: Sage.

Liu, W., and Lu, D. (2002). Rethinking the development of economic geography in mainland China. *Environmental and Planning A*, **34**, 2107–26.

Mann, G. (2007). *Our Daily Bread: Wages, Workers, and the Political Economy of the American West*. Chapel Hill, NC: UNC Press.

Nasar, S. (2011). *Grand Pursuit: The Story of Economic Genius*. New York: Simon and Schuster.

Pollard, J., McEwan, C., and Hughes, A., eds. (2010). *Postcolonial Economies*. London: Zed Books.

Scott, A. J. (2000). Economic geography: the great half-century. *Cambridge Journal of Economics*, **24**, 485–504.

Sheppard, E. (2006). The economic geography project. In S. Bagchi-Sen and H. Lawton-Smith, eds., *Economic Geography: Past, Present and Future*, pp. 34–46. New York: Routledge.

Stern, N. (2007) *The Economics of Climate Change: The Stern Review*. Cambridge: Cambridge University Press.

Tang, W. S. (2009). Chinese-language geography. In R. Kitchin and N. Thrift, eds., *International Encyclopedia of Human Geography*. Amsterdam: Elsevier.

Thrift, N. (2005). *Knowing Capitalism*. London: Sage.

Walker, R. (2012). Geography in economy: reflections on a field. In T. Barnes, J. Peck, and E. Sheppard, eds. *The Wiley-Blackwell Companion to Economic Geography*, pp. 47–60. New York: Wiley-Blackwell.

INDEX

Aalbers, M. B., 217
absolute space, 6
accumulation
  by dispossession, 138
  primitive accumulation, 138
Acid Rain Program, 141–142
Actor-Network Theory (ANT), 147
actually-existing neoliberalisms, 103
administrative principle, of hexagonal
    central place theory, 337
advertising industry, 361–365
  branding, geographies of, 362–363
  in Fordist era, 361
  importance, 361
  leading global communications groups
    and their advertising agencies, 363;
    Dentsu (Tokyo), 363; Havas (Paris),
    363; Interpublic (New York), 363;
    Omnicom (New York), 363; Publicis
    (Paris), 363; WPP (Dublin), 363–364
agglomeration economies, 376–380, 449
  binding clusters together, 376–380
  core competencies, 378
  vertical disintegration, 378
Akihabara, Tokyo (electronics shopping
    district), 354
Al-Hindi, K. F., 429
Allen, J., 77
Alvstam, C. G., 257
Amin, A., 120

Andean Common Market (ANCOM), 114
anglosphere, economic geography beyond,
    511–512
anti-globalization movements, TNC and,
    326–327
Apple iPhone 3G, 313
ASEAN Free Trade Agreement (AFTA),
    114, 320
Asheim, B., 399
Asia
  consumption processes in, 472–473;
    bottom of pyramid markets,
    473–474; China, 475; India, 473;
    McDonald's in Asia, 483; regulatory
    barriers, 475; Taiwan, 474
  high-tech clusters in, 387–388
ASOS (As Seen on Screen), online clothing
    retailer, 352
asset-based securities (ABS), 189, 215
Association of Southeast Asian Nations
    (ASEAN) grouping, 320
Atlantic Fordism, 281
authoritarian states, 101
automobile manufacturers, workplace
    employment conditions in, 172–174
  Canadian Auto Workers union (CAW),
    172
  Chrysler, 172–174
  Ford, 172–174
  General Motors, 172–174

automobile manufacturers, workplace
employment conditions in (*cont'd*)
Toronto–Windsor corridor, 172
unions role, 172–174

Bair, J., 257
Bakker, K., 151
Bank for International Settlements (BIS),
203
banking, 193–200
European countries, 197
evolution of, 193–200
finance role in production, 194–195;
consumer finance, 194; corporate
finance, 194; money of account, 195
local banking system, 197–198
private banking facilities, 199
regional banking system, 197–198
subprime loans, 200
U.S. banking system, 197
see also national banking systems
Bardhan, A., 77
Barnes, T., 25, 77, 399, 517–518
Barnett, C., 494
basic economic processes, 38–43
complexity, 41
consumers, 40
demand, 38
firms, 38
money as a universal measure of value,
41
ownership or organizational forms, 39
perfect competition, 40–41
producers, 40
supply, 38
supply and demand curves, 39
see also division of labor
Bathelt, H., 399
Beaverstock, J. V., 330
Benko, G., 517
Bergene, A. C., 183
Beugelsdijk, S., 330
BHP Billiton, 93
bid-rent curves, 8–9
Billah, M., 257
Birch, K., 77
Blake, M., 463

blocked mobility thesis, 448
BMW cars, 304–306, 310–311
body as commodity, 146–150
extraction and use of physical material
from a living body, 148
genetic material, 148
selling organs, 149–150
body scale, 18
Bonacich, Edna, 463
Bonnett, A., 25
Borderless World, The, 297
bottom of pyramid markets, 473–474
Bourdieu, Pierre, 480, 512
Bowen, J., 291
*Box: How the Shipping Container Made
the World Smaller and the World
Economy Bigger, The*, 269
branding, geographies of, 362–363
Brenner, N., 120
Bridge, G., 151
Brunn, S. D., 366
Burns, Ursula, 402–404
business owner, state as, 94–95
government-linked corporations (GLCs),
94–95
private corporations, 94
state-owned enterprises (SOEs), 94
business service clusters, 390
buyer-driven commodity chain, 234–239
characteristics of, 237

call centers, 273
cap-and-trade system, 142
capital markets, 189
capitalism/capitalist system, 55–78,
225–229
beyond national capitalism, 72–76, *see
also* global California
conceptual foundations, 60
consumption processes within, 468
contemporary capitalism, fundamental
logics that drive, 59
contradictions in, 61–64
crises in, 61–64
driving capitalism, 59–62; creative
destruction, 59–61; exploitation,
59–61; profit, 59–61

fundamentals of capitalist system, 58–64; structures of economic life, 58–59; value creation, 58–59
global capitalist system, sustaining, 74–75
new international division of labor, 68
Philippines, 56
placing, 67–72; waves of industrialization, 69–70
recovery in, 61–64
scaling, 67–72; regional scale, 68; urban scale, 70
United Kingdom, 55
United States, 56
see also inherent uneven geographies of capitalism
Caribbean Community (CARICOM), 114
Cartier, C., 463
Castree, N., 52, 77, 183
central place theory, 336–338, 435
  hexagonal central place theory pattern, Christaller's, 336–337
China, 513
  ethnic Chinese business, 459–460
Christaller, Walter, 7, 335–336, 503, 512
churning process, 201
circulating capital, 211–216
  see also financialization
Clark, G. L., 96, 217
Clean Clothes Campaign (CCC), 177
Clean Development Mechanism (CDM), 143
clusters, 372–400
  business service clusters, 390
  consumption clusters, 390
  critiques involving, 392–393; capitalist imperatives, 392; chaotic conceptions, 393; cluster interactions, 393; overprioritizing 'the local', 393; slippery scales, 392; time dimension, 392; underplaying entrepreneurship, 393
  design-intensive craft production clusters, 389
  ethnic businesses and, 445–453, *see also under* ethnic economies

flexible production hub-and-spoke clusters, 389
  high-technology innovative clusters, 389
  labor-intensive craft production clusters, 389
  production satellite clusters, 390
  project working, 395
  state-anchored clusters, 390
  temporary clusters, 394
  typology of clusters, 389–391
  see also location
  proximity
codified knowledge, 380
Coe, N. M., 77, 257, 330, 366, 399
Cohen, J., 463
collateralized debt obligations (CDOs), 187, 189
collective economies, 59
collective ownership, 49
commercial jet aircraft, 267–268
commercial subcontracting, 314–315
commodification, 126, 128–146
  creating a commodity, 131–134
    human nature, body as commodity, 146–150
    natural resource extraction, characteristics of, 133–134
    nature's commodification, variables determining, 132; economic circumstances, 132; technological and scientific knowledge, 132
    pollution, commodifying, 141–145, *see also individual entry*; rise of a green economy, 145–146
commodity chains, 224–258
  catfish commodity chain, 231
  coffee commodity chain, 230–231, 242–243
  ending of, 255–256
  geographical structures, 230–234
  global commodity chains, upgrading strategies in, 233; functional upgrading, 233; intersectoral upgrading, 233; process upgrading, 233; product upgrading, 233
  global production networks (GPNs) and, 298–299

commodity chains (*cont'd*)
  governance processes, 234–240;
    buyer-driven, 234–236; institutional
    contexts, 240–241; producer-driven,
    234–236
  input-output structure, 230
  interplace competition, 231
  producers and consumers, linking,
    229–244
  re-regulating, 244–254
common market, 113
communications systems, 266
communications technologies, 270–275
  call centers, 273
  electronic mass media, 272–273
  internet, 270–271
  mobile telecommunications, 271
  offshoring of services, 273; India,
    274–275
  optical fiber technology, 270
  satellite technology, 270
complexity, 41
computer-aided design (CAD), 279
conceptual foundations, capitalism, 60
  capitalist get profit by selling the
    produced commodity, 60
  capitalist system is based on economic
    transactions, 60
  immense dynamism and creativity, 60
consumer finance, 194
consumers, 40, 225–229
consumption, 466–496
  within capitalism, 468
  changing global consumption landscape,
    471–476; Asia, 472–473; China, 475;
    Taiwan, 474
  clusters, 390
  consumer sovereignty view, 470
  consumers-as-dupes viewpoint, 470
  consuming places, travel and tourism,
    487–493
  cultures of, 476–483
  ethical consumption politics, 484–487,
    see also separate entry
  interpreting, 468–471
  local consumption cultures, 467

  nature of, 468
  sociocultural perspective, 471
  space in Las Vegas, 469–470
containerization, 268
contemporary capitalism, fundamental
  logics that drive, 59–60
contradictions in capitalist system, 61–64
  in internal imperative for growth and
    profit, 61
  wages, 62
Conway, D., 463
Cook, I., 257
cooperative agreements, 312–313,
  322–324
cooperative economies, 59
cooperative joint venture, 322
Cornebise, M., 463
corporate and regional headquarters unit
  of TNC, 304
corporate cultures, 302–303
  operating ways, 303; corporate and
    regional headquarters, 304; material
    practices, 303; power relations, 303;
    research and development (R&D)
    facilities, 304; social relationships,
    303; transnational operating units,
    304; ways of thinking, 303
corporate finance, 194
cost-based methods, 129
Cotula, L., 151
craft-based production, 283
Crang, P., 52
Cravey, A., 429
creative centers, 381
creative destruction, 59–61
credit default swaps (CDS), 189
crises in capitalist system, 61–64
  'blockage point', 62
  over-accumulation, 62
critical isodapane, 375
Crump, J. R., 217
cultural limits and global reach, question
  of, 324–329
cultural proximity, 394
'Cultural Turn' in economic geography,
  507–508

cultures of consumption, place, and
    identity, 476–483
  Bourdieu's cultural capital, 480
  consumption work, 478–479
  mass consumption and post-Fordist
    consumption, 477
Cumbers, A., 399
cumulative causation, 10
customer relationship management (CRM)
    technologies, 279
customs union, 113

Dana, L. P., 463
Dasgupta, P., 52
Daviron, B., 257
Debbage, K. G., 494
demand, 38
dependent states, 106
deregulation as a driver of global finance,
    203–206
  state role in, 205
derivatives, 189
Derrida, Jacques, 512
Desbiens, C., 517
design-intensive craft production clusters,
    389
devaluation in capitalist system recovery,
    63–63
developmental states, 101, 104
Dicken, P., 113, 120, 257, 291, 330, 399
discourse concept, 509
discrimination and and ethnicity, 442
distance, 6–11
  space as, 11
  location and, 6–11
  importance, 10–11
Distinction: A Social Critique of the
    Judgment, of Taste, 480
distribution centers (DCs), 275–276
division of labor, 32, 42
  ethnic, 42
  gender, 42
  international, 42
  new international, 42
  spatial, 42
  technical, 42
Dolan, C., 257

domination, 74–75
Dore, R., 217
downscaling, 116
Dunn, Edwina, 333–334
Dunnhumby, 333–334
Durand, C., 330
Duval-Diop, D. M., 257
Dymski, G., 217

Eastman, George, 14
e-commerce, 276
econometrics, 35
economic flows regulation, state in, 90
economic union, 113
economies of scale, 280, 376
economy, 27–53
  beyond the assumptions of economics,
    43–50; beyond calculability, 46–49;
    beyond private property, 49–50;
    beyond rationality, 43–46; biological
    metaphors, 48–49; economic iceberg
    and the submerged non-economy, 47;
    ethnic sorting, 46; irrational behavior,
    45; labor market, 45; miscounting of
    economic processes, 47;
    religious/cultural belief system, 44
  gross domestic product (GDP), 30
  history of 'the economy', 30–37, see
    also individual entry
  meaning, 27–53
  as an organic entity, 29
  taken-for-granted economy, 29–30
  see also basic economic processes
ecosystem services approach, evaluating
    nature through, 129
  cost-based methods, 129
  cultural services, 129
  provisioning services, 129
  regulating services, 129
  revealed preference, 129
  stated preference, 129
  supporting services, 129
ecotourism, 492
education services by state, 98
Ekinsmyth, C., 429
Elden, S., 25
electronic data interchange (EDI), 276

electronic mass media, 272–273
electronics manufacturing service (EMS),
    315
elite labor migration, 178–179
    Silicon Valley's ethnic networks, 179
Emissions Trading System (ETS), 142–143
    problems addressed by, 143–144;
        allocation of permits when markets
        are first created, 143; avoiding
        leakage, 144; carbon offsets, 144;
        conflicts of interest in market
        institutions, 144; difficulties of
        continuous monitoring, 143; ensuring
        additionality, 144; market volatility,
        144; reliability of baseline measures,
        143
End of the Nation State, The, 86
Engelen, E., 217
English-language tradition in economic
    geography, 511–512
enterprise resource planning (ERP), 279
entrepreneurial process, 445
entrepreneurship and livelihood strategies,
    423–426
    women in formal entrepreneurial
        ventures, 424
environment and economy, 123–152
    costs of the disaster, 124
    nature counted in economy, 126–128;
        commodity exchanges, 127;
        conventional economic analysis, 127;
        price determination in market
        exchange, 126; supply-and-demand
        equation, 127
    pollution consequences, 124
    see also natural environment and
        economy pollution, commodifying
environmental certification, 246–248
Environmental Protection Agency's (EPA)
    Acid Rain Program, U.S., 141
epistemology, 502
e-tailers, 350
ethical commitments, 426–427
ethical consumption politics, 484–487
    'Buy Nothing Day', 484–485
    societal debates, 486

uneven geographies of ethical
    consumerism, 486
United Students Against Sweatshops
    (USAS) campaign, 485
ethnic divisions of labor, 42
ethnic economies, 432–464
    'color blind' economics, 434
    discrimination and stereotypes, 442;
        ethno-national space, 442
    economic ties binding immigrants in
        global cities, 435
    ethnic business clusters, 445–453;
        agglomeration economies, 449; ethnic
        enclave economy, 451; formation of,
        450; intraethnic economic relations,
        451; location of, 449; networks, 450;
        spatial clusters, 449, 451
    ethnic sorting in workforce, 436–445;
        Asian–American, 437; black, 437;
        distribution in various occupations,
        Los Angeles, 437–438; Han migrants,
        439; Hispanic, 437; in Urumqi,
        439–440; white, 437
    ethno-national identities, 444
    geographical dimensions and, 439
    geographical processes and, 433
    home–work linkages, 442–443
    institutional barriers, 441; Canada, 441;
        China, 441; United States, 441
    limits to ethnicity, 460–463; complexity,
        461; contradictions, 461; intra-ethnic
        cooperation, 461; statistical data, 461
    location of economic activities, 435
    qualifications and skills, 439
    segmentation, 443
    segregation, 443
    social capital and job search, 444
    spatial mismatch, 442
    transnational ethnic enterprises,
        457–460; circuit firms, 457; cultural
        enterprises, 457; ethnic Chinese
        business, 459–460; ethnic enterprises,
        457; ethnic Turks in Germany, 458;
        return migrant micro-enterprises, 457
ethnic enclave economy, 451
ethnic entrepreneurship, 446–449
    causes for, 452

co-ethnic networks, 448
general outcomes, 452
networks; and labor, 452; and markets, 452; and organizational structure, 452
Ontario stores, South Korea, 1447
ethnic resources, 447
ethnicity, definition, 436
ethnoburbs, 453
European Free Trade Association (EFTA), 114
EuropeanUnion (EU), 114–115
evolutionary perspective of technological change, 265
Explanation in Geography, 505
exploitation, 59–61
export-oriented industrialization (EOI), 104
external sphere, economy as, 36
externalized transactions in transnational corporation (TNC), 300–301

Facebook, 13, 16
in China, 16
as an economic phenomenon, 3–5
as geographical story, 3–5
in Japan, 17
at Prineville location, 20–22
failed states, 106
Fairtrade Labelling Organizations International, 249
feminist economic geography, 426–428, 506
dimensions of, 426; capitalist system, 427; ethical commitments, 426; linkages between productive jobs in workforce and reproductive labor in the home, 426; openness to role of difference and identity in economic life, 426
see also under gendered economies women
feudal economies, 59
Fields, G., 291
finance, importance of, 187–218
'end of geography' conception of finance, 192
role in production, 194–195

see also global finance
financial remittances, 454
financialization, 211–216
collateralized debt obligations (CDOs), 213
financial crisis from 2008, nation-states efforts to rescue, 214–215; massive rescue packages, 214–215
financing production, 193–200
see also banking
firms, 38
Fisher, Irving, 35
flexibility, 168
functional flexibility, 166
numerical flexibility, 166
and product technology, 280; forming economic activity clusters, 286; Italy, 286 Japanese flexible production systems, 285–286
temporal flexibility, 166
flexible production hub-and-spoke clusters, 389
flexible production, 283–285
Florida, Richard, 381
Fordism, 281, 285
Fordist production system, 280–281
Atlantic Fordism, 281
foreign direct investment (FDI) strategies, 92
formal entrepreneurial ventures, women in, 424
Foucault, Michel, 509, 512
Franchising, 312–313, 322–324
Franz, M., 257
free-trade area, 113
French, S., 217
'friction of distance', 5, 7
fully owned subsidiaries, 325
functional flexibility, 166
functional upgrading, 233
future horizons in economic geography, 499–518
changing field, 501–512
changing world, 513–517
discourse, 509
new environmental challenges, 515–517
new forms of integration, 514

future horizons in economic geography
(*cont'd*)
new technologies, 515
science and quantitative economic
geography, 502–504
structural approaches, 504–508

G20 summit, 112–113
gendered economies, 402–430
creating gendered workplaces, 419–421;
creating work environment, 420
devaluation of women's work, 418
entrepreneurship and livelihood
strategies, 423–426; in the developed
world, 425; women in formal
entrepreneurial ventures, 424
jobs and workplaces, gendering,
410–421; developed industrial
countries, 410; disciplining
mechanism in the workplace, 412;
industrial employment for women,
413; manufacturing jobs, 411; newly
industrializing countries, 410, 412;
women and the paid workforce,
410–413
nature of work and, 403–405
redundant masculinities, 421
'replacement cost' approach, 406
role of gender in economic life, 404–405
underrepresented women, 403; in upper
echelons of workforce, 405
unpaid work, gendered patterns of,
406–410, *see also individual entry*
valuing gendered work, 418–419
gendering of jobs, 42, 413–418
part-time work, 415
segmentation of workforce, 415
third world woman, devaluing, 414
in United States, 415–416
unpaid works, 415
General Agreement on Tariffs and Trade
(GATT), 111
*General Theory of Employment, Interest
and Money*, 36
generic management systems standards,
251

geographical structures of a commodity
chain, 230–234
geographies of labor, 159–177
different national labor conditions, 165;
education and training, 165;
employment practices, 165; industrial
relations, 165
firms approach to, 164
local labor control, 160, 166–167, *see
also* local labor control regime (LLCR)
workers as agents of change, 170–177;
coping strategies employed by women,
171; defending place, worker actions
in situ, 172–174; micro-politics,
action within the workplace,
170–171; worker–manager relation,
171
see also labor markets: shaping
geopolitics, 74–75
Geref?, G., 257
Germany, 116, 458
Gibbon, P., 257
Gibson, Katherine, 508
Gibson-Graham, J. K., 52
Gidwani, Vinay, 510
Glassman, J., 77, 120
Glass-Steagall Act of 1933, 197
global California, 72–76
key themes, 73; geopolitics, 74;
informationa and communications
technology (ICT), 73; labor migration,
73
global cities, 207
global commodity chains, 230
upgrading strategies in, 233
global consumption landscape, 471–476
see also under consumption
global convergence and corporate culture,
327–328
global economic power, new geography of,
513–514
global finance, 201–211
in 2007–2008
changing regimes in, 204; 1945–1973,
204; 2008 onward, 204; late 1980s to
late 2000s, 204–205; mid-1970s to
mid-1980s, 204–205

deregulation as a driver of, 203–206
finance for production to global finance shift, 202
financial centers of, 206–208
local mortgage lending and, 213
offshore currency market, 205
petrodollars, 205
as placeless, question of, 191–193
rise of, 201–211; bank-related institutions, 201; churning process, 201; global securitization and financial trading, 201–202
Shanghai's position in, 210
United Kingdom, 205
United states, 205
Global Financial Integration: The End of Geography, 191
global production networks (GPN), 298
from commodity chains to, 298–299
global scale, 17, 241
globalization and state economic processes, 86–88
globalization of retailing, 338–344
in 1990s, 340
coordination in, 342
early 2011, 340
failure in, 344
in late 1800s, 340
latest phase of, distinguished characteristics, 341; impacts, 341; scale, 341; speed, 341
leading transnational retailers, 338–339
mid-1990s, 340
scale of international retailing, 338
scope of international retailing, 338
success in, 344
globally concentrated production of TNC, 307–308
globally mobile capital and labor power, 157–159
Glückler, J., 399
gold farming, 262
Goodman, M. K., 494
Goss, J., 366
governance processes in commodity chain approach, 234–240
Japanese *sogo shosha*, 238–239

government-linked corporations (GLCs), 94–95
Grabher, G., 330
graduated sovereignty, 106–107
Graham, Julie, 508
green economy, 145–146
Gregory, D., 77
Gregson, N., 257
Grimes, J. R., 257
gross domestic product (GDP), 30, 406–407
components of, 31
guanxi capitalism, 461
Gutierrez, M., 151

Hall, C. M., 494
Hall, S., 330
Han migrant workers, 439
Hanson, S., 429, 463
Harrison, B., 330
Harvey, D., 77, 151, 214, 505
Hassler, M., 257, 330
headquarters location complexity in TNC, 304
health services by state, 98
hedge funds, 189
Heiman, M., 151
Henderson, J., 330
Henry, N., 399
heritage tourism, 490
Herod, A., 183
Hess, M., 330, 517
Hewlett Packard (HP), 301
hexagonal central place theory, Christaller's, 336–337
administrative principle, 337
marketing principle, 337
transport principle, 337
Hiebert, D., 463
high-technology innovative clusters, 389
history of 'the economy', 30–37
1930s, 35
by 1940s, 36
18th century, 31
complex mathematical techniques, 35–36

history of 'the economy' (cont'd)
    economy as independent of social, political, and cultural processes, 37
    end of 19th century, 32–33
    external sphere, economy as, 36
    individual economic decisions, 33
    Industrial Revolution in England, 31
    metaphors of economy, 34
    national political economy, 33
    national scale, 37
    new perspective by Irving Fisher, 35, 35
    1870s onwards, 33
    political economy, 32
hollowing-out the state, question of, 117–119
    business media, 118
    corporate scandals, 118
    QUANGOS (Quasi-Autonomous Non-Governmental Organizations), 117
    quasi-private institutions, 118
    regulatory regimes, 117
home–work linkages, and ethnicity, 442–443
homo economicus, 44
Hongkong and Shanghai Banking Corporation (HSBC), 296
host-market production structure of TNC, 308
Hudson, A. C., 217
Hudson, R., 25
Hughes, A., 494
humanity and nature, 146–150
    body as commodity, 146–150
    separation of, 148
Humby, Clive, 333–334
Humphrey, J., 257
hybrid cultures, 481–482
hyperglobalist position, 86–87

identity
    consumption and, 476–483; lipstick use, 480–481; shopping, 481
Immortal Life of Henrietta Lacks, The, 149
import-substitution industrialization (ISI) strategies, 104

incremental innovations technological change, 265
Indian consumers, 473–474
industrial employment for women, 413
industrial location theory, 373–376
    critical isodapane, 375
    Hollywood film production, 378–379
    internal economies, 376
    isodapanes, 375
    least cost model of industrial location, 374
    main factors, 377; dedicated infrastructure, 377; face-to-face contact, 377; intermediate industries, 377; skilled labor, 377
    Weber's, 375
Industrial Revolution, 37
industry strategies, 92
informal retail spaces, 358–361
    formal retail spaces versus, 359
    occasional markets, 359
    spazas, 360
infrastructure services by state, 98
inherent uneven geographies of capitalism, 64–67
    spatial forms of complexes, identifying, 65; city-satellite systems, 65; clusters of towns, 65; gender relations, 66; large cities or metropolises, 65; regional complexes, 65; spatial divisions of labor, 66; spatial fix, 65; territorial production complexes, 65
institutional barriers, and ethnicity, 441
institutional context of commodity chain approach, 240–244
    formal versus informal, 240
    global scale, 241
    at macro-regional scale, 241
    multi-scalar, 241
    national scale, 242
institutional proximity, 393
institutional thickness notion, 384
intellectual journeys in economic geography, 499–518
inter-firm networks of transnational corporation (TNC), 296, 312–324

cooperative agreements, 312–313, 322–324
franchising, 312–313, 322–324
joint ventures, 312, 318–322
in service sector, 323
strategic alliances, 312–313, 318–322
subcontracting, 312–313, *see also* international subcontracting
intermediate goods, 336
internal economies, 376
internalized transactions in transnational corporation (TNC), 300
International Accounting Standards Board (IASB), 119
International Accounting Standards Committee (IASC), 119
international division of labor, 42
transnational corporation (TNC) and, 317–318
international economic treaties and state, 89
International Monetary Fund (IMF), 110, 203, 241, 281
International Organization for Standardization (ISO), 251
ISO14001, 251
ISO9000 standards, 251–253
ISO9001:2008, 251
international organizations, 108
international retailing, 338–344
see also globalization of retailing
international scales, 108
international subcontracting, 314–318
commercial subcontracting, 314–315
electronics manufacturing service (EMS), 315
geographical implications of, 316–317
industrial subcontracting, 314–315
MacBook, 316
original design manufacturing (ODM), 315
original equipment manufacturing (OEM), 315
personal digital assistants (PDAs), 316
in Southeast Asia, 319–320
internationally integrated R&D lab, 306
internet, 7, 9, 270–272

interplace competition, 231
intersectoral upgrading, 233
intraethnic economic relations, 451
intra-firm networks of transnational corporation (TNC), 296, 302–312
corporate cultures, 302–303
spatiality complex in, 302, 304; BMW example, 304–306, 310–311; headquarters, 304; production operations, 307; R&D labs location, 304–306
investor, state as, 95–97
huge resource endowments in home countries, 95
longstanding conservative fiscal management, 95
significant trade imbalances through exports, 95
Isaksen, A., 330
isodapanes, 375
Ivarsson, I., 257

Jackson, P., 494
Jacobs, Jane, 353
Japan
auto industry, 287–288
flexible production systems, 285–286
Jayne, M., 494
job search, social capital and, 444
jobs and workplaces, gendering, 410–421
see also under gendered economies gendering of jobs
Johansson, O., 463
joint ventures, 312, 318–322
cooperative joint venture, 322
Jones, A., 77, 330
'just-in-case' systems, 283
'just-in-time' systems, 283
Justice for Janitors (JfJ) campaign, 175

Kalsaas, B. T., 330
Kaneko, J., 291
Kaplan, D., 463
Keil, R., 120
Kelly, P. F., 52, 120, 463
Keynes, J. M., 36, 500, 518
Keynes: The Return of the Master, 500

Keynesian welfare state, 107
Klein, Naomi, 362
knowledge economy, 421
Kondratiev waves, 265–266
Krippner, G., 217
Krugman, Paul, 10–11
  and geographical economics, 10–11
Kyoto Protocol, 142

labor geographies, *see* geographies of labor
labor-intensive craft production clusters, 389
labor markets, 45, 422–423
  home in, 422–423
  segmentation, 415, 417; theories of, 417
  shaping, 159–170, *see also* Upscaling worker action; by market intermediaries, 162; Chicago's prisons as labor market institutions, 162; governments regulating, 161; influencing factors, 161; institutional intervention in, 159–160; local labor market regulation, 160; public sector employment, 160; temporary staffing industry, 163
  space in, 422–423
  strategies, 92
  women's versus men's market entry, 422
  work in, 422–423
labor power, 154–184
  beyond capital versus labor, 180–183, *see also* Noncapitalist economies
  globally mobile capital and, 157–159
  competetion at international and/or subnational scales, 157; households, 158; labor reproduction, 158; local cultures of work and credential recognition, 158; place attachments, 158; regulation, 158
  Queen Mary College example, 154–155
  see also geographies of labor
  migrant labor
labor standards and conditions, 250
Lai, K. P. Y., 217
land use, 8–9
Langley, P., 217
Larner, W., 429

Las Vegas, 469–470
latitude, 6
lean distribution systems, 275
  distribution centers (DCs), 275–276
  electronic data interchange (EDI), 276
  transportation terminals, 275
lean production, 166
Lee, Y. S., 330
Leinbach, T. R., 291
Lepawsky, J., 257
Levinson, Marc, 269
Ley, D., 463
Leyshon, A., 25, 183, 217, 518
Li, W., 217, 463
lifecycle engineering, 279
Light, Ivan, 463
Liquor, Hospitality and Miscellaneous Workers Union (LHMU), 176
Liu, W., 517
Lloyd, P. E., 399
local banking system, 197–198
local labor control regime (LLCR), 166–167
  mobility, 167
  in Southeast Asia, 169
  stability, 167
  strategies of, 168; individual level, 168; industrial estate, 168; national regulatory policies, 168; spaces of migration, 168; workplace, 168
local scale, 18
locally integrated R&D lab, 306
location, 5
  and distance, 6–11
  effect on economic activities, 7
  of ethnic business clusters, 449
  importance of, 372–400, *see also* agglomeration economies
  industrial location theory; Sand Hill Road, Silicon Valley, 371–373
  locational analysis, 7
  see also industrial location theory place proximity, importance of
logistics revolution, 275–278
  integrated logistics systems, 276; air cargo services, 276;

temperature-controlled logistics operations, 276
lean distribution systems, 275; distribution centers (DCs), 275–276; e-commerce, 276; electronic data interchange (EDI), 276; transportation terminals, 275
longitude, 6
Lösch, August, 337
Lowe, M., 366
Lu, D., 517

Ma, L. J. C., 463
MacBook, 316
Mackinnon, D., 399
macro-economic management, in capitalist system recovery, 63
macro-regional groupings, 108, 113–114
macro-regional scale, 17
macro-regionalization of global economy, 319–320
Mahon, R., 120
making money, 187–218
  see also Finance, importance of
Malecki, E. J., 291, 366
Malls for retailing, 356–357
  promoting, 356–357
Mansvelt, J., 494
map scale, 17
maquiladoras, 318, 414
  of northern Mexico, 309–310
market regulation and state, 90
marketing principle, of hexagonal central place theory, 337
Marshall, Alfred, 377
Martin, R., 217
Marx, Karl, 33
mass market and product technology, 280
Massey, D., 15, 77, 429
Mccloskey, D., 52
Mcdowell, L., 429
McFadden Act of 1927, 197
Mcgrath-Champ, S., 183
mega-event tourism, 490–491
mergers and acquisitions (M&As), 318
metaphors of economy, 34
methodology, 502

micro-politics, action within the workplace, 170–171
migrant labor, 177–180
  mobile elites, 178–179; Silicon Valley's ethnic networks, 179
  temporary migrant workers, 179–180
Miles, S., 494
Miller, D., 494
Mitchell, K., 463
mobile telecommunications, 271
Molloy, M., 429
money as a universal measure of value, 41
Moriset, B., 291, 366
mortgage-backed securities (MBS), 189, 211
Moss, P., 429
Mullings, B., 429
multi-scalar institutional contexts, 241
'mumpreneurship', 424–425
Murphy, A. J., 366
Murphy, J. T., 330
Mykhnenko, V., 77

Nadvi, K., 257
Nasar, S., 518
national banking systems, 195–198
  changing geography of, 195–198; presence of significant stocks of wealth, 196; rationale, 195; role of the state, 196
  reorganization of, 198–200
national economy manager, state as, 91–94
  see also under state
national scale, 17, 37, 108, 242
natural environment and economy, 123–152
  economy's demand upon environment, 130
  ecosystem services approach, evaluating nature through, 129
  incorporating nature, 128–141, see also commodification ownership
  see also environment and economy
Neilson, J., 257
Nelson, L., 429

neoliberal states, 100, 102–103
  actually-existing neoliberalisms, 103
  neoliberalization, 103
  proto-neoliberalism, 103
  roll-back neoliberalism, 103
  roll-out neoliberalism, 103
networks in ethnic business clusters, 450
New international division of labor
      (NIDL), 42, 68, 317
New York Stock Exchange (NYSE),
      209–210
newly industrialized economies (NIEs),
      104, 317
Nike, 294–295
No Logo, 362
Nojiri, W., 291
Nokia, 301
noncapitalist economies, 180–183
  alternative formal employment spaces,
      181
  alternative informal employment spaces,
      181
  challenge for economic geographers, 181
  Mondragon Cooperative Corporation
      (MCC), 182
non-equity modes (NEM) of cross-border
      activity by TNCs, 312, 314
North American Free Trade Agreement
      (NAFTA), 114, 309–310
numerical flexibility, 166

O'Brien, Richard, 191–192
O'neill, P., 120
offshore currency market, 205
Offshore Financial Center (OFC), 212
offshoring of services, 273
  India, 274–275
Ohmae, Kenichi, 86, 297
Olds, K., 52
online spaces of retailing, 350–353
  ASOS (As Seen on Screen), online
      clothing retailer, 352
  challenge to, 351; store-based retailers,
      351–352
  e-tailers, 350
ontology, 502
optical fiber technology, 270

Order of Things, The, 509
Organization of Petroleum Exporting
      Countries (OPEC), 127
organizational proximity, 394
original design manufacturing (ODM), 315
original equipment manufacturing (OEM),
      233, 315
Oro, K., 257
Ouma, S., 257
outsourcing, see subcontracting
own-brand manufacturing (OBM), 233
own-design manufacture (ODM), 233
ownership, 128–141
  establishing ownership, 134–135;
      communal access, 135; private
      ownership and private exploitation,
      135–137; state ownership and private
      exploitation, 135; state ownership and
      state exploitation, 135
  state-owned resource companies,
      137–139; accumulation by
      dispossession, 138; primitive
      accumulation, 138

Page, S. J., 494
paid workforce, women and, 410–413
  see also unpaid work, gendered patterns
      of
Patel, Campillo, A., 257
path dependency, 265
patriarchy, 409
  cultural institutions, 409
  household production, 409
  sexuality, 409
  the state, 409
  violence against women, 409
  waged work, 409
Peake, L., 429
peasant economics, 59
Peck, J., 120
Peet, R., 77
perfect competition, 40–41
personal digital assistants (PDAs), 316
petrodollars, 205
Philippines, 56, 456
Pietrobelli, C., 399
Pike, A., 217

Pinch, S., 399
pink collar ghettoes, 423
place, 6, 14–17
  characteristics, 14
  consumption and, 476–483, *see also*
    tourism industry
  differences between, 15
  differences in human society between, 14
  distinctive features of, 14
  historical place-making, 15
  placeless capital, 191–193
  shaping economic opportunities, 16
  see also location
  proximity, importance of
placing capitalism, 67–72
political economy, 32
Pollard, J., 217, 518
pollution, commodifying, 141–145
  cap-and-trade system, 142
  Emissions Trading System (ETS),
    142–143
  Environmental Protection Agency's
    (EPA) Acid Rain Program, U.S., 141
  Kyoto Protocol, 142
Ponte, S., 257
Portes, Alex, 463
positivism in economic geography,
  502–504
  quantitative economic geography,
    502–504
  scientific modes, 502–504
post-Fordist production system, 280–281
post-structural economic geography,
  508–512
  circumstances that led to, 510–511
  individual identities in economic
    processes, 510
  power of representation and discourse
    in, 509–510
  power to produce knowledge, 510
  understanding economic processes,
    508–509
PotashCorp, 93
Pratt, G., 463
primitive accumulation, 138
Pritchard, B., 257
private corporations, 94

private equities, 189
private ownership, expanding, 135–137
private standards, 250
privatization
  of collectively controlled resources,
    139–141; problems in, 140
process technology, 278–288
process upgrading, 233
producer-driven commodity chain,
  234–239
  characteristics of, 237
producers, 40
product specialization for a global or
  regional market of TNC, 308
product technology, 278–288
  contemporary production systems, 282;
    geographical tendencies, 282; labor
    force, 282; production volume and
    variety, 282; supplier relationships,
    282; technology, 282
  and their geographies, 280–288;
    economies of scale, 280; flexibility,
    280; mass market, 280
  Japan's auto industry, 287
  'just-in-case' systems, 283
  'just-in-time' systems, 283
  kinds of industrial system, 281–282;
    craft-based production, 283; flexible
    production, 283; Fordism, 281;
    geographical configurations of, 285
  product innovation, 278–280;
    characteristics of contemporary
    product innovation, 279; product life
    cycle, 278–279
product upgrading, 233
production chain, 229
  see also commodity chains
production networks, of transnational
  corporation (TNC), 298–302
'production of scale', 22
production operations complexity in TNC,
  307
production process innovation, 278–280,
  284–285
  Dell Computer (example), 284–285
  flexible specialization, 285
  geographical configurations and, 285

production satellite clusters, 390
profit, 59–61
property rights and state, 89
proto-neoliberalism, 103
proximity, importance of, 391–397
  cultural proximity, 394
  institutional proximity, 393
  organizational proximity, 394
  regional cultures of production,
    380–388
  relational proximity, 394
  spatial clustering, 381
  untraded interdependencies of
    production, 380–388
  see also location
  place
public goods and services provider, state
    as, 97–100
  education services, 98
  health services, 98
  infrastructure services, 98
  national business systems, 99–100;
    business formation and management
    processes, 99; ownership patterns and
    corporate governance, 99; work and
    employment relations, 99–100
  transport services, 97–98

quantitative economic geography,
    502–504
Quasi-Autonomous Non-Governmental
    Organizations (QUANGOS), 117

Rabellotti, R., 399
radical innovations in technological
    change, 265
radio frequency identification devices
    (RFID), 276
recovery in capitalist system, 61–64
  devaluation, 63–63
  macro-economic management, 63
  spatial displacement of capital, 64
  temporal displacement of capital, 63
redundant masculinities, 421
regional banking system, 197–198
regional cultures of production, 380–388

Regional Development Agencies (RDAs),
    116
regional scale, 18, 68
  Asia, high-tech clusters in, 387–388
  untraded interdependencies at,
    384–388; inter-firm relationships,
    385; learning region, 385; regional
    cultures, 385; Silicon Valley firms,
    collaboration of, 386–387
regulation theory, 63, 505
regulatory regimes, 117
relational proximity, 394
remittances, transnational, 454–457
  into Philippines, 456
'replacement cost' approach, 406
re-regulating commodity chains, 244–254
  economic returns, 249
  strategic policy initiative, 248
  UN Global Compact, 248
  world of standards, 244–254, see also
    individual entry
rescaling the state, 108–117
  common market, 113
  customs union, 113
  downscaling, 116
  economic union, 113
  free-trade area, 113
  G20 summit, 112
  Germany, 116
  International scales, 108
  macro-regional groupings, 113–114;
    Andean Common Market (ANCOM),
    114; ASEAN Free Trade Agreement
    (AFTA), 114; Caribbean Community
    (CARICOM), 114; European Free
    Trade Association (EFTA), 114;
    EuropeanUnion (EU), 114; North
    American Free Trade Agreement
    (NAFTA), 114; Southern Cone
    Common Market (MERCOSUR), 114
  national scale, 108
  subnational scales, 108
  United States, 116
  upscaling, 106
  within states, 116
research and development (R&D) facilities
    unit of TNC, 304–306

internationally integrated R&D lab, 306
locally integrated R&D lab, 306
support lab, 306
retail spaces, configuration of, 353–361
  informal retail spaces, 358–361
  malls, 356–357
  occasional markets, 359
  stores, 357–358
  streets, 353–356, *see also* streets for
    retailing
retailing, 333–367
  central place theory, 336–338
  retail geographies, 336–338
  shifting geographies of, 338–353, *see*
    also globalization of retailing
online spaces of retailing
urban scale: retailing at
  see also advertising industry
revealed preference, 129
Rieker, M., 429
Rise of the Creative Class, The, 381
River, The, 28
Roberts, S. M., 217
Rodrigue, J. P., 291
Rogerson, C. M., 366
roll-back neoliberalism, 103
roll-out neoliberalism, 103
Rothenberg, Aalami, J., 330
rule of the law and state, 89

satellite technology, 270
scale, 6, 17–23
  body scale, 18
  as economic process frameworks, 18
  global scale, 17
  local scale, 18
  macro-regional scale, 17
  map scale, 17
  multiple scales, economic processes at,
    19
  national scale, 17
  not hierarchical, 18–19
  not naturally occurring, 22
  'production of scale', 22
  regional scale, 18
  urban scale, 18
scaling capitalism, 67–72

Schindler, S., 330
Schoenberger, E., 330
Schoenberger, L., 366
Schumpeter, Joseph, 61
Scientific Certification Systems (SCS), 254
scientific modes in economic geography,
    502–504
Scott, A. J., 399, 517
Seager, J., 429
Securities, 189
Securitization, 190
segmentation
  and ethnicity, 443
  of workforce, 415, 417–418
segregation, and ethnicity, 443
Sen, Amartya, 52
Service Employees International Union
    (SEIU), 175
Sharia law, 44
Shaw, G., 494
Sheppard, E., 77, 517
shock therapy, 110
shopping as a social activity, 481–482
shopping malls, 356–357
Silicon Valley firms, collaboration of,
    386–387
Silvey, R., 429
Silvia Nasar, 52
Skinner, C., 366
Smith, Adam, 32
Smith, N., 77
social capital and job search, 444
social process, technological
    change as, 264
Solomon, B., 151
Southern Cone Common Market
    (MERCOSUR), 114
sovereign wealth funds (SWFs), 95–96
  global landscape of, 97
space, conceptions of, 5–6
  location, 5
  place, 6
  absolute space, 6
  space as distance, 11
  scale, 6, 17–23, *see also individual entry*
  territory, 5, 12–14, *see also individual*
    entry

spaces of sale, 333–367
   see also retailing
space-shrinking technologies, 264–265
   see also communications technologies
   logistics revolution
   transportation technologies
spatial clustering, 381, 449, 451
spatial displacement of capital, in capitalist
      system recovery, 64
spatial divisions of labor, 42, 66
spatial fix, 65
spatial mismatch, 442
spatial organization of intra-TNC
      networks, 302, 304
spatial scales, 240
spazas, 360
Springer, S., 120
Springsteen, Bruce, 28, 36
standards, *see* world of standards
state, 83–122
   as business owner, 94–95;
      government-linked corporations
      (GLCs), 94–95; private corporations,
      94; state-owned enterprises (SOEs), 94
   in global finance deregulation, 205
   globalization and, 86–88; hyperglobalist
      position, 86–87
   hollowing-out, question of, 117–119
   as investor, 95–97
   as national economy architect, 88–89
   as national economy manager, 91–94;
      BHP Billiton, 93; exports, 92; foreign
      direct investment (FDI) strategies, 92;
      imports, 92; industry strategies, 92;
      labor market strategies, 92; trade
      strategies, 91–92
   as public goods and services provider,
      97–100, *see also separate entry*
   as regulator, 90–91; economic flows, 90;
      market regulation, 90
   role in contemporary global system, 84;
      as political-economic organizations,
      84; as ultimate guarantor of economic
      system within its territory, 84
   role in governing the modern economy,
      87–88
state-anchored clusters, 390

state-owned enterprises (SOEs), 94, 104,
   117
state-owned resource companies,
   137–139
   as ultimate guarantor, 88–89; dealing
      with financial crises, 88; guaranteeing
      national economic instruments,
      88–89; property rights and the rule of
      the law, 89; securing international
      economic treaties, 89
   unpacking the state, 85; multi-scalar
      organization, 85; nationhood, 85;
      sovereignty, 85; territorial
      organization, 85
   varieties of states, 100–108, *see also*
      authoritarian states
   developmental states
   neoliberal states
   welfare states
   see also rescaling the state
stated preference, 129
stereotypes and and ethnicity, 442
Stiglitz, Joseph, 52
store-based retailers, 351–352
stores for retailing, 357–358
   geography use in, 357–358
Storper, M., 65, 399
strategic alliances, 312–313, 318–322
   function-specific strategic alliances, 321;
      One World Alliance, 321; SkyTeam,
      321; Star Alliance, 321
   mergers and acquisitions (M&As), 318
   Wal-Mart and Li&Fung, 321
streets for retailing, 353–356
   Akihabara, Tokyo, 354
   Central Manchester, 354–355
   Market Street, 354
   Northern Quarter, 354
Structural Adjustment Programs (SAP),
   110
structural approaches in economic
   geography, 504–508
   anglophone Economic Geography, 506
   'Cultural Turn' in, 507–508
   feminist geography, 506
   Marxian thought in, 505–506

post-structural economic geography, 508–512

structures of economic life, 58–59

cooperative or collectivized economies, 59

feudal economies, 59

peasant economics, 59

subcontracting, 312–313

see also international subcontracting

subnational scales, 108

subprime finance, 200, 215

suburban retails, 345–349

festival marketplaces, 349

Sunley, P., 399

supply, 38

support lab of R&D, 306

Swyngedouw, E., 120

Tacconelli, W., 330

tacit knowledge, 380

taken-for-granted economy, 29–30

Tang, W. S., 517

technical division of labor, 42

techno-economic paradigm, 265

technological change, 261–292

geographical impacts of, 264; different kinds of technology and, 264; different levels of technology, 265; evolutionary perspective, 265; as a social process, 264

incremental innovations, 265

radical innovations, 265

techno-economic paradigm, 265

technology systems, 265

uneven geography of technology creation, 288–290; air travel underpinning global economy, 288; information and communication technologies underpinning internet, 289; robot industry, 289–290

see also process technology

product technology

space-shrinking technologies

universalization of technology

temporal displacement of capital, in capitalist system recovery, 63

temporal flexibility, 166

temporary clusters, 394, 396

temporary migrant workers, 179–180

territory, 5, 12–14

city governments, 12

control over, 12

definition, 12

governments, 12

national states, 12

Terror And Territory, 25

theme park tourism, 491–492

Theory of the Location of Industries, The, 374

thinking geographically, 3–26

third world woman, devaluing, 414

Thrift, N., 52, 77, 217

Tickell, A., 120

Tokatli, N., 257, 399

tourism industry, 487–493

ecotourism, 492

heritage tourism, 490

international tourism receipts and expenditure, 489

large-scale tourism, 490

mega-event tourism, 490–491

'package' holidays combining travel, 490

popularity of, 488

theme park tourism, 491–492

urban tourism, 490

world tourist flows, 2010, 489

Trade-Related Intellectual Property Rights agreement (TRIPS), 148

trade strategies, 91–92

transnational corporation (TNC), 294–331

and anti-globalization movements, 326–327

Borderless World, The, 297

complexity and operating risks in, 295

demonstration effect, 295

geographies of, 307; globally concentrated production, 307–308; host-market production structure, 308; product specialization for a global or regional market, 308; transnational vertical integration, 308–309

Hewlett Packard (HP), 301

transnational corporation (TNC) (*cont'd*)
  inter-firm networks, 296, 312–324, *see also individual entry*
  intra-firm networks, 296, 302–312, *see also individual entry*
  in *maquiladoras* of northern Mexico, 309–310
  myth of being everywhere, 296–298; HSBC, 296
  and the new international division of labor, 317–318
  Nike, 294–295
  Nokia, 301
  organizing forms, costs and benefits, 325
  production network, coordination at, 300
  production network, transactions in, 300; externalized transactions, 300–301; internalized transactions, 300
  value activity and production networks, 298–302
transnational operating units of TNC, 304
transnational remittances, 454–457
  into Philippines, 456
transnational vertical integration of TNC, 308–309, 312
transnationalism, 454
  economic geographies of, 453–460
  transnational ethnic enterprises, 457–460, *see also under* ethnic economies
transport principle, of hexagonal central place theory, 337
transport services by state, 97–98
transportation technologies, 266–270
  commercial jet aircraft, 267–268
  containerization, 268
transportation terminals, 275
Troubled Asset Relief Program (TARP), 83
Turner, S., 366

uneven economic growth, reasons, 57–58
  environmental determinism, 57
  see also inherent uneven geographies of capitalism
United Kingdom, 55

United Nations Global Compact, 248
United States, 56, 116
  gendering of jobs in, 407, 415–416
universalization of technology, 263–266
  communication, 264
  space-shrinking technologies, 264–278, *see also individual entry*
  transportation, 264
unpaid work, gendered patterns of, 406–410
  Canada, 407
  national economic output represented by, 407
  patriarchy, 409
  Scandinavian countries, 408
  tasks and functions, 406
  time spent, 407–408
  uneven global geography of, 408
  United States, 407
  see also paid workforce, women and
untraded interdependencies
  at regional scale, 384–388
  at work, 382–384; Motorsport Valley example, 382–384
untraded interdependencies of production, 380–388
upgrading strategies in global commodity chains, 233
upscaling worker action, 106, 174–177
  Clean Clothes Campaign (CCC), 177
  international trade unions, 175
  Justice for Janitors (JfJ) campaign, 175
  Liquor, Hospitality and Miscellaneous Workers Union (LHMU), 176
  local trade unions, 175
  national trade unions, 175
  organizing across localities, 174–177
  Service Employees International Union (SEIU), 175
urban entrepreneurialism, 116
urban scale, 18, 70
  retailing at, 345–350; by 1974, 345; Chicago's suburban shopping centers, 1949–1974, 346; from 1950 to mid-1970s, 345; United Kingdom, 347; United States, 346–347; until 1950s, 345

urban tourism, 490
Urumqi, labor market at, 439–440
'utility' concept, 33

value activity, of transnational corporation (TNC), 298–302
value creation, 58–59
valuing nature, 141–146
  see also commodification
varieties of states, 100–108
  conceptualizing, caveats about, 105;
    capacities difference among states to
    implement policies, 106; individual
    states that defy easy categorization,
    105; policy positions across different
    domains, 105; state as a dynamic set
    of institutions, 107; variation within
    each category, 105
  dependent states, 106
  failed states, 106
  weak state, 106
  see also authoritarian states
  developmental states
  neoliberal states
  welfare states
venture capital firms on Sand Hill Road,
  Silicon Valley, 372–373
von Hayek, Friedrich, 49
von Thünen, Johan, 7–9, 503, 512
  land use model by, 8–9

Walby, Sylvia, 409
Waldinger, Roger, 463
Walker, R., 65, 77, 517
Walks, R. A., 217
Walton, Roberts, M., 463
Waltring, F., 257
Wang, Q. F., 463
Watson, J. L., 494
weak state, 106
Wealth of Nations, The, 32
Webber, M., 151
Weber, Alfred, 374, 503, 512
Welfare states, 101, 103–104
Williams, A. M., 494
Williams, S., 494

Wills, J., 463
Wojcik, D., 217
women and the paid workforce,
  410–413
women labor, coping strategies employed
  by, 171
women workforce, 402–430
  see also gendered economies
workers shaping economic geographies,
  question of, 154–184
  see also labor power
workforce, ethnic sorting in, 436–445
  see also under ethnic economies
workplaces, gendering, 410–421
  see also under gendered economies
world of standards, 244–254
  certification process, 245
  coverage, 245
  environmental certification, 246–248
  field of application, 245
  form, 245
  generic management systems standards,
    251
  geographical scale, 245
  International Organization for
    Standardization (ISO), 251;
    ISO14001, 251; ISO9000 standards,
    251; ISO9001:2008, 251
  key drivers, 245
  labor standards and conditions, 250
  limits to, 254
  private standards, 250
  regulatory implications, 245
World Trade Organization (WTO), 91,
  111–112, 241, 326
Wright, M., 429
Wrigley, N., 330, 366

Xerox company, 402–404

Yeung, G., 217
Yeung, H. W. C., 120, 330, 463

Zook, M., 291
Zuckerberg, Mark, 3, 16
Zukin, S., 494